Outstanding Contributions to Logic

Volume 7

More information about this series at http://www.springer.com/series/10033

Heinrich Wansing
Editor

Dag Prawitz on Proofs and Meaning

Springer

Editor
Heinrich Wansing
Bochum
Germany

ISSN 2211-2758 ISSN 2211-2766 (electronic)
ISBN 978-3-319-11040-0 ISBN 978-3-319-11041-7 (eBook)
DOI 10.1007/978-3-319-11041-7

Library of Congress Control Number: 2014955373

Springer Cham Heidelberg New York Dordrecht London

Printed on acid-free paper

Springer is part of Springer Science+Business Media (www.springer.com)

Preface

The idea for the present volume goes back, as far as I remember, to a brainstorming among Ryszard Wójcicki, Jacek Malinowski, and myself at the *Trends in Logic VII* conference in Frankfurt/Main, September 2009. On the occasion of a meeting with a representative from Springer, Ryszard Wójcicki, quite enthusiastically, presented his thoughts on a new book series within the *Studia Logica Library*, an ambitious series meant to present Outstanding Contributions to Logic with volumes dedicated to eminent living logicians. It certainly comes as no surprise at all that during the brainstorming several important logicians immediately came to our minds and that Dag Prawitz was one of them.

After this meeting some time elapsed before the first steps were taken to start concrete book projects, and the division of labor had it that I should act as the coordinator of the volume for Dag Prawitz. Eventually, I approached Dag Prawitz in order to suggest the project, to invite him to think about a suitable topic and title, and to propose a possible volume editor. A most natural volume editor, it seemed to me, would have been Peter Schroeder-Heister, whose doctoral dissertation Dag Prawitz had co-supervised, who is very familiar with Prawitz's work, and who would have been a natural choice anyway because of his great expertise in proof theory and proof-theoretic semantics. However, since there is a long-standing plan for a joint monograph by Prawitz and Schroeder-Heister, the idea came up that I might edit the book, a plan I finally approved.

Soon after, a suitable title for the volume was found: *Dag Prawitz on Proofs and Meaning*. Moreover, a list of prospective contributors was compiled, and the response to the invitations for the volume was very positive and encouraging. In September 2012, a book workshop was held at Ruhr-University Bochum. During this highly productive event, Dag Prawitz not only presented a first version of his own contribution to the envisaged book, but also carefully commented on most of the other papers for his *Outstanding Contributions to Logic* volume. Various contributors to the volume, including Dag Prawitz himself, met at the *2nd Conference on Proof-Theoretic Semantics* at the University of Tübingen, March 2013, and at the workshop *Meaning: Models and Proofs* in Munich, October 2013. By the end of 2013, the volume was almost complete, awaiting some final corrections and polishing.

Meanwhile, it took a long period of time for the first (other) volumes of the new book series to appear, a period in which Ryszard Wójcicki resigned as the Editor-in-Chief of the *Trends in Logic* book series, in which I agreed to take over responsibility for *Trends in Logic*, and in which Sven Ove Hansson agreed to become the Editor-in-Chief of *Outstanding Contributions to Logic*.

Now that the volume *Dag Prawitz on Proofs and Meaning* is finally ready for production, I have the privilege to thank, first of all, professor Dag Prawitz for his kind and thoughtful cooperation. It was a great pleasure to exchange thoughts with him in person and via e-mail, and like many other contributors to the volume, I benefited form Dag's careful and constructive comments. It was a pleasure, too, to see the volume developing, and I am grateful to all the other contributors for their kind cooperation as well as their patience. In addition, I would like to thank Andrea Kruse, Caroline Willkommen, Roberto Ciuni, and Judith Hecker for their general support (in particular also during the workshop in 2012) and Tobias Koch for his LaTeX-nical assistance.

Bochum, March 2014 Heinrich Wansing

Contents

About the Contributors

Cesare Cozzo studied at Rome, Florence and Stockholm. He is an Associate Professor of Logic at the Department of Philosophy of the University of Rome "La Sapienza." His main research has been in theory of meaning and philosophy of logic. He is author of three books: *Teoria del significato e filosofia della logica* (Clueb, Bologna 1994), *Meaning and Argument* (Almqvist & Wiksell, Stockholm 1994), *Introduzione a Dummett* (Laterza, Roma 2008), and of several articles. A significant part of his work has been focused on the idea that the sense of a linguistic expression is given by some rules for its use in arguments: he has developed a fallibilist and nonholistic version of this idea.

Michael Detlefsen is the McMahon-Hank II Professor of Philosophy at the University of Notre Dame, where he has also served as the editor of the *Notre Dame Journal of Formal Logic* since 1984. The main area of his scholarly interest is the philosophy of mathematics. His current focus is freedom as an ideal of mathematical inquiry.

Kosta Došen is a Professor of Logic at the Faculty of Philosophy of the University of Belgrade. He works also at the Mathematical Institute of the Serbian Academy of Sciences and Arts, where he chairs the Seminar for General Proof Theory. He used to be a professor at the University of Toulouse III. He studied in Belgrade, and then in Oxford, under Michael Dummett's supervision. He is the author of the books *Cut Elimination in Categories* (Kluwer 1999), *Proof-Theoretical Coherence* (College Publications 2004) and *Proof-Net Categories* (Polimetrica 2007), of which the last two are co-authored by Zoran Petrić. He published many papers in logic and some related areas of mathematics—a few are in philosophy. He worked in modal logic and intuitionistic logic, and pioneered the unified investigation of substructural logics (he named them so). In general proof theory, following Joachim Lambek and Saunders Mac Lane, he works in categorial proof theory, in particular, on identity of deductions.

Roy Dyckhoff began his academic career, after a year of Operating Systems programming and two mathematics degrees, as a Fellow of Magdalen College (Oxford). He moved to the University of St Andrews (Scotland) as a Lecturer in Pure Mathematics in 1975. In 1982 he transferred to Computer Science, in which his Senior Lectureship became Honorary on his retirement in 2011. His research has covered general topology, category theory, proof theory, typed lambda calculus, proof-theoretic semantics and other topics in philosophical logic. He is best known for 1987–1989 work on a pedagogical proof assistant, MacLogic, and for a 1992 article on Vorob'ev's work on the decidability of intuitionistic propositional logic. He edited Springer volumes on *Extensions of Logic Programming* (1994, 1996) and on *Analytic Tableaux* (2000). He is a *Former Editor* of the journal *Studia Logica*, a member of the *Centre for Interdisciplinary Research in Computational Algebra* (at St Andrews) and (until retirement) of the *Center for Logic, Algebra, and Computation* (at CUNY). He has worked as Visiting Professor (or its equivalent) in Bern, Bologna, Dresden, Helsinki, Nancy, Paris and Tübingen.

Edward Hermann Haeusler is an Associate Professor of Computer Science at Pontifical Catholic University of Rio de Janeiro. He received his DSc in Computing from PUC-Rio. He has been working in Logic and Proof Theory, with applications in Computer Science.

Peter Milne has been a Professor of Philosophy at the University of Stirling since 2007. He joined Stirling from the University of Edinburgh; previously he had taught at Birkbeck College (University of London), and held lecturing positions at the University of Liverpool and Heythrop College (University of London), and a post-doctoral research position at Massey University (New Zealand). He has published numerous articles in professional journals and edited collections; his research has focused on philosophy of logic, foundations of probability, the interpretation of indicative conditionals, and the history of logic and analytic philosophy.

Grigori Mints born 06/07/1939 in St. Petersburg, Russia. BA and MS 1961, Ph.D. 1965, Sc.D. 1989, all in mathematics, Leningrad State University. Interests: mathematical logic, proof theory, constructive mathematics, automated deduction, foundations of mathematics. Researcher in Steklov Institute of Mathematics, St. Petersburg 1961–1979, Nauka and Mir Publishers 1979–1985, senior researcher in Institute of Cybernetics, Tallinn 1985–1991. Professor of Philosophy (1991–2014), Computer Science (1992–2000) and Mathematics (1997–2014), Stanford University. Visiting UC Berkeley (2004), LM University Munich (2005), Max Planck Institute of Mathematics (2007). Editor of 10 books, Author of 3 books, Translator of 8 books, more than 200 published papers, more than 3000 published reviews. *Professor Mints passed away on May 29, 2014. We are proud that his paper is part of the present volume.*

Sara Negri is Docent of Logic at the University of Helsinki and a Fellow of the Helsinki Collegium for Advanced Studies. She has visited various universities (Amsterdam, Chalmers, St Andrews, Toulouse), has been a Research Associate at the Imperial College in London, a Humboldt Fellow in Munich, and a visiting scientist at the Mittag-Leffler Institute in Stockholm. Her research contributions range from mathematical logic and constructivism, with a Ph.D. from the University of Padova in 1996, to proof theory and its applications to philosophical logic and formal epistemology. She is best known for her work on labeled sequent calculi for non-classical logics. Her publications include two monographs co-authored with Jan von Plato, *Structural Proof Theory* (2001) and *Proof Analysis: A Contribution to Hilbert's Last Problem* (2011) and about 50 research articles.

Peter Pagin is a Professor of Theoretical Philosophy at Stockholm University (Sweden). He received his Ph.D. there in 1987, with Dag Prawitz as supervisor. His outlook on the field, areas of interest, and philosophical standards were strongly shaped by this education. His main area is philosophy of language, where he has worked on topics such as assertion, compositionality, semantic holism, intensional contexts, vagueness, and general meaning theory. He has done work on natural language semantics and pragmatics, as well as on abstract semantics and in the philosophy of logic. Among the articles that strongly connect with the work of Dag Prawitz are 'Knowledge of proofs', *Topoi* 1994; 'Bivalence: meaning theory vs metaphysics', *Theoria* 1998; 'Intuitionism and the anti-justification of bivalence', in Lindström et al (eds.), *Logicism, intuitionism, formalism. What has become of them?*, Springer 2009; 'Compositionality, understanding, and proofs', *Mind* 2009; and 'Assertion, inference, and consequence' *Synthese* 2012. Pagin is Subject Editor (2010–2014) for logic and language of *Philosophy Compass*, associate editor of *Linguistics and Philosophy*, and editor of, among other things, a special issue of *Theoria* (LXIV 1998), on the philosophy of Dag Prawitz.

Luiz Carlos Pereira is an Assistant Professor of Philosophy at Pontifical Catholic University of Rio de Janeiro (PUC-Rio) and an Associate Professor of Philosophy at the State University of Rio de Janneiro (UERJ). He received his Ph.D. in Philosophy from the University of Stockholm. His main areas of interest are Logic and Analytic Philosophy.

Dag Prawitz See the scientific autobiography and the survey *Prawitz, proofs, and meaning* in this volume.

Stephen Read is currently Honorary Professor of History and Philosophy of Logic at the Arché Research Centre for Logic, Language, Metaphysics and Epistemology at the University of St Andrews, Scotland. He was originally appointed Lecturer in Logic and Metaphysics at St Andrews in 1972. His research and publications lie mostly in contemporary philosophy of logic and in the history of logic, in particular, logic and philosophy of language in the fourteenth century. His most recent book was an edition and English translation of Thomas Bradwardine's *Insolubilia* (Dallas Medieval Texts and Translations 10), Peeters 2010. His new translation of John Buridan's *Treatise on Consequences* (Fordham University Press) has appeared in 2014. His main interests lie in the concept of logical consequence, proof-theoretic semantics and the logical paradoxes.

Tor Sandqvist, an Associate Professor of Philosophy at the Royal Institute of Technology in Stockholm, has published papers dealing with belief revision, counterfactual conditionals, and the meaning-theoretical credentials of classical logic. He regards normative and inferentialist approaches to semantics with sympathy, and also takes an interest in meta-ethics and computability theory.

Peter Schroeder-Heister is a Professor of Logic and Philosophy of Language at the University of Tübingen. In 1981 Dag Prawitz was the external examiner of his doctoral dissertation on an extension of natural deduction. One of his main research interests is what he has called Proof-Theoretic Semantics.

William Tait (who has also published under the names "W.W. Tait", "William W. Tait" and "Bill Tait") is a Professor Emeritus of Philosophy at the University of Chicago. He served on the faculty of the philosophy departments at Stanford University (1958–1965), the University of Illinois-Chicago (1965–1971) and the University of Chicago (1972–1996). He has worked and published in the fields of logic (primarily proof theory), the foundations of mathematics and the history of the foundations of mathematics. He has published two books: *Early Analytic Philosophy: Frege, Russell, Wittgenstein. Essays in honor of Leonard Linsky* (Open Court Press 1997) and a collection of his own philosophical essays, *The Provenance of Pure Reason: Essays in the Philosophy of Mathematics and Its History* (Oxford: Oxford University Press 2005).

Neil Tennant is Humanities Distinguished Professor in Philosophy, Adjunct Professor of Cognitive Science, and Distinguished University Scholar at The Ohio State University, Columbus. He took his Ph.D. in logic, and his B.A. in mathematics and Philosophy at the University of Cambridge. He is a Fellow of the Academy of Humanities of Australia, an Overseas Fellow of Churchill College, Cambridge, and an Associate of the Center for Philosophy of Science at the University of Pittsburgh. His research interests include logic, philosophy of mathematics, philosophy of science (especially biology), and philosophy of language. He has written *Natural Logic* (Edinburgh University Press 1978), *Philosophy, Evolution and Human Nature*, with Florian Schilcher (Routledge & Kegan Paul 1984), *Anti-Realism and Logic* (Clarendon Press, Oxford 1987), *Autologic* (Edinburgh University Press 1992), *The Taming of The True* (Oxford University Press 1997) and *Changes of Mind* (Oxford University Press 2012).

Gabriele Usberti is a Professor of Philosophy of Language at the Università degli Studi Siena (Italy). He is the author of *Logica, verità e paradosso* (Feltrinelli 1980), *Significato e conoscenza*, (Guerini 1995), and articles in professional journals, among which 'Towards a semantics based on the notion of justification' (*Synthese* 2006), 'Anti-Realist Truth and Truth-Recognition' (*Topoi* 2012). He works in philosophical logic and the theory of meaning, trying to extend an intuitionistic meaning theory to empirical statements.

Jan von Plato is Swedish Professor of Philosophy at the University of Helsinki, Finland. His books are *Creating Modern Probability* (1994), *Structural Proof Theory* (with Sara Negri 2001), *Proof Analysis* (with Sara Negri 2011), and most recently *Elements of Logical Reasoning* (2013). Next to systematic work on proof systems and their application in geometry, algebra, and arithmetic, he has done extensive research on the historical sources of logic and foundational study, especially on Gentzen, with the book *Saved from the Cellar: Gerhard Gentzen's Shorthand Notes on Logic and Foundations of Mathematics* scheduled to appear in 2015.

Heinrich Wansing is a Professor of Logic and Epistemology at the Ruhr-University Bochum (Germany). Before that he was a Professor of Philosophy of Science and Logic at Dresden University of Technology (1999–2010). He took his M.A. and his Ph.D. in Philosophy at the Free University of Berlin and his Habilitation in logic and analytical philosophy at the University of Leipzig. He is the author of *The Logic of Information Structures* (Springer 1993), *Displaying Modal Logic* (Kluwer 1998), *Truth and Falsehood. An Inquiry into Generalized Logical Values* (with Y. Shramko, Springer 2011), and numerous articles in professional journals. Heinrich Wansing has been working mainly on philosophical logic, including the semantics and proof theory of modal, constructive, paraconsistent, many-valued, and other nonclassical logics. Moreover, he is the Editor-in-Chief of the book series *Trends in Logic* (Springer), a managing editor of the journal *Studia Logica*, and a member of a number of other editorial boards of logic and philosophy journals.

Chapter 1
Prawitz, Proofs, and Meaning

Heinrich Wansing

Abstract In this paper central notions, ideas, and results of Dag Prawitz's investigations into proofs and meaning are presented. Prawitz's seminal work on natural deduction proof systems and normalization of proofs is presented as a component of his development of general proof theory. Moreover, Prawitz's epistemic theory of meaning is introduced in the context of proof-theoretic semantics. The perspective is partly historical and mainly systematical and aims at highlighting Prawitz's fundamental and extremely influential contributions to natural deduction and a proof-theoretic conception of meaning, truth, and validity.

Keywords General proof theory · Gerhard Gentzen · Dag Prawitz · Natural deduction · Normalization · Proof-theoretic semantics · Anti-realism

1.1 Introduction

"Meaning approached via proofs" is the title of Prof. Dag Prawitz's contribution to a special issue of the journal *Synthese* devoted to proof-theoretic semantics (Prawitz 2006a),[1] and the analysis of proofs and their relation to linguistic meaning may be seen as a guiding theme running through Dag Prawitz's numerous and seminal publications in formal and more philosophical logic. Whereas originally there was a focus on analyzing meaning and validity in terms of proofs and, more generally, in

[1] An excellent overview of proof-theoretic semantics is (Schroeder-Heister 2013), see also (Schroeder-Heister 2006) and (Wansing 2000). As far as I know, the term 'proof-theoretic semantics' was coined by Peter Schroeder-Heister, who used it during lectures in Stockholm in 1987. The first appearance in print of the term 'proof-theoretic semantics' seems to be in (Schroeder-Heister 1991). The very idea of a proof-theoretic semantics has been clearly spelled out already by Prawitz (1971) and can certainly be traced back to Gentzen (1934/35). The term 'Gentzensemantik' (Gentzen semantics) is used in (Kutschera 1968, 1969).

H. Wansing (✉)
Department of Philosophy II, Ruhr-University Bochum, Universitätsstraße 150,
44780 Bochum, Germany
e-mail: Heinrich.Wansing@rub.de

H. Wansing (ed.), *Dag Prawitz on Proofs and Meaning*, Outstanding Contributions to Logic 7, DOI 10.1007/978-3-319-11041-7_1

terms of conclusive grounds for asserting the conclusions of inferences, more recently and also in the present volume, Prawitz enquires into the definition of proofs in terms of meaning. It is clearly beyond the scope of this paper to give a comprehensive survey of Dag Prawitz's work on philosophical and mathematical logic, automated reasoning, and theoretical as well as practical philosophy.[2] Prawitz's outstanding contributions to logic include his work on

- Mechanical proof procedures (Prawitz 1960; Prawitz et al. 1960),
- Normalization and strong normalization for systems of natural deduction (Prawitz 1965, 1971), in particular also the recent normalization theorem with regard to natural deduction for classical first-order Peano arithmetic using Gentzen's ordinal assignment (Prawitz 2012c),
- Translations between intuitionistic and classical logic (Malmnäs and Prawitz 1968),
- The correspondence between natural deduction and typed λ-calculus (Prawitz 1970a, 1971),
- Proof-theoretic validity concepts (Prawitz 1971, 1973, 1974), and
- Cut-elimination (Hauptsatz) for second-order logic and simple type theory (Prawitz 1967, 1969, 1970b).

This note cannot address all of these topics, nor can it deal with any of them in the depth and detail they indubitably deserve. Instead, the idea is to consider and to put into a slightly wider perspective Dag Prawitz's contributions to natural deduction and to a proof-theoretic understanding of meaning, truth, and validity.

1.2 General Proof Theory

Dag Prawitz's investigations into linguistic meaning emerged out of his conception of a general theory of proofs and his fundamental work on the normalization of derivations in various systems of natural deduction. The latter is a major achievement in its own right that is in no way subordinate to any semantical investigations. Moreover, the development of a proof-theoretic semantics requires a proof-theoretic foundation, and Prawitz's work in proof theory has provided such a basis.

Theories of linguistic meaning can take quite different forms. It is a widely shared Fregean view that linguistic meaning rests in a fundamental way upon *truth conditions*. Moreover, the development of model theory has led to defining truth conditions of sentences from rather expressive languages in suitable semantical models. Model-theoretic semantics expresses semantical realism if the models are regarded as either

[2] Moreover, it is next to practically impossible and, therefore, not intended to do full justice in this paper to the widespread and ramified secondary literature on Prawitz's writings on proofs and meaning. Thus, whereas it is projected to point out key contributions Prawitz has made to proof-theoretic semantics, the presentation will be partial and to some extent idiosyncratic in several respects. A journal special issue dedicated to Prawitz's work is issue 2–3 of *Theoria* 64 (1998). Another volume besides the present one dedicated to Dag Prawitz's work is (Pereira et al. 2014). Information on Dag Prawitz's life and intellectual development can be found in his scientific autobiography, which is part of the present volume.

being or representing parts of a mind-independent reality. Whereas model-theoretic semantics still may be regarded as the predominant paradigm in semantics, in philosophical logic rule-based, procedural approaches to linguistic meaning have become increasingly important over the past decades, to a very large extent through the pioneering contributions of Dag Prawitz and Michael Dummett. The general idea in this case is that meaning manifests itself in rules for the correct use of linguistic expressions. In order to base the analysis of linguistic meaning on the notion of proof or a concept closely related to the notion of proof and to make this approach intelligible and promising, first of all a thorough philosophical understanding and rigorous account of proofs in general and of derivations in formal languages is needed.

It was Dag Prawitz (1971, 1973, 1974) who suggested to use the term 'general proof theory'[3] to refer to

> a study of proofs in their own right where one is interested in general questions about the nature and structure of proofs (Prawitz 1974, p. 66).

Prawitz's work on proofs and meaning therefore is to be viewed in the broader context of his development of *general proof theory*,[4] a field which Prawitz contrasts with *reductive proof theory*.[5] Reductive proof theory accrued from Hilbert's program and from Gentzen's and Gödel's work on consistency proofs for formalized fragments of arithmetic.[6] Prawitz (1972, p. 123) characterizes reductive proof theory "as the attempt to analyze the proofs of mathematical theories with the intention of reducing them to some more elementary part of mathematics such as finitistic or constructive mathematics." Whereas reductive proof theory may be seen as part of mathematical logic, general proof theory has a more philosophical concern. In (Prawitz 1971, p. 237) he set the agenda of this area, by listing what he took to be the obvious topics in general proof theory:

---

[3] Georg Kreisel at that time used the term 'theory of proofs', cf. (Mints 1992, p. 123).

[4] The term 'structural proof theory' or 'structural theory of proofs' is also used nowadays, for instance in (Avigad and Reck 2001), (Mints 1992), (Negri and Plato 2001), (Plato 2009). Kosta Došen (2003), however, hesitates to employ this term to designate Prawitz's conception of general proof theory. Other useful and important references for proof theory and, in particular, general proof theory include (Buss 1998a, b), (Negri and Plato 2011), (Takeuti 1975), (Troelstra and Schwichtenberg 2000). There is also categorial proof theory, a discipline Kosta Došen and Zoran Petrić (2007) describe as:

> a field of general proof theory at the border between logic and category theory. In this field the language, more than the methods, of category theory is applied to proof-theoretical problems. Propositions are construed as objects in a category, proofs as arrows between these objects, and equations between arrows, i.e. commuting diagrams of arrows, are found to have proof-theoretical meaning.

[5] In (Prawitz 1981a), he offers a discussion of different directions in proof theory and their philosophical significance. Survey articles on the history of proof theory are, for example, (Avigad and Reck 2001), (Avigad 2011), (Plato 2008a).

[6] Information on Hilbert's program and pointers to the literature on reductive proof theory and its philosophical significance can be found in (Feferman 1988, 2000), (Sieg 2013), (Zach 2009).

2.1. The basic question of defining the notion of proof, including the question of the distinction between different kinds of proofs such as constructive proofs and classical proofs.

2.2. Investigation of the structure of (different kinds of) proofs, including e.g. questions concerning the existence of certain normal forms.

2.3. The representation of proofs by formal derivations. In the same way as one asks when two formulas define the same set or two sentences express the same proposition, one asks when two derivations represent the same proof; in other words, one asks for identity criteria for proofs or for a "synonymity" (or equivalence) relation between derivations.

2.4. Applications of insights about the structure of proofs to other logical questions that are not formulated in terms of the notion of proof.

These topics are, of course, closely interrelated. The structural analysis of proofs, for example, is relevant to the problem of the identity of proofs. Prawitz not only posed the problem of formulating identity criteria for proofs sharing the same conclusion and finite set of premises, but also suggested to account for the identity of proofs in systems of natural deduction in terms of reducibility to a unique normal form. The idea, going back to Gentzen's work, is to remove detours and other redundancies from derivations so as to lay bare their inferential and hence meaning-theoretical core content. Prawitz (1965) considers immediate reductions of different kinds. The primary reductions remove occurrences of *maximum formulas*, i.e., formulas that are both the conclusion of an introduction and the major premise of an elimination. These "proper" reductions remove obvious detours from derivations in natural deduction. The derivation on the left, for example, one-step reduces to the derivation on the right:

$$\dfrac{\begin{array}{c}\mathcal{D}\\ A\end{array} \quad \dfrac{\begin{array}{c}[A]\\ \mathcal{D}'\\ B\end{array}}{(A \rightarrow B)}}{B} \qquad\qquad \begin{array}{c}\mathcal{D}\\ A\\ \mathcal{D}'\\ B\end{array}$$

Here, following Prawitz,

$$\begin{array}{c}\mathcal{D}\\ A\end{array}$$

indicates that the derivation $\mathcal{D}$ ends with A and the notation

$$\begin{array}{c}A\\ \mathcal{D}'\\ B\end{array}$$

indicates that the derivation $\mathcal{D}'$ that ends with B depends on A as an open, undischarged assumption. The square brackets indicate that the assumption A has been discharged in the inference to $(A \rightarrow B)$. In the case of an empty discharge, where A has not been used to derive B in the inference to $(A \rightarrow B)$, the reduced derivation is $\mathcal{D}'$. A formula A is said to be derivable from a set of assumptions Δ if and only if there exists a derivation $\mathcal{D}$ of A with all undischarged assumptions in Δ, and the derivation $\mathcal{D}$ is then also called a derivation of A from Δ. As another example, consider the derivations on the left, where $i \in \{1, 2\}$, which reduce to the derivations on the right:

$$\dfrac{\dfrac{\begin{matrix}\mathcal{D}\\ A_i\end{matrix}}{(A_1 \vee A_2)} \quad \begin{matrix}[A_1]\\ \mathcal{D}_1\\ C\end{matrix} \quad \begin{matrix}[A_2]\\ \mathcal{D}_2\\ C\end{matrix}}{C} \qquad\qquad \begin{matrix}\mathcal{D}\\ A_i\\ \mathcal{D}_i\\ C\end{matrix}$$

Oftentimes a proof is said to be normal or in normal form if it is detour-free, but stronger notions of normal form based on further reductions besides detour removals have been considered in the literature as well. A *normal form theorem* for a system of natural deduction shows that for every derivation of a formula A from a set of assumptions Δ there exists a normal derivation of A from Δ. A *normalization theorem* provides a procedure for reducing any given derivation to a normal one with the same conclusion and set of premises, and a *strong normalization theorem* establishes that the iterated application of reduction steps in any order to a given derivation terminates in a normal derivation.[7] *Uniqueness* of the normal derivation arrived at is usually concluded from the confluence or Church-Rosser property of the reducibility relation: if a derivation $\mathcal{D}$ reduces to derivations $\mathcal{D}'$ and $\mathcal{D}''$, then $\mathcal{D}'$ and $\mathcal{D}''$ both reduce to a derivation $\mathcal{D}'''$ with the same conclusion and set of premises.

For the conjunction, implication, and universal quantification fragment of intuitionistic predicate logic, it can be proved that every formula in a natural deduction derivation without maximum formulas is a subformula of the conclusion of that derivation or of one of its undischarged assumptions. As explained in (Girard 1989), (Prawitz 1965), in order to obtain this subformula property for a richer vocabulary that includes disjunction, existential quantification, and absurdity, one has to consider permutative reductions. The schematic elimination rules for these connectives have an *arbitrary* conclusion C which, of course, need not be a subformula of the major premise of the elimination inference nor a subformula of an open assumption. As a result, the inductive argument used to prove the subformula property for the $\{\wedge, \rightarrow, \forall\}$–fragment cannot be applied. The problem can be avoided by removing further redundant parts of proofs that (Prawitz 1965) calls *maximum segments*. If the conclusion C of an elimination of $\vee$, $\exists$, or $\bot$ is followed by an elimination of the main logical operation of C, then the two elimination steps can be permuted. Consider the derivation

$$\dfrac{\begin{matrix}\mathcal{D}\\ C\end{matrix} \quad \dfrac{\begin{matrix}\mathcal{D}'\\ (A \vee B)\end{matrix} \quad \begin{matrix}[A]\\ \mathcal{D}_1\\ (C \rightarrow D)\end{matrix} \quad \begin{matrix}[B]\\ \mathcal{D}_2\\ (C \rightarrow D)\end{matrix}}{(C \rightarrow D)}}{D}$$

7 As pointed out by Prawitz in the preface to the Dover edition of (Prawitz 1965), the term 'normalization' was suggested by Georg Kreisel.

It reduces in one step to:

$$\dfrac{\begin{array}{c}\mathcal{D}'\\(A\vee B)\end{array}\quad \dfrac{\begin{array}{c}\mathcal{D}\\C\end{array}\quad\begin{array}{c}[A]\\\mathcal{D}_1\\(C\to D)\end{array}}{D}\quad \dfrac{\begin{array}{c}\mathcal{D}\\C\end{array}\quad\begin{array}{c}[B]\\\mathcal{D}_2\\(C\to D)\end{array}}{D}}{D}$$

In addition to detour reductions and permutative reductions, which are also called detour and permutative conversions, respectively, Prawitz (1971) considers immediate simplifications that remove redundant eliminations of disjunction or the existential quantifier in which some assumptions that may be used and discharged are absent in the derivation of the minor premise. The derivation on the left, for instance, where it is assumed that A is not used as an assumption in the derivation $\mathcal{D}'$ of C, can be immediately simplified to the derivation on the right:

$$\dfrac{\begin{array}{c}\mathcal{D}\\(A\vee B)\end{array}\quad\begin{array}{c}\mathcal{D}'\\C\end{array}\quad\begin{array}{c}[B]\\\mathcal{D}''\\C\end{array}}{C}\qquad\qquad\begin{array}{c}\mathcal{D}'\\C\end{array}$$

The reducibility relation between derivations is the reflexive-transitive closure of the one-step reduction relation, and equivalence between derivations is the reflexive, symmetric, and transitive closure of one-step reduction. Prawitz suggested to consider two derivations as representing the same proof if and only if they are equivalent. As Kosta Došen (2003) pointed out, this identity conjecture (or normalization conjecture, as he calls it) "is an assertion of the same kind as Church's Thesis: we should not expect a formal proof of it." In this respect it may seem appropriate to refer to the identity conjecture as Prawitz's Thesis or as the Prawitz–Martin-Löf Thesis.[8]

[8] In (Prawitz 1971, p. 261) and (Prawitz 1972, p. 134) Prawitz remarks that the conjecture about the identity of proofs is due to Per Martin-Löf and acknowledges an influence by ideas of William Tait (1967). In (Martin-Löf 1975, p. 104), however, he gives credit to Prawitz for the identity conjecture. An interesting consideration of the identity of proofs and logics may also be found in (Straßburger 2007). Došen (2003) considers a relation of equal generality of proofs that leads to a "generality conjecture": two derivations represent the same proof if and only if they have the same generality. In his doctoral dissertation (Widebäck 2001), Filip Widebäck showed that for a particular system one direction of the identity thesis provably holds: identical proofs in minimal implicational logic are encoded by $\beta\eta$-equivalent typed lambda-terms. As Widebäck (2001, p. 4) explains, "there is only one schematic extension of $\beta\eta$-equivalence, the trivial equivalence relation that identifies all proofs of a theorem. It is then argued that the identity relation on proofs is non-trivial, i.e. that there are non-identical proofs. This proves the completeness part of the conjecture."

1.3 Natural Deduction

Dag Prawitz's investigations into natural deduction may be seen to comprise his core contributions to proof theory. Actually, it was Prawitz who gave natural deduction a central position in proof theory, and often natural deduction is presented as introduced by Gentzen (1934/35) but studied by Prawitz (1965). Independently of Gentzen's work, systems of natural deduction have been developed also by Stanisław Jaśkowski (1934) in a linear representation and as nested systems of subordinate proofs that later have become known as Fitch-style natural deduction (Fitch 1952), see also the *Studia Logica* special issue on "Gentzen's and Jaśkowski's Heritage. 80 Years of Natural Deduction and Sequent Calculi", edited by Andrzej Indrzejczak (2014). According to Jaskowski (1934, p. 5), in 1926 Jan Łukasiewicz called attention to the fact that mathematicians in their reasoning employ suppositions and posed the problem to develop this reasoning method in terms of inference rules. In a paper on the history of natural deduction proof systems (Pelletier 1999), Pelletier comments on the publication of (Gentzen 1934/35) and (Jaśkowski 1934) by writing that "[i]n 1934 a most singular event occurred. Two papers were published on a topic that had (apparently) never before been written about, the authors had never been in contact with one another, and they had (apparently) no common intellectual background that would otherwise account for their mutual interest in this topic."[9] As pointed out by Gentzen (1934/35), it was an analysis of informal proofs in mathematics that resulted in his development of formal proof systems of natural deduction. These proof systems define the notion of a logical proof in intuitionistic and in classical first-order logic. An inspection of properties of these natural deduction proof systems then led Gentzen to a general theorem, which he called "Hauptsatz" (main theorem). The Hauptsatz says that every purely logical proof can be brought into a detour-free normal form. Gentzen proved this result for another kind of calculi, namely the sequent calculi for classical and intuitionistic predicate logic, proof systems he developed in order to prove the Hauptsatz. Sequent calculi manipulate derivability statements of the form $\Delta \vdash \Gamma$, called sequents, where Δ and Γ are finite sequence, multi-sets, or sets of formulas.[10]

In a translation of Gentzen's own words (Gentzen 1969, p. 289):

> A closer investigation of the specific properties of the natural calculus have finally led me to a very general theorem which will be referred to below as the "*Hauptsatz*."... In order to

[9] Jan von Plato's opinion on Jaśkowski's paper is rather critical. In (Plato 2012, p. 331) he writes:

> The calculus is classical and the derivations fashioned in a linear form. The latter feature, especially, makes it practically impossible to obtain any profound insights into the structure of derivations. Consequently, there are no results in this work beyond what, more or less, Aristotle could have said about the hypothetical nature of proof.

Differences between Jaśkowski's and Gentzen's formulations of natural deduction and, in particular, advantages of Jaśkowski's approach are considered in (Hazen and Pelletier 2014).

[10] Gentzen (1934/35) considered statements of derivability between finite sequences of formulas and assumed the following structural sequent rules that do not exhibit any logical operations:

> be able to enunciate and prove the *Hauptsatz* in a convenient form, I had to provide a logical calculus especially suited to the purpose. For this the natural calculus proved unsuitable. For, although it already contains the properties essential to the validity of the *Hauptsatz*, it does so only with respect to its intuitionistic form.

As is well-known, Gentzen showed that the cut-rule

$$\frac{\Gamma \vdash \Theta, D \qquad D, \Delta \vdash \Lambda}{\Gamma, \Delta \vdash \Theta, \Lambda}$$

is admissible, that is, its use has no effect on the set of sequents provable in the calculi under consideration. The formula D in the succedent of the left premise sequent and the antecedent of the right premise of the cut rule is excised, hence the name "Schnitt" (cut). With the development of the sequent calculus for classical predicate logic and for intuitionistic predicate logic obtained by a restriction to sequents with at most one formula in the succedent so as to block the provability of the double negation elimination law, there was no pressing reason for Gentzen to include a proof of the normalization theorem for natural deduction in intuitionistic predicate logic in his thesis.

Dag Prawitz' thesis *Natural Deduction: A Proof-Theoretical Study* (Prawitz 1965) was published only one year after the first part of a translation of Gentzen's doctoral dissertation into English appeared in *The American Philosophical Quarterly* and thereby became accessible to a much wider audience. The philosophical significance of Gentzen's systems of natural deduction had not at all been generally recognized at that time, and the normalizability of natural deduction for intuitionistic predicate logic was unknown, although Gentzen, as quoted above, in the published version of his thesis remarked that his system of natural deduction for intuitionistic logic contains the properties essential to the validity of the Hauptsatz. As Samuel Buss (1998a, p. 47) explains, "[t]he importance of natural deduction was established by the classical result of Prawitz [1965] that a version of the cut elimination holds also for natural deduction." Moreover, although Gentzen explained that a closer inspection of the natural deduction systems led him to the cut-elimination theorem for his sequent calculi, the relation between natural deduction and Gentzen's sequent calculi and, in particular, between cut-elimination and normalization became clear only through Prawitz's work. Prawitz (1965, Appendix A, §3) showed that normal derivations in the natural deduction calculi for minimal and intuitionistic predicate logic and for classical predicate logic without $\vee$ and $\exists$ can be mapped into cut-free sequent derivations, see also (Plato 2003). Together with the converse direction, observed in Prawitz (1965, Appendix A, §2), this allows one to derive the cut-elimination theorems for the corresponding sequent calculi. Prawitz's investigations of natural deduction in (Prawitz 1965) go beyond Gentzen's work also in extending natural deduction to

(Footnote 10 continued)

$$\frac{\Gamma \vdash \Theta}{D, \Gamma \vdash \Theta} \quad \frac{\Gamma \vdash \Theta}{\Gamma \vdash \Theta, D} \; \textit{Thinning} \qquad \frac{D, D, \Gamma \vdash \Theta}{D, \Gamma \vdash \Theta} \quad \frac{\Gamma \vdash \Theta, D, D}{\Gamma \vdash \Theta, D} \; \textit{Contraction}$$

$$\frac{\Delta, E, D, \Gamma \vdash \Theta}{\Delta D, E, \Gamma \vdash \Theta} \quad \frac{\Gamma \vdash \Theta, D, E, \Delta}{\Gamma \vdash \Theta, E, D, \Delta} \; \textit{Exchange}$$

(simple) second-order logic, ramified second-order logic, systems of modal logic based on minimal, intuitionistic, and classical propositional logic, to relevant and strict implication, Fitch's set theory, and Nelson's systems of constructive logic with strong negation.

Prawitz's normalization results advanced natural deduction into a mature field, and his *Natural Deduction: A Proof-Theoretical Study* has become a modern classic, re-published in 2006, which is still the standard reference to Gentzen-style natural deduction. Until the year 2005 it was generally believed that Dag Prawitz, together with Andres Raggio (1965), was the first to prove the normalization theorem for intuitionistic predicate logic. But then von Plato (2008b) discovered such a proof in an unpublished handwritten version of Gentzen's doctoral thesis (Gentzen 2008), so that the proof probably has been obtained by Gentzen by early 1933. This discovery notwithstanding, it is clear that together with Gerhard Gentzen, Dag Prawitz with full justification may be regarded as the most important researcher in the area of natural deduction. Without their seminal contributions, natural deduction would not be what it is today, namely one of the central paradigms in general proof theory and proof-theoretic semantics. Moreover, the proofs-as-programs or formulas-as-types interpretation of natural deduction derivations, to which Prawitz has made contributions in (Prawitz 1970a, 1971), has been essential for the development of functional programming languages and Martin-Löf Type Theory (Martin-Löf 1984). With his dissertation Prawitz accomplished the breakthrough of natural deduction. In a recent paper, von Plato (2012, p. 333f) puts an emphasis on Prawitz's normalization result for classical first-order logic and writes:

> The real novelty of Dag Prawitz' thesis *Natural Deduction: A Proof-Theoretical Study*, was not the normalization theorem for intuitionistic logic, but the corresponding result for classical natural deduction for the language of predicate logic that does not have disjunction or existence. His deft move was to limit indirect proof to atomic formulas, and to show that it is *admissible* for arbitrary formulas in the $\vee, \exists$-free fragment.

Normalization for *classical first-order logic* presented as a natural deduction calculus indeed has been discussed for many years. Whereas Prawitz, building on the definability of disjunction and existential quantification in classical predicate logic, excluded $\vee$ and $\exists$ from consideration and restricted the classical reductio ad absurdum rule

$$\dfrac{\begin{matrix}[\neg A]\\ \vdots \\ \bot\end{matrix}}{A}$$

where $\neg A$ is defined as $(A \to \bot)$, to atomic conclusions, various later proofs for the full vocabulary impose restrictions on the instantiation of elimination rules or require global proof transformations, such as in (Seldin 1986). Tennant (1978) considered applications of the classically but not intuitionistically valid dilemma rule (notation adjusted):

$$\frac{\begin{array}{cc}[A] & [\neg A] \\ \vdots & \vdots \\ B & B\end{array}}{B}$$

Here A is atomic. Moreover, Tennant takes into account only detour conversions.

Von Plato and Siders (2012) note that Prawitz's approach of making sure that a major premise of an elimination rule in a normal derivation is not derived by an application of classical reductio is enough to prove normalization of classical first-order logic in its standard vocabulary. A derivation is now said to be normal whenever all major premises of elimination rules are open assumptions, and the subformula property states that all formulas in normal derivations of a formula A from a finite set of open assumptions Γ are subformulas of $\neg C$, where C a subformula of A or the formulas from Γ. A normalization result with respect to a natural deduction proof system for classical first-order logic that does not use classical reductio ad absurdum but instead adds the rule

$$\frac{\begin{array}{c}[\neg A] \\ \vdots \\ A\end{array}}{A}$$

to a natural deduction proof system for intuitionistic logic is presented in (Seldin 1989). The addition of this rule to intuitionistic logic results in classical logic; classical reductio already gives classical logic when it is added to Johansson's minimal logic.

Von Plato and Siders (2012) explain their proof of the normalization theorem for classical predicate logic first with respect to a system of natural deduction with what is known as *general elimination rules*. In the literature these rules are also sometimes referred to as generalized elimination rules. The general elimination rule for conjunction can be found in Dag Prawitz's paper on functional completeness (Prawitz 1978) as an instance of a general elimination schema. General elimination rules for higher-level natural deduction proof systems that allow the discharge of rules have been introduced by Schroeder-Heister (1981, 1984a) and for ordinary natural deduction also by Dyckhoff (1988), Tennant (1992), López-Escobar (1999), and von Plato (2000, 2001). General higher-level elimination rules and general standard-level elimination rules are compared to each other by Schroeder-Heister (2014a).

As in the standard elimination rule for disjunction and the familiar elimination rule for the existential quantifier

$$\frac{\exists x A(x) \quad \begin{array}{c}[A] \\ \vdots \\ C\end{array}}{C}$$

where x is not free in C, or in any formula C depends on, other than $A(x)$, the general elimination rules for conjunction, implication, and the universal quantifier repeat as the conclusion of the elimination inference the minor premise obtained by inferences that allow the discharge of assumptions in applications of the general elimination rule:

$$\frac{(A \wedge B) \quad \begin{matrix}[A]\,[B] \\ \vdots \\ C\end{matrix}}{C} \qquad \frac{(A \rightarrow B) \quad A \quad \begin{matrix}[B] \\ \vdots \\ C\end{matrix}}{C} \qquad \frac{\forall x A(x) \quad \begin{matrix}[A(t)] \\ \vdots \\ C\end{matrix}}{C}$$

Here the term t is free for x in A. As in the case of the elimination rules for $\vee$ and $\exists$, permutative conversions can be defined for the general elimination rules for $\wedge$, $\rightarrow$, and $\forall$.

As already mentioned, another topic in general proof theory is the relationship between natural deduction and sequent calculi. In (Prawitz 1965, Appendix A, §2) Prawitz points out that a cut-free sequent calculus derivation may be regarded as "an instruction how to construct a corresponding natural deduction." An axiomatic sequent $A \vdash A$ corresponds to the atomic formula A, and an application of a sequent rule that introduces a logical operation φ into succedent (antecedent) position gives rise to the addition of an application of a natural deduction introduction (elimination) rule for φ at the bottom (top) of the constructed derivation. This procedure gives a many-to-one correspondence, because different sequent calculus derivations are translated by the same natural deduction derivation. An influential paper in this area has been (Zucker 1974), see also (Urban 2014) and Roy Dyckhoff's contribution to the present volume (Dyckhoff 2014). An advantageous aspect of using general elimination rules is that they admit defining a one–one correspondence between natural deduction derivations and derivations in a suitably chosen version of Gentzen's sequent calculus for intuitionistic predicate logic, namely a calculus with independent contexts, see (Negri and Plato 2001, Chap. 8). There are one–one translations from natural deduction derivations into sequent derivations and vice versa, such that cut-elimination is a homomorphic image of normalization and normalization a homomorphic image of cut-elimination, respectively.

Gentzen's sequent calculus for classical predicate logic is symmetrical: it is a multiple-premise and multiple conclusion proof system. Multiple-conclusion natural deduction calculi have also been considered in the literature, see, for instance (Shoesmith and Smiley 1978), (Ungar 1992). Tranchini (2012) has defined a multiple-conclusion natural deduction proof system for dual intuitionistic propositional logic, see also (Goré 2000). The dualization of natural deduction for intuitionistic propositional logic involved in Tranchini's calculus proceeds by inverting the introduction and elimination rules for intuitionistic logic and replacing the intuitionistic connectives by their duals, thereby obtaining downward branching multiple-conclusion derivation trees. Another dualization procedure that leads to a constructive falsification logic different from dual intuitionistic logic can be found in (Wansing 2013). Here the natural deduction rules for both proofs and their duals are single-conclusion rules.

Although proof-theoretic semanticists also investigate the meaning of expressions from natural languages [see, for example, (Francez and Dyckhoff 2010), (Francez et al. 2010)], the proof-theoretic characterization of the meaning of the logical operations and the semantical notion of validity of proofs or, more generally, inferences or arguments have been the central subject and main application area of proof-theoretical meaning analyses. This central enterprise in the development of a proof-theoretic semantics requires the availability of a suitable proof-theoretical framework that establishes a general schema for inference rules for the connectives and the quantifiers in predicate logic and possibly richer vocabularies. Dag Prawitz takes natural deduction as developed by Gentzen (1934/35) to be the most suitable rule-based framework for a semantical investigation of the logical operations. Natural deduction calculi, however, are not the only type of proof system that have been suggested for the development of a proof-theoretic semantics. Prawitz (1971, p. 246) emphasizes that "Gentzen's systems of natural deduction are not arbitrary formalizations of first order logic but constitutes a significant analysis of the proofs in this logic" and he adds in a footnote that it "seems fair to say that no other system is more convincing in this respect." On page 260 of (Prawitz 1971) he comments that the main advantage of natural deduction is

> that the significance of the analysis… seems to become more visible. Furthermore it has recently been possible to extend the analysis of first order proofs to the proofs of more comprehensive systems when they are formulated as systems of natural deduction . . ., while an analogous analysis with a calculus of sequents formulation does not suggest itself as easily. Finally, the connection between this Gentzen-type analysis and functional interpretations such as Gödel's Dialectica interpretation becomes very obvious when the former is formulated for natural deduction.

Although Gentzen introduced sequent calculi for purely *formal* reasons with a view on proving cut-elimination and the consistency of first-order Peano arithmetic, sequent calculi have also been proposed as a kind of proof systems suitable for the development of proof-theoretic semantics. Schroeder-Heister (2009, p. 237), for example, considers Gentzen's sequent calculus as a "more adequate formal model of hypothetical reasoning," because it "does more justice to the notion of *assumption* than does natural deduction." Sequent calculi have other advantages as well. They allow one, for example, to neatly draw the distinction between internal and external consequence relations (Avron 1988). In (Mares and Paoli 2014), Edwin Mares and Francesco Paoli argue that this distinction can be applied to obtain solutions to certain variants of the semantical and set-theoretical paradoxes. They also present a natural deduction single-conclusion introduction rule for multiplicative (intensional) disjunction:

$$\frac{\begin{array}{c}[\neg A]\\ \vdots\\ B\end{array}}{(A \oplus B)}.$$

This rule is impure; as a meaning assignment it makes the meaning of $\oplus$ dependent on the meaning of negation. Its multiple-conclusion multi-set sequent calculus counterpart avoids exhibiting negation (notation adjusted):

$$\frac{\Gamma \vdash \Delta, A, B}{\Gamma \vdash \Delta, (A \oplus B).}$$

A classic source for suggesting sequent calculi as a framework for meaning-theoretical analyses is (Hacking 1979)[11]; a more recent reference is (Paoli 2007). Moreover, generalizations of Gentzen's sequent calculus, such as, for example, hypersequent calculi (Avron 1996; Baaz 2003) and display sequent calculi (Belnap 1982, 1995; Goré 1998; Postniece 2010; Restall 2000; Wansing 1998) have been investigated; a survey of both frameworks can be found in (Ciabattoni et al. 2014). In particular, display calculi have been motivated as proof systems especially useful for a proof-theoretic analysis of the meaning of a broad range of logical operations, including modal operators.

It should also be noted that Prawitz is not at all at odds with sequent calculi and, moreover, classical semantics. On the contrary, he has made contributions to the theory of cut-elimination by proving *Takeuti's conjecture*. Takeuti's conjecture is normally stated as the claim that cut-elimination holds for simple type theory and sometimes as the less general claim that second-order classical logic is closed under cut. Schütte (1960) had shown that Takeuti's conjecture for simple type theory is equivalent to a semantical property, viz. the extendibility of every partial valuation to a total valuation. Takeuti's conjecture was proved for second-order logic by Tait (1966) and Prawitz (1967). For higher-order logic the conjecture was verified by Prawitz (1969) and Takahashi (1967). In these papers Schütte's semantical equivalent to cut-elimination is proved non-constructively.

1.4 The Relation Between Proofs and Meaning

1.4.1 Natural Deduction Rules and the Meaning of the Logical Operations

There is a now famous and often-quoted passage from (Gentzen 1934/35) to which Prawitz has drawn wide attention. Since this passage may be regarded as the conceptual basis of the whole enterprise of a proof-theoretic semantics of the logical

[11] Hacking's paper has been criticized by Sundholm (1981) inter alia for comments on the relation between cut-elimination and the conservativeness of introducing new logical operations in sequent calculi.

operations, its repeated quotation is, therefore, next to unavoidable for the present purposes[12]:

> The introductions represent, as it were, the 'definitions' of the symbols concerned, and the eliminations are no more, in the final analysis, than the consequences of these definitions. This fact may be expressed as follows: In eliminating a symbol, we may use the formula with whose terminal symbol we are dealing only 'in the sense afforded by the introduction of that symbol' (Gentzen:1969, [p. 80]).

In this paragraph, Gentzen first of all suggests that the natural deduction introduction rules for the logical operations are *meaning constitutive*: they define these symbols. Since the rules come in pairs of sets of introduction and elimination rules, it is natural to enquire about the relationship between the elimination rules and the introduction rules, and Gentzen addresses this issue as well. The relation is a semantical one: the formula that contains the logical operation to be eliminated as its main logical operation may be used only as what it *means* on the basis of the introduction of that operation. However, Gentzen also surmises that rendering this thought more precise should make it possible to establish the elimination rules on the basis of certain requirements as unique functions of their corresponding introduction rules. One may thus wonder whether the elimination rules for a logical operation φ are indeed uniquely determined by the introduction rules for φ. Is there a procedure that, given the introduction rules for φ, returns the unique set of elimination rules for φ? What does such a procedure achieve? Does it *justify* the resulting elimination rules? Does it make sure that the introduction rules indeed satisfy certain requirements usually imposed on a definition, in particular, non-creativity? Moreover, do the elimination rules for φ have an impact on *the meaning* of φ, after all? As to the latter question, there are at least three options:

1. Verificationist theories of meaning: The meaning of a logical operation φ is completely determined by φ's introduction rules. The elimination rules are derivative and do not contribute to the meaning of φ.
2. Pragmatist theories of meaning: The meaning of a logical operation φ is, contrary to what Gentzen writes in the above short paragraph, completely determined by the elimination rules for φ. The introduction rules are derivative and do not contribute to the meaning of φ.
3. "Equilibriumist" theories of meaning: The meaning of a logical operation φ is, again contrary to Gentzen's explanation in the above short paragraph, determined by both the introduction and the elimination rules for φ. The introduction rules and the elimination rules mutually depend on each other and are in balance.

Dag Prawitz (1965) explains Gentzen's statement concerning the relationship between the natural deduction introduction and the elimination rules by using the term 'inversion principle' that was introduced by Lorenzen (1950) to express a closely related idea.[13] An informal explanation of the inversion principle is that

---

[12] The paper by Sara Negri and Jan von Plato in the present volume, (Negri and Plato 2014), contains additions from manuscript sources to this famous passage.

[13] The relation between Prawitz's inversion principle and Lorenzen's slightly more general conception of an inversion principle is explained in (Schroeder-Heister 2008).

> an elimination rule is, in a sense, the inverse of the corresponding introduction rule: by an elimination rule one essentially only restores what had already been established by the major premise of the application of an introduction rule (Prawitz 1965, p. 33).

According to Prawitz, the elimination rules in natural deduction are indeed semantically *justified* by their corresponding introduction rules, the latter being self-justifying. Moreover, in (Prawitz 1985, p. 159f.), for example, he emphasizes that this justification of the elimination rules is closely related to the normalizability of proofs:

> The way in which the elimination rules are justified by the introduction rules can be seen to be what makes possible normalizations of proofs in natural deduction or, equivalently, elimination of cuts in the calculus of sequents as established in Gentzen's Hauptsatz.

The elimination rules are thus semantically justified in virtue of the detour conversions, whereas within a pragmatist theory of meaning the introduction rules are justified on the basis these conversions. As already mentioned, detour conversions are possible not only with Gentzen's elimination rules, but also with the general elimination rules that follow the pattern of the elimination rules for $\vee$ and $\exists$. In the case of the general elimination rule for conjunction, for example, the derivation on the left reduces in one step to the derivation on the right:

$$\dfrac{\dfrac{\begin{matrix}\mathcal{D} & \mathcal{D}'\\ A & B\end{matrix}}{(A \wedge B)} \qquad \begin{matrix}[A]\,[B]\\ \mathcal{D}''\\ C\end{matrix}}{C} \qquad\qquad \begin{matrix}\mathcal{D}\,\mathcal{D}'\\ \mathcal{D}''\\ C\end{matrix}$$

According to meaning theoretical equilibriumism, the relation between the introduction and elimination rules for a given logical operation is a mutual dependence, and a now often used notion that seems to be appropriate to a symmetric conception of this relation is the notion of *proof-theoretic harmony*.[14] This notion of harmony was introduced by Michael Dummett, who in (Dummett 1973, p. 396) discusses

> different aspects of 'use', and the requirement of harmony between them. Crudely expressed, there are always two aspects of the use of a given form of sentence: the conditions under which an utterance of that sentence is appropriate, which include, in the case of an assertoric sentence, what counts as an acceptable ground for asserting it; and the consequences of an utterance of it, which comprise both what the speaker commits himself to by the utterance and the appropriate response on the part of the hearer, including, in the case of assertion, what he is entitled to infer from it if he accepts it.

On p. 454 f. he then remarks:

> [i]n the case of a logical constant, we may regard the introduction rules governing it as giving the conditions for the assertion of a statement of which it is the main operator, and the elimination rules as giving the consequences of such a statement: the demand for a harmony

[14] Schroeder-Heister (2013) notes that "[t]his terminology is not uniform and sometimes not even fully clear. It essentially expresses what is also meant by 'inversion'." See also Schroeder-Heister's contribution to the present volume (Schroeder-Heister 2014b).

> between them is then expressible as the requirement that the addition of the constant to a language produces a conservative extension of that language.[15]

In (Tennant 1978, p. 74), Neil Tennant defines a "Principle of Harmony", the satisfaction of which is guaranteed by proof normalization. Steinberger (2013) distinguishes between three types of proof-theoretic harmony: harmony-as-conservative extension (a conception associated with Belnap and Dummett), harmony-as-deductive equilibrium (developed by Tennant), and harmony-as-levelling procedure (a conception Steinberger locates in the writings of Prawitz and of Dummett, who refers to detour conversions as a levelling procedure, because it levels maximum formulas as local peaks). In any case Dag Prawitz's inversion principle is a principle of semantical justification of the elimination rules. It suggests the normalization theorem, the proof of which is "based on this principle" Prawitz (1965, p. 34). Moreover, harmony as conservative extension has been repeatedly criticized by Prawitz, who observed that in the case of higher-order concepts, harmony does not guarantee the conservative extension property. In Prawitz (1994, p. 375) he explains that Dummett's explication of proof-theoretic harmony in terms of conservative extensions

> can hardly be correct, however, because from Gödel's incompleteness theorem we know that the addition to arithmetic of higher order concepts may lead to an enriched system that is not a conservative extension of the original one in spite of the fact that some of these concepts are governed by rules that must be said to satisfy the requirement of harmony.

The general elimination rules in natural deduction that take their pattern from the familiar elimination rules for $\vee$ and $\exists$ have given rise to the notion of "general-elimination harmony", cf. (Read 2010, 2014). As Read (2010) emphasizes, the possession of this property by a natural deduction proof system implies neither normalizability, nor the conservative extension property, nor consistency. An in-depth study of various inversion principles and their relation to introduction rules can be found in Peter Milne's contribution to the present volume, (Milne 2014).

In (Prawitz 2007), Dag Prawitz in addition to discussing verificationist and pragmatist theories of meaning, briefly considers a third kind of theories of meaning, namely *falsificationist* theories of meaning as suggested by Dummett (1993). Prawitz distinguishes between "obvious rules for the falsification of a sentence" and "the standard way of falsifying a compound sentence," the latter consisting of inferring a contradiction from the sentence under consideration. The first-mentioned rules result in a notion of falsity assumed in David Nelson's constructive logics with strong negation [see (Almukdad and Nelson 1984), (Kamide and Wansing 2012), (Nelson 1949), (Odintsov 2008), (Wansing 1993b)], logics for which Prawitz (1965, p. 97) has presented natural deduction proof systems. In these logics a falsification of a strongly negated formula A amounts to a verification of A. A falsification of a strongly negated implication $(A \rightarrow B)$ can therefore be understood as a falsification of the strong negation of A together with a falsification of B. As to the second-mentioned way of falsifying a compound sentence, Prawitz notices a problem. A

[15] In a footnote on p. 455 of (Dummett 1973), Dummett however, qualifies this statement by writing that "this is not to say that the character of the harmony demanded is always easy to explain, or that it can always be accounted for in terms of the notion of a conservative extension."

falsification of an implication $(A \rightarrow B)$ now requires in addition to a falsification of B also a *verification* of A, which is to be distinguished from a falsification of A's negation. Prawitz draws the conclusion that a "falsificationist meaning theory seems thus to have to mix different ideas of meaning in an unfavorable way" (Prawitz 2007, p. 476). These considerations are taken up in my paper (Wansing 2013), where a natural deduction calculus is defined from two primitive sorts of derivations, namely proofs and dual proofs, that can be combined with each other to obtain more complex proofs and dual proofs. If introduction rules are meaning constitutive, as Gentzen suggested, then now both the rules for introducing a logical operation φ into proofs and the rules for introducing φ into dual proofs define the meaning of φ.

In order to explicate the way in which the elimination rules of natural deduction are semantically justified, Prawitz developed various notions of proof-theoretic validity. The idea is that the elimination rules are justified because they preserve proof-theoretic validity. I will turn to this issue in Sect. 1.4.2, before I will come back to the question whether the introduction rules uniquely determine the elimination rules in the context of comments on proof-theoretic notions of functional completeness in Sect. 1.4.3.

1.4.2 Proof-Theoretic Validity Concepts

From a logical point of view, the notion of valid inference (entailment) is perhaps the most important semantical notion, and if proof-theoretic semantics is a semantics at all, it ought to provide a notion of valid and hence justified inference. In the tradition of denotational semantics and, in particular, in Tarski's model theoretic conception of semantical consequence, the notion of entailment is defined in terms of truth preservation from the premises to the conclusion(s) or, less standardly, in terms of falsity preservation from the conclusion(s) to the premises. An inference from a set of assumptions Δ to a single conclusion A is said to be valid if and only if for every model $\mathcal{M}$, if all elements of Δ are true in $\mathcal{M}$, then A is true in $\mathcal{M}$ (or, whenever for every model $\mathcal{M}$, if A is false in $\mathcal{M}$, then some element of Δ is false in $\mathcal{M}$, which on the classical understanding of falsity as non-truth amounts to the same notion). In Dag Prawitz's opinion, however, a reasonable notion of valid inference cannot be defined in terms of truth preservation because this conception cannot explain why logical inferences may endow an epistemic subject with knowledge. Accordingly, in Prawitz's opinion, denotational semantics lacks a notion of valid inference.

A proof-theoretic notion of validity of inferences has been developed by Prawitz (1971, 1973, 1974) and this proof-theoretic conception of validity, which is meant to provide an epistemically reasonable notion of valid inference, has played a very prominent role in proof-theoretic semantics, see also (Prawitz 2006b). The central concept is that of a valid closed proof. A closed proof is a proof from the empty set of assumptions, and a proof that is not closed is said to be open. Moreover, the concept of validity is defined relative to a given *atomic system*. An atomic system S comprises inference rules for generating atomic formulas built up from some descriptive constants, and an atomic formula is defined to be valid with respect to *S* if and only if it is derivable in *S*. Since the validity of a proof depends on the meaning of the logical operations occurring in that proof and since the meaning of a logical operation is assumed to be defined by its introduction rules, the introduction rules play an essential role in the definition of proof-theoretic validity. In (Prawitz 1973, p. 234), Prawitz explains that

> [t]he main idea is this: while the introduction inferences represent the form of proofs of compound formulas by the very meaning of the logical constants when constructively understood and hence preserve validity, other inferences have to be justified by the evidence of operations of a certain kind.

These justifying operations are the detour reductions used to obtain proofs in normal form. Closed proofs that end with an application of an introduction rule are called *canonical* proofs. A closed valid proof of a compound formula *A* should then either itself be a canonical proof of *A* or it should be reducible to a canonical proof of *A*.

The notion of a valid proof relative to a given atomic system *S* and a given notion of reducibility to canonical proofs is then defined as follows:

1. Every closed proof in *S* is valid.
2. A closed canonical proof is valid if its immediate subproofs are valid.
3. A closed non-canonical proof is valid if it reduces to a valid closed canonical proof or to a closed proof in *S*.
4. An open proof is valid if every closed proof obtained by substituting valid closed proofs for its undischarged assumptions is valid.

If the language under consideration contains variable-binding operators, it is required that an open proof is considered valid if every closed proof obtained by first replacing its free variables with closed terms and then its open assumptions with valid closed proofs is valid.[16] A natural deduction rule is valid with respect to an atomic system *S* and the assumed reduction procedures if it leads from valid proofs to valid proofs. Detour reducibility can be used to show that the elimination rules for $\wedge$, $\vee$, $\rightarrow$, $\exists$, and $\forall$ are sound; they preserve proof-theoretic validity.

In (Prawitz 2014a), Prawitz points out that

> [a] limitation of the notion of validity ... is that it is defined only for deductions in a given formal system. In contrast, a notion like truth is defined for sentences in general but singles out a subset of them. Similarly, one would like to have a notion of validity defined for a broader range of reasoning, singling out a subdomain of correct reasoning that could properly be called proofs.

---

[16] In (Prawitz 1971, p. 289), Prawitz explains that "[s]ince the introductions and eliminations are inverses of each other, Gentzen's idea to justify the eliminations by the meaning given to the constants by the introductions may be reversed. ... A derivation will then be valid when it can be used to obtain certain valid derivations of the subformulas." This remark is related to the already mentioned *pragmatist* theories of meaning, which Prawitz critically discusses in (Prawitz 2007). In a pragmatist theory of meaning "the meaning of a sentence is determined by what counts as its direct consequences and the meanings of the immediate subsentences" (Prawitz 2007, p. 469). As already noted, if a pragmatist theory of meaning is developed in the context of natural deduction, the meaning of a logical operation is laid down by its elimination rules instead of its introduction rules.

Moreover, he continues to explain that in (Prawitz 1973), the goal was

> to extend the notion of validity to a broader class of reasoning, what I called *arguments*. The simple idea is to consider not only deductions that proceed by applications of a given set of rules of inferences but trees of formulas built up of arbitrary inferences. . . . Such an arbitrary tree of formulas (in the language of first order logic) with indications of how variables and hypotheses are bound was called an *argument skeleton*.

The idea is that an argument is built up from an argument skeleton and reductions assigned to inferences that are not meaning constitutive.

The notion of valid proof defined above agrees with the notion defined in (Prawitz 1973, Definition 6.1) if 'proof' is replaced by 'argument schema'. There are schematic proofs that are valid with respect to any atomic system S for a given set of reductions, and that are logically valid in this sense. Indeed, intuitionistic predicate logic is sound with respect to proof-theoretic validity; given the familiar set of one-step detour conversions, every proof in intuitionistic predicate logic is logically valid. Prawitz (1973) conjectured that the converse holds as well, i.e., that if a derivation in the language of first-order logic is valid for every atomic system, then there exists a corresponding derivation in intuitionistic predicate logic. In (Prawitz 2014a), Prawitz states this *completeness conjecture* as follows:

> the conjecture that there are no stronger justifiable elimination rules within first order logic than the ones formulated by Gentzen can now naturally be formulated as follows: Every valid inference rule that can be formulated within first order languages holds as a derivable inference rule within the system of natural deduction for intuitionistic logic.

If the completeness conjecture turned out to be true, it would, of course, designate intuitionistic logic from the standpoint of Prawitz's conception of proof-theoretic semantics. As Schroeder-Heister (2013) explains, there are, however, "considerable doubts concerning the validity of this conjecture for systems that go beyond implicational logic. In any case it will depend on the precise formulation of the notion of validity, in particular on its handling of atomic systems."[17]

One may wonder how the proof-theoretic notion of validity compares to the model-theoretic notion of valid consequence when it comes to demonstrating that a given consequence fails to be valid. There cannot, of course, be a counter-model construction. In (Prawitz 1985), Prawitz defines a sentence A to be a valid consequence of sentences $A_1, \ldots, A_n$ whenever there exists a justifying reduction procedure with respect to which the one-step argument

$$\frac{A_1 \ldots A_n}{A}$$

is logically valid. A further detailing is required if the notion of validity of proofs is supposed to be persistent (or monotonic) with respect to extensions of a given atomic system and set of reduction procedures, cf. (Schroeder-Heister 2013).

Ignoring the set of reductions, for the propositional language containing conjunction, disjunction, and implication, Schroeder-Heister (2013) rephrases the definition

---

[17] A critical discussion of proof-theoretic conceptions of validity can also be found in (Read 2013).

of validity of proofs with respect to an atomic base S as a definition of *S-validity of formulas*, so as to obtain clauses that are analogous to the evaluation clauses in the denotational Kripke semantics for positive intuitionistic logic. The clause for implication, for example, is:

$(A \to B)$ is S-valid whenever for every extension S' of S, if A if S'-valid then B is S'-valid.

The atomic systems thus play the role of information states that are partially ordered by the expansion relation between atomic systems. A formula A fails to be a valid consequence of formulas $A_1, \ldots, A_n$ whenever there is an atomic system S such that while $A_1, \ldots, A_n$ are S-valid, A is not. Schroeder-Heister (2013) notes that Prawitz's conjecture is correct for this "validity-based" semantics with respect to the $\{\to, \wedge\}$-fragment of positive intuitionistic logic if the atomic systems may contain rules that discharge assumptions. The presence of disjunction, however, gives rise to problems. De Campos Sanz and Piecha (2014) have observed that Mints's rule

$$\frac{((A \to B) \to (A \vee C))}{(((A \to B) \to A) \vee ((A \to B) \to C))}$$

which is underivable in positive intuitionistic logic, provides a counterexample. However, it is questionable whether this observation with respect to Schroeder-Heister's validity-based semantics delivers a counterexample to Prawitz's conception of valid consequence.

According to Dag Prawitz's present standpoint, the earlier conception of proof-theoretic validity has not attained an epistemically reasonable notion of valid proofs regarded as representations of arguments. In (Prawitz 2014a), Prawitz expresses doubts about his notion of validity of arguments and reformulates the above definition of validity of proofs in two steps, thereby arriving at a notion of validity for "interpreted proof terms" from an extended typed lambda calculus. This notion of validity of interpreted proof terms comes quite close to the well-known *Brouwer-Heyting-Kolmogorov (BHK) interpretation* of the intuitionistic connectives and quantifiers in terms of (direct) proofs, see, for example, (Troelstra and van Dalen 1988). The crucial clause in the BHK interpretation is that for implication: a proof of $(A \to B)$ is a construction c that transforms any proof π of A into a proof $c(\pi)$ of B. The reason for Prawitz's revision of his earlier notion of validity is that in the previous conception of arguments,

> the justifications are operations defined on argument skeletons rather than on arguments, i.e. skeletons together with justifications, and furthermore that the value of the justifying operations consist of just argument skeletons instead of skeletons with justifications. It is the skeletons with justifications that represent arguments, valid or invalid ones, and when an argument step is to be justified it is conceivable that one wants the justification to depend on the entire arguments for the premisses and not only on their skeletons. ... [T]o accomplish that the justifications operate not just on argument skeletons, but, as it were, on skeletons with justifications, requires a more radical change in the approach. We need then conceive of the valid arguments, i.e. proofs, as built up of operations defined on proofs and yielding proofs as values.

A proof term of type A stands for a proof of A, and the proof-term forming operation $\rightarrow I$ that comes with the implication introduction rule is the following:

> $\rightarrow I(\alpha^A(P; B/(A \rightarrow B))$ is a proof term of type $(A \rightarrow B)$ and binds free occurrences of the proof variable α^A of type A in P, if P is a proof term of type B.

As a result, valid interpreted proof terms may be seen to stand for proofs in intuitionistic predicate logic, and the elimination rules for intuitionistic predicate logic emerge as valid in the newly defined sense. Moreover, Prawitz (2014a) restates the completeness conjecture with respect to this notion of proof-theoretic validity.

In (Prawitz 1971), Prawitz also uses a different notion of validity in proving normalization and distinguishes between the semantical understanding of validity based on the natural deduction introduction rules and validity used in proofs of normalizability. In particular, for proving normalizability, normal proofs are defined straightforwardly as valid, so that normal non-canonical proofs emerge as valid, although they are not, or at least not entirely, semantically justified by introduction inferences, cf. (Schroeder-Heister 2006). There is thus a notion of validity of proofs that is based on viewing the introduction rules as meaning-constitutive and a validity concept used in normalizability theory. Prawitz's work on proof normalization was inspired by Martin-Löf's notion of computability (Martin-Löf 1971), who adjusted William Tait's (1967) notion of convertibility for terms with combinators to natural deduction. Girard (1971) referred to Tait's convertibility predicate as "reducibilté". These days, validity concepts used in normalizability theory are often referred to as computability predicates, and Prawitz now agrees that it is unfortunate to call a computability predicate a validity notion. In order to prove *strong normalization* for the natural deduction proof systems for minimal and intuitionistic predicate logic, taking into account detour conversions and permutative conversions for $\vee$ and $\exists$, and classical predicate logic in the vocabulary without $\vee$ and $\exists$, Prawitz strengthened the validity notion of normalizability theory to a concept of *strong validity* by requiring that when a proof ends with the application of an elimination rule, then *every* derivation to which the given one immediately reduces is valid. Strong validity implies strong normalizability, and the proof of strong normalization consists in demonstrating that every proof is strongly valid. Prawitz (2014a) notes that "[m]odifying Martin-Löf's notion, I used it to prove strong normalization for various systems of natural deductions." As Anne Troelstra and Helmut Schwichtenberg (2000, p. 224) remark:

> Strong normalization was strongly put on the map by Prawitz [1971], who proved strong normalization for a natural-deduction version of intuitionistic second-order logic, using Girard's extension of Tait's method.[18]

[18] Prawitz's proof of strong normalization for second-order intuitionistic logic with respect to both detour and permutative conversions has been completed by Tatsuta (2005), who noted counterexamples to Theorem 2.2.1 (Prawitz 1971, p. 302), that is used in Prawitz's proof. A prominent reference to strong normalization is (Joachimski and Matthes 2003). A textbook presentation of the application of a computability predicate in proofs of strong normalization can be found in (Girard 1989), and a detailed discussion of various proof-theoretic validity concepts is presented by Schroeder-Heister (2006).

In (Prawitz 1981b), Prawitz then also proved strong normalization for classical second-order logic.

1.4.3 Proof-Theoretic Notions of Functional Completeness

Less prominent than proof-theoretic validity concepts but nevertheless important to proof-theoretic semantics is the idea of proof-theoretic conceptions of functional completeness, an idea emphasized, for example, in (Wansing 2000). The earliest proofs of functional completeness with respect to a proof-theoretic semantics are the functional completeness results for intuitionistic propositional logic and David Nelson's constructive propositional logics with strong negation (Almukdad and Nelson 1984; Kamide and Wansing 2012; Nelson 1949; Odintsov 2008) obtained by Franz von Kutschera (1968, 1969) using higher-level sequent systems.[19]

The problem of functional completeness for a given logic $\mathcal{L}$ consists of finding a preferably finite set op of logical operations from $\mathcal{L}$ such that *every* logical operation of $\mathcal{L}$ is explicitly definable by a finite number of compositions from the elements of op. The logical operations of $\mathcal{L}$ can be given just by the set of primitive operations of the language of $\mathcal{L}$, which is then trivially a set of operations with respect to which $\mathcal{L}$ is functionally complete. It may, however, also be possible to define a class of permissible or meaningful logical operations such that the set of primitive operations of $\mathcal{L}$ or some subset of it suffices to define every permissible operation. The singleton set containing the Sheffer stroke, for example, is not only functionally complete for classical propositional logic in any of its standard vocabularies, but the Sheffer stroke suffices to explicitly define *every* finitary truth function on the set of classical truth values. In the context of the proof-theoretic semantics of the logical operations based on natural deduction, the basic question is which shape natural deduction introduction and elimination rules may take for the introduction rules, or the elimination rules, or both to serve as meaning constitutive rules, and whether all operations the rules of which instantiate the general introduction and elimination schemata can be explicitly defined by a finite number of compositions from the elements of some finite set of operations.[20]

If, with Gentzen, it is assumed that the introduction rules are meaning constitutive, the definition of a set of permissible connectives requires (i) a specification of the

---

[19] It seems that von Kutschera was not aware of Nelson's work and independently developed constructive propositional logics with strong negation under the names 'direct propositional logic' ("direkte Aussagenlogik") and 'extended direct propositional Logic' ("erweiterte direkte Aussagenlogik").

[20] In correspondence, Prawitz suggested a more general notion of functional completeness, namely to call a set op of operations "complete in a language L with respect to a property P of operations, if all operations with the property P are definable in L in terms of op." The property P may then, for example, be the property of having its meaning given by an explicit schematic introduction rule or the property of being a Booelan function.

permissible introduction rules and (ii) a way of uniquely determining the elimination rules from the introduction rules. Within an equilibriumist approach, it is clear that some constraints *must* be imposed on the introduction and elimination rules, since on pain of triviality not any pair of introduction and elimination rules can assign meaning to a logical operation. The pertinent counterexample is Arthur Prior's binary tonk. Assuming reflexivity and transitivity of inference, the pair of rules is:

$$\frac{A}{(A\ \mathtt{tonk}\ B)} \qquad \frac{(A\ \mathtt{tonk}\ B)}{B}$$

It trivializes the notion of inference, and an obvious response to this problem is to require detour reducibility (or cut-eliminability in the sequent calculus), cf., for example, (Avron 2010), (Belnap 1962), (Humberstone 2013), (Wansing 2006a). Even in the absence of a transitive derivability relation, arbitrary rule schemata are problematic. The following left and right sequent rules for the unary operation $\bullet$ allow one to derive $\bullet A \vdash \bullet B$ for arbitrary formulas A and B, which makes it difficult to consider $\bullet$ as meaningful:

$$\frac{\Gamma, A \vdash B}{\Gamma, \bullet A \vdash B} \qquad \frac{}{\Delta, A \vdash \bullet B.}$$

In (Prawitz 1978), Prawitz considers the set of all finitary connectives defined by natural deduction introduction rules that instantiate what he calls explicit general introduction schemata, which come with associated general elimination schemata. He shows the set of intuitionistic connectives $\{\bot, \wedge, \vee, \rightarrow\}$ to be functionally complete with respect to the former set, in the sense that for every n-place connective φ with permissible introduction rules, there is a formula schema $\phi(p_1, \ldots, p_n)$ built up from the propositional variables $p_1, \ldots, p_n$ and connectives from $\{\bot, \wedge, \vee, \rightarrow\}$ such that for any formulas $A_1, \ldots, A_n$,

$$\varphi(A_1, \ldots, A_n) \leftrightarrow \phi(A_1, \ldots, A_n)$$

is provable in intuitionistic propositional logic together with the introduction and elimination rules for φ.[21]

Whereas (Prawitz 1978), like Zucker and Tragesser (1978), considers natural deduction rule schemata the instantiations of which are ordinary insofar as these rules allow the *discharge of formulas*, Schroeder-Heister (1984a), inspired by (Kutschera

[21] Zucker and Tragesser (1978) consider an interpretation of propositional connectives in terms of what they take to be the most general form of a natural deduction introduction rule. They show that in their framework for every permissible connective φ one can find a finite combination of connectives from $\{\bot, \wedge, \vee, \rightarrow\}$ with the same set of introduction rules and thus with the same meaning as φ. It is unclear, however, whether the existence of shared introduction rules implies replaceability in all deductive contexts, which follows from explicit definability. A proof of functional completeness of $\{\bot, \wedge, \vee, \rightarrow\}$ and $\{\neg, \wedge, \vee, \rightarrow\}$ with respect to evaluation clauses generalizing Kripke's relational semantics for intuitionistic propositional logic can be found in (McCullough 1971). Model-theoretic proofs of functional completeness along these lines for various constructive modal propositional logics with strong negation are presented in (Wansing 2006b).

1968), proved $\{\bot, \wedge, \vee, \rightarrow\}$ to be functionally complete for intuitionistic propositional logic with respect to a generalized natural deduction framework that allows assumptions of arbitrary finite level, that is, allows the *discharge of rules*.[22] This result is extended to a proof of functional completeness of the set $\{\bot, \wedge, \vee, \rightarrow, \exists, \forall\}$ for intuitionistic predicate logic by Schroeder-Heister (1984b). The higher-level sequent systems of von Kutschera are generalized to obtain functional completeness results for substructural subsystems of intuitionistic propositional logic and substructural subsystems of Nelson's constructive propositional logics with strong negation in (Wansing 1993a, b).

1.4.4 Some Other Philosophical Aspects

The themes of proof normalization, identity criteria for proofs, and other topics from Prawitz's agenda of general proof theory may be regarded to deal with derivations as mind-independent entities. The work of Prawitz on proofs and meaning has made it clear that a structural analysis of proofs is needed for developing an *epistemic* theory of the meaning of the logical operations and an understanding of logical consequence suitable to explain its procedural aspects and its epistemic accessibility. After all, proofs are regarded as entities that may provide epistemic subjects with knowledge. Prawitz examines the question, why some inferences confer evidence on their conclusions when applied to premises for which one already has evidence, as a fundamental problem of deductive inference and, one might add, therefore also as a basic problem of the epistemology of logic in general, see also (Prawitz 2009, 2011, 2012b).

Originally Prawitz concentrated on analyzing meaning in terms of proofs. More recently, however, and in his contribution (Prawitz 2014b) to the present volume, Prawitz enquires into the definition of proofs in terms of *legitimate inferences*. This approach may be seen as radically different from standard proof-theoretic semantics. Instead of defining the validity of inferences in terms of valid proofs (or arguments), now proofs are defined in terms of legitimate inferences. A legitimate inference is an inference that can be used to obtain a justification (or conclusive ground, or warrant) for asserting its conclusion. Since the notion of a ground for the assertion of a conclusion is an epistemic concept and since meaning is explained in terms of conclusive grounds, Prawitz's theory of meaning is indeed epistemic in contrast to the classical theory of meaning defined in terms of truth conditions. According to Prawitz, the notion of a legitimate inference is *conceptually prior* to the notion of proof: a proof is a chain of legitimate inferences that provides a conclusive ground for asserting the conclusion of the proof, whereas there does not exist a general concept of proof that one could use to elucidate the notion of a conclusive ground for an assertion. The classical, model-theoretic conception of valid inference clearly

---

[22] This higher-level approach avoids a problem that arises with applying Prawitz's elimination schema to implication, cf. footnote 2 of the translation of (Prawitz 1978) into German.

falls short of guaranteeing legitimacy of an inference: a model-theoretically valid one-step inference, for example, does not in general provide a conclusive ground for asserting its conclusion.

Epistemic theories of meaning are often associated with anti-realistic conceptions of truth and validity, and one may wonder whether proof-theoretic semantics comes with a commitment to anti-realism. Schroeder-Heister (2013) notes that "[f]ollowing Dummett, major parts of proof-theoretic semantics are associated with anti-realism," and Dubucs and Marion (2003, p. 235) even speak of "the traditional anti-realism of Dummett and Prawitz." Customarily, realism is understood as the doctrine that there is an objective reality of entities, or entities of a certain kind, that exists independently of any conscious beings. According to Dummett (1978), realism can be characterized semantically as assuming *bivalence*, namely the view that every meaningful declarative sentence from a certain discourse is either true or false but not both true and false and not neither true nor false, provided it is neither vague nor ambiguous. In (Dummett 1993, p. 75), Dummett explains that "[a] theory of meaning in terms of verification is bound to yield a notion of truth for which bivalence fails to hold for many sentences which we are unreflectively disposed to interpret in a realistic manner." Dummett and Prawitz, however, disagree about the notion of truth. For Prawitz, a sentence is by definition true just in case it is provable, and provability is, in his opinion, a tenseless and objective notion. According to Prawitz, identifying truth with the *actual* existence, in the sense of possession, of a proof is a "fatal flaw" (Prawitz 2012a). In his reply to (Dummett 1998), Prawitz (1998, p. 287) points out that

> a sentence is provable is here to mean simply that there is a proof of it. It is not required that we have actually constructed the proof or that we have a method for constructing it, only that there exists a proof in an abstract, tenseless sense of exists. . . . [T]ruth is something objective: it is in no way we who make a sentence true. To ask whether a mathematical sentence is true is to make an objective question, whose answer, if it has an answer, is independent of time. However, since the theory of meaning that I have in mind is constructive, it does not follow that every such question has an answer; we have indeed no reason to assume that.

It is important to note that by a proof Prawitz in this context means an *act product*. In his reply to Sundholm (1998), Prawitz (1998, p. 323) explains that

> a proof (i.e. a proof-object, but since we do not use the word proof for the act, the suffix "object" is less necessary here) I take to be the act product that results from an act of proving. The realist and the verificationist can agree on this analysis. The difference between them is that the verificationist but not the realist takes what counts as a direct proof or verification as being constitutive of meaning.

Moreover, for Prawitz, proofs as act products are epistemological entities insofar as the possession of a proof results in knowledge.

In a recently published paper on anti-realism and universal knowability, Michael Hand maintains that in view of the Knowability Paradox and unknowable true Fitch conjunctions of the form $A \wedge \neg KA$, where KA is to be read as "A is known by someone at some time," anti-realism must not only countenance true propositions that, as a matter of fact, remain unknown, but also true propositions the verification of which is unperformable. According to Hand (2010, p. 36), true Fitch conjunctions provide a "kind of innocuous recognition transcendence, and to acknowledge it we need Prawitz's ontological approach to antirealistic truth." Pagin (1994, p. 99) comes to the conclusion that if one accepts the existence of unknowable proofs, it will be difficult to find reasons against classical logic that are not also reasons against intuitionistic logic. In any case for Prawitz the truth of a sentence amounts to "the tenseless existence of a proof or ground" (Prawitz 2012a).

Usually the different co-existing approaches to linguistic meaning (model-theoretic, proof-theoretic, dynamic, game-theoretic, etc.) are viewed as rivals of each other. If a sharp distinction is drawn between syntax and semantics and if proof theory is seen as belonging to syntax, the expression 'proof-theoretic semantics' even appears to be a contradiction in terms.[23] If the different paradigms of meaning theories are seen as contenders, proof-theoretic semantics would be a methodology incompatible with realism as represented by model-theoretic semantics. However, this is not exactly Prawitz's attitude. According to him, "the picture produced by model theory is incomplete in essential respects" (Prawitz 1972, p. 131). Pointing to the epistemic dimension of proofs and the conception of proofs as processes[24] that reveal logical consequences, Prawitz writes:

> [i]n model theory, one concentrates on questions like what sentences are logically valid and what sentences follow logically from other sentences. But one disregards questions concerning how we know that a sentence is logically valid or follows logically from another sentence. General proof theory would thus be an attempt to supplement model theory by studying also the evidence or the process - i.e., in other words, the proofs - by which we come to know logical validities and logical consequences (Prawitz 1974, p. 66).

1.5 Conclusion

The above comments on an impressive œuvre that has developed over more than 50 years are bound to be incomplete. Nevertheless, it should have become clear that Dag Prawitz's contributions to logic are indeed outstanding. Prawitz's investigations into systems of natural deduction, at which we have glimpsed, have established natural deduction as a central paradigm in general proof theory. Moreover, Prawitz's contributions to a rule-based epistemic theory of meaning, truth, and validity form the centrepiece of an important and very active meaning-theoretical research programme that reveals the philosophical impact of Gentzen's ideas, namely proof-theoretic semantics.

Acknowledgments I wish to thank Dag Prawitz for his very kind and thoughtful co-operation during the edition of the *Outstanding Contributions to Logic* volume devoted to him and his work on proofs and meaning. In particular, I am grateful for his critical and helpful comments on earlier

---

[23] Such a remark has been made also by Prawitz (2006a, p. 507) and Schroeder-Heister (2006, p. 526).

[24] Proofs as acts in comparison to proofs as objects are discussed in (Sundholm 1998), and Dag Prawitz's comments on this discussion can be found in (Prawitz 1998, p. 318 ff.)

versions of the above paper. Moreover, I would like to thank Peter Schroeder-Heister for some useful comments and Jan von Plato for his very detailed and helpful remarks on the penultimate version of this paper.

References

Almukdad, A., & Nelson, D. (1984). Constructible falsity and inexact predicates. *Journal of Symbolic Logic*, *49*, 231–233.

Avigad, J. (2011). Proof theory. To appear in S.O. Hansson & V. Hendricks (Eds.), *Handbook of formal epistemology*. Dordrecht: Springer.

Avigad, J., & Reck, E. H. (2001). Clarifying the nature of the infinite: The development of metamathematics and proof theory. Carnegie-Mellon technical report CMU-PHIL-120. Available at http://www.andrew.cmu.edu/user/avigad/Papers/infinite.pdf.

Avron, A. (1988). The semantics and proof theory of linear logic. *Theoretical Computer Science*, *57*, 161–184.

Avron, A. (1996). The method of hypersequents in the proof theory of propositional non-classical logics. In W. Hodges, M. Hyland, C. Steinhorn, & J. Truss (Eds.), *Logic: From foundations to applications* (pp. 1–32). Oxford: Oxford Science Publications.

Avron, A. (2010). Tonk—A full mathematical solution. In A. Bletzki (Ed.), *Hues of philosophy: Essays in memory of Ruth Manor* (pp. 17–42). London: College Publications.

Baaz, M., Ciabattoni, A., & Fermüller, C. (2003). Hypersequent calculi for Gödel logics—A survey. *Journal Logic and Computation*, *13*, 1–27.

Belnap, N. D. (1962). Tonk, plonk and plink. *Analysis*, *22*, 130–134.

Belnap, N. D. (1982). Display logic. *Journal of Philosophical Logic*, *11*, 375–417. Reprinted with minor changes as Sect. 62 of Anderson A.R., Belnap N.D., and Dunn J.M. (1992). *Entailment: The logic of relevance and necessity* (Vol. 2). Princeton: Princeton University Press.

Belnap, N. D. (1995). The display problem. In H. Wansing (Ed.), *Proof theory of modal logic* (pp. 79–92). Dordrecht: Kluwer Academic Publishers.

Buss, S. R. (1998a). An introduction to proof theory. In (Buss 1998b) (pp. 1–78).

Buss, S. R. (Ed.). (1998b). *The handbook of proof theory*. Amsterdam: North-Holland.

Ciabattoni, A., Ramanayake, A. R., & Wansing, H. (2014). Hypersequent and display calculi—A unified perspective, in *Studia Logica*, published online July 2014. doi:10.1007/s11225-014-9566-z.

de Campos Sanz, W., Piecha, T. (2014). A critical remark on the BHK interpretation of implication. *Philosophia Scientiae*, *18*(3).

Došen, K. (2003). Identity of proofs based on normalization and generality. *The Bulletin of Symbolic Logic*, *9*, 477–503.

Došen, K. (2007). *Proof-theoretical coherence*. London: College Publications, 2004, revised version 2007. Available at http://www.mi.sanu.ac.rs/~kosta/coh.pdf

Dubucs, J., & Marion, M. (2003). Radical anti-realism and substructural logics. In A. Rojszczak et al. (Eds.), *Philosophical dimensions of logic and science: Selected contributed papers from the 11th international congress of logic, methodology, and philosophy of science* (pp. 235–249). Dordrecht: Kluwer Academic Publishers.

Dummett, M. (1973). *Frege. Philosophy of language*. New York: Harper & Row.

Dummett, M. (1978). *Truth and other enigmas*. London: Duckworth.

Dummett, M. (1993). *The seas of language*. Oxford: Clarendon Press.

Dummett, M. (1998). Truth from the constructive standpoint. *Theoria*, *64*, 122–138.

Dyckhoff, R. (1988). Implementing a simple proof assistant. In: *Proceedings of the workshop on programming for logic teaching (Leeds, 6–8 July 1987)* (pp. 49–59). Centre for Theoretical Computer Science, University of Leeds.

Dyckhoff, R. (2014). Cut elimination, substitution and normalisation, this volume.
Feferman, S. (1988). Hilbert's program relativized: Proof-theoretial and fondational reductions. *Journal of Symbolic Logic*, *53*, 284–364.
Feferman, S. (2000). Does reductive proof theory have a viable rationale. *Erkenntnis*, *53*, 63–96.
Fitch, F. (1952). *Symbolic logic. An introduction*. New York: The Ronald Press Company.
Francez, N., & Dyckhoff, R. (2010). Proof-theoretic semantics for a natural language fragment. *Linguistics and Philosophy*, *33*, 447–477.
Francez, N., Dyckhoff, R., & Ben-Avi, G. (2010). Proof-theoretic semantics for subsentential phrases. *Studia Logica*, *94*, 381–401.
Gentzen, G. (1934/35). Untersuchungen über das logische Schließen. *Mathematische Zeitschrift*, *39*, 176–210, 405–431. English translation in *American Philosophical Quarterly* 1 (1964), 288–306 and *American Philosophical Quarterly* 2 (1965), 204–218, as well as in *The collected papers of Gerhard Gentzen*, M.E. Szabo (Ed.), Amsterdam: North Holland, 1969, 68–131.
Gentzen, G. (2008). The normalization of derivations (title added by Jan von Plato). *The Bulletin of Symbolic Logic*, *14*, 245–257.
Girard, J.-Y. (1971). Une extension de l'interprétation de Gödel à l'analyse, et son application à l'elimination des coupures dans l'analyse et la théorie des types. In J. E. Fenstad (Ed.), *Proceedings of the Second Scandinavian Logic Symposium* (pp. 63–92). Amsterdam: North-Holland Publishing Company.
Girard, J.-Y., Lafon, Y., & Taylor, P. (1989). *Proofs and types*. Cambridge: Cambridge University Press.
Goré, R. (1998). Substructural logics on display. *Logic Journal of the IGPL*, *6*, 451–504.
Goré, R. (2000). Dual intuitionistic logic revisited. In R. Dyckhoff (Ed.), *Automated reasoning with analytic tableaux and related methods* (pp. 67–252). Springer Lecture Notes in AI 1847. Berlin: Springer.
Hacking, I. (1979). What is logic? *Journal of Philosophy*, *76*, 285–319.
Hand, M. (2010). Antirealism and universal knowability. *Synthese*, *173*, 25–39.
Hazen, A. P. & Pelletier, F. J. (2014). Gentzen and Jaśkowski natural deduction: Fundamentally similar but importantly different. *Studia Logica*, published online June 2014. doi:10.1007/s11225-014-9564-1.
Humberstone, L. (2013). Sentence connectives in formal logic. *The Stanford encyclopedia of philosophy* (Summer 2013 Edition), E. N. Zalta (Ed.), http://plato.stanford.edu/archives/sum2013/entries/connectives-logic/
Indrzejczak, A. (2014). *Studia Logica* special issue on Gentzen's and Jaśkowski's Heritage. 80 years of natural deduction and sequent calculi, to appear.
Jaśkowski, S. (1934). On the rules of supposition in formal logic. In J. Łukasiewicz (Ed.), *Studia Logica, Vol. 1*, published by the philosophical seminary of the faculty of mathematics and natural sciences, Warsaw, 1934, 5–32. Reprinted in S. McCall (Ed.), *Polish Logic 1920–1939*, pp. 232–258. Oxford: Oxford UP, 1967.
Joachimski, F., & Matthes, R. (2003). Short proofs of normalization for the simply-typed lambda-calculus, permutative conversions and Gödel's T. *Archive for Mathematical Logic*, *42*, 59–87.
Kamide, N., & Wansing, H. (2012). Proof theory of Nelson's paraconsistent logic: A uniform perspective. *Theoretical Computer Science*, *415*, 1–38.
López-Escobar, E.G.K. (1999). Standardizing the N systems of Gentzen. In X. Caicedo & C. H. Montenegro (Eds.), *Models, algebras and proofs* (Lecture Notes in Pure and Applied Mathematics, Vol. 203, pp. 411–434). Marcel Dekker, New York.
Lorenzen, P. (1950). Konstruktive Begründung der Mathematik. *Mathematische Zeitschrift*, *53*, 162–202.
Malmnäs, P.-E., & Prawitz, D. (1968). A survey of some connections between classical, intuitionistic and minimal logic. In H. A. Schmidt (Ed.), *Contributions to mathematical logic* (pp. 215–229). Amsterdam: North-Holland.
Mares, E., & Paoli, F. (2014). Logical consequence and the paradoxes. *Journal Philosophical Logic*, *43*, 439–469.

Martin-Löf, P. (1971). Hauptsatz for the intuitionistic theory of iterated inductive definitions. In J. E. Fenstad (Ed.), *Proceedings of the Second Scandinavian Logic Symposium* (pp. 179–216). Amsterdam: North-Holland Publishing Company.
Martin-Löf, P. About models for intuitionistic type theories and the notion of definitional equality. In S. Kanger (Ed.), *Proceedings of the Third Scandinavian Logic Symposium* (pp. 81–109). Amsterdam: North-Holland.
Martin-Löf, P. (1984). *Intuitionistic type theory*. Naples: Bibliopolis.
McCullough, D. P. (1971). Logical connectives for intuitionistic propositional logic. *Journal of Symbolic Logic*, *36*, 15–20.
Milne, P. (2014). Inversion principles and introduction rules, this volume.
Mints, G. E. (1992). *Selected papers in proof theory*. Amsterdam: Bibliopolis.
Negri, S., & von Plato, J. (2001). *Structural proof theory*. Cambridge: Cambridge UP.
Negri, S. & von Plato, J. (2011). *Proof analysis. A contribution to Hilbert's last problem*. Cambridge: Cambridge UP.
Negri, S. & von Plato, J. (2014). Meaning in use, this volume.
Nelson, D. (1949). Constructible falsity. *Journal of Symbolic Logic*, *14*, 16–26.
Odintsov, S. P. (2008). *Constructive negations and paraconsistency*. Dordrecht: Springer.
Pagin, P. (1994). Knowledge of proofs. *Topoi*, *13*, 93–100.
Paoli, F. (2007). Implicational paradoxes and the meaning of logical constants. *Australasian Journal of Philosophy*, *85*, 553–579.
Pelletier, F. J. (1999). A brief history of natural deduction. *History and Philosophy of Logic*, *20*, 1–31.
Pereira, L. C., Haeusler, E. H. H., & da Paiva, V. (Eds.). (2014). *Advances in natural deduction. A celebration of Dag Prawitz's work*. Dordrecht: Springer.
Postniece, L. (2010). *Proof theory and proof search of bi-intuitionistic and tense logic*. Ph.D. thesis, Australian National University, Canberra.
Prawitz, D. (1960). An improved proof procedure. *Theoria, 26*, 102–39, Reprinted (with comments) in J. Siekmann & G. Wrightson (Eds.), *Automation of reasoning 1: Classical papers on computational logic 1957–1966* (pp. 162–201). Berlin: Springer, 1983.
Prawitz, D. (1965). *Natural deduction. A proof-theoretical study*. Stockholm: Almqvist and Wiksell (Reprinted with Dover Publications, Mineola/NY, 2006).
Prawitz, D. (1967). Completeness and Hauptsatz for second order logic. *Theoria*, *33*, 246–258.
Prawitz, D. (1969). Hauptsatz for higher order logic. *Journal of Symbolic Logic*, *33*, 57–452.
Prawitz, D. (1970a). Constructive semantics. In *Proceedings of the 1st Scandinavian Logic Symposium Åbo 1968*. Filosofiska Föreningen och Filosofiska Institutionen vid Uppsala Universitet, Uppsala, 96–114.
Prawitz, D. (1970b). Some results for intuitionistic logic with second order quantifiers. In J. Myhill, et al. (Eds.), *Intuitionism and Proof Theory, Proceedings of the Summer Conference at Buffalo* (pp. 259–269). Amsterdam: North-Holland.
Prawitz, D. (1971). Ideas and results in proof theory. In J. E. Fenstad (Ed.), *Proceedings of the Second Scandinavian Logic Symposium* (pp. 235–307). Amsterdam: North-Holland Publishing Company.
Prawitz, D. (1972). The philosophical position of proof theory. In R. E. Olson (Ed.), *Contemporary philosophy in Scandinavia* (pp. 123–134). Baltimore: The John Hopkins Press.
Prawitz, D. (1973). Towards a foundation of general proof theory. In P. Suppes, et al. (Eds.), *Logic, methodology and philosophy of science IV* (pp. 225–50). Amsterdam: North-Holland Publishing Company.
Prawitz, D. (1974). On the idea of a general proof theory. *Synthese*, *27*, 63–77.
Prawitz, D. (1978). Proofs and the meaning and completeness of the logical constants. In J. Hintikka et al. (Eds.), *Essays on mathematical and philosophical logic* (pp. 25–40) Dordrecht: Reidel, translated into German as Beweise und die Bedeutung und Vollständigkeit der logischen Konstanten, *Conceptus, 26*, 3–44, 1982.

Prawitz, D. (1981a). Philosophical aspects of proof theory. In G. Fløistad (Ed.), *Contemporary philosophy. A new survey*. Volume 1: Philosophy of Language (pp. 77–235). The Haugue: Martinus Nijhoff Publishers.

Prawitz, D. (1981b). Validity and normalizability of proofs in 1st and 2nd order classical and intuitionistic logic. In S. Bernini (Ed.), *Atti del congresso nazionale di logica* (pp. 11–36). Naples: Bibliopolis.

Prawitz, D. (1985). Remarks on some approaches to the concept of logical consequence. *Synthese*, *62*, 153–171.

Prawitz, D. (1994). Review of M. Dummett, *The logical basis of metaphysics*. London: Duckworth, 1991. *Mind*, *103*, 373–376.

Prawitz, D. (1998). Comments on the papers. *Theoria*, *64*, 283–337.

Prawitz, D. (2002). Problems for a generalization of a verificationist theory of meaning. *Topoi*, *21*, 87–92.

Prawitz, D. (2006a). Meaning approached via proofs. *Synthese*, *148*, 507–524.

Prawitz, D. (2006b). Validity of inferences, In *Proceedings from the 2nd Launer Symposium on Analytical Philosophy on the Occasion of the Presentation of the Launer Prize at Bern 2006*. Available at http://people.su.se/~prawd/Bern2006.pdf.

Prawitz, D. (2007). Pragmatist and verificationist theories of meaning. In R. E. Auxier & L. E. Hahn (Eds.), *The philosophy of Michael Dummett* (pp. 455–481). Chicago: Open Court.

Prawitz, D. (2009). Inference and knowledge. In M. Pelis (Ed.), *The logica yearbook 2008* (pp. 175–192). London: College Publications.

Prawitz, D. (2011). To explain deduction, to appear in a volume published in connection with the Lauener Prize 2010 given to Michael Dummett.

Prawitz, D. (2012a). Truth as an epistemic notion. *Topoi*, *31*, 9–16.

Prawitz, D. (2012b). The epistemic significance of valid inference. *Synthese*, *187*, 887–898.

Prawitz, D. (2012c). A note on Gentzen's 2nd consistency proof and normalization of natural deductions in 1st order arithmetic, draft.

Prawitz, D. (2014a). An approach to general proof theory and a conjecture of a kind of completeness of intuitionistic logic revisited. In L. C. Pereira, E. H. Haeusler, & V. de Paiva (Eds.), *Advances in natural deduction. A celebration of Dag Prawitz's work* (pp. 269–279). Dordrecht: Springer.

Prawitz, D. (2014b). Explaining deductive inference, this volume.

Prawitz, D., Prawitz, H. & Voghera, N. (1960). A mechanical proof procedure and its realization in an electronic computer, *Journal of the Association for Computing Machinery 7*, 102–128. Reprinted (with comments) in J. Siekmann & G. Wrightson (Eds.), *Automation of reasoning 1: Classical papers on computational logic 1957–1966* (pp. 28–202). Berlin: Springer, 1983

Raggio, A. (1965). Gentzen's Hauptsatz for the systems NI and NK. *Logique et Analyse*, *8*, 91–100.

Read, S. (2010). General-elimination harmony and the meaning of the logical constants. *Journal of Philosophical Logic*, *39*, 557–576.

Read, S. (2013). Proof-theoretic validity. To appear in C. Caret & O. Hjortland (Eds.), *Foundations of logical consequence*. Oxford: Oxford University Press.

Read, S. (2014) General-elimination harmony and higher-level rules, this volume.

Restall, G. (2000). *An introduction to substructural logic*. London: Routledge.

Schroeder-Heister, P. (1981). *Untersuchungen zur regellogischen Deutung von Aussagenverknüpfungen*. Ph.D. thesis, University of Bonn.

Schroeder-Heister, P. (1984a). A natural extension of natural deduction. *Journal of Symbolic Logic*, *49*, 1284–1300.

Schroeder-Heister, P. (1984b). Generalized rules for quantifiers and the completeness of the intuitionistic operators & $\vee$, $\supset$, $\forall$, $\exists$. In M. M. Richter, et al. (Eds.), *Computation and proof theory, Proceedings of the Logic Colloquium Held in Aachen, July 18–23, 1983, Part II*. Springer Lecture Notes in Mathematics (Vol. 1104, pp. 399–426). Berlin: Springer.

Schroeder-Heister, P. (1991). Uniform proof-theoretic semantics for logical constants (Abstract). *Journal of Symbolic Logic*, *56*, 1142.

Schroeder-Heister, P. (2006). Validity concepts in proof-theoretic semantics. In R. Kahle & P. Schroeder-Heister (Eds.), *Special issue on proof-theoretic semantics* (*Synthese 148*, 525–571).
Schroeder-Heister, P. (2008). Lorenzen's operative justification of intuitionistic logic. In M. van Atten, P. Boldini, M. Bourdeau & G. Heinzmann (Eds.), *One hundred years of intuitionism (1907–2007)* (pp. 214–240). Basel: Birkhäuser.
Schroeder-Heister, P. (2009). Sequent calculi and bidirectional natural deduction: On the proper basis of proof-theoretic semantics. In M. Peliš (Ed.), *The logica yearbook 2008* (pp. 237–251). London: College Publications.
Schroeder-Heister, P. (2013). Proof-theoretic semantics. In E. N. Zalta (Ed.), *The Stanford encyclopedia of philosophy* (Spring 2013 Edition), http://plato.stanford.edu/archives/spr2013/entries/proof-theoretic-semantics/
Schroeder-Heister, P. (2014a). Generalized elimination inferences, higher-level rules, and the implications-as-rules interpretation. In L. C. Pereira, E. H. Haeusler, & V. de Paiva (Eds.), *Advances in natural deduction. A celebration of Dag Prawitz's work* (pp. 1–29). Dordrecht: Springer.
Schroeder-Heister, P. (2014b). Harmony in proof-theoretic semantics: A reductive analysis, this volume.
Schütte, K. (1960). Syntactical and semantical properties of simple type-theory. *Journal of Symbolic Logic*, *25*, 305–326.
Seldin, J. (1986). On the proof theory of the intermediate logic MH. *Journal of Symbolic Logic*, *51*, 626–647.
Seldin, J. (1989). Normalization and excluded middle I. *Studia Logica*, *48*, 193–217.
Shoesmith, D. J., & Smiley, T. J. (1978). *Multiple conclusion logic*. Cambridge: Cambridge University Press.
Sieg, W. (2013). *Hilbert's programs and beyond*. Oxford: Oxford University Press.
Steinberger, F. (2013). On the equivalence conjecture for proof-theoretic harmony. *Notre Dame Journal of Formal Logic*, *54*, 79–85.
Straßburger, L. (2007). What is a logic, and what is a proof? In J.-Y. Béziau (Ed.), *Logica Universalis* (2nd ed., pp. 135–152). Basel: Birkhäuser.
Sundholm, G. (1981). Hacking's logic. *The Journal of Philosophy*, *78*, 160–168.
Sundholm, G. (1998). Proofs as acts and proofs as objects: Some questions for Dag Prawitz. *Theoria*, *64*, 187–216.
Szabo, M. E. (Ed.). (1969). *The collected papers of Gerhard Gentzen*. Amsterdam: North Holland Publishing.
Tait, W. (1966). A nonconstructive proof of Gentzen's Hauptsatz for second order predicate logic. *Bulletin of the American Mathematical Society*, *72*, 980–983.
Tait, W. (1967). Intentional interpretation of functionals of finite type I. *The Journal of Symbolic Logic*, *32*, 198–212.
Takahashi, M. (1967). A proof of cut-elimination theorem in simple type-theory. *Journal of the Mathematical Society of Japan*, *19*, 399–410.
Takeuti, G. (1975). *Proof theory, studies in logic and the foundations of mathematics,* Vol. 81. Amsterdam: North-Holland. (2nd ed. from 1987 reprinted with Dover Publications, Mineola/NY, 2013).
Tatsuta, M. (2005). Second order permutative conversions with Prawitz's strong validity. *Progress in Informatics*, *2*, 41–56.
Tennant, N. (1978). *Natural logic*. Edinburgh: Edinburgh University Press.
Tennant, N. (1992). *Autologic*. Edinburgh: Edinburgh University Press.
Tranchini, L. (2012). Natural deduction for dual-intuitionistic logic. *Studia Logica*, *100*, 631–648.
Troelstra, A., & van Dalen, D. (1988). *Constructivism in mathematics* (Vol. 1). Amsterdam: North-Holland.
Troelstra, A., & Schwichtenberg, H. (2000). *Basic proof theory* (2nd ed.). Cambridge: Cambridge UP.

Ungar, A. (1992). *Normalization, cut elimination, and the theory of proofs*. CSLI Lecture Notes No. 28, Stanford.
Urban, C. (2014). Revisiting Zucker's work on the correspondence between cut-elimination and normalisation. In L. C. Pereira, E. H. Haeusler, & V. de Paiva (Eds.), *Advances in natural deduction. A celebration of Dag Prawitz's work* (pp. 31–50). Dordrecht: Springer.
von Kutschera, F. (1968). Die Vollstäandigkeit des Operatorensystems $\{\neg, \wedge, \vee, \supset\}$ für die intuitionistische Aussagenlogik im Rahmen der Gentzensemantik. *Archiv für Mathematische Logik und Grundlagenforschung, 11*, 3–16.
von Kutschera, F. (1969). Ein verallgemeinerter Widerlegungsbegriff für Gentzenkalküle. *Archiv für Mathematische Logik und Grundlagenforschung, 12*, 104–118.
von Plato, J. (2000). A problem of normal form in natural deduction. *Mathematical Logic Quarterly, 46*, 121–124.
von Plato, J. (2001). Natural deduction with general elimination rules. *Archive for Mathematical Logic, 40*, 541–567.
von Plato, J. (2003). Translations from natural deduction to sequent calculus. *Mathematical Logic Quarterly, 49*, 435–443.
von Plato, J. (2008a) The development of proof theory. *The Stanford encyclopedia of philosophy* (Fall 2008 Edition), E. N. Zalta (Ed.), http://plato.stanford.edu/archives/fall2008/entries/proof-theory-development/
von Plato, J. (2008b). Gentzen's proof of normalization for natural deduction. *The Bulletin of Symbolic Logic, 14*, 240–244.
von Plato, J. (2009). Proof theory of classical and intuitionistic logic, Chapter 11 in L. Haaparanta (Ed.), *History of modern logic* (pp. 499–515). Oxford: Oxford University Press.
von Plato, J. (2012). Gentzen's proof systems: Byproducts in a work of genius. *The Bulletin of Symbolic Logic, 18*, 313–367.
von Plato, J., & Siders, A. (2012). Normal derivability in classical natural deduction. *Review of Symbolic Logic, 5*, 205–211.
Wansing, H. (1993a). Functional completeness for subsystems of intuitionistic propositional logic. *Journal of Philosophical Logic, 22*, 303–321.
Wansing, H. (1993b). *The logic of information structures*. Springer Lecture Notes in Artificial Intelligence (Vol. 681). Berlin: Spinger.
Wansing, H. (1998). *Displaying modal logic*. Dordrecht: Kluwer Academic Publishers.
Wansing, H. (2000). The idea of a proof-theoretic semantics and the meaning of the logical operations. *Studia Logica, 64*, 3–20.
Wansing, H. (2006a). Connectives stranger than tonk. *Journal of Philosophical Logic, 35*, 653–660.
Wansing, H. (2006b). Logical connectives for constructive modal logic. *Synthese, 150*, 459–482.
Wansing, H. (2013). Falsification, natural deduction, and bi-intuitionistic logic. *Journal of Logic and Computation*, published online: July 17. doi:10.1093/logcom/ext035.
Widebäck, F. (2001). *Identity of proofs*. Ph.D. thesis, University of Stockholm, Stockholm: Almqvist and Wiksell International.
Zach, R. (2009). Hilbert's program. *The Stanford encyclopedia of philosophy* (Spring 2009 Edition), E. N. Zalta (Ed.), http://plato.stanford.edu/archives/spr2009/entries/hilbert-program/
Zucker, J. (1974). Cut-elimination and normalization. *Annals of Mathematical Logic, 7*, 1–112.
Zucker, J., & Tragesser, R. (1978). The adequacy problem for inferential logic. *Journal of Philosophical Logic, 7*, 501–516.

Chapter 2
A Short Scientific Autobiography

Dag Prawitz

2.1 Childhood and School

Being born in 1936 in Stockholm, I have memories from the time of the Second World War. But Sweden was not involved, and my childhood was peaceful. One notable effect of the war was that even in the centre of Stockholm, where I grew up, there was very little automobile traffic. Goods were often transported by horse-drawn wagons. At the age of six we children could play in the streets and run to the nearby parks without the company of any adults.

One memory from this time happens to illustrate a theme that I was to be quite concerned with as an adult: the difference between canonical and non-canonical methods and the power of the latter. Children in Sweden normally do not begin school until the year they reach seven. Many learn to count before that. Some of us could count to one hundred and were even able to add small numbers together. We did this orally in the canonical way: To get the sum of 3 and 5, we counted 4, 5, 6 … and kept track of the number of steps until they were 5, whereupon we proudly announced 8. To add big numbers was of course out of question, even if we understood how it could be done in principle: to add 30 and 50, we would have to start counting 31, 32, 33, and would loose count of the number of steps long before reaching 80. My father then told me of a non-canonical way: to add 30 and 50 is to add 5 tens to 3 tens, just like adding 5 apples to 3 apples. I still recall the faces of my friends when I started to practice this trick, a mixture of suspicion and admiration: how could I do so big sums so fast?

Otherwise, I was slow as a child. My father tried to teach me how to read, but I never managed words longer than one syllable before entering school. Having absorbed what I learned in school, I showed some pedagogic ambitions, however. At home I set up my own school where I took the role of the teacher of my three years younger sister Gunilla, trying to teach her to read and write. This was successful enough to get her accepted in the second form when she entered school at seven; it is another story that skipping the first form turned out to have many drawbacks.

D. Prawitz (✉)
Department of Philosophy, Stockholm University, 10691 Stockholm, Sweden
e-mail: dag.prawitz@philosophy.su.se

H. Wansing (ed.), *Dag Prawitz on Proofs and Meaning*, Outstanding Contributions to Logic 7, DOI 10.1007/978-3-319-11041-7_2

Summers were spent on an island in the Stockholm archipelago, a different world I loved very much. A renowned Hungarian mathematician, Marcel Riesz, professor at a Swedish university, often came to stay with us for a while. He had been the teacher of my father, and they used to discuss mathematics while sunbathing on the rocks at the seaside. When I entered secondary (middle) school at the age of eleven, he wanted to teach me something and tried to prove Pythagoras' theorem in a way that should enable me to take it in. I understood the general idea of proving theorems from axioms, but I questioned the axioms and grasped little of the proof. He also explained to me why a natural number is divisible by 3 if, and only if, the sum of the numbers that occur in its Arabic notation is divisible by 3. I understood his demonstration, I thought, but when he gave me the task of writing down the proof in my own words, it was too much for me. My mother was afraid that I would become overworked by such attempts, and asked Marcel to stop these exercises.

My early childhood seems to me now to have been rather dull. I entered gymnasium (high school) at 15, which was like taking a step into the world of adults; teachers showed us a new kind of respect, addressed us in a more formal way, and had more interesting things to say—this gave us more self-respect, and we wanted to do things that adults do. In my class, we were a group of friends who started to publish a magazine. It was stencilled and sold well within the whole school. As editor of the magazine, I was summoned to the headmaster one day. A poem written by another pupil that we had published was offensive when some of the words were taken in a sexual sense. The fact was that I was completely ignorant of that reading.

I was also active in several clubs that flourished at the gymnasium. I came to chair several of them: the school's rifle-club, its literary society, which then celebrated its 50th anniversary, and the pupils' council, then a fairly new invention. I was also a little involved in an association of similar councils in the Stockholm area that was formed at this time. In this connection, the idea of a joint school magazine came up. Together with some pupils from other schools, I started such a magazine, and it came to be sold in schools throughout the whole of Sweden. It was printed at one of the leading Swedish newspapers, and its layout was that of a newspaper. Soon I spent most of my time as editor in the composing room of this newspaper, as a reporter at different events, or as canvasser for advertisements to finance the project. It was a huge undertaking and filled up more than my spare time. The headmaster of my school (now another one) encouraged my work, and allowed me to take time off from school. My schoolwork was minimized accordingly.

The school subjects to which I devoted most energy were history and philosophy. Most of our teachers were very competent, and many were qualified to teach at the university. My philosophy teacher was especially qualified and had written a book about Nietzsche. He was also an inspiring teacher. Philosophy was an optional subject the last two years of gymnasium, and my interest in philosophy dates from this time.

2.2 Undergraduate Studies

When I finished school, I got a temporary job for the summer as journalist at a provincial newspaper. For the next year, I had been awarded a very generous scholarship to study at the University of Wisconsin, where I was accepted as a junior student (to have completed a Swedish high school at this time was considered comparable to having studied two years at an American university). There I studied journalism, rhetoric, and psychology. My plans for the future were quite unclear. I wanted to study at university for a few years. As possible careers afterwards I was thinking of journalism or politics. What I did not have the slightest idea of at the time was that the University of Wisconsin had a most eminent logician Stephen Kleene, though I would become very aware of this two years later.

Back at Stockholm I wanted to allow myself a short time of luxury, during which I would study philosophy, psychology, and mathematics for a first degree, before doing something more useful that could earn me a living. The normal time for graduating from Swedish universities was three years, but I was eager to get it done faster.

I started with theoretical philosophy, which is a separate subject at Swedish universities. It was taught at Stockholm University by only one professor, Anders Wedberg, and an assistant, Stig Kanger. Wedberg was an historian of philosophy, internationally known for his book *Plato's Philosophy of Mathematics*. But he had studied logic at Princeton at its glorious period at the end of the 30s, and had written two small booklets on modern logic intended for a general audience. They were written in a fluent style and explained not only the main ideas of sentential and predicate logic in an intuitive way but touched also on more advanced and fascinating themes such as Gödel's completeness and incompleteness theorems.

In my first semester, Wedberg lectured on logic. As a teacher, he taught us scrupulously the language of first order logic. This combination of intuition and exactitude fascinated me. It was thrilling to demonstrate rigorously by intuitive reasoning that a conclusion necessarily follows from given premisses, and it was mysterious how this could be possible at all. It seemed that modern logic was able to explain why such necessary connections hold, and it was impressive that this could be done so exactly and in such details that within certain areas, a machine could in principle always take over and find a proof, if the connection did hold.

In the second semester we were ready for Hilbert and Ackermann's book *Grundzüge der theoretischen Logik*. It was then in its third edition from 1949 (not to be confused with the completely changed fourth edition). Wedberg told us that the principle of substitution for second order logic was wrongly formulated in the first edition and that even the attempted emendation in the second edition had gone wrong. Let us now see if it has come right in the third edition, he suggested. I wondered myself how one was to decide such a question when even the experts had got it wrong. I imagined that one would have to read a lot in order to compare different formulations of the rule that could be found in the literature. Wedberg's intention was of course different: we should think out for ourselves how the rule must be formulated in order to be correct. He taught us how to reason, and gave us faith in our ability to do so.

At the end of this semester, Kanger was to defend his doctoral thesis at the kind of public disputation that is still in use in Sweden; I followed it with fascination. One of his results was a new completeness proof for first order logic that offered a seemingly feasible method for proving theorems. Even a computer could use it, Wedberg suggested, and so the possibility of a machine proving theorems could be realized in practice.

I now got the idea of devoting the approaching summer to such a project. I withdrew to a lonely country house in Denmark, bringing with me Kanger's dissertation *Provability in logic* and Beth's essay on semantic tableaux, *Semantic entailment and formal derivability*, which I had been told contained a completeness proof similar to Kanger's. The idea was to work out an algorithm for proving theorems that could be implemented on a computer. There were no programming languages at that time. Programs had to be written directly in the machine code for a particular machine. I knew no such codes, but I understood enough of the principles of mechanical manipulations to invent my own programing language. I defined a number of specific operations for syntactic transformations of formulas. In terms of these, I laid down instructions for how, given a sequence of formulas in a first order language as input, one was to make syntactic manipulations in a certain order and store the result in various specified memories. The outcome of following the instructions would be a proof that the last one of the input formulas was a logical consequence of the preceding ones, if this was in fact the case.

I presented and discussed my algorithm in an essay I submitted as a graduation paper when I returned to Stockholm in the autumn. My studies in psychology and mathematics had run parallel with the ones in theoretical philosophy, and later in the autumn I obtained my first university degree (corresponding to a BA). I wanted now to go in for theoretical philosophy. To my delight Wedberg proposed that I should continue as a graduate and offered me a scholarship. In January 1958, I enrolled and started to study Kleene's *Introduction to Metamathematics*.

2.3 Mechanical Proof Procedures

The first Swedish computer had been developed in the 50s, and for a short while it held the world record for speed. A successor to it now filled several big rooms in the former building of the Royal Institute of Technology in Stockholm. It so happened that my father used it for certain tasks and knew how to program it. He was kind enough to use part of his summer vacation to translate into the code for this machine the algorithm that I had written in my homemade programing language.

My father was a mathematician and did not know any modern logic; his contact with the subject amounted to having heard Hilbert delivering the lecture "Über das Unendliche" at the celebration in honour of Weierstrass in 1925 at Münster—having been sent there by Mittag-Leffler, whose assistant he was at that time. Strangely enough we seldom spoke about such things. He was usually quite busy, and after having translated my algorithm into a machine program he did not have more time to devote to it. His work was continued in the autumn by a fellow-student of mine, Neri Voghera, who was an assistant at the institute responsible for the machine for which the program was written. He tested and modified the program, and in 1958 he could run it to prove simple theorems of predicate logic. It seems to have been the first full-scale theorem-prover for predicate logic implemented on a computer. I presented our work at the First International Conference on Information Processing at UNESCO in Paris in June 1959, and a joint paper by the three of us was published in 1960.

It was clear to me already in the summer of 1957 that the proof procedure I was working on was not very efficient after all. One of its main limitations was that instances of quantified sentences that had to be formed in the proof search were generated more or less at random. In the essay I wrote that summer, I discussed how to improve the procedure in that respect: the substitutions for quantified variables should be postponed to a stage at which one could see that the substitutions were useful for finding a proof. When I presented the essay in the autumn, it turned out that Stig Kanger had a similar idea.

A year later, I was able to work out this idea in the form of a method that did not make any substitution until it was guaranteed that the substitution instances would generate a proof. The method was presented in a rather unwieldy paper called "An improved proof procedure" published in 1960. It would have benefited from some guidance, but to supervise graduate students was not the custom in Sweden at that time, at least not in philosophy. Anyhow, the ideas of this paper came to some use in the soon growing field of automated theorem proving, as was generously acknowledged by Martin Davis when the paper was later republished in the volume *Automation of Reasoning*. It also became my thesis for a second university degree (*filosofie licentiatexamen*), which was then a required step before one could go on to a doctorate. I now left this field for other interests that I wanted to pursue in a doctoral dissertation, and returned to it only a couple of times at the end of the 60s, when I described the main idea in a more mature and compressed way and also developed it somewhat further.

2.4 Towards the Doctoral Thesis—Normalizations of Natural Deductions

Before I could take up doctoral studies, I had to complete my military service. I had so far been conscripted for a few summer months as a recruit, and had been able to postpone the rest of the service, but now it had to be fulfilled. I was placed as a psychologist because of my earlier psychology studies. My duties were not without interest. I conducted hundreds of interviews with soldiers as part of the military selection process, and worked for several months at a psychiatric clinic with the aim of learning how to sort neurosis in the case of war. But I doubt that I should have been able to function very well as a military psychologist if it had really come to it.

In the summer of 1961, I returned to studies. I had been awarded a scholarship for four years in order to write a doctoral dissertation. Again I withdrew to a lonely place, this time renting a summerhouse in the Stockholm archipelago. There I started to read works by logicians who had philosophical ideas about what matters in deductions.

One reason for leaving the field of automated theorem proving was that I had a wish to understand philosophically what it really was that made something a proof. I took for granted that this question must somehow be connected with the meaning of the sentences occurring in the proof. Stig Kanger, who was the closest senior logician in my surroundings, was sure that all such questions concerning meaning and all that really matters in logic were to be approached model-theoretically. To begin with, this seemed plausible to me. At least, it was not contradicted by my experience of how a completeness proof could generate a proof algorithm. It also seemed to be confirmed by other facts such as the one that Kanger had obtained in his dissertation a three-line proof of Gentzen's main result, the Hauptsatz, from a completeness proof. For a while, I was led to study model theory, but it soon left me unsatisfied, and when resuming studies in 1961, I turned to approaches that linked questions of meaning to deductive matters.

I began reading works by Paul Lorenzen and Haskell Curry. Curry had made several attempts at giving a kind of inferential interpretation of the logical constants, which attracted me, but I was repelled by his formalist outlook and his general way of writing, and never looked at his work on combinatory logic. Lorenzen related the meaning of the logical constants to statements about underlying logic free calculi like Curry, but he was more inspiring to read, although I was not yet quite ready for his constructivist attitude.

Then I reread Gerhard Gentzen's "Untersuchungen über das logische Schliessen". I had read this work before and like most people I had mostly paid attention to his sequent calculus. This time I was immediately struck by the depth of his idea that the introduction rules of the system of natural deduction determine the meaning of the logical constants and that other rules are justified by reduction procedures that eliminate certain uses of them. It seemed to give some kind of answer to my question what it is that makes something to a proof. I saw in a flash that by iterating these reductions it was possible to put the deductions in a certain normal form, which gave Gentzen's idea a sharper form.

The discovery of this normalization procedure filled me with great joy that summer in the Stockholm archipelago. Soon I also realized that this procedure corresponded to eliminating cuts in the sequent calculus, which gave the latter enterprise a greater significance. I was surprised that Gentzen had not stated this beautiful normalization theorem for natural deduction, but it seemed clear from his writings that he had seen the possibility of such a theorem and that his Hauptsatz had its source in this insight. Accordingly, I was not surprised when Jan von Plato a few years ago told me that he had found a proof of the normalization theorem for intuitionistic logic in Gentzen's unpublished Nachlass.

When I returned to the mainland in the autumn, I presented in outline the proof of the normalization theorem at a joint Stockholm-Uppsala seminar, which had just then been initiated. Wedberg pointed to a problem concerning the inductive measure that I employed. I could remove the problem shortly afterwards, and used the rest of the year to work out other details connected with normalization.

The next spring I went to the Institute for Mathematical Logic at Münster and stayed there for the summer. I attended lectures by Hans Hermes, Gisbert Hasenjaeger, and Wilhelm Ackermann. I was glad to see that the author of my old textbook of logic was still alive; Ackermann was now a retired schoolteacher, and came to the institute every second week to lecture on his notion of "strenge Implikation" and to stay for the joint seminars, which were led by Hermes. They gathered all at the institute and were its highlights. I was invited to present my results about normalization of natural deduction at one of its sessions. My presentation was well received on the whole, but Hasenjaeger wondered why I bothered with natural deduction when Gentzen had been so happy to leave it after having found his sequent calculus.

Hasenjaeger's reaction was not uncommon. It took some time before the normalization theorem was accepted as a significant way of formulating the main idea behind Gentzen's Hauptsatz. Certainly, for certain aims and from certain perspectives, sequent calculi are preferable to systems of natural deduction. But the philosophical essence of Gentzen's result appears most clearly in natural deduction: here one sees how applications of introduction rules give rise to canonical forms of reasoning and how other forms of reasoning which proceed by applications of elimination rules are justified to the extent that they can be reduced to canonical form; the prerequisite of this justification by reduction was that the eliminations were the "inverses" of the introductions, to use a term that I had found in Lorenzen, and when this condition was satisfied, the prerequisite for proving the Hauptsatz for the corresponding calculus of sequents was also at hand. As is clear from the introduction to his second consistency proof, Gentzen himself saw this as a key to the understanding of his result. I am glad that Gentzen's high appreciation of his normalization result has been further confirmed by von Plato's recent investigations of Gentzen's Nachlass; it appears that Gentzen even wanted to make it a cornerstone of a book on the foundations of mathematics that he was planning.

Most of what was to become my doctoral dissertation *Natural Deduction* three years later was ready for publication now after my first year as candidate for the doctorate. But my plans for the dissertation were more ambitious. There were several other themes that I was working on. One was the question of the best way to present classical logic as a natural deduction system. I wanted to have a more dual system than Gentzen's, and experimented with several different ideas, for instance, letting the deduction trees branch downwards at disjunction eliminations or taking the nodes of the deduction trees as disjunctions.

Another theme was the extension of the normalization theorem to second order logic. I knew that Takeuti had conjectured that the Hauptsatz held for second order logic, but I had not yet digested his rather long proof of the claim that this result would yield the consistency of first order Peano arithmetic.

A third theme was the relation between classical logic and intuitionistic logic. A fourth was the interpretation of intuitionistic logic. I saw that one could map natural deductions into an extended typed lambda calculus. To each provable sentence in predicate logic I assigned a term in an extended lambda calculus, and took it as an interpretation of intuitionistic logic in terms of constructions. It intrigued me that this mapping of two natural deductions of which one reduced to the other resulted in two terms in the lambda calculus that denoted the same construction. I presented this idea at a seminar in Stockholm in 1963 attended by Wedberg, who to my surprise turned out to be well versed in Church's lambda calculus; recalling his period of study at Princeton in the 30s, this should not have been surprising after all.

On the whole, however, there were few persons with whom I could speak about logical matters; supervision was still something unheard of. However, in 1963 I got to know Christer Lech, the only mathematician at Stockholm with an interest in logic at that time. We discussed various things including themes around my planned dissertation such as alternative systems of natural deduction for classical logic, and he then came up with helpful ideas about how to formulate the reductions if one based the system on the axiom of the excluded middle. Christer held a position as docent, which gave him plenty of time for research and the right to teach what he wanted. One semester he chose to lecture on Gentzen's second consistency proof. I felt that I understood the proof better than he because of knowing how to normalize natural deductions. This boosted my self-confidence, but the feeling was not really well grounded—it was not till a year ago that this understanding materialized in the form of a proof of a normalization theorem for Peano arithmetic.

For the spring term of 1964, I was invited to UCLA as visiting assistant professor. I was to be a substitute for David Kaplan who was on leave. The invitation came from Richard Montague, whom I had met at a conference at Åbo (Turku) in the summer of 1962. Some of the participants had attended the World Congress of Mathematics held immediately before in Stockholm and went together by boat from Stockholm to Åbo. In this connection, I met several persons whom I knew before only because of having read their works, like Church, Tarski, Mostowski, and Curry; the latter introduced me to Kripke, who accompanied him.

The invitation to teach at UCLA flattered me—I was only 27 and had not yet got my doctorate. My duties there were to give an introductory course in logic and to run a more advanced seminar in philosophy of logic. The textbook for the course was written by Kalish and Montague and used a system of natural deduction that they had recently developed. It had some pedagogical merits, but the authors had understood nothing of the proof-theoretical potentialities of a system of natural deduction.

As theme for my seminar, I chose to discuss various interpretations of intuitionistic logic, starting with the ones by Kolmogorov and Heyting, and continuing with my imbedding of intuitionistic natural deductions into an extended lambda calculus. Intuitionism was a rare thing at UCLA; Montague who attended a few of the sessions complained that some of the interpretations were too informal.

Every second week there was a joint logic colloquium for the whole Los Angeles area in which a number of very competent logicians participated; the most well known was perhaps Abraham Robinson. I was invited to present a paper and chose to talk about my ideas of a normalization theorem for a system of classical natural deduction where the deduction trees branch downwards at disjunction eliminations. The idea is quite natural but the precise formulation is delicate, as several people have discovered who have later tried to develop such a system. I had a simple solution, I believed, but was worried that it might be too simplistic. I started my presentation with a warning, saying that there might be something wrong with the proof since it used only usual induction over the length of the deduction tree, which meant that the proof worked also for second order logic, but such a proof should not be possible in view of Takeuti's claim.

After a few minutes of lecturing, I felt very uneasy and started to see the situation from the outside: who am I, standing here in front of all these competent people, to present an obviously fallacious proof? I blushed, my heart was throbbing, and I felt as if my tongue was stuck in the mouth. But the lecture went on, and no one could see anything wrong with the proof. The presentation was generally considered to have been quite clear. My tension seemed not even to have been noticed; this was confirmed by some students who had attended the seminar and whom I told how I had felt during the talk. Montague was of the opinion that I should not worry about Takeuti's claim, which opinion he backed up by saying that people often make unfounded claims that so and so cannot be proved because of Gödel's result.

Not long afterwards I found the error in my proof, and I also understood how Takeuti's claim was obtained as an immediate corollary of three very easy lemmata; the full proof was later put in my dissertation on less than a page.

My stay in Los Angeles was otherwise quite pleasant. I rented a small cottage in Brentwood with an orange tree outside. Socially I saw especially Don Kalish quite often, who among other things opened my eyes to what was happening in Vietnam. He also introduced me to Rudolf Carnap, who had recently retired from UCLA and now lived in Santa Monica.

This was a time when in my opinion it was still preferable to cross the Atlantic by boat, and in connection with the two voyages I stayed in New York for quite a while. I also went to New York to attend a meeting of the Association for Symbolic Logic, at which I presented a proof of the normalization theorem for classical logic, choosing this time a formulation with the law of the excluded middle as an axiom. I included some applications to ideas suggested by Fitch. In this connection, Church taught me a lesson: never *postulate* proofs to be normal—what use could there be of a system without free access to modus ponens; I think he was very right.

In New York I was kindly invited to Martin Davis's house. There I also met Raymond Smullyan, who told me enthusiastically that he was very taken with my modification of Beth's semantic tableaux used in my first paper on mechanical proof procedures and intended to develop it further in the text book that he was planning. This topic was now very far from my mind, and I was sorry that I could not really share his enthusiasm.

Back in Stockholm in the autumn, it was high time for me to publish my dissertation and to get my degree—there was only a year left of my scholarship. I put together what I had ready, including several extensions of my results to other languages than the ordinary first order ones. The dissertation had to be available in printed form in good time before it was to be defended at the public disputation, and it would take some time to get it printed with the technique of that day. But I was still hesitating about what form I should choose for the system of classical logic. Shortly before the deadline for handling in the manuscript to the printer, I changed the system from one where the nodes of the deduction tree were assigned sequences of formulas interpreted disjunctively to the one that appears in the monograph. This was certainly a wise move, although it does not do justice to the duality of the logical constants when interpreted classically.

The disputation took place on the very last day of the spring term of 1965. Kanger was the faculty opponent. To reduce the tension of this public event, he told me the main points of his opposition the day before. But he left one surprise. In summing up his evaluation, he made the complaint that I had not seen, or had chosen not to see, that my main result could very easily be obtained from Gentzen's Hauptsatz for the sequent calculus. At this point, my future first wife, who was sitting with Christer Lech in the audience at the first row, pointed emphatically at the beginning of the dissertation, obviously wanting to encourage me to refer to it. But I was dumb with astonishment. Not only had I indeed written on the first page of the preface that the normalization theorem was equivalent to the Hauptsatz, but in an appendix, I had even shown in sufficient detail how it could be derived from the Hauptsatz. Kanger's accusation was too absurd, and in addition, it revealed that he had missed or did not want to accept what I considered to be the main point of my thesis. It was not the custom to respond to what the opponent said in his summing up. So I remained silent. But sometimes I still regret that I did not give a quick, crushing reply.

2.5 Docent at Stockholm and Lund—Visiting Professor in US

The Swedish doctorate at this time corresponded to a German Habilitation or a French doctorat d'État. Provided that the dissertation was considered to be sufficiently good, it gave the doctor the desirable position as docent for six years, if there was such a vacant position at the time in question. The docentship in theoretical philosophy was already occupied by Jan Berg, who had got this position for his dissertation on Bolzano a couple of years earlier. But the one in practical philosophy was free, and I was allowed to "borrow" it as long as there was no one qualified for the position in that subject.

I could do so for two years. Then my good friend Lars Bergström had completed his dissertation in practical philosophy. He asked me to be his opponent at the disputation; according to an old tradition there should be, besides the faculty opponent, a second opponent chosen by the author. This was the immediate reason for my engagement in a critical study of utilitarianism, which resulted in several papers in the next years. He got his doctorate and took over my position. To my luck, Sören Halldén, who had recently been appointed to a professorship at Lund University, then offered me to become a docent there.

The position gave "protected space", to use a vogue term in education policy for naming a phenomenon that is now rare. The holder of the docentship could engage in research at will without the need to write any further applications or reports. The teaching load was small and could be fulfilled as seminars connected with the research.

This was an opportunity to move to the countryside. I married Louise Dubois, whom I had known for many years and to whom I had dedicated my dissertation.

We bought a country house located in a rural area south of Stockholm at a distance of less than two hours travel by train or car, and settled down there. It was a grand villa, built as the main building of an estate, but now partitioned from it. There were a few neighbours, a lake, which the villa had a splendid view of, open fields at one side, and deep forests at the back—a perfect place for a peaceful life as a docent in philosophy. I needed to go to Stockholm only for one day a week. When I got attached to Lund, to which the distance was greater, I went there occasionally for short periods.

I now took up themes that I had worked on as a doctoral student but had not been able to include in my dissertation. I first turned to questions concerning various relations between classical and intuitionistic logic that I found useful to approach proof-theoretically. As for the translation of intuitionistic logic into classical logic, it seemed likely that the Gödel-McKinsey-Tarski interpretation of intuitionistic sentential logic in classical S4 could be extended to predicate logic by using the normal form theorem that I had established in my dissertation for classical S4 with quantifiers. I gave this problem as a task to Per-Erik Malmnäs, an able student whom I had got to know when I taught logic courses as a graduate student. He wanted to write a graduation paper, and solved the problem quite quickly. We decided to write a joint paper "A survey of some connections between classical, intuitionistic, and minimal logic"; it should be said that Per-Erik did the hard work. I presented part of the paper at a logic colloquium at Hannover in August 1966 organized by Kurt Schütte, who invited us to include the full paper in a volume that he edited. From the conference I otherwise recall Peter Aczel, the only person there of my age, with whom I was glad to speak.

Next I returned to the problem of extending the normal form theorem to second order logic. Having no idea of how the reductions could be shown to terminate, I chose to approach the problem model-theoretically, which was most conveniently done by trying to prove the Hauptsatz for second order sequent calculus; in other words, Takeuti's conjecture. As had been noted by Kanger and Schütte, the Hauptsatz for first order logic follows immediately from their kind of completeness proof. They had both tried to extend their results to second order logic but had not succeeded in that attempt so far (Kanger had announced such a result, but no proof had been forthcoming). In a vaguely nominalist spirit, I had interested myself in models where the second order variables range not over all relations between individuals but only over definable ones. In these terms, the problem that appeared when trying to extend Kanger's or Schütte's result to second order logic could be described by saying that the counter model that one gets for an unprovable sentence by taking an infinite branch in an attempted cut-free proof might not be closed with respect to definability. It had seemed to me for some time that the natural way to overcome this problem should be to extend the second order domains of the obtained counter model by adding the relations that could be defined in terms of the relations already contained in the domain. This operation might have to be repeated a number of times along an initial segment of ordinals, but eventually the domain must become closed under definability.

When I finally tried this idea in the spring of 1967, it worked out fairly immediately to my surprise; I had expected greater difficulty in view of the fact that Takeuti's conjecture had been open for quite a long time. First I was afraid that I had overlooked something, but soon I had a complete proof, and in June I had written it up in a paper "Completeness and Hauptsatz for second order logic", which I submitted to *Theoria.*

The question now arose whether this result could be extended to higher logic. Later in the summer, I took up this question, and it turned out that the same construction did not quite work for higher types. To overcome the new problems, I made the interpretation non-extensional, and found a way to extend the given model in order to make it closed under definability by a more complicated operation performed once, instead of repeating an operation a transcendental number of times as in the previous proof. With some modifications like this, the proof went through. Quite satisfied with this, I wrote a new paper, "Hauptsatz for Higher Order Logic", which I sent to Schütte and submitted to the *Journal of Symbolic Logic*.

Shortly afterwards, I went to the third International Congress in Logic, Methodology and Philosophy of Science (LMPS), which was held in Amsterdam at the end of August. There I found out that in the autumn a proof of the Hauptsatz for second order logic, using a method different from mine, had appeared in a paper by William Tait. Furthermore, I met Schütte who told me that a few months earlier he had received an unpublished paper from a young Japanese logician Moto-o Takahashi containing essentially the same proof as mine of the Hauptsatz for higher order logic. Schütte nevertheless recommended my paper for publication in view of some sufficiently interesting differences between the proofs. He also arranged that I could present my proof at a session that had a gap.

The entire Congress took place at a hotel in the centre of Amsterdam, Krasnapolsky, and most of the participants also stayed in that hotel, which facilitated exchanges and created a nice feeling of homeliness. There I met Georg Kreisel who urged me to investigate whether a similar proof of the Hauptsatz could be worked out for second order intuitionistic logic. I was only moderately interested in this problem but looked at it anyway when I had returned home. It turned out to be fairly easy to give a proof that followed the same pattern as the one for classical logic, using Beth's interpretation of intuitionistic logic generalized to second order, which turned out to be a special case of Kripke's interpretation when similarly generalized. I presented the result at the Conference on Intuitionism and Proof Theory, held at Buffalo in the United States in August 1968.

I was more interested in taking up a third theme from my pre-doctoral period, the idea of interpreting intuitionistic logic in terms of constructions denoted by terms in an extended lambda calculus. I had not worked on it after presenting it at my seminar series in Los Angeles in 1964. But now in September of 1968, there was yet another conference, the First Scandinavian Logic Symposium, held at Åbo in Finland; one of Stig Kanger's many good initiatives. It seemed to be a suitable occasion to present this idea, but I regretted that I had not paid more attention to it in previous years and that I had not succeeded in giving it the presentation I had wanted and that it deserved. I returned to the subject in a little more detail two years later at the Second Scandinavian Logic Symposium held at Oslo, but then it would be another 30 years before I took it up again in a different context.

The second semester of the academic year 1968–69, I spent as visiting professor at Michigan University in Ann Arbor. One may wonder why, and I have no real answer to that question. The invitation came from Irving Copi, whose system of natural deduction I had written a very short note about. But this was a very minor thing, and the department at Ann Arbor did certainly not have its strength in logic. Its staff had a good reputation though, especially for their competence in moral philosophy, which I had some interest in. The philosophical dedication of the staff was impressive, demonstrated among other things by their gathering at the house of a member of the faculty every Sunday afternoon to listen to a paper one of them had newly written.

The subsequent academic year, I was to spend at Stanford by invitation of Kreisel. I had bought a car in Ann Arbor, and at the beginning of the summer I drove from there to Palo Alto. In Chicago I picked up Per Martin-Löf. I had met Per in December 1965 for the first time on the initiative of Christer Lech, who had asked me to attend a session of his seminar where Per was to make a presentation. After that we saw each other very occasionally to begin with. But Per had a growing interest in logic. We went together to the LMPS Congress in Amsterdam and to the Buffalo conference, and we soon exchanged ideas quite frequently. Now he accompanied me on the drive to Stanford, and stayed with me there for the summer.

Per had been a year at the University of Illinois in Chicago by invitation of Bill Tait. There he had also got to know Bill Howard and had learned about his idea of formulas-as-types. Howard had first connected his idea with Gentzen's sequent calculus, but Per saw that it was more fruitful to relate it to my work on natural deduction that I had told him about. By slightly modifying Howard's idea, one got an isomorphic imbedding of intuitionistic natural deductions into an extended typed lambda calculus, whereas mine was homomorphic.

From this time, Per was a great supporter of my perspective on Gentzen's work, and he soon came to develop it further. One of his first contributions was to carry over Tait's notion of convertibility to natural deduction. It gave a new method for proving that the reductions that I had defined terminate in normal deductions. A second contribution was that he extended the idea of introduction and elimination rules to concern not only the logical constants but also the concept of natural number and more generally concepts defined by induction or by generalized induction, and obtained a normalizability result for the theory of such concepts. He came to present these results the next summer at the Second Scandinavian Logic Symposium in Oslo. A third idea of Per's concerned questions about the identity of proofs that we discussed together. If a deduction reduced to another, the corresponding lambda terms denoted the same construction, as I had already noted, and it seemed reasonable to think that the deductions then denoted the same proof. Per suggested that something like the converse of this was also true, more precisely, that two deductions represent the same proof only if one could be obtained from the other by a sequence of reductions and their converses. I liked this idea and came to refer to it as a conjecture of Martin-Löf.

At Stanford I was glad to get to know Solomon Feferman, who attended a series of seminars on my work on natural deduction that I gave during the summer term. Georg Kreisel came to Stanford in the autumn. My contacts with him had been limited, but we were soon on very good terms, had a joint seminar, and saw each other regularly for hours several times a week. He, too, now became convinced of my perspective on

Gentzen's work and became an enthusiastic supporter. As he put it when he wrote a long, critical review in the *Journal of Philosophy* of Szabo's translation into English of Gentzen's collected papers: "I was slow in taking in this part of Gentzen's work. ...As I see it now, guided by D. Prawitz's reading of Gentzen, the single most striking element of Gentzen's work occurs already in his doctoral dissertation". This striking element consisted in "informal ideas on a theory of proofs, where proofs are principal objects of analysis, and not a mere tool".

He was especially interested in the significance of the reductions by which a deduction d is transformed to a normal deduction $|d|$ and in the ideas about the identity of proofs that I had discussed with Per. To give what he called "the flavour of the potentialities of a theory of proofs", he referred to my result about normalizability, which he wanted to give the following "succinct formulation":

> To every deduction d there is a normal deduction $|d|$ that *expresses the same proof as d.* (Thus normal derivations provide canonical representations, roughly as the numerals provide canonical representations for the natural numbers.)

Of course, Kreisel also enriched my perspective and fused it with his own elaborated logical views. His reflections on these themes were likewise to be presented at the Second Scandinavian Logic Symposium in a long paper, "A Survey of Proof Theory II". Here he expressed certain reservations concerning the idea of the introduction rules as meaning constitutive, but nevertheless a main theme was the significance of the reductions in natural deduction.

These discussions with Martin-Löf and Kreisel were a great stimulus to me and made me again focus on the theme of my doctoral dissertation. During my time at Stanford I started to write a paper that would bring that theme up to date and give a survey of what I was now calling general proof theory, as opposed to reductive proof theory. It came to be called "Ideas and results in proof theory", and it too came to be presented at the Second Scandinavian Logic Symposium. The greater part of it was finished when I left Stanford in March to go to my country house in Sweden.

Kreisel came there at the beginning of June to stay with me before we went on to Oslo for the symposium. He brought news about a problem that we had much discussed at Stanford and that I had also discussed with Per: how, if at all, could one establish for second order logic, not only that to any deduction there is one in normal form that proves the same, but also that this normal deduction is obtainable by the standard reduction procedure—to use a terminology that Kreisel suggested, one wanted a *normalization theorem* not only a *normal form theorem.* Kreisel's good intuition had led him to think that a graduate student in Paris, Jean-Yves Girard, whom he had been supervising in the spring, was working on something that was relevant to this question and now had a result that may yield a positive solution of our open problem. He had actually sent me a paper by Girard in advance, which I had misplaced and could not now find. Unfortunately, Kreisel could not explain Girard's idea, so we could not settle its applicability.

The symposium in Oslo, organized by Jens-Erik Fenstad, was given a wider format than the First Scandinavian Logic Symposium since several non-Scandinavians were also invited. It became an important event in proof theory because of the many contributions in that field. Girard was not coming to the symposium but another of his supervisors, Iegor Reznikoff, was present and volunteered to explain Girard's idea. Unfortunately no one understood much of the explanation, except that Girard had made some kind of construction using impredicative quantification which by instantiation gave a computability predicate of the right kind for the system of terms that he had constructed. When I returned to my country house after the symposium, Iegor accompanied me, and staying with me for a while, he tried to explain Girard's idea a little further. The idea was still mysterious to me, but the idea of an impredicative construction was suggestive. When Iegor had left I sat down and was able to work out a proof of the normalization theorem. Very satisfied that this old problem had now been solved, I wrote a letter to Per at the beginning of July telling him what I had found. He was making a sailing trip, but back at home he replied that he had found the same proof. I included my proof as an appendix to my paper in the proceedings from the Oslo symposium, while Per presented his proof in a separate paper that was also included in the proceedings. Girard contributed a paper to the proceedings where the application of his idea to natural deduction was explicitly worked out.

Among the achievements in general proof theory during these years, I valued most highly the new method for proving normalizability of natural deductions that was obtained by Martin-Löf carrying over Tait's notion of convertibility to them. I modified the method so as to obtain strong normalizability, that is, the result that any sequent of reductions regardless of the order in which they are made terminate in a normal deduction. What I especially liked about this notion, which Martin-Löf had called computability, was that it seemed to allow one to make precise the idea that reasoning which proceeds by introductions preserves validity in virtue of the meaning of the logical constant in question, while reasoning by eliminations is justified or valid because of being reducible to valid reasoning as defined by introductions. In accordance with this view, which was presented in an appendix to my paper to the Oslo symposium, I renamed the notion and called it validity.

In my Oslo paper, validity, or rather several different notions called validity, was defined for derivations of various formal systems. However, this was not what I really wanted. The desideratum was to define validity for reasoning in general—to define it for derivations, all of which then turned out to be valid, was like defining truth for provable sentences instead of for sentences in general.

In the academic year after the Oslo symposium I was back as docent at Stockholm University; Jan Berg had got a professorship at the Technische Hochschule in München, and I could take over his position as docent. I now wanted to extend the definitional domain of the notion of validity so as to cover reasoning in general, which I represented by what I called arguments. They were made up of arbitrary inference steps arranged in tree form. To each inference step that was not claimed to be meaning constitutive there was to be assigned an alleged justification in the form of an arbitrary reduction procedure. The notion of validity was now worked out for such arguments in my paper "Towards a foundation of a general proof theory". It was presented the next summer at the 4th LMPS Congress, which was held in August-September at Bucharest, and to which I had been invited to give a lecture.

That summer my term of six years as docent was at an end. My future as an academic philosopher would then have been very uncertain, if I had not just won a permanent position as professor of philosophy at Oslo University.

2.6 Professor of Philosophy at Oslo University—Fellow at Oxford

The appointment of a professor at a Scandinavian university was a serious affair at that time, particularly in Norway, and especially if it was a professorship in philosophy. The procedure started with appointing a committee, which worked for about a year to rank the applicants. Then the whole faculty, in this case the faculty of Humanities, was to give its opinion, the Rector of the university was to make his decision, and finally the government was to make the appointment. In my case, the committee was deeply divided about whom to rank first. When it had delivered its publicly available report, thick as a book, a general dispute broke out, which took place among other things in the main daily newspapers of the country.

The position to which I had applied was vacant after Arne Næss, who had been *the* professor of philosophy at the university for 30 years; he was a kind of icon, held in high esteem. A number of people were of the opinion that I was not worthy to succeed him. I was too specialized and too young; in one newspaper it was said that a quickly raised broiler should not succeed Næss. Altogether there were 21 long newspaper articles that discussed the issue. Of them 18 were against and 3 were in favour of my appointment. Nevertheless, the faculty put me first and I was appointed. Many were indignant. I had a former girl friend, now living in Oslo, who warned me about accepting the position in view of the general opinion.

I ignored the warning, and in the summer I went to a big meeting held for several days in the Norwegian mountains and organized by students. Most of them had been active in the student revolts at the end of the 60s, and most of them were against my appointment. I survived unhurt, and when I later took office at the philosophy department all were friendly. As I happened to find out later by reading a protocol, one professor had suggested at a meeting of the department held after I had been appointed that they should write to the faculty to say that they had no room for me because Næss's office had to be converted to a tearoom. But soon he was to propose that we should run a series of seminars together, and so we did. My great support at the department was Dagfinn Føllesdal, who had proposed that I should apply for the position, and with whom I had many joint seminars in philosophy of language.

I liked living in Oslo but it was a new life for me in many respects. My wife did not accompany me to Oslo, and we were eventually to separate and to divorce. As professor I had several new obligations. Philosophy was a bigger and more important subject in Norway than in Sweden, which had its ground in the fact that all students in all faculties had to study philosophy in their first year at the university. Being a non-native of Norway, I was not expected to know how things were run at the university, and I cannot say that I was burdened with many administrative duties. But to come up to expectations I found it important to cultivate other interests that I had in philosophy besides logic. One such interest was the concept of cause, about which I wrote some essays, stimulated by writings of Georg Henrik von Wright, who even visited the department for a month. I also wrote long reviews of philosophy books for a daily newspaper; for instance, I recall devoting much time to John Rawls, *A Theory of Justice*.

For all that, I still pursued logic but now with less intensity. In 1974, I gave a course in proof theory at the Summer Institute and Logic Colloquium at Kiel. I also presented a paper at a concurrent symposium in honour of Kurt Schütte in which I looked at Gentzen's symmetrical sequent calculus from a classical semantic perspective. The point was that the calculus could be seen as defining a semantic notion in such a way that the completeness property could be expected as a feature of this notion. But I tried to do too many things in the paper, and the point was lost in a multitude of definitions and results. Part of the paper had been stimulated by discussions with Georg Kreisel, with whom I continued to have good contact. Our roads sometimes more or less crossed in various parts of Europe, and he then suggested that I visit him. Occasionally, we spoke German with each other. Since as a Swede I found it strange to use the formal *Sie* when knowing each other so well, we agreed to use the singular *du* and our first names when addressing each other, an informality that he practiced with very few persons, he said. We continued reading and commenting on each other's papers. But it was clear that his former enthusiasm, not to say over-enthusiasm, about my work gradually cooled off. Some time after my six years at Oslo, his comments started to be rude. Knowing his character, I was not offended, but I found it better to drop the contact. We seldom saw each other after that. An exception occurred much later when he came up to Oxford a couple of times to visit me when I was staying there for the summer of 1995, borrowing Daniel Isaacson's house.

Two new contacts outside of Norway that I made during these years were especially important to me. One was with Oxford and Michael Dummett, and the other was with Italy and Italian logic. I was invited to give a talk at a meeting at Santa Margherita Ligure in June 1972, where I presented in a more accessible form some of the ideas of a general proof theory that I had set out at the LMPS Congress in Bucharest the year before. It was a small Italian conference—the other foreign guests were Arend Heyting, Charles Parsons, and Gert Müller—and was my first real contact with Italian philosophy, besides having got to know Ettore Casari in my visits to Münster ten years earlier. Casari came to the meeting with some of his students, among them Daniela Montelatici and Giovanna Corsi. From Florence there was also Marisa Dalla Chiara, and from Rome came Carlo Cellucci. All of them I came to know very well in the succeeding years. I had not been aware before of the great interest in logic within Italian philosophy. It was an interest that continued growing, and it embraced proof theory of the kind that engaged me. I came often to return to Italy for this reason.

The conference in Santa Margherita was one of these nice, relaxed, and yet fruitful meetings that Italian philosophers often organize so well according to my experience: there was generous time for presentations and discussion, but there was also time for excursions at sea to Portofino, for walks, for hearing Marisa singing together with Giuliano Toraldo di Francia, and for a splendid dinner at the home of Evandro Agazzi, who was the organizer of the meeting.

But the meeting is most memorable to me for a personal reason. There I met my future second wife, Daniela Montelatici, who at that time was at the end of her university studies. She was the main reason for me to return to Italy frequently. Three years later we went together to Oxford for a year.

I was allowed to take a sabbatical from Oslo for the academic year 1975–76, and chose to spend the year at Oxford. Michael Dummett and Robin Gandy had very warmly welcomed me coming there when I told them about my intention. I had met Robin Gandy at the conference in Hannover in 1966. He was to me then just a tall man with a strong Oxford accent—when passing me in the hallway of the venue, he said very briefly but encouragingly enough: "hello, I liked your *Natural Deduction*, that is the way to present Gentzen". Later he invited me to Oxford several times, and now we got to know each other very well.

Michael Dummett was a new acquaintance. I had first met him at a conference at Oberwolfach in 1974. At that time I did not know anything about his work, but I soon understood that we had converging interests. Some of his ideas of a theory of meaning could be seen as a generalization of Gentzen's idea of natural deduction; in particular, his talk of two aspects of language, the conditions for appropriately making a statement and the conclusions that could be appropriately drawn from a statement, that must be in harmony with each other, as Dummett put it. This idea amounts within logic to nothing other than what I had called the inversion principle (following Lorenzen) for introduction and elimination rules, on which I had based the proof of the normalization theorem for natural deduction. My reason for wanting to go to Oxford was especially that I wanted to learn more about what I saw as Dummett's extension to language in general of a logical theme that I had been engaged in.

It was arranged that I should be a fellow at Wolfson College since it was considered important to have an affiliation of this kind. To begin with we rented Philippa Foot's charming three-store house at Walton Street while she was in California for most of the year. When she returned, Robin Gandy was going away for the rest of the academic year and offered us his apartment at Wolfson College.

The time for my stay in Oxford was very appropriate. Some of Dummett's main contributions to the field that I was interested in, "What is a theory of meaning? (II)" and "The philosophical basis of intuitionistic logic", were published that academic year. Furthermore, Per Martin-Löf also came to Oxford for the autumn of 1975 with a fresh and intensive interest in theory of meaning. He gave a series of well attended seminars at All Souls at which he rejected the idea that introduction rules or conditions for making a statement are meaning constitutive. Instead, it is the elimination rules or the rules for drawing conclusions from a statement that are meaning constitutive, he argued, and outlined an entire meaning theory for a substantial part of mathematics on this basis. Thus, there were a lot of issues to discuss at Oxford.

I was very impressed by Dummett's papers although I did not agree with all of his ideas. My occupation with them gave my philosophical thinking a new direction. Per's seminars were also quite impressive, but I stuck to the opposite view that it is the introduction rules that determine the meaning of the logical constants and that this idea was the one that was to be extended to mathematics and language in general. I started to work on my own version of this view. A first result was the paper "Meaning and proofs", which was finished before I left Oxford. Per came a little later to realize that the meaning theory that he had presented in his seminars did not work. When he worked out his type theory in detail, he concurred with me that it is the introduction rules that have to be meaning constitutive in a meaning theory for mathematics.

Michael Dummett had discussed two possible meaning theories, one that identified the meaning of a sentence with what it takes to verify its truth, and one that identified it with what conclusions can be drawn from its truth, calling the first verificationism and the second pragmatism. In the spring of 1976 we ran a series of joint seminars discussing this and other meaning theoretical issues. Our discussions were to continue in subsequent years at a number of conferences, in proceedings, in three books devoted to Dummett's philosophy, and in one volume devoted to philosophical ideas of mine. I visited Oxford again many times, and Michael accepted several invitations to come to Stockholm after I had returned there from Oslo. As to the choice between verificationism and pragmatism, he came to lean on the whole to the former, later relabeling it justificationism in which the basic concept was justification of an assertion rather than verification of the truth of a sentence. In *The Logical Basis of Metaphysics* published 1991, he devoted some chapters to what he called proof-theoretical justifications of logical laws, which he wrote originally soon after my stay in Oxford. There he discussed critically, but in the end mainly with approval, meaning theories of the kind that I had tried to base on the notion of a valid argument taking introduction rules as meaning constitutive. But he also returned to a notion of valid argument based on elimination rules, somewhat surprisingly referring to Martin-Löf, not noticing that he had given up that project because the idea does not work if one tries to extend it beyond first order predicate logic.

When I was in Oxford, Anders Wedberg was granted a pension a few years before the regular retirement age because of bad health, and his chair was advertised vacant. I applied and at the end of 1976 I was appointed. The appointment procedure was less dramatic than the one at Oslo, although Per Martin-Löf, who had also applied for the position, was a strong candidate. I stayed in Oslo for the academic year 1976–77, not to leave my position there too abruptly. Daniela stayed in Florence. In the spring we visited Belgium and France together. A group at the University of Leuven had invited me to give a series of lectures and treated us with great hospitality, showing us around in Belgium. We also saw the newly erected Université de Louvain at Louvain-la- Neuve and were shocked by the effects of language conflicts. We then went to Paris. Jules Vuillemin had invited me to give a lecture at Collège de France. I had not understood before what a prestigious institution this was. At the reception afterwards, at which I was given a medal, I felt that some of the faculty members considered me too young for this honour; if I had known before of this medal, I might have been able to give a better lecture in order to feel worthy of the medal.

This year I engaged myself in the on-going discussion about atomic energy, which was especially intensive in Sweden because of a decision that was to be made about whether to build further nuclear power stations. Together with Jon Elster I arranged a series of seminars about the philosophy of risks. I came to be member of a subgroup of the Energy Commission that the Swedish government appointed and that had the task to give a foundation for a national policy on current issues about energy production.

Our task was to study principles of balancing different kinds of risks, especially when it was a question of events whose occurrence had low probability but whose effects would be catastrophic. We were impressed by a principle for decision-making referred to in insurance business as the notion of Maximum Probable Loss, which we understood as the principle that one should not always choose the alternative with the highest expected utility but should refrain from such an action if there were non-negligible probabilities for negative consequences of a magnitude above a certain limit, for instance leading to the non-survival of the agent. In my opinion, an application of this principle, which I found most reasonable, gave the result that one should not build nuclear power stations of the kind available at that time. Arguing for this view, I came to participate a little in the subsequent political debates in Sweden. Today I am less sure of this view because of the risk of climate catastrophes when using alternative energy sources.

2.7 Back to Stockholm—Chairman of the Department

I moved to Stockholm in the autumn of 1977. Daniela made the big decision to leave Florence and to move to Stockholm with me. We married, and after some years we had three children in rapid succession: Camilla, Erik, and Livia. Daniela decided that it was enough with one philosopher in the family and switched her subject to psychology. Parallel with giving birth to three children, she started a new, long university education to get a degree in psychology. To this she added a considerable period of training to become authorized as a psychoanalyst. In the 90s she could open her own psychoanalytical clinic, in which she is still working. I am glad to have been able thanks to my previous acquaintance with psychology to follow her professional career a little, and that she has been able to understand something of what I have been doing thanks to her previous studies in philosophy.

The Department of Philosophy at Stockholm University, in particular Theoretical Philosophy, had suffered a period of great weakness in the years before I got back there. No one had taken a doctorate in Theoretical Philosophy after I had done so in 1965. Anders Wedberg had been absent for long periods because of illness and because of a research project that he was engaged in. The number of students had grown a lot in the 70s. However, most of the teaching was done by a group of graduate students who had been doctoral students for years without seemingly getting any closer to a degree, much to the amazement of the rest of the Faculty of Arts; even the director of studies was such a graduate student. In addition there had been difficult conflicts between these students and the few permanent teachers. The latter tried to stop the graduate students from governing the department, which was not easy, among other reasons because of the so-called student democracy that had been introduced in the 70s. Students had been given a great say at all levels of the university, in particular at the departmental level, and together with the representatives of the administrative staff, they could even have a majority vote on the board of the department. I was told horror stories about meetings of the board, which could go on for more than ten hours with a short adjournment for dinner. The director of studies, who was in fact sometimes governing the department, and who had a seat at the faculty level too, even tried to stop my appointment, proposing that the position should be replaced by less expensive positions at a lower level. Gaining no hearing for that proposal, he left the department before I arrived.

Obviously I had to take on the responsibility for revitalizing theoretical philosophy at Stockholm University. I took on the task as chairman of the department for the first ten years. Everyone at the department was tired of the long period of conflicts, and the department meetings now ran fairly smoothly; after a while the problem was rather to get the elected representatives to participate in the meetings. The administrative duties consumed time but were not otherwise a problem. The challenge was to extend the teaching staff and to get the doctoral students to take their doctorates.

The regular teaching staff in theoretical philosophy was still very small. There was one position as docent, held by Alexander Orlowski, who had succeeded me, and a half position as lecturer besides my own position. Orlowski had been docent for almost six years and got another position when he reached this time limit. The position as docent came to be filled by a succession of very competent people from other universities, one after the other. They were in order Mats Furberg, Dick Haglund, Göran Sundholm, and Dag Westerståhl. After some years in Stockholm, each of them became professor at another university. The eminent institution of docentship was then unfortunately abolished, but at that time our staff had grown considerably.

The system of higher education was reformed in Sweden in the 70s. Among other things the former doctoral degree was gradually replaced with a doctoral exam, which was meant to be less demanding than the former degree. Many teachers were critical of the reform and saw it as a lowering of standards. However, another important ingredient in the reform was that dissertations for the doctoral exam were to be assessed only with respect to whether they were accepted or rejected, while the ones for the degree had been graded essentially along the old Latin scale of honours. The requirement for getting a low, passing grade had not been very high, and it is doubtful whether it had been higher than the requirement for passing the new doctoral exam. But to get such a low grade had been considered to be a failure, worse than not to take the degree at all, and this was certainly a major reason why candidates postponed the presentation of their dissertations. In view of the situation at my department, the reform seemed to me reasonable.

But most graduate students are self critical, and in spite of the reform, it was not very easy to get candidates to take their doctorates. New candidates with grants sometimes succeeded in taking their exam in almost the expected time of four years. Among the first of them was Luiz Carlo Pereira, who came from Brazil to study proof theory with me, and then returned to Brazil where he later got a professorship. Another was Lars Hallnäs, whose dissertation extended the normalization theorem to a version of set theory, and a third was Torkel Franzén, who wrote a dissertation defending realism in mathematics. The two got positions at a newly started research institute for applied computer science, where I was also engaged as a consultant for some years. Slightly later Peter Pagin wrote a doctoral dissertation on rules and their place in the theory of meaning. He stayed at the department and became lecturer.

For older candidates who had already used their grants it was often more difficult to find time to complete the dissertation at the side of other work. One of the first of them to take his exam was my old student Per-Erik Malmnäs, who had switched from proof theory to the philosophy of probability while I was in Oslo. Closely following him came Gunnar Svensson, whose dissertation was on Wittgenstein's later philosophy. Both of them got positions as lecturers in the department. Gunnar Svensson was also appointed director of studies and assumed the responsibility for a lot of administrative business, which was a great help to me.

At about the end of my first 10 years at the department there had been altogether ten doctoral exams; not a great number perhaps, but a definite improvement. At this stage the teaching staff had grown considerably, and gradually I could share supervision of graduate students with colleagues. My own teaching was mostly within logic and philosophy of language in the form of courses at undergraduate level as well as seminars at graduate level. I was especially pleased to be able to gather most of the people working in theoretical philosophy at a joint higher seminar once a week. It was satisfying to see the department growing and to regain health, but it did not give me much time for other things.

2.8 New Tasks

My old friend and fellow-student Lars Bergström now came back to the department as professor of Practical Philosophy and took over my task as chairman. Numerous other duties and undertakings were added instead. The department had for a long time been quite isolated within the faculty and had been looked upon with suspicion. To change the situation I engaged myself at the faculty level too, and was elected Vice Dean for two periods; I steered away from being a candidate for the position as Dean, but was instead on the Board of the University for one period. A more demanding task that I took on was to be a member for six years of the Swedish Research Council for the Humanities and Social Sciences. I was the chairman of two priority committees that had to evaluate applications for research projects within certain areas, and for a period I was vice president of the entire council.

Another demanding undertaking was to organize the 9th International Congress of Logic, Methodology and Philosophy of Science in 1991. Stig Kanger had volunteered to organize this congress and this was confirmed at the 8th LMPS Congress held in Moscow in 1987, but he died tragically in 1988 without having taken any measures to make such a congress possible. I had not much choice but to take over the responsibility—I had participated in all the LMPS Congresses since the one in Amsterdam in 1967 except one, and on the whole it seemed to me a worthwhile institution. The first challenge was to raise the considerable amount of money that was needed. I was fairly successful in this—it helped that I had become a member of the Royal Swedish Academy of Sciences and of the Royal Swedish Academy of Letters, History and Antiquities, and of course my affiliation with the Research Council also helped. It remained to actually organize the congress, a task that I was glad to share with Dag Westerståhl, who did much of the hard work as secretary of the organization committee.

Among the most time consuming tasks were commissions as expert in conjunction with appointments. Swedish philosophy and the Swedish university system in general were in rapid expansion for several decades in the last century, but the number of professors in most subjects, in any case in philosophy, was small and constant. The demanding procedure that was applied in the Nordic countries for filling academic appointments (and is still more or less in place) required a committee of professors who evaluated all the works done by the applicants, an evaluation that was to be openly accounted for in a lengthy, public document. There was always a new or vacant position to be filled, and the few professors were constantly called upon as experts. Even when I was abroad on a sabbatical, large boxes full of books and papers could arrive that had to be evaluated for some appointment.

There were two extensive tasks that I quite willingly took on during these years. One was to be the director of the publishing house Thales. Swedish philosophy had been given a quite generous donation to be used for translations of philosophical literature into Swedish. Money was allotted to translators by decision of all professors of philosophy in the country at a joint meeting. We had found however that in spite of the translation being paid for in this way, commercial publishing houses were often unwilling to publish the book in an acceptable way. We were also dissatisfied with their failure to keep books in stock that we wanted to use at the universities. To improve the situation we decided bravely to take the matter in our own hands. On the proposal of Stig Kanger, we formed a foundation that was to carry on a non-profit-making publishing house. It was named Thales, and I assumed the task as its director.

Thales started in 1985 with a negligible initial capital. To register a foundation a donation is necessary, so we six present when the foundation was formed donated 100 crowns each, and this was our whole capital! Business was accordingly very slight to begin with. But we were able to get some small grants from other sources and could publish some of the translations that were already paid for. Furthermore, we were allowed to take over from commercial publishing house the rights of some translations that they did not want to republish. Some of these continued to sell very well, such as Wittgenstein's *Tractatus* and *Philosophical Investigations*, which Anders Wedberg had translated in the 60s and 70s. Thanks to a bank loan we were also able to take over philosophy books from another publishing house that had gone bankrupt. After ten years we had published around 50 titles and were selling 40 additional titles that we had taken over.

I stayed as director for an additional ten years, and the publishing house continued to grow. We translated a great number of classical texts, from Aristotle to Frege and Husserl. My last year as publisher coincided with a Kant anniversary, 200 years having passed since his death, at which we came out with translations of all of Kant's critical works; they had not appeared in Swedish before. But we also translated contemporary philosophers like Derrida, Dummett and Davidson, brought out original works by Swedish philosophers, and published three journals.

In the last years my work as director really amounted to a part time job. I liked the varied activities as a publisher of philosophical books. In a small publishing house like Thales they were often of a very practical nature. It was not only a question of accepting manuscripts and finding good translators and editors. It was equally important to choose suitable printers, to develop a good system of distribution, and above all to make both ends meet. Economically this non-profit-making enterprise was very successful: the year I left, its net income was 1,000 times the initial capital. The point of the activities was of course to make classical and modern philosophy available in the Swedish language, and thereby contribute to maintaining and developing a Swedish philosophical vocabulary. In a small language area like the Swedish-speaking one, this does not come about by itself, but is of great importance, it seems to me, since thinking is most efficiently done in one's native language.

Another task that I quite willingly took on was to be president of the Rolf Schock Foundation. Rolf Schock and I were fellow-students from 1961. Both of us worked on doctoral dissertations in logic in the Stockholm philosophy department, but he presented his dissertation at Uppsala University some years after I got my degree. We were on quite good terms. He liked to take up philosophical questions of a logical kind for discussion. I was the Faculty Opponent when he was to defend his dissertation. Then our ways parted, and when I came back to Stockholm I had the unpleasant task of assessing his work when he applied for various positions, which he never got, and which always led to his appeal to higher levels, and to a certain bitterness against me from his side. He lived a quite simple life, taking various jobs, and I felt bad about not being able to offer him an academic position. In 1986 he died in an accident. To the surprise of most of us it turned out that he had a considerable fortune, half of which he wanted to donate to prizes that were to be given by Swedish academies, among them the Swedish Academy of Sciences. A foundation was formed for this aim and I was asked to represent the Academy on the board of the foundation. In this capacity I also became president of the foundation in accordance with its rules.

The first thing we had to do was to bring the capital, which came from the sale of real estates in Berlin, to Sweden. However, before we were able to do so, Rolf's mother succeeded in freezing the money in Germany. She was very rich and was one of the heirs according to Rolf's will, but she demanded a greater share of the inheritance. She sued the foundation and the other heirs, and this started a long lawsuit in two courts. The mother won in the first court, and when the outcome in the second court seemed to go in the same direction, there was a settlement, essentially on her conditions.

A big part of the fortune was thereby lost, but the returns on the capital that came to the foundation were still enough to award four prizes of 400,000 crowns each, every second year. According to Rolf's will, one prize was to be given in logic and philosophy, one in mathematics, one in art, and one in music. Rolf wanted the first two prizes to be awarded by the Royal Swedish Academy of Sciences, the third by the Royal Academy of Fine Arts, and the fourth by the Royal Swedish Academy of Music. The question was how this was to be organized. Some were in favour of letting each academy arrange this as it pleased, but to me it seemed much nicer to let the prizes lend lustre to each other and to arrange a joint prize ceremony. This also became the decision of the board, and the first four prizes were awarded in 1993 at a an elegant, yet fairly relaxed ceremony, at which the King of Sweden handed over the prizes, followed by a buffet supper. Rolf had often seemed quite rebellious, but he had asked very established institutions to award the prizes, and I think that he would have liked the form of the prize ceremony that we decided on.

Of greater importance was of course which laureates were chosen. For the prize in logic and philosophy there was a committee of five persons, which I chaired—the other members were Lars Bergström, Dagfinn Føllesdal, Per Martin-Löf, and Georg Henrik von Wright. Our first problem was how to interpret the phrase "logic and philosophy". To understand it as an intersection would make the field very narrow compared to the fields of the other three prizes. But it did not seem right, and it was probably not the intention of Rolf, to understand it simply as a union. Our solution was to say that the prize should be given for a contribution to logic that was of philosophical relevance or to philosophy that had some bearing on logic. There were obviously a great number of people who had made important contributions that fitted this description. We invited a group of people to make nominations and our choice was finally W.V. Quine. He arrived in Stockholm very alert, pleased to get the prize, and delivered a lecture about the notion of object, which attracted a big audience. Even Davidson and Dreben came to Stockholm to honour Quine on this occasion and we arranged seminars with all three of them on the day before the prizes were awarded. Two years later Michael Dummett was awarded the prize. I stayed in charge of the Foundation for 10 years and as chairman of the committee for the price in logic and philosophy for four years. Then others took over.

I found it difficult to combine concentrated work in logic or philosophy with all these activities. The summers were more relaxed and were spent with the family. For many years we spend the summers in Italy. We then bought a small farm south of Stockholm. I liked to grow vegetables, and did a lot of that, supplying our own and some other family's entire needs for the whole year. A former student of mine, Gunnar Stålmarck, who had started a quite successful business based on a patent he had got for a hardware used to solve problems in sentential logic very efficiently, and who for a while employed the majority of the Swedish logicians, introduced me to sheep-breeding. It required a lot of fencing, because one had to change ground quite often to avoid intestinal worms, but otherwise the sheep took care of themselves. We started in a modest scale, slaughtered the lamb rams in the autumn, but kept the ewes. Soon they were quite a flock of sheep—numbering 52 in all, when we sold the farm since the children had grown up and were less interested in spending all their free time there.

2.9 Philosophy in Spare Time

Most of the papers in logic and philosophy that I did produce in these years were written for conferences in Italy or in connection with longer stays in Italy. For the Italian National Congress of Logic in 1979, which was held for a whole week in Montecatini, I worked out more systematically proofs of the normalization theorems for first and second order intuitionistic and classical logic using the notion of validity.

The paper was published in the proceedings of the congress, which was not very well circulated, and seems to have remained rather unknown. I learned the other day that several otherwise very competent and knowledgeable logicians thought that the problem of normalizing deductions in second order classical logic remained open.

In 1981, shortly after our first child Camilla was born, we spent six months at a vineyard in Chianti. At that time Marisa Dalla Chiara organized a conference on the foundations of mathematics. It was a good and quite lively conference. Among those who came from abroad were Kreisel, Putnam, Takeuti, Feferman, and Girard. I have a vivid memory of the conference among other things because we invited all its participants to a supper on our terrace of the vineyard; a risky undertaking for a great thunderstorm was looming up, which would have completely spoiled the whole supper—it came on shortly after the guests had left.

At the conference, I presented a paper where I compared and discussed some approaches to the concept of logical consequence. Besides the classical one by Tarski and the one that I tried to develop on the basis of my notion of valid argument from 1971, now revised in important respects, I discussed a notion based on ideas in a dissertation by Peter Schroeder-Heister. I had recently been asked to assess his dissertation, which I found interesting. It led later on to his staying for some time in Stockholm followed by further cooperation and to a long friendship. Among other things, he came to contribute to the further development of the notion of valid argument.

I was able to take leave from Stockholm University also for two other longer periods spent in Italy. In 1983, when our second child Erik was still a baby, we spent ten months in Rome at the invitation of Carlo Cellucci. I was attached to Rome University, La Sapienza, as "Professore a contratto". There I gave a course in proof theory with lecture notes, which were carefully edited in Italian by Cesare Cozzo; my intention was that they could be the germ to a book on general proof theory.

I did not do much to implement that idea. When my dissertation was brought out, I had taken the risk on my own expense to print it in a greater number of copies than demanded by the university, but it had anyway been sold out fairly quickly. The question then arose whether to print a new edition. I preferred to write instead a new book that would bring the material of the dissertation up to date. But since this idea was not realized, the question of printing a new edition of the dissertation came back now and then, and when Dover Publications asked for the permission to make reprint as late as 2005, I finally decided to agree to that idea.

During my stay in Rome, I also took up once more a study of Dummett's theory of meaning, and contributed a paper to the first volume that came out about his philosophy. While in Italy, I also gave some lectures in Siena, and at a meeting there I presented a further development of Hallnäs's result on normalizing deductions in set theory, which was included in another virtually unknown proceedings.

In 1990–91, I had another sabbatical leave, and we then stayed in Florence for the whole academic year, sending our children to Italian schools. It was not the best time for leave since Dag Westerståhl and I were still preparing the details for the LMPS Congress that was to be held in August. Communications were not yet by e-mails but took place by fax. I recall how I was constantly going back and forth to my brother in-law's office that had a fax machine.

That year I managed to took up a problem I had lectured on in Stockholm in 1979 and in Oxford in 1980 concerning how to prove a normalization theorem for arithmetic by transfinite induction. It seemed likely that the ideas used by Gentzen in his second consistency proof could be used for this purpose. I had specific ideas about how this could be done by adding a rule of explicit substitution to natural deduction. Making some progress in this project, I presented the ideas at Ettore Casari's Saturday seminars, which I often visited that year; my presentation was in Italian—it was high time, it seemed to me, that I started speaking Italian on such occasions. But I was not able to bring the project to a successful conclusion.

Among other memorable conferences in Italy that I have participated in there were two held in the small Sicilian town of Mussomeli. They were memorable not only because of their philosophical content but also because of the friendliness, interest, and hospitality shown by officials and ordinary people. I do not know what they got out of the very specialized conferences that were held in their town. The first one, held in September 1991, was organized by Brian McGuinness and Gianluigi Oliveri and was devoted to "The Philosophy of Michael Dummett", which was also the title of the book that came out as a result. Dummett was honorary president of the second conference held in 1995, which had the title "Truth in Mathematics". In my paper to the latter conference I discussed Dummett's view of truth and presented my own, different view.

Both Dummett and I presented papers discussing the notion of truth at a conference held in Santiago de Compostela in January 1996, and the discussion between us continued at another conference "Logic and Meaning: Themes in the work of Dag Prawitz" held in June the same year in Stockholm. The latter was a conference arranged by Peter Pagin and Göran Sundholm as a celebration of my 60th birthday. It resulted in a special issue of *Theoria* with papers partly different from the ones presented at the conference but addressing themes that I had worked on. I was asked to write responses to the papers, which took me considerable time. Several of the papers seemed to me to require sophisticated replies, and I struggled with them for several summers to the frustration of Peter who was editing the volume; the issue came out as part 2–3 of *Theoria* for the year 1998, but was actually printed in 2000.

From the middle of the 90s, I led an interdisciplinary research project "Meaning and Interpretation". A group of people from philosophy, theory of literature, and linguistics had taken the initiative in this. It came to involve 22 researchers altogether from different disciplines and ran for six years, partly financed by grants from a research foundation. We arranged a large number of seminars and several conferences with invited guests from abroad. Several publications came out as a result. It gave me a little time for research, during which I wrote some minor papers, but it was not enough for more concentrated work on my part.

Thanks to this research project and other research grants, we could employ a greater number of philosophers than our ordinary university budget allowed. At the end of the 90s there were altogether seven lecturers in theoretical philosophy. The grants were sufficient to support a number of doctoral students too. In the 90s there was also another group of students interested in phenomenology, and I was able to arrange matters so that we received a grant from a private foundation that could be used for the support of some of them. The group was led lead by Alexander Orlowski, and we also had sufficient resources to enable us to invite Dagfinn Føllesdal to participate in some of the activities. Among the doctors emanating from the group were Hans Ruin, who would form his own group at another university in the Stockholm area, and Daniel Birnbaum, who would come to have a career as a director of art museums in Germany and Sweden.

Among the dissertations that I supervised myself during the 90s, I especially appreciated the ones by Cesare Cozzo and Filip Widebäck. Widebäck's dissertation *Identity of Proofs* took up the conjecture that I had been discussing with Martin-Löf and Kreisel in 1969 and showed, independently of some similar results established by others at about the same time, that in the case of the implication fragment of sentential logic the proposed identity criterion is maximal in the sense that adding some further identities between proofs makes the relation collapse, that is, all proofs would become identical.

Cesare Cozzo's dissertation *Meaning and Argument* was a thoughtful contribution to what soon came to be known as inferentialism, a view that advocated a theory of meaning based on inference rules, taken in a wide sense. Ever since I met Cesare during my stay in Rome in 1983, I have greatly appreciated discussing various things with him, and it was a joy that he, after taking his doctorate at Florence, decided to write a second doctoral dissertation with me.

Superficially the kind of theory of meaning that I had tried to develop starting from Gentzen's idea of introduction rules as meaning constitutive could seem to be a kind of inferentialism. But when the matter is considered more closely an essential difference is obvious. The notion of validity of an argument that I had made the basis of my attempted theory of meaning is in general not a recursively enumerable predicate, whereas to be provable according to a fixed set of inference rules is such a predicate. This is an important difference, because in view of Gödel's incompleteness theorem we must be ready to extend our inference rules, for instance by bringing in concepts of higher order. We shall then be able to prove assertions in the original language that were not provable before the extension. If the meaning of asserted sentences is tied to a set of inference rules and their truth is tied to what can be proved by these rules, it is difficult to avoid the conclusion that the meaning and the significance of the assertions change when we extend the language, but this is counter intuitive because it clashes with our natural inclination to take the assertions to remain the same when the language is extended. In contrast, the notion of the validity of an argument and a concept of truth based on this notion are not defined with reference to a particular formal system and are consequently not affected by this kind of problem. (By the way, Cozzo escapes the problem by defining the meaning of a sentence by reference not to all inference rules that are in force, but only to some of them that concern the sentence in a qualified sense, and by explicating truth not in terms of what is provable in a language but in a more complicated way in terms of possible rational extensions of languages).

My meaning-theoretical use of the notion of valid argument, which I did not adequately separate from the proof-theoretical use until the beginning of the 80s, did not attract much attention to begin with. I was glad to be followed later in my endeavour to explicate a notion of valid reasoning in this way by Michael Dummett in his book the *Logical Basis of Metaphysics* from 1991 and by Peter Schroeder-Heister at the conference *Proof-Theoretic Semantics* that he arranged in 1999 and in the subsequent volume with the same title; there Peter made very clear how and why the notion of validity must differ depending on whether it is used as a basic semantic notion or for the end of proving normalizability. However, I always considered this project as a tentative one, and at a conference on natural deduction that Luiz Carlo Pereira arranged in Rio de Janeiro in 2001, I considered a notion of valid proof term that had a greater affinity with the usual notion of intuitionistic proof. More recently my interest has turned to the question what it is that gives a proof its epistemic force, and since this question cannot be answered in terms of valid arguments or valid proof terms, I am now more concerned with a notion of legitimate inference, which I consider more basic. But his belongs to another chapter of my life.

2.10 Retirement

At the turn of the century there were three more semesters before I turned 65, which was the stipulated age of retirement. There were several things I wanted to do before I was pensioned off. Among other things, I regretted that I had never lectured on the history of philosophy, and I planned to do so in the autumn of 2000. As usual there were too many things that had to be done. I was preparing the lectures, I was reading proofs of a second edition of a Swedish textbook in logic that I had written long ago and that was now published by Thales, I had promised the Research Council evaluations of some applications for research projects, and so on. Daniela said that I was leading an especially hectic existence at this time. Anyway, on the 11th of September, I bicycled from my home in the centre of Stockholm to the university campus, situated a few kilometres north of the city centre; usually a pleasant ride, the end of which went through woodland scenery. At an earlier point where the path crossed a two-lane street, I was hit by a car. It was a clearly marked bicycle crossing, in which the traffic should give way to bicycles, at least those coming from the right, and the cars in the first lane rightly stopped. The driver of a small lorry in the second lane, seeing a gap in front of him, speeded up, and hit me when I came to that lane. From a legal point of view the driver was to blame, but I should of course have been more careful.

My first reaction was "how silly to fall down like this, now I just have to continue". But people around me came running, saying that I should just stay where I was, and in fact I could not move very much, having received a double fracture to my pelvis. I was taken by ambulance to the hospital. Before being operated on, I made two phone calls, one to Ulf Jacobsen, who was the main editor of Thales and with whom I had almost daily contact, telling him that I had a small problem but would call him again soon, and one to Daniela, to whose answering machine I said that I was in the hospital and asked her to inform the Research Council that I would be a few days late with my report. The operation went well anatomically but it gave rise to internal bleeding which was difficult to stop. I had been quite conscious up to the time of the operation and had felt relatively well under the circumstances. But since the internal bleeding did not want to stop, the doctors continued to give me anaesthetics after the operation, and I remained anaesthetized, deeply unconscious, in the intensive care unit for two weeks. The bleeding soon threatened the kidneys and the lungs, and the situation was critical for a while. It was a strange feeling to wake up after two weeks. The situation improved gradually but slowly. After about half a year, my first sick leave as an employee, I was back at work, and in late spring, shortly before my retirement in June, my colleagues said that they recognized me as I used to be.

In the summer I went to Brazil for one month with Daniela and the two of our children who wanted to go there with us. The conference that Luiz Carlo Pereira arranged in Rio lasted a week. The rest of the month we spent at various places. We were especially fond of the coast in northern Brazil, east of Fortaleza, where a colleague lent us a hut for a week.

The next two years I had a part time appointment in the department, which allowed me to finish in a less abrupt way various things in which I had been engaged, such as the research project "Meaning and Interpretation". I also continued as director of Thales to the autumn of 2004, at which point my real retirement started, but one could also say that I retired in September 2001 at the date of my accident.

At this time there was a reform at the Swedish universities allowing competent lecturers to be promoted as professors. Six lecturers in theoretical philosophy became professors in this way at about the time I retired. Three of them I have already mentioned: Peter Pagin, who had been a main participant in the project Meaning and Interpretation and with whom I share many interests, Per-Erik Malmnäs, who had also taken up an old interest in Greek philosophy which I would soon benefit from, and Gunnar Svensson, who had taken over as chairman of the department and in whose care it has continued to thrive. The other three had been recruited externally: Staffan Carlshamre, who also took over my work as publisher and has continued to develop Thales, Dugald Murdoch, who had written his dissertation at Oxford on philosophical aspects of the Copenhagen interpretation of quantum mechanics and with whom I shared an interest in the epistemic aspects of inferences, and Paul Needham, who also has a British background and is working mainly in the philosophy of science. It was considered to be enough with six professors of theoretical philosophy. They could be said to be my successors, since my position was not advertised as vacant.

At home there were also big changes. The children left home and went to live for themselves, eventually with partners, and now there are even grandchildren. I am glad to have survived my accident to see my children enter on their careers.

I see my retirement as a very happy change. I am now free to engage in whatever I like, and can for instance concentrate again more whole-heartedly on philosophical questions of my choice. It is like being back at the time when I was a student or docent, but now being less anxious to produce something. I still have a room at the department and go there for some seminars and colloquiums—for instance, in the Logic, Language, and Mind Seminars that Peter Pagin organizes—or when I feel like it, but I have no obligations.

From the autumn of 2004, I have declined administrative commissions with a few exceptions. I led a committee for assessing the Swedish philosophy departments, commissioned by the National Agency for Higher Education, took part in a second round of such assessments, and participated in a committee for assessing Norwegian philosophy. I now enjoy enormously the freedom from all such duties.

Instead I can accept invitations to speak at conferences or at universities, which previously I often had to decline. In 2006, I was invited to give what is called the Kant lectures at Stanford. It was a pleasure to see again several old friends: Solomon Feferman, whom after my time in Stanford I had seen on numerous occasions in Stockholm and other places in Europe, Patrick Suppes, who was as vital as ever and had more energy to discuss my lectures than I had myself, and Gregori Mints, whom I had first met in Moscow in 1974.

That same year, I gave a talk at a memorial symposium in honour of Georg Henrik von Wright in Åbo on "Logical Determinism and The Principle of Bivalence". An extensive series of seminars where Per Martin-Löf went through Aristotle's logical work had inspired me to take up Aristotle's Sea Battle. Per-Erik Malmnäs, also active at these seminars, inspired me to study the ancient commentaries to Aristotle's work—a new kind of experience for me.

Still the same year, I was glad to speak on "Validity of Inferences" at a symposium in Bern, when Dagfinn Føllesdal was awarded the Lauener Prize, and on the same topic at another symposium arranged in Stockholm by Peter Pagin to celebrate my birthday.

For a couple of months in 2007, I was fellow at the Institute of Advanced Studies at Bologna on the invitation of Giovanna Corsi, a friend since my first Italian conference in 1972. I was even given the Medal for Science that the Institute awards each year, and gave also a course at the Philosophy Department of Bologna University.

Among other events in the last years, I gave the lecture at Michael Detlefsen's inauguration ceremony as "Professeur d'Excellence" in Paris 2008. I was back in Paris for a longer visit in 2009 to lecture at several conferences and at Collège de France, this time at the invitation of Ann Fagot, a friend since we were both at Stanford in 1969–70. I was in Italy to give talks at two conferences in 2010, "Logic and Knowledge" in Rome and "Anti-realistic Notions of Truth" in Siena, and a course at a summer school for doctoral students in Siena organized by Gabriele Usberti in 2011. Later that year I replaced Saul Kripke as a plenary speaker at the LMPS Congress at Nancy, speaking about "Is there a general notion of proof?". In the autumn the same year, I gave the Burman lectures at Umeå by the invitation of Sten Lindström, and gave a talk at a conference on "Evidence in Mathematics" arranged by Dagfinn Føllesdal. Last summer I was again in Rio de Janeiro to give a series of lectures.

Many of the talks on these occasions have had to do with the epistemic force of deductive inferences. Perhaps I have spoken too often of this problem, but it is one that intrigues me and that has been neglected in contemporary logic. Modern logic has on the whole ignored dynamic aspects. I am not fond of vogue words like "dynamic", but I think it is appropriate here; an inference is first of all a matter of getting to know something that one did not know before. This epistemic phenomenon cannot be explained in terms of truth preservation under various interpretations in the way logical consequence is usually explicated. It is not easy to say how it is to be explained. Aristotle had at least a name, perfect syllogism, for an inference that justifies its conclusion. For an inference to be legitimately used in a deductive proof it must provide us with a ground for the conclusion given that we had grounds for the premisses. Although the premisses of such a legitimate inference cannot assert a true proposition while the conclusion asserts a false on, it is obvious that truth preservation is not a sufficient condition for being legitimate. Nor can legitimacy be explained in terms of my notion of valid argument, as I have already remarked. It seems to me that to account for what makes an inference legitimate we have to rethink what an inference is. This is the topic of my paper in this volume, and I shall not speak more about it here.

Another question that has engaged me in the last one and a half years is one that I left in 1991, namely, how to prove a normalization theorem for first order arithmetic using transfinite induction. It had remained an open problem, and I took up it up again when asked if I could contribute something to a volume planned in connection with the 100th anniversary of Gentzen's birth. We are many who expected it to be possible to carry over Gentzen's second consistency proof to natural deduction and to extend it by using his methods so as to get a normalization theorem. Gentzen had obviously intended to prove this more general result, but had met with difficulties and had then confined himself to proving that cuts could be eliminated from an imagined proof of a contradiction in the sequent calculus. With the knowledge that we have now, more than half a century later, it should be possible to establish the more general result. But those of us who have tried to do so have found it surprisingly difficult. In my new, more relaxed situation after retirement, I was able to concentrate on the problem for some weeks, and then I saw where my previous attempt had gone wrong. After some additional weeks, when I was swinging between thinking that I knew how to do it and being in despair of finding a solution, I was able to modify the strategy in the right way so that all the pieces fell into place. At least, so it seemed, and it was very satisfying that this old problem now seemed to be solved. It remained to polish the proof and correct various oversights, but I now trust that the paper, having been read by several colleagues, is in order.

It is a curious thing to write an autobiography. In a way it is fun to recall old memories. I have written down some episodes as they have come to mind without any real plan, making selections with respect to whether they really mattered to me and were relevant in the "short scientific biography" that I have been asked to write for this volume. But when I check them, I find that my memory is not always as accurate as I thought and would like it to be. And when I look at the result, I sometimes wonder what person I am picturing. Needless to say, there are many other people than those mentioned here who have been important to me personally or professionally, and probably there are many other stories that I would have found more relevant to retell if I had thought about them.

Chapter 3
Explaining Deductive Inference

Dag Prawitz

Abstract We naturally take for granted that by performing inferences we can obtain evidence or grounds for assertions that we make. But logic should explain how this comes about. Why do some inferences give us grounds for their conclusions? Not all inferences have that power. My first aim here is to draw attention to this fundamental but quite neglected question. It seems not to be easily answered without reconsidering or reconstructing the main concepts involved, that is, the concepts of ground and inference. Secondly, I suggest such a reconstruction, the main idea of which is that to make an inference is not only to assert a conclusion claiming that it is supported by a number of premisses, but is also to operate on the grounds that one assumes or takes oneself to have for the premisses. An inference is thus individuated not only by its premisses and conclusion but also by a particular operation. A valid inference can then be defined as one where the involved operation results in a ground for the conclusion when applied to grounds for the premisses. It then becomes a conceptual truth that a valid inference does give a ground for the conclusion provided that one has grounds for the premisses.

Keywords Inference · Deduction · Proof · Ground · Meaning · Logical validity · Inferentialism · Intuitionism · Proof-theoretic semantics

To justify deduction in order to dispel sceptical doubts about the phenomenon is not likely to succeed, since such an attempt can hardly avoid using deductive inference, the very thing that is to be justified. This should not prevent us from trying to explain

An early draft of this paper was presented at a seminar arranged by professor Wansing to discuss the contributions to this volume and at another seminar at the Department of Philosophy at Stockholm University. I thank the participants for many valuable comments, many of which have influenced the final version of this paper; special thanks to Per Martin Löf for remarks about references and to Cesare Cozzo for carefully commenting a late version of the paper.

D. Prawitz (✉)
Department of Philosophy, Stockholm University, SE-10691 Stockholm, Sweden
e-mail: dag.prawitz@philosophy.su.se

H. Wansing (ed.), *Dag Prawitz on Proofs and Meaning*, Outstanding Contributions to Logic 7, DOI 10.1007/978-3-319-11041-7_3

why and how deductive inference is able to attain its aims. Deduction should thus be explained rather than justified.[1]

For some time, I have been concerned with trying to explain deductive inference in a quite specific sense. Not being wholly satisfied with what I have written on this theme, I take this opportunity to give what I hope will be a more satisfactory presentation of the problem and the directions for its possible solution, as I see it today; some repetition of what I have written earlier will be unavoidable.

In Sect. 3.2, I make precise what in my opinion ought to be explained about deductive inference and the form the explanation should take. Before entering into that, it is useful to narrow down the phenomenon that I want to explain, which I do in Sect. 3.1 by considering among other things the different elements in an inference and the aim of a deductive inference. The question that I am raising is in brief: Why do some inferences confer evidence on their conclusions when applied to premisses for which one already has evidence? What is it that gives them this epistemic power? I do not claim originality for raising this question, but I hope the readers will agree with me that it is a fundamental problem with no obvious solution.

The solution that I propose in the rest of the paper is however bound to be controversial. I am myself dubious about some points. Others may find better solutions by following other paths, or even by following the same general path that I do. In Sect. 3.3, I argue that linguistic meaning must be given in epistemic terms if one is to account for how inferences yield evidence. Section 3.4 discusses three proposed epistemic theories of meaning. One of them is rejected, while the other two are found to contain ideas that suggest a direction for answering the raised question. They form a starting point for Sects. 3.5 and 3.6, in which a notion of ground is developed and the concept of inference is reconstructed in terms of which the explanation sought for is given. The proposal is discussed in the last Sect. 3.7.

3.1 The Phenomenon of Deductive Inference

Drawing conclusions from what one holds to be true is an everyday activity of great importance not only for humans but also for animals. It can be studied from various angles such as evolutionary, historical, sociological, psychological, philosophical, and logical points of view.

Philosophers and logicians distinguish among other things between inductive, abductive, and deductive inferences. Philosophers and psychologists make a quite different distinction between what they call intuitive and reflective inference. To make a reflective inference is to be aware of passing to a conclusion from a number of premisses that are explicitly taken to support the conclusion. Most inferences that we make are not reflective but intuitive, that is, we are not aware of making

[1] Dummett (1973 and 1991, Chap. 8) speaks of the justification of deduction, but gives the reason stated here for why it is an explanation rather than a justification that one may hope to obtain. However, our views of what the explanation should amount to differ.

them, and most of them are of course not deductive either. The two distinctions cross each other, however: for instance, some intuitive inferences are deductive, and even animals make them; Chrysippus' dog is a conceivable example.[2] Humans certainly often behave like Chrysippus' dog: taking for granted the truth of a disjunction 'A or B', and getting evidence for the truth of not-A, we start to behave as if we held B true without noticing that we have made an inference.

On the other hand, when we deliberate over an issue or are epistemically vigilant in general, we are conscious about our assumptions and are careful about the inference steps that we take, anxious to get good reasons for the conclusions we draw. Such reflective inferences are presumably confined to humans. The systematic use of reflective deductive inferences seems to be a recent phenomenon, dating back to the Greek culture; the Babylonian mathematicians were quite advanced, for instance knowing Pythagoras' theorem in some way, but, as far as we know, they never tried to prove theorems deductively.

I am here interested in philosophical aspects of deductive inferences, particularly in those of logical relevance. The logical relevance of the distinction between intuitive and reflective inferences may be doubted, but the distinction turns out to be significant also for logic. Trying to say what a deductive inference consist in, I begin now with the reflective ones.

3.1.1 A First Characterization of Inferences—Inferential Transitions

Although there is no generally accepted view of what an inference is, I think that it is right to say as a first approximation that a reflective inference contains at least a number of assertions or judgements made in the belief that one of them, the conclusion, say $\mathcal{B}$, is supported by the other ones, the premisses, say $\mathcal{A}_1, \mathcal{A}_2, \ldots, \mathcal{A}_n$. An inference in the course of an argument or proof is not an assertion or judgement to the effect that $\mathcal{B}$ "follows" from $\mathcal{A}_1, \mathcal{A}_2, \ldots, \mathcal{A}_n$, but is first of all a transition from some assertions (or judgements) to another one. In other words, it is a compound act that contains the $n+1$ assertions $\mathcal{A}_1, \mathcal{A}_2, \ldots, \mathcal{A}_n$, and $\mathcal{B}$, and in addition, the claim that the latter is supported by the former, a claim commonly indicated by words like "then", "hence", or "therefore".

This is how Frege saw an inference, as a transition between assertions or judgements.[3] To make an assertion is to use a declarative sentence A with assertive force, which we may indicate by writing $\vdash A$, using the Fregean assertion sign. We may

---

[2] Chrysippus (c. 280–207 BC) considers a story of the following kind: Running along a road following his master, who is ahead out of sight, a dog comes to a fork. The dog sniffs one of the roads. Finding no trace of the master, he suddenly sets off along the other road without sniffing.

[3] Among contemporary logicians, Martin-Löf (1985) has especially taken this point of view. For an instructive exposition of the difference between Frege's view and the later view that logic is not concerned with inference, but with logical consequence, see Smith (2009).

also say with Frege that a sentence A expresses a thought or proposition p, while $\vdash A$, the assertion of A, is an act in which p is judged to be true, not to be confused with ascribing to p the property of being true.[4]

Frege saw a judgement as an inner, mental act and an assertion as the outward, linguistic manifestation of such an act. Accordingly, the transition that takes place when we make an inference is either a mental act or a linguistic act (a speech act), depending on whether we take it as a transition between judgements or assertions. In any case, it is an act of a more complex kind than that of judgement or assertion, since it contains a number of judgements or assertions and in addition a claim that they stand in a certain epistemic relation to each other. I shall call such an act an *inferential transition.*

It does not matter for what I am interested in here whether we speak of judgements or assertions in this context and hence whether we take an inferential transition to be mental or linguistic. In the sequel, I shall usually speak of assertions, thus suggesting that the inference has a linguistic component, but most of what I shall say applies equally well if this component is lacking.

However, unlike Frege, I want to account for the fact that an inference may occur in a context where assumptions have been made, and a qualification of what has been said is therefore needed. It should be understood that the assertions in an inferential transition need not be categorical, but may be made under a number of assumptions. In such a case, I shall speak of an *assertion under assumptions*, and the term "assertion" is to cover also assertions made under assumptions, sometimes called hypothetical assertions. I write

$$A_1, A_2, \ldots, A_n \vdash B,$$

to indicate that the sentence B is asserted under the assumptions $A_1, A_2, \ldots, A_n$.

To make an assumption is a speech act of its own, and one may allow an inference to start with premisses that are assumptions, not assertions. It is convenient however not to have to reckon with this additional category of premisses, but to cover it by the case that arises when A is asserted under itself as assumption, that is, an assertion of the form $A \vdash A$.

Furthermore, we have to take into account that a sentence that is asserted by a premiss or conclusion need not be closed but may be open. For instance, to take a classical example, we may assume that $\sqrt{2}$ equals a rational number n/m and infer that then 2 equals n^2/m^2. In such cases, I shall speak of *open assertions*, and the term assertion is to cover also such assertions.

Does an inference contain something more than an inferential transition, and if so, what? This will be a crucial issue in this paper. On the one hand, when we make inferences in the course of a proof, all that we announce is normally a number of inferential transitions. On the other hand, there seems to be something more

---

[4] Like Frege, and unlike Martin-Löf (1985), I do not take an expression of the form "it is true that ...", where the dots stand for a declarative sentence, to be the form of a judgement. To assert a sentence of the form "... is true", where the dots stand for the name of a sentence A, is to make a semantic ascent, as I see it, and is thus not the same as to assert A.

that goes on at an inference, some kind of mental operation that makes us believe, correctly or incorrectly, that we have a good reason to make the assertion that appears as conclusion. For the moment, I leave the question open and concentrate on the inferential transitions. They may be indicated by figures like:

$$\frac{\Gamma_1 \vdash A_1 \qquad \Gamma_2 \vdash A_2 \qquad \ldots \qquad \Gamma_n \vdash A_n}{\Delta \vdash B}$$

where Γ_i and Δ stand for (possibly empty) sequences of (closed or open) sentences; thus, $\Gamma_i \vdash A_i$ and $\Delta \vdash B$ stand for assertions. Although an inference may occur in the context of assumptions, some of which may be discharged by the inference, new assumptions are not introduced by the inference, and hence, each sentence of Δ belongs to some Γ_i $(i \leq n)$. The long horizontal line indicates that the assertion below the line is claimed to be supported by the assertions above the line. I shall call such a figure an *inference figure*. This is of course how inferences are commonly indicated, although often with a different interpretation.

If we thought of an inference as consisting not of a transition between assertions but of just an explicit claim about an inferential relation between the premisses and the conclusion, it would remain to account for the successive assertions that do occur in a proof as well as for how the claim is to be represented; it would certainly be inadequate to represent it as the assertion of an implication.[5] As I see it, the claim comes to an expression when making an inference only in the form that the conclusion is an assertion made with reference to some premisses; "because" of them, as one sometimes says.[6]

3.1.2 Inference Acts, Inference Figures, and Inference Schemata

When we characterize an inference as an act, we may do this on different levels of abstractions. If we pay attention to the agent who performs an inference and the occasion at which it is performed, we are considering an *individual* inference act. By abstracting from the agent and the occasion, as we usually do in logic, we get a *generic* inference act. We may further abstract from the particular premisses and the conclusion of a generic inference and consider only how the logical forms of

---

[5] One of the lessons of Carroll's (1895) regress is that we never get to a wanted conclusion if we see an inference as the assertion of an implication (see further fn 12 and 13).

[6] It is sometimes argued against this view that an inference may result not in the acceptance of the conclusion but in the rejection of a former belief. But such a belief revision is better analysed as consisting of a series of acts, the first of which is an inference in the present sense, resulting in an assertion being made or a belief being formed, which is then found to be in contradiction to another belief already held. Instead of using these two beliefs as premisses in another inference that would result in the categorical assertion of a contradiction, some of the former beliefs are reclassified as assumptions. Under these assumptions, a contradiction is inferred, and the resulting hypothetical assertion is the premiss of a last inference (a reductio ad absurdum) in which the negation of one of the assumptions is inferred.

the sentences involved are related (as we do when we say that modus ponens is an inference with two premisses such that for some sentences A and B, one of the premisses is the assertion of $A \supset B$, while the other is the assertion of A and the conclusion is the assertion of B). I shall call what we then get an *inference form*.

Making the same distinctions with respect to inferential transitions, we have already seen that a generic inferential transition is individuated by an *inference figure* of the kind displayed in the above, in other words, by its premisses and its conclusion. Similarly, the transition contained in an inference form is individuated by what I shall call an *inference schema*, following common usage (in some contexts it may be called an *inference rule*); here the sentence letters do not stand for particular sentences but are to be seen as parameters.

We usually take for granted that everything logically relevant about inference acts can be dealt with when an inference is identified with a set of premisses and a conclusion, in other words, with what individuates a generic inferential transition or the form of such a transition. We can then direct our attention solely to inference figures or schemata and disregard the acts that they represent, which means that we can see premisses and conclusions as sentences instead of assertions; the Fregean assertion sign may be taken just as a punctuation sign that separates a sentence in an argument from the hypotheses on which it depends.

When we take this point of view, the distinction between intuitive and reflective inferences is of no interest, nor is there any room for it. The premisses and the conclusion of an intuitive inference act are implicit beliefs ascribed to the agent when we interpret her behaviour as involving such an act, while the premisses and the conclusion of a reflective inference are explicit assertions or judgements. But the difference is of no concern, if what matters are the sentences that are asserted or believed to be true.

But even from this point of view, a general characterization of inferences must contain something more than what has been said so far, because nothing has been said about which inferences are the good or interesting ones.

3.1.3 The Aim of Inference

Which then are the good inferences? The standard answer is of course that they are the (logically) *valid* ones, where validity of an inference is defined as logical consequence, either in the traditional way to mean that the truth of the premisses necessarily implies the truth of the conclusion, or in the way of Bolzano and Tarski to mean that under all variations of the meaning of the non-logical vocabulary the inference preserves truth. Since this definition of validity is with reference to the truth of the sentences that individuate a generic inferential transition, the answer is still within the framework of inference figures and inference schemata supposed to be what matters in logic.

To discuss whether this is an adequate answer, we must say something about the point of inferences, why we are interested in making and studying them, and now the inferences must be seen as acts. In view of the fact that our beliefs are founded, among other things, on intuitive or reflective inferences, we may say that the function of inferences in general is to arrive at new beliefs with a sufficient degree of veracity. For an inference to have this function, validity does not seem to be what really matters. An inference must preserve truth, but that it preserves truth necessarily or under all variations of the non-logical vocabulary seems to be an unnecessarily strong requirement. Furthermore, it may not be enough to preserve truth in contexts where the premisses are only probably true; what then matters is rather likelihood.

However, it is the point of reflective inferences that I want to discuss especially, and in this context we can speak of aims. The personal aims of subjects who make inferences may of course differ, but we can speak of an aim that should be present in order that an inference is to count as reflective. As already said, it is an ingredient in what it is to make such an inference that the conclusion is held, correctly or incorrectly, to be supported by the premisses. In view of this, reflective inferences must be understood as aiming at getting support for the conclusion. This may be articulated in different ways. We may say that the primary aim is to get a good *reason* for the assertion that occurs as conclusion. Since the term reason also stands for cause or motive, another and better way to express the same point is to say that the aim is to get adequate *grounds* for assertions or sufficient *evidence* for the truth of asserted sentences. Since assertions are evaluated among other things with respect to the grounds or evidence the speakers have for making them, we may also say that the aim of reflective inferences is to make assertions *justified* or *warranted*.

The terms that I have used here have of course many uses. I am speaking of evidence and grounds for an assertion and am using the terms synonymously to refer to what a subject should be in possession of in order to be justified or warranted in making the assertion. Already saying this is hopefully sufficient to help the reader to distinguish my use of these terms from some other common uses, but it is an objective of the paper to clarify the terms further. Evidence and ground are notions on a par with notions like truth that may not be possible to define in terms of more basic notions, but have to be explained by exploring their relations to other notions and by describing the form that evidence or grounds take for assertions of different kinds. These are questions that I shall return to in the paper. To avoid a possible misunderstanding, I emphazise already now that I understand evidence or a ground for an assertion as something objective, not to be confused with the psychological notion of subjective conviction.[7]

Most of what has been said up to now applies not only to deductive inferences. What is particular about them is that they aim at *conclusive evidence* or *conclusive grounds* when applied to premisses for which one already has conclusive evidence or

[7] Some writers, e.g. Chateaubriand Filho (1999), point out that a flawless sequence of inferences may fail to carry conviction while a geometrical drawing may convince us completely of the truth of an alleged theorem. As long as this is a psychological question about what subjectively convinces us, I am not concerned with it here. If the drawings are regimented in such a way that it can be claimed that we get conclusive evidence for some logically compound assertions by observing the drawings, it may amount to a theory of what it is to have a ground for an assertion, alternative to the one developed in Sect. 3.5.

grounds. It is definitely a crucial element of our ideas about deductive inferences that they aim at something more than making the conclusion plausible or giving reason to think that the asserted sentence is probably true. This aspect must therefore be included when we characterize deductive inferences. The important point to note is that the aim of a reflective inference cannot be described just in terms of truth. The aim has not been attained when the sentence that is asserted by the conclusion just happens to be true.[8] Nor can it be sufficient that, as a matter of fact, the sentence is true in all possible worlds, or under all variations of the meaning of the non-logical vocabulary, such that the sentences asserted by the premisses are true. In an act of reflective inference, the conclusion is held to be supported epistemically, and the inference would be unsuccessful if such a support was not really obtained. The point could also be made by saying that we expect an inference to afford us with knowledge in a Platonian sense, which is again to say that it should give us not only a true belief, but also a ground for the belief.[9]

Since I here restrict myself to deductive inferences applied to premisses for which one claims conclusive evidence or conclusive grounds, the evidence or ground that one seeks to obtain by an inference will always be intended to be conclusive. I shall therefore normally leave out the attribute "conclusive".

One could have a terminology according to which it is built into the notion of inference that it attains its aim, but this is not how we usually use the term inference, allowing for the possibility that an inference is invalid, and I shall follow common usage here.

3.1.4 Deductive Proofs

Unlike inference, the term proof is normally used to indicate epistemic success. It is not only that we cannot say, without being ironical, that we have proved a sentence but it turned out to be false. An alleged proof that starts from premisses for which one has no grounds or turns out to have a gap has to be withdrawn; a proof with gaps is a contradiction in terms.

What is a gap in a proof? It occurs exactly at a point where an inferential transition is made such that there are grounds for the premisses, but the inference does not succeed to carry this evidence further so as to confer evidence on the conclusion. In other words, it appears when an inference is made that is not successful with respect to its primary aim as described above. In sum, it is required of a proof that all its inferences are successful.

---

[8] Pagin (2012) develops a view of what has to be required of a good or valid inference that is different from mine, but he makes essentially the same point that truth is not enough; as he puts it, it has to be arrived at by a reliable method.

[9] As Cellucci (2013) points out we are often interested in new proofs of what is already known to be true, which he sees as an argument against the idea that knowledge is the aim of inference. New proofs are interesting when they give new grounds, so this observation reinforces the idea that the primary aim of inference is to acquire grounds or evidence.

To characterize a proof as a chain of inferences, as we usually do, we thus need this notion of successful inference. It is convenient to have a term for this, that is, for inferences that can be used legitimately in a proof, and I have called them *legitimate inferences* (Prawitz 2011). Accordingly, a generic inference is said to be legitimate, if a subject who makes the inference and has evidence for its premisses thereby gets evidence for the conclusion; or more precisely, it should follow that she has evidence for the conclusion from the assumptions that she performs the inference and has evidence for the premisses. We can now say that a deductive proof is a chain of legitimate inferences.

3.2 A Fundamental Problem

That deductive proof delivers conclusive evidence is thus a conceptual truth when the concepts involved are understood as in the preceding section. But the fact that we use the notions of proof and evidence in this way does not of course imply that there are proofs. Even less does it explain why and how a chain of deductive inferences can yield evidence for its last conclusion. To explain this, I take to be a fundamental problem of logic and the philosophy of logic. Using the notion introduced at the end of the preceding section, we can say that the problem is to explain how there can be legitimate inferences.

That there is something of fundamental importance to explain here seems to me obvious, but the need of an explanation may be most apparent in the case of sentences whose uses are first learned in other contexts than inferences. We learn how to verify an arithmetical identity $t = u$ by making the relevant computations. To assert such an identity is then naturally taken to be a claim to the effect that a computation of the two terms t and u would yield the same value; we get evidence for the claim by making the computations and observing the result. Having learned what is meant by $t = u$ in this way, it is not at all obvious why one should expect to get evidence for the truth of $t = u$ by making instead a deductive proof, for instance by mathematical induction. Similarly, we learn how to verify an assertion about the length of a distance by measuring it. How is it that we can get evidence for such an assertion by instead measuring some other lines, for instance, the legs of a right-angled triangle where the distance appears as hypotenuse, and then inferring the assertion via a proof of Pythagoras' theorem?

For some sentences like universal quantifications, we expect that evidence for asserting them can only be got by some kind of inferences, but the problem is equally significant for them, since not all inferences yield evidence. The general problem is, metaphorically expressed: Why do certain inferences have the epistemic power to confer evidence on the conclusion when applied to premisses for which there is evidence already?

We take for granted that some inferences have such a power, and there is no reason to doubt that they have. But what is it that gives them this power? This should be explained.

Some may think that it is already explained or could easily be explained in terms of the notion of valid inference as commonly defined. Although this does not seem likely in view of the fact that the definition of validity refers to truth and not to any epistemic notion concerned with how truths are established, this possibility should not be excluded off hand and will be discussed below (Sect. 3.2.2).

Others may think that it is utopian to hope for an explanation of such fundamental facts. In any case, the problem why certain inferences are legitimate is not often formulated seriously. There are exceptions such as Boghossian (2001, 2003, 2012) and Sundholm (1998a, 2012), who have taken up the problem in somewhat different terms.

It is to be noted however that the problem can be formulated in terms that Aristotle already introduced. He saw that the use of what he calls syllogisms may not give evidence. This is apparent from the fact that immediately after having given his often quoted definition of syllogism, which is a forerunner to how the validity of an inference became defined traditionally, he differentiates between *perfect* and *imperfect* syllogisms, where the notion of perfect seems to have the same drift as legitimate. In Aristotle's words, as translated by Ross (1949, p. 287), a perfect syllogism is one "that needs nothing other than the premisses to make the conclusion evident". Aristotle did not try to explain what makes an inference perfect, but he took some syllogisms as perfect, and much of his work is concerned with reducing imperfect syllogisms to perfect ones.[10]

3.2.1 The Form of an Explanation

What can be expected of an explanation why some inferences are legitimate? As for explanations in general, one would expect to be given some properties of the entities in question that imply the facts to be explained, or, more precisely in this case, one wants to find a condition on a deductive inference that is sufficient and necessary for it to be legitimate, and that by and large is satisfied by inferences that we use as if they were legitimate.

The task to show that a condition C on generic inferences is sufficient for legitimacy can be spelled out as the task of establishing for any generic inference I and subject S that from the three facts

(1) the inference I satisfies the condition C,
(2) the subject S has evidence for the premisses of I
(3) S performs I,

it can be derived that

(4) S gets evidence for the conclusion of I.

To find such a derivation is the business of the philosopher who seeks an explanation. This may be described as taking place on a meta-level, where the subject's

[10] As argued by Corcoran (1974), this is a central aim of Aristotle's logical work.

activities on the object-level is explained. The subject is not to do anything except performing the inference I; the point is that thereby, without doing anything else, she gets evidence for the conclusion.

When deriving (4) from (1)–(3), one may very well use a generic inference of the same kind as the one that is shown to be legitimate in this way. The point is not to convince a sceptic who doubts that the inference is legitimate, but to explain why it is legitimate, and to this end there can be no objection to use this very inference.

To be able to derive (4) from (1)–(3), the concepts that occur in these clauses must of course be made more precise, and it is now that the question whether an inference consists of something more than an inferential transition becomes crucial. There is an apparent risk that to identify an inference with such a transition gives a concept that is too impoverished to allow an explanation of why certain inferences provide us with evidence.

3.2.2 Can Legitimacy be Explained in Terms of Validity?

We should now consider the question whether we do not already have the wanted condition C in the form of validity of an inference. Many introductory textbooks in logic intimate this more or less explicitly when they ask what an epistemically good inference is and thereafter seem to answer the question by defining the notion of validity, either in the traditional way or in the way of Bolzano and Tarski.

When this notion of validity is defined, no presumption is made about an inference being identified by something above its premisses and conclusion, so let us assume that an inference involves nothing more than an inferential transition. Then it is obvious however that applying a valid inference to premisses for which one has evidence does not in general give one evidence; just think of all cases of logical consequence that are difficult to establish or that no one has established, or alternatively, think of how short proofs could be if all valid inferences were legitimate—in fact, the proofs would never need to contain more than one inference.

Some nevertheless insist upon the relevance of validity, arguing that although validity in general is not a sufficient condition for legitimacy, simple forms of validity are, for instance, the ones met in textbooks. If validity in general is not a sufficient condition, it can of course not be the *validity* of those simpler inferences that is responsible for their legitimacy—it must somehow be there simplicity or something related that is responsible. One suggestion could be that the property 'known to be valid' is responsible for the inferences to be legitimate.

This is to suggest that the explanation should take a somewhat different form from how it is described above. Instead of a condition C on inferences, there should be a relation $C(S, I)$ between a subject S and an inference I, namely that S knows I

to be valid, which can be shown to be a sufficient and necessary condition for an inference to be legitimate.[11]

To show the condition to be sufficient, we should thus demonstrate that (4) follows when the condition $C(I)$ in clause (1) is replaced with $C(S, I)$, saying that S knows I to be valid; call this new clause (1′). It is important at this point not to confuse the subject and the philosopher who seeks an explanation. Already from (1) (C standing for validity) and (2), the philosopher can easily derive that the sentence occurring in the conclusion of I is true, and has then evidence for the conclusion of I in view of this derivation; the philosopher argues on the meta-level and is not begging the question, because the issue is not whether he or she has evidence, but whether the subject has. The latter does not follow from (1) and (2). The suggestion is now that from (1′), (2), and (3), it follows that the subject has evidence for the conclusion of I.

However, it is not clear how this follows. Indeed, the idea that it follows was made problematic by Lewis Carroll in his tale about Achilles and the Tortoise.[12] It may be suggested that the subject could derive from (1′) and (2) that she has evidence for the conclusion $\mathcal{A}$ in the same way as the philosopher showed that he or she could get such evidence. But, of course, the last step where one looks at a derivation of $\mathcal{A}$ and concludes that one has evidence for $\mathcal{A}$ cannot be taken now when the very question is whether the performance of an inference gives evidence; it has first to be shown that the steps in the derivation preserve evidence.[13] Furthermore, to argue that the subject can get evidence for $\mathcal{A}$ by further reasoning is to depart from the idea that it is by the original inference that the subjects gets evidence for $\mathcal{A}$.

The idea that one must know an inference to be valid in order to get evidence for the conclusion is problematic also in other respects. One must ask how the knowledge is to be acquired. If the subject has to prove the inference to be valid, the inferences

---

[11] Etchemendy (1990) argues in effect that if validity is defined in the way of Bolzano and Tarski, then knowing an inference to be valid is not sufficient for it to be legitimate, while if validity is defined in the traditional way it is sufficient. As will be seen, I do not think that there is such a difference.

[12] Carroll's is thereby a philosopher who has raised the problem why a subject gets conclusive evidence for the conclusion (or as he puts, gets forced to accept the conclusion) of an obviously valid inference whose premisses she accepts; his point being that it does not help that she accepts the validity of the inference. See also fn 5.

[13] The dialectical situation that we are in, after having conceded that the subject S does not get evidence for the conclusion $\mathcal{A}$ merely because of the inference I being valid, is quite similar to the one that Carroll describes, where S (the Tortoise) asks why she should accept the conclusion $\mathcal{A}$ of an inference I whose premisses she accepts. The advocate of the relevance of validity (Achilles) asks S to accept that the sentence occurring in $\mathcal{A}$ must be true if the sentences occurring in the premisses are, in other words, that I is valid, as defined traditionally. S agrees to this, and let us say that she now knows that I is valid. She has thus an extra premiss, which coincides with our clause (1′), from which to infer that she has evidence for $\mathcal{A}$, but she still asks why it follows that she has evidence for $\mathcal{A}$. Admittedly, there is a valid inference to $\mathcal{A}$ from the now available premisses, but there was one already from the original premisses, and it has been conceded that the mere validity of an inference is not enough in order to infer that a subject who performs the inference gets evidence for the conclusion. It could now be argued that the new inference is not only valid but is surely known by S to be valid, and to argue so (which is what Achilles does) is to take a second step in the regress that Carroll describes.

used in that proof must be proved valid too in order to have any force. Another regress than the one noted by Carroll then emerges. Clearly, if in general we had to prove an inference to be valid before using it legitimately, the use of inferences would never be able to get off the ground.

To take C to be instead the condition that the validity of the inference is self-evident, say in view of the meaning of the involved sentences, is again to require too much. In some cases, such as the inference schema of conjunction introduction, the validity of the schema is a straightforward part of the truth-condition for the sentence occurring in the conclusion. But such cases are exceptional. Even in the case of very simple inferences, the validity is not in general self-evident but can be seen to be valid by a simple derivation.

Although it cannot be excluded that (4) could be derived from (1′), (2), and (3) by explicating adequately the involved concepts, the prospects do not seem promising, regardless of whether validity is defined in the traditional way or in the way of Bolzano and Tarski, and in any case no cogent argument has so far been given. It may be thought that validity is nevertheless a necessary condition for an inference to be legitimate. Of course, the sentences occurring in the premisses of an inference cannot be true while the sentence occurring in conclusion is false if the inference is to be legitimate. However, to require that the same hold for all variations of the meaning of the non-logical parts of the sentences involved rules out the legitimacy of inferences that are regularly used in mathematics. For instance, an inference by mathematical induction passes from the assertion of true sentences to the assertion of a false one when some non-logical constant (such as natural number) is suitably re-interpreted. Whether validity defined in the traditional way is necessary for legitimacy depends on how the modal notion that occurs in the definition is analysed.

3.2.3 The Intertwining of Evidence and Proofs

To get any further with our problem we must first of all inquire about what it is to have evidence for an assertion. As already remarked, for logically compound sentences there seems to be no alternative to saying that evidence comes from inference. In view of this, it is easy to understand how the problem of explaining why inferences give evidence has come to be overlooked. On the other hand, since not any inference gives evidence, one cannot account for evidence by referring to inferences without saying which inferences one has in mind, and to use the notion of legitimate inference at this point would of course be patently circular.[14]

How can we avoid to get entangled into circles when trying to explain what evidence is? It may have to be accepted that it is in the nature of the meaning of some

[14] From this one may be tempted to draw the conclusion that proof and evidence are equivalent notions, as Martin-Löf (1985) affirms, saying, "proof is the same as evidence". But against this speaks that evidence is a much more general and basic notion. As already noted, there are kinds of assertions for which evidence is got first of all by other means than proofs.

types of sentences that evidence for them can only be explained in terms of certain kinds of inferences. The legitimacy of these inferences is then a datum that has to be accepted as somehow constitutive for the meaning of these sentences. This idea will be developed in the next section.

But not all cases of accepted inferences can reasonably be seen as constitutive of the meaning of the involved sentences, as I shall argue (Sect. 3.4.1). For them, it remains to explain why they are legitimate. One possibility may be to define the notion of proof in another way than as a chain of inferences. I shall discuss some attempts in this direction in Sect. 3.4. If one succeeds in that, it may be possible to account for evidence in terms of proof without circularity. The legitimate inferences could then be taken to be the ones that give rise to a new proof when attached to a proof. This is to turn the usual conceptual order between inferences and proofs upside down, but it has seemed to me a fruitful idea. I do not any longer adhere to it, but the approach that I now want to follow is nevertheless an offshoot of it.

3.3 Meaning and Evidence

It is a truism to say that what counts as evidence for an assertion must depend on the meaning of the asserted sentence. It is equally obvious that the legitimacy of an inference must likewise depend on the meanings of the sentences involved. We are familiar with how in classical semantics inferences are shown to be valid by appealing to the truth-conditions that define the meanings of sentences of different forms. To show that an inference is legitimate we must similarly derive this fact from the meanings of the sentences involved. The difference is however that in this latter case we want to derive, not that the inference is truth-preserving, but that one gets evidence for the conclusion by performing the inference, given evidence for the premisses. For it to be possible to derive the latter it is not sufficient, I want to argue, that the meanings of sentences are given in the form of bare truth-conditions, that is, conditions in terms of a notion of truth that is not explained in terms of something else. It seems that meaning must instead be given in epistemic terms.

Consider a simple example like conjunction introduction, where the assertion of a conjunction $A\&B$ is inferred from two premisses, asserting A and asserting B. Given that a subject has (conclusive) evidence for the truth of A and for the truth of B, we can infer that A is true and that B is true, and hence it follows from the meaning of a conjunction, given in terms of its truth-condition, that the conjunction $A\&B$ is true. (Note again that it is not a question of the subject inferring this; cf Sect. 3.2.1). But we want to show not only that the conjunction is true, but that the subject gets evidence for this by making the inference of conjunction introduction, and this does not follow from the meaning of conjunction given in terms of its truth-condition and from what we have said so far about inferences.

It may retorted that it is just an obvious fact that an agent who has evidence for the truth of A and for the truth of B has evidence for the truth of the conjunction $A\&B$. But this is not a fact—indeed, it is a fact that one sometimes has evidence for two conjuncts but does not bring these pieces of evidence together so as to get evidence for the conjunction. Our knowledge is unfortunately not closed under logical consequence, nor is the set of assertions for which we have evidence closed under the operation of legitimate inference (in contrast to the set of assertions for which we can acquire evidence). Having evidence for the assertion of A and for the assertion of B do not together *constitute* having evidence for the assertion of $A\&B$.[15] The step to go from having the first two pieces of evidence to having also the third piece of evidence is small, but it is still a step that has to be taken. To leave all such small steps out of account would be to say that one already has all evidence that one can get by making inferences of the kind that we are familiar with.

It is not to be denied that if there is evidence for asserting the two conjuncts, there is evidence for the truth of the conjunction, too. Nor is it to be denied that a person who has the first two pieces of evidence can easily have or get the third piece of evidence, provided that she knows the meaning of conjunction. This one may call an undeniable fact, but it is more or less what we want to explain, hopefully by deriving it from the meaning of conjunction.

One may doubt that this simple fact can be derived from something more basic, and claim it to be a brute fact that just has to be accepted. But as said, it must depend on the meaning of conjunction; it would not be a fact, if conjunction meant something else. Thus, instead of saying that it is a brute fact, one should say that it is a fact that is constitutive of the meaning of conjunction, or at least a part of what constitutes that meaning.

It cannot be excluded that it can be derived from some other principle about evidence that a person who knows the meaning of conjunction and has evidence for asserting two conjuncts can get evidence for asserting the conjunction. But the question then arises why this other principle holds. That so and so counts as evidence for an assertion A must in some cases, for instance the one just considered, be explained in the end by referring to some conventions, it seems, and to refer to conventions in this context is reasonable if it is a question of the meaning of A. It is therefore difficult to see how it can be avoided that some basic principles about evidence are formulated in terms of meaning and conversely that meaning is thereby to some extent accounted for in epistemic terms.

Dummett, inspired by Wittgenstein, argues more generally that to account for how language functions, linguistic meaning must be connected with our use of language, and in particular with what justifies our assertions or counts as grounds for them. For him the argument is a part of a more comprehensive argument in favour of constructivism or anti-realism, which eventually leads to discarding classical logic. These possible consequences are not something that I shall be concerned with here. It may be true that an epistemic theory of meaning is more easily given if the language is understood intuitionistically, and such a usage is presupposed in the main proposals that I shall consider in the next sections. But I leave it open here whether similar

---

[15] This point has been argued for by Detlefsen (1992). Even if evidence for asserting $A\&B$ is said to consist of a pair whose elements are evidence for asserting A and evidence for asserting B, we have still to form this pair to get evidence for asserting the conjunction.

epistemic theories of meaning could not be developed that are in accordance with the classical understanding of logical constants.

3.4 Three Proposals for How Linguistic Meaning Can be Explained in Epistemic Terms

I shall consider three different proposals, which can be dated back to ideas put forth by Carnap, Gentzen, and Heyting in the 30's, for how linguistic meaning can be explained in epistemic terms.

Carnap and Gentzen made the suggestion that the meanings of sentences are determined by inference rules concerning these sentences, an idea that has become known as inferentialism. Carnap (1934) made the sweeping suggestion that the meanings of the sentences of a formal language are determined by all the inference rules of the language. This is a form of what I shall call *radical inferentialism*, which I shall discuss briefly in Sect. 3.4.1 and find reason to reject.

Gentzen (1934–1935) made the more sophisticated proposal that certain specific inferences, instances of what he called introduction rules for logical constants, determine the meanings of these constants, while other inferences are justified in terms of these meanings. There are various ways in which this idea may be developed into a theory of meaning in which inference or proof rather than truth is a central concept, what Peter Schroeder-Heister has called *proof-theoretic semantics*. In Sect. 3.4.2, I shall consider one way based on defining a notion of valid argument.

Heyting (1930, 1931, 1934) did not make a suggestion along such inferential lines, but explained the intuitionistic meaning of propositions in terms of what he called intended constructions. To find the intended construction was considered to be the required ground for asserting the proposition. The constructions in terms of which Heyting explained meaning had in this way epistemic significance. This idea will be considered in Sect. 3.4.3.

Neither the notion of valid argument nor the notion of intuitionistic construction was aimed at answering the question that I am raising in this paper, and alone they do not suffice for this aim, but they will be a starting point for proposals that I shall make in Sect. 3.5.

3.4.1 Radical Inferentialism

The most radical form of inferentialism maintains that all accepted inferences should be seen as determining the meaning of the involved sentences. This view is indeed very radical and breaks completely with a traditional view of inferences. A less implausible but still quite radical form of inferentialism acknowledges that inferences are accepted because of being decomposed into chains of other accepted inferences, and counts only inferences that cannot be so decomposed as meaning constitutive; the latter ones may be called *primitive*.

Questions of meaning may quite generally be approached from a third or a first person perspective, that is, we may ask questions about a language that people already use or about a language that we shape. Carnap's inferentialism is of the latter kind: the meanings of sentences of a formal language that we set up are to determined by its primitive inference rules. A radical inferentialism of the former kind says that the meaning attached to a sentence by a linguistic community is determined by the (primitive or totality of) inferences that involve the sentence and are accepted by the linguistic community.[16]

That a generic inference is accepted means that it is accepted as legitimate, in other words, as giving evidence for the conclusion when applied to premisses for which one already has evidence. In this way, inferentialism is also a view of what gives evidence for assertions, and the fundamental problem gets thereby an immediate and trivial solution, build into what counts as evidence.

It is not quite fair to paraphrase this view as saying: An inference gives evidence because this is how we reason in our language—there is nothing more to say. There is a rationale for postulating the accepted (primitive) inferences as legitimate based on the view that they are meaning constitutive: the conclusions of the inferences are understood as not saying anything more than that evidence for them is acquired by means of the accepted deductive practice. However, there are reasons to object to this view of meaning and of our deductive practice.

One main reason may be illustrated as follows. There are cases where the adoption of an inference rule may reasonably be seen as just an expression of what a notion means. Conjunction introduction discussed in the previous section may be one such case. The adoption of the rules that 0 is a natural number and that the successor of a natural number is again a natural number may similarly be seen as an expression of what we mean by natural number. But this cannot be taken as a reason for saying the same about all primitive inference rules. For instance, an inference by mathematical induction gives evidence for the conclusion because of what we mean by natural number, not the other way around: the principle of induction is a primitive inference rule (unless we are in impredicative second order logic), yet is not adopted as a convention, as a part of what we mean by natural number, but is seen to be legitimate in virtue of that meaning.

Furthermore, there are well known objections to the adoption of an arbitrary set of inference rules as constitutive of the meaning of a sentence A formulated by Prior (1960), showing that it may lead to catastrophic consequences for how other sentences may be used deductively, and by Dummett (1991, p. 206), who illustrates by examples that there may be no way of figuring out the significance of asserting A; as he concludes, the rules may tell us how one is allowed to operate with A in arguments but not what A means.

---

[16] Cesare Cozzo (1994) has worked out an inferentialism according to which the meaning of an expression is determined only by those primitive inferences that *concern* the expression in a genuine (rigorously defined) sense. To its advantage the resulting meaning theory becomes compositional.

The situation is very different in all these respects when we take a specific kind of inference rules such as Gentzen's introduction rules as meaning constitutive.

3.4.2 Proof-Theoretic Semantics

An introduction rule in Gentzen's system of natural deduction has the property that sentences that occur as premisses or as assumptions discharged by the inference are constituents of the sentence that occurs as conclusion. The condition that has to obtain for inferring the assertion of a sentence A by such a rule is thus stated in terms of the constituents of A, just as a classical truth-condition for a sentence is stated in terms of its constituents. Therefore, when the meaning of A is taken to be given by its introduction rule, the significance of asserting A is quite clear: the condition that has to obtain in order for a speaker to get a ground for the assertion by an introduction inference is understood if the constituents of A are understood.

When only the introduction rules are taken to be meaning constitutive, reasoning which in its final step proceeds by applying an introduction rule gets a particular position, marked by saying that such a piece of reasoning is in *canonical form*. According to the meaning of a closed assertion A, it is reasoning in canonical form that gives evidence for A, provided of course that the reasoning has provided evidence for the premisses of the final step. The question then arises how reasoning that is not in that form can be valid at all. For this to be the case it must be clear how the reasoning can be rewritten in canonical form. In other words, an inference that is not an instance of an introduction rule but is used in a piece of reasoning for arguing for a closed assertion A must be justified by there being an operation, usually referred to as a *reduction*, that transforms this reasoning into canonical form.

Gentzen (ibid) exemplified how his elimination rules are justified in terms of the meanings given by his introduction rules by describing some of these reductions. To generalize this line of thought to reasoning in general, we may consider arbitrary arguments and define what it is for them to be valid. Let us call a chain of arbitrary inference figures an *argument structure*, and let us say that it is *closed* if its last assertion is closed and categorical, and that it is *open* otherwise. By an *argument* is understood a pair (S, J) where S is an argument structure and J is a set of assignments of operations j to occurrences of inference figures in S that are not instances of introduction rules. An operation j assigned to an occurrence of an inference figure is to be a partial operation on argument structures that end with the inference figure in question. When defined for an argument structure for an assertion A under assumptions Γ, j is to produce another argument structure for A under assumptions Δ such that $\Delta \subseteq \Gamma$. J can be seen as an alleged justification—the definition of validity below imposes restrictions that have to be satisfied for J to be a real justification. An argument (S, J) is said to reduce to the argument (S^*, J^*), if S^* is obtained from S by successively replacing initial parts S_1 with the result of applying to S_1 the reduction in J assigned to the last inference figure of S_1, and J^* is J restricted to S^*.

The ideas about valid reasoning stated above can now be turned into a definition of what it is for an argument to be *valid*, where an open argument is seen as parametric with respect to free individual variables and assumptions. Validity is defined relative to a given set of individual terms and a given set of closed arguments for atomic sentences, none of which is to be an argument for the atomic sentence $\bot$ standing for falsehood. The basic clause of the definition says that the given arguments for atomic sentences, which also count as canonical, are valid. The inductive clauses are:

(1) A closed argument in canonical form for the assertion of a compound sentence is valid if and only if its immediate sub-arguments are valid.
(2) A closed argument that is not in canonical form is valid if and only if it reduces to a valid argument in canonical form.
(3) An open argument is valid if and only if its closed substitution instances are valid, obtained by replacing first individual variables with closed individual terms and then each initial assertion of the form $A \vdash A$ with a closed valid argument for $\vdash A$ (of course, cancelling the associated assumptions A further down in the argument).[17]

The idea is that a valid argument for an assertion $\Gamma \vdash A$ as now defined by recursion over sentences represents a proof. It is to be noted that the definition does not presuppose the notion of legitimate inference. This opens for the possibility mentioned above (Sect. 3.2.3) that the latter notion can be accounted for in terms of proofs instead of the other way around.

However, it must be discussed whether it is right that a valid argument represents a proof in a sense of interest here, that is, whether to have constructed a valid argument for an assertion really amounts to having evidence for the assertion. Assume that the answer to the question is affirmative in the case of canonical arguments and let us consider what knowledge one has when one knows a valid argument for a closed categorical assertion $\mathcal{A}$ that is not in canonical form. One then knows *how* to get in possession of evidence for $\mathcal{A}$: one applies the reductions and given that the argument is valid, one will eventually be able to reduce it to a closed argument for $\mathcal{A}$ in canonical form. But this is not the same as knowing *that* applying the reduction rules will produce an argument in canonical form. If the latter is requested to have evidence for $\mathcal{A}$, then it is not enough to know or to be in possession of a valid argu-

[17] A notion of validity along these lines was first defined for natural deductions (Prawitz 1971, Appendix A1), and was then generalized to arbitrary arguments (Prawitz 1973); the definition given here is essentially as stated there except for letting the justifications be assignments to occurrences of inference figures instead of inference forms. Other variations occur in the definitions of validity given by Dummett (1991) and Schroeder-Heister (2006). None of these variations has relevance to the main question discussed in this section. I investigate in a forthcoming paper (Prawitz 2014) how these different notions of validity are related to each other and to the notion on intuitionistic proof discussed in the next section. It should be mentioned that previously I have also considered a variant form of validity that followed Tait's (1967) definition of convertible terms and Martin-Löf's (1971) definition of computable deduction in taking all normal derivations to be valid, which I used to prove normalizability or strong normalizability. But I now concur with Peter Schroeder-Heister saying that this notion does not explicate the idea that the introduction rules is meaning constitutive and it is better not called validity.

ment for $\mathcal{A}$, one must also know that what one is in possession of is a valid argument. I shall return to this question in the next section, since a similar question will occur there.

3.4.3 Intuitionistic Constructions

When Heyting explained the intuitionistic meaning of a sentence as expressing the intention or problem of finding a construction that satisfies certain conditions, he emphasized that the construction was thought of as possible in principle to be experienced by us. As Heyting put it "the intention … does not concern the existence of a state of affair imagined as independent of us, but an experience imagined as possible".[18]

The "realization of the required construction" or "fulfilment of the intention" expressed by a proposition is, according to Heyting, exactly what is required in order to assert the proposition intuitionistically. This realization, which in other words means that one has found or got in possession of the intended construction, thus amounts to having what I have called a ground for asserting the proposition. Heyting's explanation of linguistic meaning has clearly an epistemic character by being phrased, not in terms of a condition concerning a world thought of as independent of us that has to be satisfied in order for the sentence to be true, but in terms of what we should experience or know to be justified in asserting the sentence.[19] The epistemic character became even more pronounced in later work where meaning is explained by "giving necessary and sufficient conditions under which a complex expression can be asserted" (Heyting 1956, p. 97).

The realization of the required construction is called a *proof* by Heyting. Although the term proof is here used in its usual epistemic sense in so far as the realization of the intended construction is the requirement for asserting the sentence, the explanation of the term does not presuppose the notion of inference. Therefore we have again (as in Sect. 3.4.2) a candidate for how to account for legitimate inferences.

To discuss this possibility in a more precise way, the meaning explanations have to be made more explicit. There are a number of proposals for how to do this.[20] Fol-

[18] "Die Intention geht … nicht auf einen als unabhängig von uns bestehend gedachten Sachverhalt, sondern auf ein als möglich gedachtes Erlebnis (Heyting 1931, p. 113)".

[19] Heyting stresses this character by contrasting his explanation with classical explanations in terms of transcendental state of affairs. (To assert a proposition is "la constatation d'un fait. En logique classique c'est un fait transcendant; en logique intuitionniste c'est un fait empirique" (Heyting 1930, p. 235)).

[20] Kleene's realizability in terms of recursive functions is the first systematic interpretation of intuitionistic sentences. Kreisel (1959) considered an interpretation that was instead in terms of effective operations of higher types introduced by Gödel in connection with his system T (used for an interpretation deviating form Heyting's). Kreisel (1962) called his interpretation "general realizability" (later renamed "modified realizability" by Troelstra (1973)). He also suggested another interpretation in terms of an abstract notion of construction (Kreisel, 1962a), deviating from Heyting's explanations in a way that is of interest for the discussion here (see fn 22). The well-known

lowing Kreisel (1959), I shall take constructions to be essentially effective operations (or computable functions) of higher types. For brevity I shall consider fragments of first order languages where the logical constants are restricted to $\bot$, &, $\supset$, and $\forall$ ($\neg A$ is short for $A \supset \bot$). All the individuals of the domain $\mathcal{D}$ that is quantified over are supposed to be named by individual terms. I assume that we are given a set of constructions of atomic sentences, none of which is to be a construction of the atomic sentence $\bot$. The *types* are then as follows:

(a) The numeral 0 is a type, namely the type of the given constructions of atomic sentences, and $\mathcal{D}$ is a type, namely the type of the given individuals.
(b) If τ_1 and τ_2 are types, then so are $\langle \tau_1, \tau_2 \rangle$ and $\tau_1(\tau_2)$; the first being the type of pairs of objects of type τ_1 and τ_2 and the second the type of effective operations or functions whose domains are sets of objects of type τ_2 and whose values are objects of type τ_1.

Relative to a given domain $\mathcal{D}$ and given constructions of atomic sentences, together constituting what I shall call a *base*, it is defined what is meant by something being a *construction* of a compound sentence A, using recursion over A:

(1) c is a construction of $A_1 \& A_2$ iff $c = \langle c_1, c_2 \rangle$ where c_i is a construction of A_i $(i = 1, 2)$.
(2) c is construction of $A \supset B$ iff c is an effective operation that applied to any construction c' of A yields as value a construction $c(c')$ of B.
(3) c is construction of $\forall x A(x)$ iff c is an effective operation that applied to any term t yields as value a construction $c(t)$ of $A(t)$.

We want to discuss to what extent a construction c of a sentence A can be looked upon as a ground for asserting A, and it must then be made clear what it is to be in possession of c, which requires that c is known under some description. We must therefore have a language of *construction terms* denoting constructions. It can be set up as an extended lambda calculus. To each of the given constructions of atomic sentences there is to be a constant that denotes it. They count as a terms of type 0. To each type τ, there is to be variables ξ^τ that range over constructions of type τ and count as terms of type τ. Other terms are built up by the rules:

(I) If T_1 and T_2 are terms of type τ_1 and τ_2, then $p(T_1, T_2)$ is a term of type $\langle \tau_1, \tau_2 \rangle$; it denotes the pair of the constructions denoted by τ_1 and τ_2.
(II) If T is a term of type $\langle \tau_1, \tau_2 \rangle$, then $p_1(T)$ and $p_2(T)$ are terms of type τ_1 and τ_2; they denote the first and second element of the pair denoted by T, respectively.
(III) If T is a term of type σ, then $\lambda \xi^\tau T$ is a term of type $\sigma(\tau)$; it denotes the operation that for a construction denoted by a term U of type τ produces as value the construction denoted by $T(U/\xi^\tau)$.

(Footnote 20 continued)
acronym *BHK* is used for two different interpretations, the *Brouwer-Heyting-Kreisel-interpretation* due to (Troelstra, 1977), and the *Brouwer-Heyting-Kolmogoroff-interpretation* due to Troelstra and Dalen (1988). They are quite informally stated, the first one being inspired by Kreisel (1962a).

(IV) If T is a term of type $\sigma(\tau)$ and U is a term of type τ, then $\{T\}(U)$ is a term of type σ that denotes the result of applying the operation denoted by T to the construction denoted by U.

If we allow ourselves of the term true, we may say that a sentence A is *true* relative to a base $\mathcal{B}$, when there is a construction of A relative to the base $\mathcal{B}$. A sentence may be said to be *logically true* if it is true relative to any base.

To each sentence A that is provable in intuitionistic predicate logic, there is a construction term T that denotes a construction of A relative to any base $\mathcal{B}$; the intuitionistic predicate logic is thus sound with respect to this interpretation, and the language of the extended lambda calculus is complete with respect to the constructions needed to establish this soundness in the sense that there are construction terms that denote them.[21]

For a particular base $\mathcal{B}$, we may of course need to add other operators to the extended lambda calculus defined above in order to get terms that denote the constructions relative to $\mathcal{B}$, and we know by Gödel's incompleteness result that if $\mathcal{B}$ is sufficient to get the expressivity of arithmetic, there is no closed extension of the lambda calculus that is sufficient to get terms that denote all constructions relative to $\mathcal{B}$.

We want to get clear about the status of the constructions with regard to the theme of this essay. There is a certain ambiguity in Heyting's saying that "a proof of a statement consists in the realization of its required construction" (Heyting 1934, p. 14). It could mean that the process of finding the required construction is a proof, or that the result of this process, that is, the required construction, is a proof. It has become standard terminology in intuitionistic contexts to use the term proof in the second sense; for instance, in the BHK-interpretations (fn 20) and in Martin-Löf (1984) type theory, what correspond to Heyting's intended construction in terms of which linguistic meaning is explained are referred to as proofs.

It must be admitted however that constructions as defined by clauses (1)–(3) are objects, viz. pairs and functions in a typed universe, not proof acts. It is true that the construction terms above can be seen as codifications of proofs in predicate logic, but as just remarked they are not sufficient in general to denote the constructions defined by (1)–(3). Furthermore, the constructions are defined by recursion over sentences, while proofs as we usually know them are generated inductively by adding inferences. Such considerations may be the reasons why Martin-Löf (1998) and Sundholm (1998) denied that the notion of proof in this intuitionistic sense is an epistemic concept. Sundholm (1994) explained the role of these proofs to be that of a *truth-maker*.

When truth is defined as the existence of a construction, the constructions can certainly be called truth-makers, although one can discuss how this is to be taken metaphysically. The crucial question in the present context is whether the constructions are grounds not only in that ontological sense, truth-grounds, so to say, but also in the epistemic sense of being what one has to be in possession of in order to

---

[21] I have here essentially followed an earlier presentation of mine (Prawitz 1970, 1971), where a homomorphic mapping of intuitionistic natural deductions into an extended lambda calculus is defined. By applying ideas of Howard (1980, privately circulated from 1969) one can get an isomorphic mapping by considering a finer type structure.

be justified in making assertions. How such a double role could be possible at all is explained by Heyting's remark quoted at the beginning of this section; the point being that intuitionistic truth-makers, unlike truth-makers from a realist point of view, are of a kind for which it makes sense to speak of being in possession of them.

In support of the view that a construction of a sentence A is a ground for asserting A, one may refer to Heyting's meaning explanations, and say: A is understood as the intention or the problem of finding a certain construction—having found the construction, the intentions is fulfilled or the problem is solved, and therefore one should also be said to have a ground for asserting A. This is also in agreement with how we have defined truth: having found a construction of A, one is in possession of a truth-maker of A, guaranteeing the truth of A.

To this one may object that it is one thing to know or to have found a construction and another to know that it is a construction of the required kind. The situation is parallel to the one in the preceding section. To be in possession of a construction of for instance $A \supset B$ is to know an effective operation that applied to a construction of A yields a construction of B, and hence it is to *know how* to find a construction of B given one of A, but it is not to *know that* there is a such an effective operation.

There have been many different positions on this issue. For instance, Kreisel and Dummett have demanded more of a proof than we have done above, thereby taking implicitly the second of the two positions discussed above. The two BHK-interpretations (fn 20) are split on this issue, while Martin-Löf (1984, 1998) seems to take the first of the two positions. Sundholm has made a suggestion that reminds of the first of the two readings of Heyting mentioned above.[22]

More must obviously be said about what constitutes evidence or grounds for assertions. Furthermore, no account has been given of how the performance of an inference produces a ground. The next sections will be devoted to these two issues.

---

[22] Kreisel (1962a) proposes that a proof of $A \supset B$ or $\forall x A(x)$ is a pair whose second member is a proof of the fact that the first member is a construction that satisfies clause 2) or 3) above. Thus, he presupposes that we already know what a proof is; it is thought that the second proof establishes a decidable sentence and that a reduction has therefore taken place. Troelstra (1977) first BHK-interpretation follows Kreisel's proposal saying that a proof of $A \supset B$ or $\forall x A(x)$ consists of a construction as required in clause 2) or 3) together with the *insight* that the construction has the required property. The second BHK-interpretation by Troelstra and Dalen (1988) drops the additional requirement of insight without any comments.

Dummett (1977, pp.12–13) maintains that clauses 2) and 3) do not correctly define what an intuitionistic proof is, and says that a proof of e.g. $A \supset B$ is "a construction of which we can recognize that applied to any proof of A, it yields a proof of B"; thus, a proof is not a pair that contains a proof, but the recognizability in question is a required property of a proof.

Sundholm (1983), who analyses in detail the difference between Heyting's and Kreisel's views of proofs, differentiates between constructions in the sense of objects and in the sense of processes, and suggests that it is not from the construction c of A but from the construction of this construction that it is to be seen that c is a construction of A.

Although Martin-Löf (1998) denies that intuitionistic proofs are proofs in an epistemic sense, knowing them are according to him what entitles us to make assertions (but see also fn 26). This is related to his view of the two kinds of knowledge that occurred in the discussion above. In his view knowledge of a truth is to be analysed in the end as *knowledge how*; *knowledge that* "evaporates on the intuitionistic analysis of truth" (Martin-Löf 1985).

3.5 Evidence and Grounds

3.5.1 Evidence Represented in the Form of Grounds

One finds something to be evident by performing a mental act, for instance, by making an observation, a computation, or an inference. After having made such an act, one is in an epistemic state, a state of mind,[23] where the truth of a certain sentence is evident, or as I have usually put it, one is in possession of evidence for a certain assertion. With a similar wording, one can say that this essay raises the question how an inference act can bring a subject into a state where she has (conclusive) evidence for the truth of a sentence in an objective sense.

I have assumed that a subject who performs a reflective inference is aware of making an inferential transition and that she believes explicitly that she has thereby got evidence for the conclusion. We cannot assume that her awareness goes much farther. She may have the impression of having performed a mental operation by which she has become convinced that the sentence asserted by the conclusion must be true, given that the sentences asserted by the premisses are true. But we cannot expect that a phenomenological analysis will reveal an awareness of the nature of this operation and explain why it gives objective evidence.

We cannot either expect that a subject who performs an inference can tell what her obtained evidence consists in. Asked about that, she may refer to the premisses of the inference and say that they are her grounds. It is common to use the term ground in this way, but this is a usage quite different from how the term has been used here. A ground for an assertion has been understood as something that justifies the assertion. The premisses are assertions, and they or the fact that they are made does not in itself justify the conclusion. What is relevant for the justification of the conclusion is that one has evidence for the premisses. But this evidence justifies the premisses and not simultaneously the conclusion. It is by having evidence for the premisses and performing the inference that evidence or a ground is attained for the conclusion. How this is brought about is what we want to explain.

Rather than trying to analyse phenomenologically the states of mind where we experience evidence—let us call them evidence states—we have to say what evidence states are possible and what operations are possible for transforming one evidence state to another. For instance, we may want to say that anyone who knows the meaning of the sentence $t = t$ has access to a state in which she has evidence for asserting it. Similarly, we may want to say that anyone who knows the meaning of conjunction and is in a state in which she has evidence for asserting a sentence A as well as for asserting a sentence B, can put herself in a state in which she has evidence for asserting $A\&B$.

To state principles like this, it is convenient to think of evidence states as states where the subject is in possession of certain objects. I shall call these objects *grounds*, and reserve now this term for that use in the sequel. I am so to say reifying evidence and am replacing evidence states with states where the subject is in possession of

[23] Whether it is *merely* a mental state, as Williamson (2000, p. 21) claims knowledge to be, need not be discussed here.

grounds. Principles concerning the possibility of a state where a subject has evidence for an assertion $\mathcal{A}$ can then be stated as principles about the existence of grounds for $\mathcal{A}$. We can say for instance: There is a ground for asserting $t = t$; if there is a ground for asserting A and a ground for asserting B, then there is also a ground for asserting $A\&B$.

We can make the latter even more articulate by saying that there is an operation Φ that applied to grounds α and β for asserting A and B, respectively, produces a ground $\Phi(\alpha, \beta)$ for asserting $A\&B$. Transitions from one evidence state to another can in this way be represented as operations on grounds. More precisely, the mental act by which a subject gets from one evidence state to another can be described as an act where the subject gets in possession of a new ground by applying a particular operation to grounds that she is already in possession of.

This is not meant as a realistic description of the mental act, but is suggested as a theoretical reconstruction of what goes on when we pass from one evidence state to another. The idea is that we can reconstruct an inference as involving an operation on grounds, and that this will allow us to explain how an inference can give evidence. This idea will be made more precise in Sect. 3.6.

In the rest of this section, I shall state principles about the existence of grounds and operations on grounds based on the idea that the meaning of a sentence is explained in terms of what counts as a ground for asserting it. The grounds will be seen as abstract entities. As such we get to know them via descriptions. To form a ground for an assertion is thus to form a term that denotes the ground, and it is in this way that one comes in possession of the ground. Simultaneously with saying what grounds and operations on grounds there are, I shall therefore indicate a language in which the grounds can be denoted.

As already foreshadowed by the example above of an operation Φ of conjunction grounding, the grounds that come out of this enterprise will be like intuitionistic constructions and the language in which they will be described will be like the extended lambda calculus already considered in the preceding section. The type structure will however be made more fine-grained by using sentences as types following Howard (1980), so that the question whether a term denotes a ground for an assertion of a sentence A coincides with the question of the type of the term. The grounds will thereby be among the objects that one comes across within intuitionistic type theory developed by Martin-Löf (1984). But what is of interest here is whether what is defined as a ground for the assertion of a sentence A is not only a truth-maker of A but is really a ground the possession of which makes one justified in asserting A. Before resuming this discussion in the final Sect. 3.7, I shall now state in detail what are taken to be grounds for asserting first order intuitionistic sentences and then how the performance of an inference can be seen as involving an operation on these grounds.

3.5.2 The Language of Grounds

I assume as before that we are given a first order language and with that individual terms and what counts as grounds for asserting atomic sentences. Among the atomic sentences there is again to be one for falsehood, written $\bot$, such that there is no given ground for asserting it. $\neg A$ is defined as $A \supset \bot$. The task is to specify in accordance with the ideas explained above what grounds there are for asserting compound sentences of the language relative to the given grounds—a relativization that will be left implicit.

As already mentioned, I shall use sentences, both open and closed ones, as types. A ground for asserting A will thus be a ground of *type* A. I shall sometimes say "ground for A" as short for "ground for asserting A". Also the terms denoting grounds will be typed.

The language of grounds contains *individual terms*, among which are individual variables $x, y, \ldots$, *ground constants* denoting the given grounds for atomic sentences, *ground variables* $\xi^A, \zeta^A, \ldots$, ranging over grounds of type A, where A is a closed or open sentence, and symbols for the following *primitive operations* that produce grounds when applied to grounds: $\&I$, $\vee I_1$, $\vee I_2$, $\supset I$, $\forall I$, and $\exists I$; they will also be used autonomously.

The *ground terms* comprise ground constants, ground variables, and what can be built up from them inductively by using the primitive operations. I shall use t, T, and U as syntactical variables, t for individual terms, T and U for ground terms. As before A and B range over sentences. The inductive clause for forming ground terms then says that $\&I(T, U)$, $\vee I_1(T)$, $\vee I_2(T)$, $(\supset I\xi^A)(T)$, $(\forall Ix)(T)$, and $\exists I(T)$ are ground terms.

The operations $\supset$I and $\forall I$ are variable binding, indicated as usual by the attached variable. The application of the operation $\forall Ix$ to a term T is restricted by the condition that the variable x is not to occur free in the type index of a bound occurrence of a variable ξ^A in T.

The ground constants and ground variables are given with their types. The primitive operations are also to be understood as coming with types, although this may be left implicit. This is harmless, except for $\vee I_1(T)$, $\vee I_2(T)$, and $\exists I$, for which the type will sometimes be indicated within parentheses. The types of the compound ground terms are as follows:

(1) $\&I(T, U)$ is of type $A\&B$, if T and U are of types A and B, respectively.
(2) $\vee I_i(A_i \rightarrow A_1 \vee A_2)(T)$ is of type $A_1 \vee A_2$, if T is of type A_i, $i = 1, 2$.
(3) $(\supset I\xi^A)(T)$ is of type $A \supset B$, if T is of type B.
(4) $(\forall Ix)(T)$ is of type $\forall x A$, if T is of type A.
(5) $\exists I(A(t/x) \rightarrow \exists x A)(T)$ is of type $\exists x A$, if T is of type $A(t/x)$.

The term t that is existentialized in clause 5) is determined by the type of the operation; it had otherwise to be made explicit as an additional argument of the operation.

The meanings of closed compound sentences of the various forms, intuitionistically understood, are given by saying that there are certain operations that produce grounds for asserting the sentences when applied to other grounds appropriately, and that all grounds there are for asserting these sentences are produced in that way. The grounds are denoted by the terms used in clauses above, but we may also say directly what grounds there are for different sentence forms. For conjunction, disjunction, and existential quantification, this may be spelled out by saying that for any closed sentences A, B, and $\exists x A$ and closed term t:

(1*) If T denotes a ground α for asserting A, and U denotes a ground β for asserting B, then $\&(T, U)$ denotes a ground for asserting $A\&B$, viz $\&I(\alpha, \beta)$.
(2*) If T denotes a ground α for A_i, then $\vee I_i(A_i \rightarrow A_1 \vee A_2)(T)$ denotes a ground for $A_1 \vee A_2$, viz $\vee I_i(A_i \rightarrow A_1 \vee A_2)(\alpha)$.
(5*) If T denotes a ground α for $A(t/x)$, then $\exists I(A(t/x) \rightarrow \exists x A)(T)$ denotes a ground for $\exists x A(x)$, viz $\exists I(A(t/x) \rightarrow \exists x A)(\alpha)$.

To this is to be added that the above is an exhaustive specification of what grounds there are for closed sentences of the forms in question; in other words, nothing is a ground for a closed sentence of one of these forms, unless it is a ground in virtue of these principles. Furthermore, it is to be added that different primitive operations generate different values, and that the same operation generates different values for different arguments. In other words, there are identity conditions like: $\&I(\alpha_1, \beta_1) = \&I(\alpha_2, \beta_2)$, only if $\alpha_1 = \alpha_2$ and $\beta_1 = \beta_2$.

Although 1*, 2*, and 5* may be acceptable as they stand even when the sentences are classically understood, taking 2* and 5* as exhaustively determining the meaning of disjunction and existential quantification is of course compatible only with an intuitionistic reading; the classical meaning of these forms have to be specified differently.

3.5.3 Operations on Grounds

Primitive operations such as $\&I$ do not only produce grounds for assertions of closed sentences when applied to such grounds, they also constitute in themselves grounds for assertions under assumptions. Knowing the meaning of conjunction, one also knows that there is an operation, namely $\&I$, such that when one is in possession of grounds for the assumptions of the closed hypothetical assertion

$$A, B \vdash A\&B,$$

one can get in possession of a ground for the categorical assertion $\vdash A\&B$ by applying the operation. Such an operation is what is to be required of a ground for a closed hypothetical assertion. Thus, the operation $\&I$, which is also denoted by the term $\&I(\xi^A, \zeta^B)$, is to count as a ground for this hypothetical assertion.

By composition of the primitive operations we get grounds for more complex hypothetical assertions. In the other direction, a ground term consisting of just a ground variable ξ^A, where A is a closed sentence, denotes an operation, namely the identity operation, that counts as a ground for asserting A under the assumption A.

This simple way of getting grounds for hypothetical assertions does not go very far, of course, and in general, one has to define new operations to this end. For instance, for any closed sentence $A_1\&A_2$ we can define two operations, which I call $\&E_1$ and $\&E_2$, both of which are to have as domain grounds for $A_1\&A_2$. The intention is that the operation $\&E_i$ is always to produces grounds for $A_i (i = 1, 2)$ when applied to grounds for $A_1\&A_2$ and that it therefore can be taken as ground for the closed hypothetical assertion

$$A_1\&A_2 \vdash A_i.$$

This is attained by letting the two operations be defined by the equations

$$\&E_1[\&I(\alpha_1, \alpha_2)] = \alpha_1 \text{ and } \&E_2[\&I(\alpha_1, \alpha_2)] = \alpha_2.$$

The fact that the operation $\&E_i$ always produces a ground for A_i when applied to a ground α for $A_1\&A_2$ is not an expression of what & means, as clause 1* is. Instead it *depends* on what & means, and has to be established by an argument: Firstly, in view of clause 1) specifying the grounds for $A_1\&A_2$ exhaustively, α must have the form $\&I(\alpha_1, \alpha_2)$, where α is a ground for asserting $A_1\&A_2$. Secondly, because of the identity condition, α_1 and α_2 are unique. Hence, according to the equations that define the operations, $\&E_i(\alpha) = \alpha_i$, where α_i is a ground for asserting A_i, given that α is a ground for asserting $A_1\&A_2$.

More generally, an operation is said to be of type $(A_1, A_2, \ldots, A_n \rightarrow B)$, where $A_1, A_2, \ldots, A_n$ and B are closed sentences, if it is an n-ary effective operation that is defined whenever its i:th argument place is filled with a ground of type A_i and then always produces a ground of type B. Such an operation is a ground for the hypothetical assertion $A_1, A_2, \ldots, A_n \vdash B$.

An operation is given by stating the types of its domain and range and, for each argument in the domain, the value it produces for that argument. Note that it follows as an extremity that for any sentence A there is an operation of type $(\bot \rightarrow A)$; the condition that for each argument in the domain, it is specified what value it produces for that argument is vacuously satisfied since the domain is empty.

So far only grounds for closed assertions have been considered. A ground for the assertion of an open sentence $A(x_1, x_2, \ldots, x_m)$ with the free variables $x_1, x_2, \ldots, x_m$ is an effective m-ary operation defined for individual terms that always produces a ground for asserting $A(t_1/x_1, t_2/x_2, \ldots, t_m/x_m)$ when applied to $t_1, t_2, \ldots, t_m$.

When $A_1, A_2, \ldots, A_n$ and B are open sentences, an operation of type $(A_1, A_2, \ldots, A_n \rightarrow B)$ is accordingly an n-ary effective operation from operations to operations of the kind just described. It is again a ground for $A_1, A_2, \ldots, A_n \vdash B$.

The language of grounds is to be understood as open in the sense that symbols for defined operations of a specific type can always be added. A closed ground term denotes a ground for a closed categorical assertion as specified by 1*, 2*, and 5* and as will be specified by 3* and 4*. An open ground term denotes such a ground when saturated by appropriate terms for the free variables, or more generally, it denotes a ground under a given assignment to its free variables. We can then say that an open ground term $T(\xi_1, \xi_2, \ldots, \xi_n)$ of type B, whose free ground variables are $\xi_1, \xi_2, \ldots,$ and ξ_n of types $A_1, A_2, \ldots,$ and A_n, denotes an operation of type $(A_1, A_2, \ldots, A_n \rightarrow B)$, viz the operation that, when applied to grounds $\alpha_1, \alpha_2, \ldots,$ and α_n of type $A_1, A_2, \ldots,$ and A_n, produces the ground of type B which $T(\xi_1, \xi_2, \ldots, \xi_n)$ denotes under the assignment of $\alpha_1, \alpha_2, \ldots,$ and α_n to $\xi_1, \xi_2, \ldots,$ and ξ_n.

The meaning of implication and universal quantification can now be spelled out by stating that there are operations such that for any closed sentences A, B, and $\forall x A(x)$:

(3*) If T denotes a ground α for asserting B under the assumption A (that is, α is an operation of type $(A \rightarrow B)$), then $(\supset I \xi^A)(T)$ denotes a ground for asserting $A \supset B$, viz $(\supset I)(\alpha)$.

(4*) If T denotes a ground α for asserting $A(x)$ (that is, α is ground of type $A(x)$), then $(\forall I x)(T)$ denotes a ground for asserting $\forall x A(x)$, viz $(\forall I)(\alpha)$.

To this should be added again that the specifications of grounds are exhaustive and that there are identity conditions. In this connection it is important to recall that the language of grounds is open; it can be seen as an extended, and always extendible, typed lambda calculus whose terms are interpreted as denoting grounds. Operations on grounds for $A \supset B$ can then also be defined. For instance, an operation $\supset E$ of type $(A \supset B, A \rightarrow B)$ is defined by

$$\supset E((\supset I \xi)(T(\xi)), U) = T(U),$$

where ξ and U are of type A, T is of type B, and $T(U)$ is the result of substituting U for free occurrences of ξ in T.

A closed ground term whose first symbol is one of the primitive operations is said to be in *canonical form*—the form used to specify the grounds there are for different assertions.

Analogously, by an operation of type $((\Gamma_1 \rightarrow A_1), (\Gamma_2 \rightarrow A_2), \ldots, (\Gamma_n \rightarrow A_n) \rightarrow (\Delta \rightarrow B))$, where Γ_i and Δ stand for sequences of sentences, is meant an n-ary effective operations from operations of the type indicated before the main arrow to operations of the type indicated after the main arrow. They are not denoted by ground terms since the language of grounds has no variables of type $(\Gamma \rightarrow A)$. But the result produced by such an operation when applied to specific arguments is an operation of type $(\Delta \rightarrow B)$ that can be denoted by an open ground term.

As the names suggest, the operations $*I$ correspond to introduction rules and $\&E_1$, $\&E_2$, and $\supset E$ correspond to elimination rules in Gentzen's system of natural deduction. I conclude this section with giving examples of two other kinds of defined operations. Let ***Barb*** be the operation of type $(\forall x(Px \supset Qx), \forall x(Qx \supset Rx) \rightarrow \forall x(Px \supset Rx))$ that is defined by the equation

$$\textbf{\textit{Barb}}[(\forall I x)(T), (\forall I x)(U)] = (\forall I x)[(\supset I \xi^{Px})[\supset E(U, \supset E(T, \xi^{Px}))]].$$

As seen it is a ground for a hypothetical assertion corresponding to Aristotle's syllogism Barbara when formulated in a first order language.

Let ***Mtp*** be the operation of type $(A \vee B, \neg A \rightarrow B)$ whose values are defined by

$$\boldsymbol{Mtp}(\vee I_2(\alpha), \beta) = \alpha$$

It is a ground for the hypothetical assertion $A \vee B, \neg A \vdash B$.[24]

3.6 Deductively Valid Inferences

There are many reasons for thinking that one should answer in the affirmative the question that has so far been left open whether an inference contains something more than an inferential transition. It is hard to see how an act that consists of only an inferential transition, essentially just an assertion claimed to be supported by certain premisses, could be able to confer evidence on its conclusion. Another observation that points to an affirmative answer is that there are examples of transitions that can be performed for different reasons and therefore seem to belong to different inferences.

If the above explication of the notion of ground is accepted, it suggests itself that an inference is essentially an operation on grounds that produces a ground. I therefore propose as a reconstruction of the notion of inference that to *perform a reflective inference* is, in addition to making an inferential transition, to apply an operation to the grounds that one considers oneself to have for the premisses with the intention to get thereby a ground for the conclusion. Accordingly, I take an *individual* or *generic inference* to be individuated by what individuates an individual or generic inferential transition from premisses $\mathcal{A}_1, \mathcal{A}_2, \ldots,$ and $\mathcal{A}_n$ to a conclusion $\mathcal{B}$, and, in addition, by alleged grounds $\alpha_1, \alpha_2, \ldots, \alpha_n$ for the premisses and an operation Φ. Conforming to usual terminology according to which an inference may be unsuccessful, no requirement is put on the alleged grounds and the operation; in other words, $\alpha_1, \alpha_2, \ldots, \alpha_n$ may be any kind of entities, and Φ may be any kind of operation.

It is now obvious how a notion of valid of inference can be defined in a way that is congenial to the problem raised in this paper. I shall say that an individual or generic inference, individuated as above, is *(deductively) valid*, if $\alpha_1, \alpha_2, \ldots,$ and α_n are grounds for $A_1, A_2, \ldots,$ and A_n and Φ is an operation such that $\Phi(\alpha_1, \alpha_2, \ldots, \alpha_n)$ is a ground for the conclusion B. More precisely, if the inferential transition is indicated by the figure

$$\frac{\Gamma_1 \vdash A_1 \qquad \Gamma_2 \vdash A_2 \qquad \ldots \qquad \Gamma_n \vdash A_n}{\Delta \vdash B}$$

[24] This answers questions raised by Cesare Cozzo in this volume and by Pagin (2012) whether one has to decompose the inference schemata Barbara and Modus ponendo tollens, respectively, into a chain of natural deduction inference schemata in order to show that they are legitimate. As will be seen (Sect. 3.6), there is no such need: when the operations ***Barb*** and ***Mtp*** are assigned to the inference schemata in question, valid forms of inferences arise.

where Γ_i and Δ stand for sequences of sentences and each sentence of Δ belongs to some $\Gamma_i (i < n)$, then Φ has to be an operation of type $((\Gamma_1 \to A_1), (\Gamma_2 \to A_2), \ldots, (\Gamma_n \to A_n) \to (\Delta \to B))$. As remarked above, the ground for the assertion B under the assumptions Δ, produced when the operation Φ is applied to grounds for the premisses, can then be denoted by a ground term. If Δ is empty and B is closed, then the ground term is closed and denotes a ground for asserting B.

Similarly, an *inference form* is individuated by the form of an inference transition and an operation, and is defined as *valid* if the operation when applied to grounds for the premisses produces a ground for the conclusion. Since an inference form does not have specific premisses and a specific conclusion, this operation has no specific type, but is specified by a term with ambiguous types; for each instance of the form, the operation is of a specific type and is denoted by the corresponding instance of that term. An *inference schema* is defined as valid when there is an operation such that the inference schema together with that operation is a valid inference form (see Sect. 3.1.3 for the terminology).

A valid generic inference is obviously legitimate and we thus have a condition C on generic inferences of the kind sought for (Sect. 3.2.1): given (i) that an inference I is valid, (ii) that a subject S has grounds for the premisses of I, and (iii) that she performs I, it follows directly from the definition of validity (iv) that S gets in possession of a ground for the conclusion of I—the premiss (iii) now means that S applies an operation to the grounds for the premisses, which she is in possession of according to (ii), and she thereby gets a ground for the conclusion because of (i).

An individual or generic inference can err in two ways: the alleged grounds for the premisses may not be such grounds, or the operation may not produce a ground for the conclusion when applied to grounds for the premisses.[25] In the second case we could say that the inference is *irrelevant*, and in the first case that it is *incorrect with respect to the premisses*.

As the notion of validity is now defined, an inference that is valid is so in virtue of the meaning of the sentences involved in the inference. One can be interested in defining a notion of validity where only the meaning of the logical constants that occur in the sentences play any role. The notion of validity defined above may be called deductive validity to differentiate it from such a narrower notion of logical validity, which is now easily defined by using the same strategy as used by Bolzano and Tarski: Inferences on various level of abstractions are *logically valid* if they are deductively valid and remain deductively valid for all variations of the meaning of the non-logical vocabulary.

A (individual or generic) proof may now be defined as a chain of valid (individual or generic) inferences. As the notions are now explicated, a proof of an assertion does not constitute a ground for the assertion but produces such a ground, and it is to be noted that two different proofs may produce the same ground for the assertion.

---

[25] As Cesare Cozzo remarks in his contribution to this volume, previous definitions of inference that I have given had the defect that they did not allow one to make this distinction.

3.7 Concluding Remarks

The question raised in this paper is why some inferences give evidence for their conclusions. The proposed answer is: Whether an inference confers evidence on its conclusion depends on the nature of the operation Φ that is a constitutive part of an inference, as this notion has been reconstructed. That Φ produces a ground for the conclusion when applied to grounds for the premisses is in some cases just an expression of what the sentence asserted by the conclusion means and is in other cases a consequence of how the operation Φ is defined.

Furthermore, it is now to be understood that when a subject performs an inference she is aware of applying the operation Φ to grounds that she considers herself to have for the premisses and takes the result obtained to be a ground for the conclusion. If the inference is valid, the result of applying Φ to the grounds for the premisses is in fact a ground for the conclusion, as this notion has now been defined.

One may ask whether this definition of ground is reasonable. Does a ground as now defined really amount to evidence? When the assertion is a categorical one, the ground is a truth-maker of the asserted sentence; since the meanings of the sentences are given by laying down what counts as grounds for asserting them, the truth of a sentence does not amount to anything more than the existence of such a ground. Nevertheless one may have doubts about whether to be in possession of a truth-maker of a sentence as understood here really amounts to being justified in asserting the sentence. The problem is that the truth-maker of a sentence may have become known under a different description than the normal form used when the meaning of the sentence is explained.

A ground is thought of as an abstract object and is known only under a certain description of it. When a subject performs a valid inference and applies an operation Φ to what she holds to be grounds for the premisses, she forms a term T that in fact denotes a ground for the drawn conclusion $\mathcal{A}$, but it is not guaranteed in general that she knows that T denotes a ground for $\mathcal{A}$.

It can be assumed to be a part of what it is to make an inference that the agent knows the meanings of the involved sentences. Since the meanings of closed atomic sentences are given by what counts as grounds for asserting them, she should thus know that T denotes a ground for asserting an atomic sentence A when this is how the meaning of A is given. Such knowledge is preserved by introduction inferences, given again that the meanings of the involved sentences are known: The term T obtained by an introduction is in normal formal, that is, it has the form $\Phi(U)$ or $\Phi(U, V)$, where Φ is a primitive operation and the term U or the terms U and V denote grounds for the premisses—knowing that these terms do so, the agent also knows that T denotes a ground for the conclusion, since this is how the meaning of the sentence occurring in the conclusion is given.

However, when Φ is a defined operation, the subject needs to reason from how Φ is defined in order to see that T denotes a ground for the conclusion. If T is a closed term, she can in fact carry out the operations that T is built up of and bring T to normal form in this way, but she may not know this fact. Furthermore, when

T is an open term, it denotes a ground for an open assertion or an assertion under assumption, and it is first after appropriate substitutions for the free variables that one of the previous two cases arises.

It is thus clear that to have come in possession of a ground for the conclusion of a valid inference, as now understood, does not imply that one knows that what one is in possession of is such a ground. But then one may ask if to make a valid inference really gives the evidence that one should expect. I shall conclude by making some remarks about this question.

To make an inference is not to assert that the inference is valid. Nor is it to make an assertion about the grounds that one has found for the conclusion of the inference. One may of course reflect over the inference that one has made, and, if successful, one may be able to demonstrate that a ground for the conclusion has been obtained and that the inference is valid. But a requirement to the effect that one must have performed such a successful reflection in order that one's conclusion is to be held to be justified would be vulnerable to the kinds of vicious regresses that we have already discussed (Sect. 3.2.2).

It goes without saying that if one asserts that an inference is valid or that α is a ground for its conclusion $\mathcal{A}$, one should have grounds for these assertions, but they assert something more than $\mathcal{A}$. Similarly, we have to distinguish between asserting a sentence A and a sentence of the form "... is true", where the dots are replaced by a name of a sentence A (cf Sect. 3.1.1 and fn 4). The latter sentence is on a meta-level as compared to the former one, and has been assumed here to be equivalent with the sentence "there is a truth-maker (or ground) of ...". To be justified in asserting it, it is of course not sufficient only to produce a truth-maker of A. One must also have a ground for the assertion that what is produced is a truth-maker of A, which has to be delivered by a proof on the meta-level of an assertion of the form "... is a ground for asserting ...".[26] This proof will in turn depend on its inferences giving evidence for their conclusions. To avoid an infinite regress it seems again to be essential that there are inferences that give evidence for their conclusions without it necessarily being known that they give such evidence.

It may be argued that the condition for having evidence for an assertion is luminous, to use a term of Williamson (2000), and that therefore it is seen directly without proof whether something constitutes evidence for an assertion. In a frequently quoted passage, Kreisel (1962a, p. 201) says that he adopts as an idealization that "we recognize a proof when we see one". In support of this dictum, it is sometimes said that if we are presented with an argument without being able to recognize that it is a proof, then it is not a proof. If it is decidable in this way whether something is a proof, the same should hold for the properties of being a legitimate inference and of constituting evidence or a ground for an assertion.

---

[26] Martin-Löf (1984) type theory contains rules for how to prove such assertions, or judgements, as they are called there, written $a \in A$ (or $a : A$ in later writings). There are also assertions of propositions in the type theory, but they have the form "A is true" and are inferred from judgements of the form $a \in A$, where a corresponds to what I am calling a ground. Thus, the assertions in the type theory are, as I see it, on a meta-level as compared to the object level to which the assertions that I am discussing belong. (But compare also fn 22, which has a more general setting than type-theory.)

It is of course an essential feature of a formal system that it is decidable whether something is a proof in that a system. For a closed language of grounds where the term formation is restricted by specifying what operations may be used, it may similarly be decidable whether an expression in the language denotes a ground. But, as already noted, we know because of Gödel's incompleteness result that already for first order arithmetical assertions there is no closed language of grounds in which all grounds for them can be defined; for any such intuitively acceptable closed language of grounds, we can find an assertion and a ground for it that we find intuitively acceptable but that cannot be expressed within that language. The crucial question is therefore if it is decidable for an arbitrary definition of an operation, which we may contemplate to add to a given closed language of grounds, whether it always produces a ground of a particular type when applied to grounds in its domain? This is what must hold if we are to say that the property of being a ground is decidable, and it seems to me that we must be sceptical of such an idea, and therefore also of the idea that the condition for something to be a proof or to constitute evidence is luminous.

By the reconstruction that makes an inference to something more than a mere inferential transition, it has been made a conceptual truth that a person who performs a valid inference is aware of making an operation that produces what she takes to be a ground for the conclusion, although the latter is not what she asserts and is not what she has a ground for. It would be preferable if one could explicate the involved notions in another way so that the performance of a valid inference resulted in an even greater awareness and further knowledge. This may perhaps be achieved by putting greater restrictions on the operations that can be used to form grounds than I have done or by following quite different paths when analysing the concepts of inference and ground. But I do not think that we can make it conceptually true in a reasonable way that a valid inference produces not only evidence for its conclusion but also evidence of its own validity and evidence for the fact that what it produces is evidence for the conclusion.

References

Boghossian, P. (2001). How are objective epistemic reasons possible? *Philosophical Studies*, *106* (1–2), 340–380.

Boghossian, P. (2003). Blind reasoning. *Proceedings of the Aristotelian Society, Supplementary*, *77*(1), 225–248.

Boghossian, P. (2012). What is an inference? *Philosophical Studies*, *169*(1), 1–18.

Carnap, R. (1934). *Logische syntax der sprache*. Vienna: Springer.

Carroll, L. (1895). What the tortoise said to achilles. *Mind*, *4*(14), 278–280.

Cellucci, C. (2013). Philosophy of mathematics: Making a fresh start. *Studies in History and Philosophy of Science*, *44*(1), 32–42.

Chateaubriand Filho, O. (1999). Proof and logical deduction. In E. H. Hauesler & L. C. Pereira (Eds.), *Proofs, types and categories*. Rio de Janeiro: Pontifícia Universidade Católica.

Corcoran, J. (1974). Aristotle's natural deduction system. In J. Corcoran (Ed.), *Ancient Logic and its Modern Interpretation* (pp. 85–131). Dordrecht: Reidel.

Cozzo, C. (1994). *Meaning and argument. A theory of meaning centered on immediate argumental role*. Stockholm: Almqvist and Wicksell International.
Detlefsen, M. (1992). Brouwerian intuitionism. In M. Detlefsen (Ed.), *Proof and knowledge in mathematics*, (pp. 208–250). London: Routledge.
Dummett, M. (1973). *The Justification of Deductions*. London: British Academy.
Dummett, M. (1977). *Elements of intuitionism*. Oxford: Clarendon Press.
Dummett, M. (1991). *The logical basis of metaphysics*. London: Duckworth.
Etchemendy, J. (1990). *The concept of logical consequence*. Cambridge: Harvard University Press.
Gentzen, G. (1934–1935). Untersuchungen über das logische schließen I. *Mathematische zeitschrift*, *39*(2), 176–210.
Heyting, A. (1930). Sur la logique intuitionniste. *Académie Royale de Belgique, Bulletin de la Classe des Sciences*, *16*, 957–963.
Heyting, A. (1931). Die intuitionistische grundlegung der mathematik. *Erkenntnis*, *2*(1), 106–115.
Heyting, A. (1934). *Mathematische Grundlagenforschung, Intuitionismus, Beweistheorie*. Berlin: Springer.
Heyting, A. (1956). *Intuitionism, An Introduction*. Amsterdam: North-Holland Publishing Company.
Howard, W. (1980). The formula-as-types notion of construction. In J. R. Hindley & J. P. Seldin (Ed.), *To H.B. Curry: Essays on Combinatory Logic, Lambda Calculus and Formalism* (pp. 479–490). London: Academic Press.
Kreisel, G. (1959). Interpretation of analysis by means of constructive functionals of finite types. In A. Heyting (Ed.), *Constructivity of Mathematics* (pp. 101–128). Amsterdam: North-Holland Publishing Company.
Kreisel, G. (1962a). Foundations of intuitionistic logic. In E. Nagel (Ed.), *Logic methodology and philosophy of science* (pp. 198–212). Stanford: Stanford University Press.
Kreisel, G. (1962b). On weak completeness of intuitionistic predicate logic. *Journal of Symbolic Logic*, *27*(2), 139–158.
Martin-Löf, P. (1971). Hauptsatz for the intuistionistic theory if iterated inductive definitions. In J. E. Fenstad (Ed.), *Proceedings of the Second Scandinavian Logic Symposium* (pp. 179–216). Amsterdam: North-Holland Publishing Company.
Martin-Löf, P. (1984). *Intuistionistic type theory*. Napoli: Bibliopolis.
Martin-Löf, P. (1985). On the meaning of the logical constant and the justification of the logical laws. *Atti degli Logica Matematica 2* (pp. 203–281). Siena: Scuola de Specializzazione in Logica Matematica: Dipartimento de Matematica, Università di Siena.
Martin-Löf, P. (1998). Truth and knowability: on the principles C and K of Michael Dummett. In H. G. Dales & G. Oliveri (Eds.), *Truth in Mathematics* (pp. 105–114). Oxford: Clarendon Press.
Pagin, P. (2012). Assertion, inference, and consequence. *Synthese*, *187*(3), 869–885.
Prawitz, D. (1970) Constructive semantics. In *Proceedings of the 1st Scandinavian Logic Symposium Åbo 1968* (pp 96–114). Uppsala: Filosofiska Föreningen och Filosofiska Institutionen vid Uppsala Universitet.
Prawitz, D. (1971). Ideas and results in proof theory. In J. E. Fenstad (Ed.), *Proceedings of the Second Scandinavian Logic Symposium* (pp. 235–307). Amsterdam: North-Holland Publishing Company.
Prawitz, D. (1973). Towards a foundation of general proof theory. In P. Suppes (Ed.), *Logic, Methodology and Philosophy of Science IV* (pp. 225–250). Amsterdam: North-Holland Publishing Company.
Prawitz, D. (2011). Proofs and perfect syllogisms. In E. R. Grosholz & E. Ippoliti & C. Cellucci (Eds.), *Logic and Knowledge* (pp. 385–402). Newcastle on Tyne: Cambridge Scholar Publishing.
Prawitz, D. (2014). On relations between Heyting's and Gentzen's approaches to meaning. In T. Piecha & P. Schroeder-Heister (Eds.), *Advances in Proof-Theoretic Semantics*, forthcoming.
Prior, A. (1960). The runabout inference ticket. *Analysis*, *21*(1), 38–39.
Ross, W. D. (1949). *Aristotle's prior and posterior analytics*. Oxford: Oxford University Press.

Smith, N. (2009). Frege's judgement stroke and the conception of logic as the study of inference not consequence. *Philosophy Compass*, *4*(4), 639–665.
Sundholm, G. (1983). Constructions, proofs and the meaning of logical constants. *The Journal of Philosophical Logic*, *12*(2), 151–172.
Sundholm, G. (1994). Existence, proof and truth-making: A perspective on the intuistionistic conception of truth. *Topoi*, *12*(2), 117–126.
Sundholm, G. (1998a). Inference versus consequence. In T. Childers (Ed.), *The LOGICA Yearbook 1997* (pp. 26–35). Prague: Filosofia Publishers.
Sundholm, G. (1998b). Proofs as acts and proofs as objects. *Theoria*, *54*(2–3), 187–216.
Sundholm, G. (2012). "Inference versus consequence" revisited: Inference, consequence, conditional, implication. *Synthese*, *187*(3), 943–956.
Tait, W. W. (1967). Intensional interpretations of functionals of finite type I. *Journal of Symbolic Logic*, *32*(2), 198–212.
Troelstra, A. S. (1973). *Mathematical Investigations of Intuistionistic Arithmetic and Analysis*. Lecture Notes in Mathematics. Berlin: Springer.
Troelstra, A. S. (1977). Aspects of constructive mathematics. In J. Barwise (Ed.), *Handbook of Mathematical Logic* (pp. 973–1052). Amsterdam: North-Holland Publishing Company.
Troelstra, A. S. & van Dalen, D. (1988). *Constructivism in mathematics*. Amsterdam: North-Holland Publishing Company.
Williamson, T. (2000). *Knowledge and its limits*. Oxford: Oxford University Press.

Chapter 4
Necessity of Thought

Cesare Cozzo

Abstract The concept of "necessity of thought" plays a central role in Dag Prawitz's essay "Logical Consequence from a Constructivist Point of View". The theme is later developed in various articles devoted to the notion of valid inference. In Sect. 4.1 I explain how the notion of necessity of thought emerges from Prawitz's analysis of logical consequence. I try to expound Prawitz's views concerning the necessity of thought in Sects. 4.2, 4.3 and 4.4. In Sects. 4.5 and 4.6 I discuss some problems arising with regard to Prawitz's views.

Keywords Logical consequence · Deductive necessity · Inferential compulsion · Valid inference · Grounds

4.1 Inference and Consequence

"For my part, of all things that are not under my control, what I most value is to enter into a bond of friendship with sincere lovers of truth." These words written by Baruch Spinoza (1995, p. 132) in a letter of the 5th January 1665, come to my mind when I think how lucky I have been to be a pupil of Dag Prawitz. Lovers of truth accept to be bound by the gentle force of good arguments. The most intense manifestation of this gentle force is the necessity of thought. So often have I experienced the cogency of compelling arguments in Dag's friendly voice that for me the necessity of thought is strictly associated with his person. Therefore it is right, among the many topics about which he has written, to choose necessity of thought. This concept appears in an essay about logical consequence (Prawitz 2005).

The pre-theoretical usage of "consequence" in everyday language is erratic and there are various attempts at making the notion precise, but many philosophers agree that we can distinguish between a more general relation of deductive consequence and a more specific relation of *logical* consequence. These are relations between truth-bearers, i.e. entities that are capable of being bearers of truth, e.g. sentences,

C. Cozzo (✉)
Dipartimento di Filosofia, Via Carlo Fea 2, 00161 Rome, Italy
e-mail: cesare.cozzo@uniroma1.it

H. Wansing (ed.), *Dag Prawitz on Proofs and Meaning*, Outstanding Contributions to Logic 7, DOI 10.1007/978-3-319-11041-7_4

statements, or propositions. If Q is a truth-bearer and Γ is a set of truth-bearers, we can distinguish a more general and a more specific way in which they can be related:

(A) Necessarily, if all members of Γ are true, then Q is true as well;
(B) fact (A) above obtains because of the *logical form* of the truth-bearers in Γ and of Q.

Q is a *deductive consequence* of Γ if, and only if, fact (A) obtains. Q is a *logical consequence* of Γ if, and only if, both (A) and (B) obtain. In short: logical consequence is deductive consequence in virtue of logical form. The specifically *logical* character of this kind of consequence is constituted by fact (B). Note that, on the prevailing reading, (A) and (B) are facts independently of human acts. Logical consequence and deductive consequence are usually conceived as relations that have nothing to do with human acts.

A valid inference, on the other hand, is an inference, and an inference is a human act. We can distinguish between mental inferences and linguistic inferences, but both kinds of inferences are acts, either mental acts or speech acts. Göran Sundholm characterizes *mental* inferences as

> acts of passage in which a certain judgement, the conclusion of the inference, is drawn on the basis of certain already made judgements, the premisses of the inference (Sundholm 1994, p. 373).

Similarly one can say that a *linguistic* inference is the act of moving from a finite set of linguistic premises to a linguistic conclusion. The conclusion is the assertion of a sentence (possibly under assumptions). The premises are assertions of sentences (possibly under assumptions) or simply assumptions of hypotheses. The movement, the passage from premises to conclusion, is an act by which one takes a responsibility, publicly to others, and mentally to oneself. One takes responsibility for some support that the premises provide for the conclusion. The strongest support is provided by deductively valid inferences. An implicit claim that the premises support the conclusion is expressed by "therefore", "hence", "because" and similar words or phrases. Linguistic inferences are complex speech acts. Arguments are concatenations of linguistic inferences; so arguments too are complex speech acts.

The practice of providing arguments in support of assertions in order to resolve disagreements or doubts is a crucial aspect of life, and since many arguments are bad arguments, logicians of all times have insisted, as John of Salisbury (2009, p. 76) did in his twelfth-century defence of logic, the *Metalogicon*, that we need to distinguish "which reasoning warrants assent, and which should be held in suspicion". Logic teachers often say to their students: "we must study logic because we need to understand the difference between good arguments and bad arguments". In saying this, they present logic as a theory of logically valid inference and suggest that a valid inference is a good inference. So the characterization of logic as a theory of valid inference takes priority in the order of motivation. Most contemporary logicians, however, would soon add that the characterization of logic as a theory of logical consequence is theoretically deeper because validity can be *analysed* or *defined* in terms of consequence: "an argument is said to be valid – strictly speaking logically valid—if [and only if] its conclusion is a logical consequence of its premises" (Beall 2010, p. 7). For John Burgess (2009, p. 2) the connection between validity and consequence is even closer: "an argument is logically valid, its conclusion is a logical consequence of its premises" are "ways of saying the same thing". This is the prevailing view, amounting to a reduction of the notion of the logical validity of an inference to the notion of logical consequence. Sundholm (1998) has criticized this reduction, which is widespread in analytical philosophy (but is older). Obviously, its tenability depends on the meaning we decide to give to the expression "logical consequence".

What then is logical consequence? If the student asks, the teacher might answer with an explanation like that given at the start. To understand this, however, we have to clarify (A) and (B). A helpful reformulation of (B) is the following:

(B1) Fact (A) obtains only in virtue of the meanings of the logical constants occurring in the members of Γ and in Q.

The idea that fact (A), the fact that Q is a deductive consequence of Γ, depends only on the meaning of the logical constants occurring in Γ and in Q can be reformulated, in the footsteps of Bolzano and Tarski, by saying that for all uniform variations of the content of the non-logical parts of Γ and Q the result of varying Q would remain a deductive consequence of the result of varying Γ. In other words, if $V(E)$ is the result of a variation V of the content of the non-logical parts of E, (B1) can be reformulated as follows:

(B2) For any variation V of the contents of the non-logical parts of Γ and Q, it is a fact that $V(Q)$ is a deductive consequence of $V(\Gamma)$.

But how should we understand the modal notion "necessarily" involved in the analysis of deductive validity, according to (A)? We can explain necessity in terms of possible worlds:

(A1) Q is true in every possible world in which every member of Γ is true.

An attempt to analyse logical consequence through (A1), however, leads to some problems. One is that of providing a plausible notion of possible world that does not presuppose the notion of logical consequence (nor the related notion of consistency). If you do not fulfil this task, your analysis will be circular. The problem would perhaps be solved if possible worlds were represented by interpretations in the sense of model theory. In this case (A1) would be equated with the usual model-theoretic analysis of logical consequence:

(A2) All models of Γ are models of Q.

Stewart Shapiro (2005, pp. 661–667) maintains that (A2) is an adequate mathematical representation of (A1), when (A1) holds only in virtue of the meanings of the logical terms. John Etchemendy (1990) illustrates some of the difficulties that beset such a view. This is a controversial issue.

Another problem concerns the relation between consequence and valid inference and the reasonable requirement that valid inferences should be good inferences in the pre-theoretical sense, i.e. inferences whose premises genuinely support the conclusion. Suppose we accept (A2) as our notion of logical consequence and adopt the usual reductive definition of logically valid inference, according to which an inference is logically valid if, and only if, its conclusion is a logical consequence of its premises. The result is that our notion of valid inference will be inadequate as a tool for clarifying the pre-theoretical distinction between good and bad deductive arguments. There are infinitely many pairs $\langle\{P_1, \ldots, P_n\}, Q\rangle$ such that all models of $\{P_1, \ldots, P_n\}$ are models of Q but we fully ignore that they are. If one is fully unaware that this relation obtains, one will not (or in any case not legitimately) take any responsibility for a support that the premises $P_1, \ldots, P_n$ provide for the conclusion Q. The passage from premises to conclusion would not be a good inference. Thus the set of logically valid inferences according to the usual reductive definition includes infinitely many possible acts that would not be good inferences. We would never perform those acts and would never treat them as good inferences, as inferences which warrant assent to the conclusion once the premises are warranted. Having detected this problem, Dag Prawitz (2005, p. 677) proposes an entirely different reading of (A):

(A3) The truth of Q follows by necessity of thought from the truth of all members of Γ.

A first explanation of the notion of "necessity of thought" is provided by the following formulations: "one is committed to holding Q, having accepted the truth of the sentences in Γ"; "one is compelled to hold Q true, given that one holds all the sentences of Γ true "; " on pain of irrationality, one must accept the truth of Q, having accepted the truth of the sentences in Γ" (Prawitz 2005, p. 677). The necessity of thought consists in the fact that a person who accepts the truth of the sentences in Γ is compelled to hold Q true. The key feature of the relation between Γ and Q is the *compulsion of the inference* from Γ to Q. One cannot be compelled if one does not feel compelled. Inferential compulsion is a power that acts upon us only in so far as we are *aware* of its force. Therefore the necessity of thought is an *epistemic* necessity: by making the inference a person *recognizes* a guarantee of the truth of Q given a recognition of the truth of the members of Γ.

Prawitz (2005, p. 677) observes that "if the Tarskian or model theoretic notion of logical consequence contains a trace of modality, then it is not of an epistemic kind". More generally, it seems that not only is (A3) profoundly different from (A2), but, at least on initial consideration, it also differs from (A1). Suppose that Q is true in every possible world in which every member of Γ is true, but John is not aware that this is so. In this case, even if John holds all the sentences of Γ true, John is not compelled to hold Q true. He is not irrational if he does not endorse or does not assert Q, because he is not aware of any epistemic connection between Γ and Q. Therefore, at least *prima facie*, (A1) does not imply (A3).

How can the necessity of thought arise? Prawitz writes:

> To develop the idea of a necessity of thought more clearly we must bring in reasoning or proofs in some way. It must be because of an awareness of a piece of reasoning or a proof that one gets compelled to hold Q true given that one holds all the sentences of Γ true (Prawitz 2005, p. 677).

Thus he proposes the following reformulation of (A3):

(A4) There is a proof of Q from hypotheses in Γ.

By "proof" he means an abstract epistemic entity represented by a valid argument, which is a "verbalised piece of reasoning" (Prawitz 2005). An alternative explication of (A3) is therefore:

(A5) There is a valid argument for Q from hypotheses in Γ.

Prawitz reverses the usual order of analysis. The validity of an inference is not analysed in terms of logical consequence, but the other way around: valid argument, or proof, is the conceptually basic notion and logical consequence is analysed in terms of it.

4.2 The Fundamental Task

In later essays Dag Prawitz has developed a notion of valid inference that is immediately connected with the necessity of thought. So far, the necessity of thought has been characterized as an inferential compulsion. But we can view it from different angles. It manifests itself as a special awareness, a special experience, we might say: the experience of a threefold power or force. We can describe it from the point of view of a person who accepts the sentences in Γ and does not endorse Q yet, but then feels compelled by the inference to accept the truth of Q on pain of irrationality. This is the experience of a power to compel a person to accept a conclusion. The same binding force connecting Γ and Q can be described from the different viewpoint of someone who is justified in holding the members of Γ true and through the necessity of the inferential step from Γ to Q acquires a justification for holding Q true. This is the experience of a "power to justify assertions and beliefs" (Prawitz 2010, p. 1). We can see these two faces of the necessity of thought (inferential compulsion and power of delivering a justification) if we imagine a dialogue between a proponent P who asserts Q and an opponent O who tries to object to Q: if both accept Γ, the inference is such that P is justified in asserting Q and thereby O is compelled to accept the assertion.

> Such an epistemic necessity comes close to Aristotle's definition of a syllogism as an argument where "certain things being laid down something follows of necessity from them, i.e. because of them without any further term being needed to justify the conclusion". It is of course right to say that there is an epistemic tie between premisses and conclusion in a valid inference – some kind of thought necessity, we could say, thanks to which the conclusion can become justified (Prawitz 2009, pp. 181–182).

The link between justification and knowledge explains a third aspect of the necessity of thought: we can view it as the feature of inferences which constitutes their "fundamental epistemic significance" in the following sense:

> The "fundamental epistemic significance", which logically valid arguments have [...] consists of course in the common use of valid arguments to acquire new knowledge: given a valid argument, the premisses of which express knowledge that we are already in possession of, we can sometimes use it to acquire the knowledge expressed by the conclusion (Prawitz 2012a).

The necessity of thought manifests itself as the experience of the power of inferential compulsion, of the power of delivering justifications and of the power of providing new knowledge. This threefold power, it seems, is brought into effect precisely through our experience of it. One would not be compelled if one had no experience of compulsion. We may say that the necessity of thought always has a phenomenal character. But the claim that the necessity of thought has a phenomenal character (i.e. that we must experience the necessity of thought) should not be taken to imply that a feeling of being compelled is sufficient for a person to be genuinely experiencing the necessity of thought: one might feel compelled by mistake, without being really compelled.

The epistemic power and compelling force of inferential acts is one of those phenomena whose problematic character we mostly fail to notice because they are always before our eyes. But the need to understand it, the need to understand the necessity of thought, clearly lies at the heart of the philosophy of logic. The old distinction between logic and rhetoric depends on the idea that there is such a thing as a distinctive kind of inferential compulsion which leads us to truth and thereby differs from the persuasive power of rhetorical speeches. According to Prawitz

> it is a task for philosophy to account for the validity of an inference in such a way that it follows from the account or is directly included in the account that [...] a valid deductive inference delivers a conclusive ground for its conclusion, given conclusive grounds for its premises (Prawitz 2010, p. 2).

In other words Prawitz proposes a condition of adequacy for a philosophical analysis of the notion of deductively valid inference: from the fact that an inference J is valid it should be possible to derive that J is endowed with necessity of thought. Let us say that "the fundamental task" is to devise an analysis of deductive validity which satisfies this condition. Prawitz (2009, 2012a) gives a more precise formulation of the task. Suppose that the following conditions hold:

(a) There is a valid inference J from the premises $P_1, \ldots, P_n$ to the conclusion Q;
(b) an agent X has grounds for $P_1, \ldots, P_n$.

The problem for the philosopher of logic is to devise a notion of validity and to specify a further condition (c) in such a way that from assumptions (a)–(c) one can derive the crucial meta-logical conclusion:

(d) Agent X has a ground for Q.

4.3 Grounds

Beware of the word "ground". One of the senses of this word in English is that *grounds* are reasons for saying, doing or believing something. Prawitz also has this sense in mind, but in his writings the word is a technical term. One must understand its role in a semantic theory, a *theory of grounds*, which provides an explication of the meanings of logical constants, of the notion of inference and of deductive validity. A first clarification is offered by Prawitz in this excerpt:

> I have used the term ground in connection with judgements to have a name on what a person needs to be in possession of in order that her judgement is to be justified or count as knowledge, following the Platonic idea that true opinions do not count as knowledge unless one has grounds for them. The general problem that I have posed is how inferences may give us such grounds (Prawitz 2009, p. 179).

This is a first approximation. We shall see that the notion is multifaceted. Before returning to the notion of ground I summarize Prawitz's way of accomplishing the fundamental task.

What further condition (c) should be added to (a) and (b) in order for it to be the case that agent X has a ground for the conclusion Q of inference J? A first candidate is "agent X knows that the inference J from $P_1, \ldots, P_n$ to Q is valid". But Prawitz discards this proposal, because if knowledge of validity is not immediate, it will lead to an infinite regress. Since he does not want to resort to an unexplained notion of immediate knowledge "without any argument" (Prawitz 2009, p. 176), he specifies condition (c) as follows:

(c) Agent X makes the inference J from $P_1, \ldots, P_n$ to the conclusion Q.

This move introduces the idea that an inference is an act. Condition (a) says that there is a valid inference. But at this stage of Prawitz's heuristic strategy the notion of inference and the notion of validity are not analysed yet. Prawitz is trying to shape the correct analyses of the two notions, which should accomplish the fundamental task. One might understand condition (a) as saying that there is a valid inference conceived as a mere pair premises-conclusion with a special property constituting validity (e.g. truth-preservation) without any reference to human acts. Adding condition (c) brings in a decisive conceptual ingredient: inferences are acts made by persons. There is something right in this way of tackling the fundamental task, Prawitz remarks, but only if we adopt a notion of inference which is different from the notion of inference as a complex speech act outlined at the beginning of this paper. If making the inference J is only to assert $P_1, \ldots, P_n$, then to say "hence" and assert Q, "then one cannot expect that conditions (a)–(c) together with reasonable explications of the notions involved are sufficient to imply that the person in question has a ground for the conclusion" (Prawitz 2009, p. 178). There are situations, Prawitz says, where the complex speech act is performed and conditions (a)–(c) interpreted as above are all satisfied, but (d) does not hold:

> a person announces an inference in the way described, say as a step in a proof, but is not able to defend the inference when it is challenged. Such cases occur actually, and the person may then have to withdraw the inference, although no counterexample may have been given. If it later turns out that the inference is in fact valid, perhaps by a long and complicated argument, the person will still not be considered to have had a ground for the conclusion at the time when she asserted it, and the proof that she offered will still be considered to have had a gap at that time (Prawitz 2009, p. 178).

According to Prawitz in the case described the person "announces an inference", but does not actually make it, because

> "to infer" or "to make an inference" must mean something more than just stating a conclusion and giving premisses as reasons. The basic intuition is, I think, that to infer is to "see" that the proposition occurring in the conclusion must be true given that the propositions occurring in the premisses are true, and the problem is how to get a grip of this metaphoric use of "see" (Prawitz 2009, p. 179).

To *see* that the conclusion follows from the premises is to be aware of the necessity of thought. In this passage Prawitz intends to capture the phenomenal character of the necessity of thought. We have an *experience* of necessity. But this experience is at the same time the experience of *performing the act of making an inference*. Therefore it is an *active experience*. To develop this basic intuition Prawitz proposes an explication of the act of inference centred on the notion of an *operation on grounds*.

For the sake of simplicity, at the beginning of this paper I gave a description of linguistic inferences which was admittedly rough. Already in "Towards a Foundation of a General Proof Theory" Prawitz (1973, p. 231) makes it clear that an adequate general characterization of an inference must take into account not only premises and conclusion, but also the arguments for the premises, the assumptions that are discharged and the variables that are bound by making the inference. Even if we add these further ingredients to the initial rough linguistic characterization, however, the new description that we get is still the description of a kind of complex speech act, the act of presenting a linguistic construction as public evidence for an assertion. The notion of inference centred on an operation on grounds is different, because an inference in this sense is the mathematical representation of a *mental* act of inference. It is important to distinguish the three levels of the picture outlined by Prawitz: a mental level, a linguistic level and a mathematical level. *First*: to the mental level belong acts of judgements and other mental epistemic acts like observation, calculation, reasoning. All these acts can remain linguistically unexpressed. The genuine active experience of making an inference also belongs to this level: it is the conscious act of moving mentally from judgements-premises to a judgement-conclusion in such a way that the agent cannot help but recognize that this movement leads the mind from correct premises to a correct conclusion. *Second*: to the linguistic level belong acts of assertions, and linguistic practices of argumentation in support of assertions (including public appeal to intersubjective non-linguistic evidence). At this level we communicate what we do at the mental level and make it verbally manifest. But it can happen that the linguistic expression is not accompanied by corresponding mental acts. For example it can happen that an announced linguistic inference is not accompanied by a genuine mental inference. *Third*: to the mathematical level belong mathematical representations of the mental level in terms of grounds and operations on grounds. This is the level of abstract entities. The logician resorts to abstract entities in order to construct a theory that makes the mental phenomenon of deduction intelligible. A mental act of inference can be mathematically represented as the application of an operation on grounds.

> The mental act that is performed in an inference may be represented, I suggest, as an operation performed on the given grounds for the premisses that results in a ground for the conclusion, whereby seeing that the proposition affirmed is true (Prawitz 2009, p. 188).

A *generic inference act*, Prawitz explains, is individuated by four items:

> If we abstract from the specific performance and even from the agent, we can say that a *(generic) inference (act)* is determined in the simplest case by four kinds of elements: 1) a number of assertions or judgements $A_1, A_2, \ldots, A_n$, called the premisses of the inference, 2) grounds $\alpha_1, \alpha_2, \ldots, \alpha_n$ for $A_1, A_2, \ldots, A_n$, 3) an operation Φ applicable to $\alpha_1, \alpha_2, \ldots, \alpha_n$, and 4) an assertion or judgement B, called the conclusion of the inference (Prawitz 2012a, p. 895).

To *make* the inference is to apply operation Φ to the grounds $\alpha_1, \alpha_2, \ldots, \alpha_n$. After providing a mathematical representation of what it is to make an inference, Prawitz explains what it is for an inference to be valid:

> Such an inference (whether an individual or generic act) is said to be valid, if $\Phi(\alpha_1, \alpha_2, \ldots, \alpha_n)$ is a ground for B (Prawitz 2012a, p. 896).

Suppose we adopt these notions of inference, of validity and of making an inference. Now assume:

(a) There is a valid inference J from the premises $P_1, \ldots, P_n$ to the conclusion Q,
(b) agent X has grounds for $P_1, \ldots, P_n$ and
(c) X makes the inference J.

It is straightforward to deduce:

(d) agent X has a ground for Q.

So Prawitz has accomplished the fundamental task: he has provided a notion of deductively valid inference from which it follows that a deductively valid inference has the special power of delivering grounds and is thus endowed with the necessity of thought.

The linchpin of Prawitz's analysis are the operations on grounds. What is an operation on grounds? What are grounds? Prawitz's basic idea is that a ground is something of which an agent is in possession when that agent is justified in making a judgement (Prawitz 2009, p. 179). By "making a judgement" Prawitz means performing the mental act of judging a proposition to be true. This act can be justified or unjustified. It is justified if, and only if, the person who makes it is in possession of a ground for the judgement. Grounds are "abstract entities" (Prawitz 2009, p. 187). But to be in possession of a ground is a mental state. A judgement can be linguistically expressed by a speech act of assertion. If the assertion is challenged, "the speaker is expected to be able to state a ground for it". So, Prawitz says, "to have a ground is [...] to be in a state of mind that can manifest itself verbally" (Prawitz 2009, p. 183). How is the possession of a ground manifested verbally? By formulating an argument: "we usually communicate a piece of reasoning by presenting a chain of arguments" (Prawitz 2012a, p. 889).

Why is possession of a ground sufficient to justify a judgement or an assertion? Because the very content of that judgement or assertion is determined by a condition fixing what constitutes a ground for it. The notion of ground is the key-notion of an epistemic conception of the meanings of sentences and of the contents of judgements:

> the meaning of a sentence is determined by what counts as a ground for the judgement expressed by the sentence. Or expressed less linguistically: it is constitutive for a proposition what can serve as a ground for judging the proposition to be true (Prawitz 2009, p. 184).

To grasp a proposition is to grasp what counts as a ground for the judgement that the proposition is true. "Grounds on this conception [are] something that one gets in possession of by doing certain things" (Prawitz 2012a, p. 893). Let us call the act of doing such things a "grounding act". When one performs a grounding act one gets in possession of a ground. To grasp a proposition is to grasp what grounding act one must perform in order to get a ground for the judgement that the proposition is true. We get a ground for a simple judgement of numerical identity by making a special calculation. We get a ground for a simple observational judgement by making an adequate observation. We get a ground for the judgement that a logically compound proposition is true by performing an adequate operation on the grounds for its constituents (Prawitz 2012a, p. 893). For example, we get a ground for judging the proposition $P \wedge S$ to be true by performing the operation of conjunction-grounding, abbreviated $\wedge G$. Conjunction-grounding brings together a ground π for P and a ground σ for S and forms a ground $\wedge G(\pi, \sigma)$ for $P \wedge S$. This operation is not only the essential ingredient of an epistemic rule for justifying a judgement and of a pragmatic rule for making a correct assertion which expresses the judgement. It is also the core of a semantic and ontological explanation. It serves to explain the *content* of the judgement. Its content is that a proposition $P \wedge S$ is true. So to explain the content of the judgement is to explain what that proposition *is*. This is done by Prawitz in terms of the operation of conjunction-grounding: it is "a proposition such that a ground for judging it to be true is formed by bringing together two grounds for affirming the two propositions [P and S]" (Prawitz 2009, p. 184). Grounds are also the key-notion of an epistemic conception of *truth*: a proposition is true if, and only if, a ground for a judgement that the proposition is true exists in a tenseless sense, independently of whether it is possessed by us (Prawitz 2012c).

Content-constitutive grounding operations like conjunction-grounding explain why certain inferences are valid in the new sense of "valid" specified by Prawitz. The inferences whose validity is explained are instances of natural-deduction introduction rules. The operation on the grounds for the premises is a mathematical representation of a transition from premises to conclusion experienced with the characteristic compulsion of the necessity of thought. Conjunction-grounding makes inferences of conjunction-introduction valid in accordance with Prawitz's analysis of validity. The mental act of moving from a justified judgement that P is true and a justified judgement that S is true to a justified judgement that $P \wedge S$ is true is mathematically represented by the application of conjunction-grounding to ground π for P and ground σ for S with ground $\wedge G(\pi, \sigma)$ as a result.

Grounds for judgements on propositions whose truth is conclusively established are treated by Prawitz as mathematical objects and are called *closed* grounds. Grounds for judgements that depend on assumptions are treated as functions and are called *open* grounds. Open grounds of this kind are functions from grounds to grounds. An open ground for a judgement depending on m assumptions $A_1, \ldots, A_m$

has m argument places that can be filled by closed grounds $\alpha_1, \ldots, \alpha_m$ for $A_1, \ldots, A_m$. When all argument places are filled the value of the function is a closed ground. Another kind of open grounds are those for open judgements on propositional functions (Prawitz 2009, pp. 185–186). Some grounding operations constitutive of compound propositions operate on open grounds. For example *implication-grounding* $\rightarrow G$ can be applied to an open ground with one argument place $\beta(\xi^p)$ for judging Q to be true under the assumption that P is true. The result $\rightarrow G\xi^p(\beta(\xi^p))$ is a closed ground for judging $P \rightarrow Q$ to be true. In this case, too, the operation in question explains why certain inferences are valid. Implication-grounding makes inferences of implication- introduction valid in accordance with Prawitz's new theoretical notion of validity.

Operations on grounds like conjunction-grounding or implication-grounding are content-constitutive, and therefore also meaning-constitutive. But there are also operations on grounds of a different sort. For example the binary operation $\rightarrow R$ is defined by the equation (Prawitz 2009, p. 189):

$$\rightarrow R(\rightarrow G\xi^p(\beta(\xi^p)), \alpha) = \beta(\alpha)$$

where $\rightarrow G\xi^p(\beta(\xi^p))$ is a ground for $P \rightarrow Q$, $\beta(\alpha)$ is the ground for Q obtained by saturating $\beta(\xi^p)$ with α, which is a ground for P. Thus operation $\rightarrow R$ applied to α and $\rightarrow G\xi^p(\beta(\xi^p))$ makes an inference of implication-elimination valid in Prawitz's sense: an inference whose premises are $P \rightarrow Q$ and P, with conclusion Q.

I hope that this outline gives the idea of Prawitz's ground-theoretical analysis of the notion of deductively valid inference. Once deductive validity is clarified, the notion of a proof can be defined. Descartes (1985, p. 15) likened deduction to a chain. Prawitz defines a *proof* as a chain of valid inferences (Prawitz 2011, p. 400). A proof is compelling because it is constituted of compelling inferences.

4.4 Recapitulation

It is a widespread conviction that a characteristic human phenomenon is the necessity of thought which compels us when we make inferences. The pivotal role of this conviction for the philosophy of logic can hardly be underestimated. The idea, however, is not unchallenged. Some philosophers have denied that we are compelled when we go the way we do in a chain of inferences (Wittgenstein 1956, Sect. 113). If we think that there is such thing as the necessity of thought, we ought to explain how this phenomenon is possible. Dag Prawitz draws our attention to this important problem. His contribution in the essays under consideration can be divided into two parts. In the first part he outlines a research agenda aimed at solving the problem, and highlights a fundamental task. I think he would agree with the following summary of the task:

(Ft) We must analyse "deductively valid" in such a way that from "inference J is valid" we can derive "J is experienced with necessity of thought".

In the second part of his contribution Prawitz proposes a semantic theory, the mathematical theory of grounds, and develops the fundamental task into a more precise requirement on a notion of validity framed in the theory of grounds:

(R) We must analyse "deductively valid" and "to make an inference" in such a way that from

  (a) there is a valid inference J from the premises $P_1, \ldots, P_n$ to the conclusion Q;
  (b) agent X has grounds for $P_1, \ldots, P_n$;
  (c) agent X makes the inference J from $P_1, \ldots, P_n$ to the conclusion Q;

  we can derive the conclusion:

  (d) agent X has a ground for Q.

Prawitz goes on by proposing an interpretation of (c): to make an inference is to apply a grounding operation to the grounds for the premises. Then he defines "deductive validity": the inference is valid if, and only if, the result of applying the grounding operation is a ground for the conclusion. Thus requirement (R) is satisfied. The crucial conceptual tool employed by Prawitz is the notion of "grounding operation". In the next section I will illustrate a difficulty concerning the second part of Prawitz's contribution.

4.5 Mistakes

Does Dag Prawitz succeed in solving the problem of the necessity of thought and providing an analysis of deductive validity that fulfils the fundamental task? A difficulty confronting his approach is that we often make inferences with mistaken premises, though we (wrongly) judge that they are true. Consider the following example. The grounding act for an observational judgement is the act of making an adequate observation. Lisa can see that the traffic light is green. In Prawitz's theoretical frame she gets a ground for the judgement

(i) it is true that *the traffic light is green.*

This judgement is correct. Lisa knows the rules of the road. Therefore, another correct judgement for which Lisa possesses a ground is

(ii) it is true that *if the traffic light is green, cars may pass.*

Thus Lisa makes an inference and concludes:

(iii) It is true that *cars may pass.*

On Prawitz's view, Lisa has applied an operation on grounds, i.e. $\rightarrow R$, and has obtained a ground for the conclusion. Very well, but sometimes our observational judgements are wrong. It can happen that Ugo judges that the traffic light is green, though in reality it is red. So, Ugo can make judgement (iii) on the basis of judgements (ii) and (i) like Lisa. Unlike Lisa, however, Ugo does not really possess a ground for (i), because (i) is not correct, and Ugo does not get a ground for (iii), because (iii) is incorrect as well. Moreover, Ugo cannot apply any operation on grounds: such operations can be applied only to the appropriate grounds and in this case one of the two grounds that are needed is not there.

An unsophisticated way to describe the two courses of thinking is that Lisa and Ugo perform mental acts of the same kind: they move from judgements of the same type (corresponding to (ii) and (i)), with the same content, to conclusions that are judgements of the same type (iii), with the same content. So Lisa and Ugo both perform inferences, and their individual inferences are of the same kind. The unsophisticated logician would also say that both inferences are equally valid: they are both instances of *modus ponens*. Moreover, given that Ugo judges that the two premises are true, he is no less compelled to judge that cars may pass than Lisa is. There is of course an important difference: Ugo is mistaken and Lisa is right. But Ugo's mistake is in one of the premises, not in the act of drawing the conclusion from the premises. It is important, the unsophisticated logician would say, to distinguish a mistake in the premises from a mistake in the inference.

From Prawitz's point of view the unsophisticated description is wholly wrong. Ugo's mental act of moving from premises to conclusion and Lisa's inference are deeply different. According to the proposed definition of an inference act (Prawitz 2012a), an inference act must involve grounds for the premises. Ugo does not have a ground for (i). So, Ugo's act is not an inference. A fortiori, it is not a valid inference. Lisa's act, on the other hand, is a valid inference. Therefore the two acts are deeply different.

The above considerations show that if the theory of grounds aims to provide a mathematical representation of our deductive activity, it faces the following objections, which indicate a clash between the ground-theoretical analysis of deduction and some pre-theoretical convictions. *First*: Pre-theoretically it seems reasonable to say that our deductive activity includes acts of inference that can be valid or invalid, but Prawitz's definition of an act of inference is such that only valid inferences are inferences. It is part of what it is for an act J to be an inference that the agent has grounds for the premises and that J involves an operation Φ applicable to the grounds for the premises. The only operations described in Prawitz's sketch are operations that yield a ground for the conclusion, when they are applied to the grounds for the premises. Therefore, unless new, different operations on grounds are introduced, if J is an inference, J is valid. However, it seems reasonable to say that a fallacy of affirming the consequent is also an inference, though a deductively invalid one. Is there an operation on grounds that corresponds to this inference? If there were and if it were applied to grounds for the premises, it would not yield a ground for the conclusion. *Second*: Pre-theoretically it seems reasonable to say that our deductive activity includes inferences (valid or invalid) with mistaken premises. But Prawitz's analysis of an act of inference implies that if the premises of J are incorrect (i.e. the agent has no grounds for them), J is not an inference. (An act of inference is determined by four elements, one of which being the grounds for the premises.) *Third*: Pre-theoretically, it seems reasonable to say that an inference can be valid even if its premises are mistaken, but on Prawitz's view this is impossible. The reason is that Prawitz's analysis of an inference blurs the distinction between a mistake in the premises and a mistake in the act of moving from the premises to the conclusion. In both cases the act is not a valid inference, because it is not even an inference. By contrast, if we adopt classical model-theoretic semantics, there is no problem with saying that an instance of *modus ponens* is logically valid, because the conclusion is a model-theoretic consequence of the premises, even though one of the premises is (or both are) false in the real world. *Fourth*: It seems reasonable to say that the experience of necessity of thought also characterizes the transition from mistaken premises to their immediate consequences. Prawitz's grounding operations are not defined for mistaken premises devoid of corresponding grounds. The notion of a grounding operation is meant as the mathematical representation of a transition from premises to conclusion experienced with necessity of thought. But the representation is inadequate, because it leaves out an important part of the phenomenon: all the compelling inferences with mistaken premises.

To the first objection Prawitz might reply that he is interested in valid compelling inferences and is thus entitled to leave aside fallacies, which are only attempted, and failed, inferences. The other objections might perhaps be dealt with by adopting the following line of thought. The ground-theoretical analyses of inference and valid inference can probably be adjusted so as to take into account mental inferences with mistaken premises. One might extend the notion of ground and introduce a mathematical representation of the results of epistemic acts underlying mistaken premises. One might say that by an epistemic act (e.g. the act of looking at the traffic light) a person can get a *ground-candidate* that can be a genuine ground or a *pseudo-ground*, which does not provide a justification. The same ground-candidate is a ground for certain judgements (e.g. judgements to the effect that the traffic light is red) and a pseudo-ground for other judgements (e.g. judgements to the effect that the traffic light is green). One can then define operations whose arguments are ground-candidates, not only genuine grounds for the premises. For example, one can define an extension of the operation $\rightarrow R$ which behaves as before when applied to grounds for the premises, but yields a pseudo-ground for the conclusion if one of its arguments is a pseudo-ground (as in the case of Ugo above). If one adopts this line of thought, the definition of an inference should be modified so as to admit that an inference has ground-candidates for the premises, not necessarily genuine grounds. The definition of validity is essentially unaltered: an inference is valid if the corresponding operation on grounds yields a ground for the conclusion whenever it is applied to grounds for the premises. A particular inference can thus be valid even if the ground-candidates for the premises are in fact pseudo-grounds and the premises are therefore mistaken. This more general ground-theoretical notion of an inference perhaps agrees better with our pre-theoretical views. However, a question remains, to which I now turn.

4.6 Epistemic Contexts

The theory of content underlying Prawitz's ground-theoretic semantics is centred on the idea that some grounding operations are content-constitutive. They are embodied in inferences that are instances of natural-deduction introduction rules. Prawitz's examples of grounding operations that are not content-constitutive correspond to natural-deduction elimination rules (conjunction-elimination, implication-elimination, arithmetical induction). My question is: can there be grounding operations specifically corresponding to inferences of a different kind? Consider the following syllogism in Barbara.

(BA) Every human is an animal.
Every philosopher is a human.
Therefore every philosopher is an animal.

Can there be a single specific grounding operation that brings together two grounds for the premises of (BA) so as to get a ground for the conclusion? Pre-theoretically (BA) is compelling, and if the mathematical representation of our active experience of the compelling power of an inference is the application of a grounding operation, then there should be a grounding operation that mathematically represents the act of making an inference like (BA). Someone will reply that this is not necessary: we do not need a special grounding operation for Barbara, because the necessity of thought pertaining to (BA) can be explained as a result of a composition of grounding operations corresponding to natural-deduction inferences. We can break up (BA) into a concatenation of applications of natural-deduction rules. Thus the compelling force of an individual inference in Barbara results from a combination of the compelling forces of many more elementary inferences.

The problem with such a reply is that the experience of a compelling inference has been known to human beings since the remote past, while natural-deduction was invented by Gentzen in the 1930s. A whole community can feel that a syllogism in Barbara is compelling, even though no one is capable of publicly showing that it can be reduced to a concatenation of more elementary applications of natural-deduction rules, simply because no one in the community knows how to present arguments in natural-deduction style. For illustration, here is a little story. Petrus, a medieval philosopher, is engaged in an honest discussion aimed at truth. The participants in the discussion constitute a small philosophical community in the Middle Ages. They rationally cooperate in a common investigation, try to be open-minded, impartial, careful, curious and are willing to retract their assertions if the evidence is against them. Petrus accepts (the Latin version of) our syllogism (BA) without hesitation, because he feels that, on pain of irrationality, he must accept the conclusion, having accepted the truth of the two premises. All the members of the philosophical community have the same attitude as Petrus. For the sake of dispute, however, one of them, Ioannes, *challenges the others to provide a proof that (BA) is valid.* Petrus replies that the challenge is unreasonable: those who doubt the validity of syllogisms like (BA) do not attach genuine meaning to their words; as Aristotle said, it is a clear sign of ignorance to demand that everything be proven; of course, nobody can provide a more articulated argument to show that (BA) is valid, but such an argument is not necessary. To accept syllogisms in Barbara, Petrus explains, is a precondition of all demonstrative reasoning: that is why, if we are rational, we must accept (BA). All the participants in the discussion agree and Ioannes too apologizes for his awkward challenge. The little story is realistic enough, I believe. How should one comment on it?

Let us call "reductionism" the view that all instances of necessity of thought in logic are combinations of grounding operations embodied in natural-deduction inferences. The reductionist thinks that when Petrus utters (BA) in Latin, by asserting "omnis homo est animal", "omnis philosophus est homo" and then "ergo omnis philosophus est animal", the complex speech act does not necessarily reveal the presence of a compelling proof. There is a compelling proof only if a series of mental acts is performed in somebody's mind: the concatenation of acts of transition from premises to conclusions mathematically represented by the appropriate chain of applications of natural-deduction grounding operations. According to *the stern reductionist*, in performing or endorsing the speech act (BA) no one in the medieval community is also performing that concatenation of mental acts. That the members of the community fail to do so is revealed by their *inability to defend the announced linguistic inference*, when it is challenged. When Ioannes dares to call into question the validity of (BA) their reaction is dogmatic. They disguise their lack of logical insight by depicting their inability as an aspect of the allegedly privileged epistemic status of syllogisms in Barbara. But, far from vindicating the speech act (BA), their dogmatic reaction is a manifestation of the fact that they are not really able to *see* that the conclusion follows from the premises. So, none of those who perform or endorse (BA) during the medieval discussion are really performing the mental act of making an inference. To make an inference means more "than just stating a conclusion and giving premises as reasons" (Prawitz 2009, p. 179). It is necessary "to see" that the proposition in the conclusion must be true, given that the propositions in the premises are true. But in the medieval community, with regard to (BA), no one really sees this necessity.

The reductionist must choose between two lines of thought: the stern interpretation or the *generous* interpretation of the little story. The generous reductionist thinks that the stern interpretation of the little story is intolerably restrictive: it leads to the conclusion that there has almost never been a compelling act of inference in the past and that there almost never is an act of inference among people who are not trained logicians. The generous reductionist opposes this exceedingly restrictive attitude and maintains that Petrus and his fellows do perform the concatenation of mental transitions corresponding to a combination of instances of natural-deduction rules. They perform the series of mental inferences in a hidden way, even though no publicly accessible manifestation of their mental activity is available apart from the fact that they assert "omnis homo est animal", "omnis philosophus est homo", and then "ergo omnis philosophus est animal" or immediately accept this linguistic move.

The reader will probably agree that the generous interpretation is implausible. It is implausible because mental inferences are conceived here as conscious acts, not as an unconscious mechanism. The generous interpretation ascribes to Petrus and his fellows a concatenation of mental inferences corresponding to a natural-deduction derivation of the conclusion of (BA) from its premises. But if mental inferences are conscious acts, and they internally perform these conscious acts, why are the members of the medieval community so reluctant to make such acts explicit? Ascribing a concatenation of mental inferences to them is at odds with their reluctance to provide an argument more articulated than the simple (BA). Thus the generous interpretation is at the least unwarranted, if not downright refuted by the community's overt behaviour.

If the generous interpretation is discarded, there are two possibilities: either to adopt the stern interpretation or to reject reductionism and admit of a specific grounding operation for Barbara. A reason for someone to reject reductionism is that the stern interpretation appears too restrictive. The non-reductionist may adopt a contextualist stance with respect to the necessity of thought.

Petrus and his fellows are honestly engaged in a discussion aimed at knowledge and truth. Presumably, they would take the issue seriously were Ioannes able to substantiate his challenge to prove the validity of (BA) by showing that it is not at all an unreasonable challenge. *We can now substantiate* the problem by explaining the meaning of sentences of the form "every X is a Y" independently of syllogisms in Barbara and by exhibiting other ways of obtaining the conclusion of (BA) from the premises by means of a more articulated argument constituted of more elementary inferences. But neither Ioannes nor anyone else in the medieval epistemic context sketched in our little story is capable of doing this. They lack the necessary logical and meaning-theoretical equipment. Therefore when Petrus and his fellows discard the problem, their behaviour does not display irrationality. In their epistemic context Petrus and his fellows feel that, if they care for truth, they must immediately accept the conclusion of (BA) once they have accepted the premises. In other words they feel compelled. They are not fully aware of why they feel this way. But it is not a mere feeling. Their feeling of compulsion *accords with the epistemic context.*

An epistemic context is a social context constituted by a truth-oriented community, shared practices, a common language, accepted reasonings and knowledge-claims, open problems, and instances of contextual rigour. An announced linguistic inference is an *instance of rigour relative to an epistemic context* if, and only if, all members of the truth-oriented community in the context expect of every member that he or she understands and accepts the act of inference and does not require further justification of it. The feeling of compulsion associated with an act of transition from premises to conclusion accords with an epistemic context if the corresponding linguistic inference is an instance of rigour relative to the context. Petrus and his fellows belong to an epistemic context where syllogisms in Barbara are instances of rigour. Therefore their feeling of compulsion associated with (BA) accords with the epistemic context. The non-reductionist concludes that *in their epistemic context* Petrus and his fellows *are* compelled. They make a genuine inference endowed with necessity of thought in that context. But in different contexts the instances of rigour are different and a similar inference would be devoid of necessity of thought. The non-reductionist concedes that if Paolo, a student of contemporary first order predicate logic, reacted to a question regarding the validity of (BA) like Petrus does in his medieval context, then Paolo would simply show his lack of logical competence. The non-reductionist maintains that the notion of necessity of thought is relative to the epistemic context.

An inference cannot have necessity of thought independently of the epistemic context. A possible development of this idea consists in saying that an act of inference is also determined by the epistemic context to which it belongs, so that a difference in the epistemic context implies that the act of inference is different. The truth condition of the sentence "Petrus' act of inference is compelling", on this view, depends on whether the act of inference is an instance of rigour relative to the epistemic context to which Petrus belongs (This kind of contextualism is therefore different from the variety of epistemic contextualism defended, among others, by DeRose 2009).

By contrast, the notion of necessity of thought adopted by the stern reductionist is non-relative. The stern reductionist may allow that Petrus and his fellows in the medieval epistemic context do their best in good faith to be rational and that they feel compelled when they perform the speech act of passing from the premises of (BA) to the conclusion. However, according to the stern reductionist, they *are not* compelled, because they do not really see that the conclusion is true given that the premises are: their logical tools and their logical penetration are inadequate. Thus behind their speech act there is not a genuine mental inference endowed with necessity of thought. If we could show them that a syllogism in Barbara can be replaced by a more articulated argument in natural-deduction style, they would acknowledge that they were not able to vindicate their speech act and agree that we instead can perform all the inferences required to really see that the conclusion follows from the premises. The stern reductionist thinks that there is such a thing as a fully articulated proof, which counts as an absolute standard of rigour for all contexts. Only possession of such a proof provides necessity of thought. We, who can provide arguments in natural-deduction style, in our epistemic context, are able to carry out an absolutely rigorous deduction that constitutes a fully articulated version of (BA). This is why we genuinely experience necessity of thought when we pass from the premises to the conclusion. Petrus and his fellows in the medieval context cannot carry out a really rigorous proof. Therefore, with regard to (BA), they do not have real experience of the necessity of thought.

Both the contextualist non-reductionist and the stern reductionist think that a mere feeling of being compelled is not sufficient to genuinely experience necessity of thought. Both believe that one can feel compelled by mistake, without being really compelled. The non-reductionist believes that Petrus is compelled in the medieval epistemic context as described above because his feeling of being compelled accords with the context: in light of the epistemic tools available in that context the need for and possibility of a further articulation of (BA) are not perceived. But a counterpart of Petrus in a more sophisticated and conceptually richer epistemic context (a contemporary course of predicate logic) might feel compelled like Petrus without being really compelled. His fellows in the richer context would be entitled to require a more articulated argument. For this kind of non-reductionist a feeling of compulsion is not sufficient: an experience of real inferential compulsion requires a suitable epistemic context. For the stern reductionist what is required is the genuine possession of a fully articulated proof.

Recourse to the idea of a fully articulated proof can explain the necessity of thought of simple deductions like (BA). But to what extent can we apply the notion of a fully articulated proof? Can we apply such an analysis of the necessity of thought to real-life mathematical proofs? Prawitz is aware that the proofs that appear on a mathematical paper are not fully articulated.

> Specialists may allow proofs with great leaps that look like gaps to those less familiar with the subject. What is a proof for the specialist may have to be expanded by insertions of smaller inference steps before it becomes a proof for the non specialist. But we expect there to be an end to such expansions, resulting in what is sometimes called fully articulated proofs, so that at the end we can say: here is something that is objectively a proof (Prawitz 2011, p. 386).

The argument that counts as a proof for the non-specialist and an argument expressing the fully articulated proof, however, usually cannot be the same argument. The argument for the non-specialist must be understandable and must therefore contain shortcuts. An argument corresponding to a fully articulated proof of a real mathematical theorem, where every logical detail is fully explicit, would be extremely long and complicated and therefore unreadable and unintelligible. As a consequence, the fully articulated proof is an idealized proof, not a real-life proof. The notion of an idealized proof is perhaps less problematic if we want to use it to develop an epistemic conception of truth (Prawitz 2012b, c). But the problematic nature of the notion of idealized proof worsens if we want to use it to explain the necessity of thought, which is the experience of concrete human beings in an epistemic context where acts of inference are performed or assessed. We are not interested in the experience of idealized beings. We are interested in an experience of real humans. The source whence an experience of real humans springs can be in their acts, the real acts of agents placed in a social context. But how can an experience of real humans spring from acts of an idealized agent independent of any contextual limitation?

References

Beall, J. C. (2010). *Logic*. London: Routledge.

Burgess, J. (2009). *Philosophical logic*. Princeton: Princeton University Press.

DeRose, K. (2009). *The case for contextualism*. Oxford: Clarendon Press.

Descartes, R. (1985). Rules for the direction of the mind. In: trans by D. Murdoch (Ed.) *The philosophical writings of Descartes* (pp. 7–78). Cambridge: Cambridge University Press.

Etchemendy, J. (1990). *The concept of logical consequence*. Harvard: Harvard University Press.

Prawitz, D. (1973). Towards a foundation of general proof theory. In P. Suppes (Ed.), *Logic, methodology and philosophy of science IV* (pp. 225–250). Amsterdam: North-Holland Publishing Company.

Prawitz, D. (2005). Logical consequence from a constructivist point of view. In S. Shapiro (Ed.), *The Oxford handbook of philosophy of mathematics and logic* (pp. 671–695). Oxford: Oxford University Press.

Prawitz, D. (2009). Inference and knowledge. In M. Pelis (Ed.), *The Logica yearbook 2008* (pp. 175–192). London: College Publications.

Prawitz D (2010) Validity of inferences. In: Proceedings from the 2nd Launer Symposium on Analytical Philosophy on the Occasion of the Presentation of the Launer Prize at Bern 2006.

Prawitz, D. (2011). Proofs and perfect syllogisms. In C. Cellucci, E. R. Grosholz, & E. Ippoliti (Eds.), *Logic and knowledge* (pp. 385–402). Cambridge: Cambridge Scholars Publishing.

Prawitz, D. (2012a). The epistemic significance of valid inference. *Synthese*, *187*(3), 887–898.
Prawitz, D. (2012b). Truth and proof in intuitionism. In P. Dybier, S. Lindström, E. Palmgren, & G. Sundholm (Eds.), *Epistemology versus ontology* (pp. 45–67). Heidelberg: Springer.
Prawitz, D. (2012c). Truth as an epistemic notion. *Topoi*, *31*(1), 9–16.
Salisbury, J. (2009). *Metalogicon*. Philadelphia: Paul Dry Books.
Shapiro, S. (2005). Logical consequence, proof theory and model theory. In S. Shapiro (Ed.), *The Oxford handbook of philosophy of mathematics and logic* (pp. 651–670). Oxford: Oxford University Press.
Spinoza, B. (1995). *The letters*. Indianapolis: Hackett Publishing Company.
Sundholm, G. (1994). Ontologic versus epistemologic: Some strands in the development of logic. In D. Prawitz & D. Westerståhl (Eds.), *Logic and philosophy of science in Uppsala* (pp. 373–384). Dordrecht: Kluwer Academic.
Sundholm, G. (1998). Inference versus consequence. In T. Childers (Ed.), *The Logica yearbook 1997* (pp. 26–35). Prague: Filosofia.
Wittgenstein, L. (1956). *Remarks on the foundations of mathematics*. Oxford: Blackwell. (3rd edition 1998).

Chapter 5
On the Motives for Proof Theory

Michael Detlefsen

Abstract The central concerns of this chapter are to (i) identify the chief motives which led Hilbert to develop his "direct" approach to the consistency problem for arithmetic and to (ii) expose certain relationships between these motives. Of particular concern will be the question of the role played by conceptual freedom (i.e. freedom in use of concepts) in this development. If I am right, Hilbert was motivated by an ideal of freedom, though it was only one of a number of factors that shaped his proof theory.

Keywords Hilbert · Frege · Bernays · Proof theory · Consistency · Model construction · Direct approach to consistency · Finitary viewpoint · Conceptual freedom · Abstract axiomatic method · Formal axiomatization · Contentual axiomatization · Frege's futility argument · Rigor · Denken.

5.1 Introduction

There was a well-known discussion among 19th and early 20th century foundational thinkers concerning the conditions under which concepts (or, equivalently for our purposes, expressions) may be justifiedly used in mathematical reasoning.

A significant issue in this discussion, and one with which I will be centrally concerned here, was a matter of freedom. Some maintained that use of concepts in mathematical reasoning is in some way(s) exceptionally free—free in such ways and/or to such an extent as to set conceptual usage in mathematics apart from conceptual usage in all other sciences and also in everyday thinking.

Others found this difficult to square with what was commonly accepted as a proper control on our use of concepts in mathematics—namely, consistency. If consistency is legitimately required of our use of concepts in mathematics, then how can our use of concepts be properly described as being free?

M. Detlefsen (✉)
Department of Philosophy, University of Notre Dame,
100 Malloy Hall, Notre Dame 46556, USA
e-mail: mdetlef1@nd.edu

H. Wansing (ed.), *Dag Prawitz on Proofs and Meaning*, Outstanding Contributions to Logic 7, DOI 10.1007/978-3-319-11041-7_5

Frege and Hilbert had a brief exchange on this matter in their correspondence. Frege's part consisted largely of a reiteration of his misgivings concerning claims (mainly by those Frege counted as formalists) that we are free to create concepts and to use those concepts in our mathematical thinking.[1]

He reminded Hilbert that those who have emphasized the importance of freedom in our use of concepts in mathematics have also acknowledged that, to be justified, all such uses must be consistent.

Frege then offered two observations which, taken together, pose a serious challenge to anyone who, like Hilbert, maintains that there is significant scope for free use of concepts in mathematics. The first was that

I. practically speaking, the only way to establish the consistency of our uses of concepts is to find "witnessing" instances of those uses (i.e. to find objects which make the axioms governing the uses mentioned patently true).

The second was that

II. to provide such witnesses is, in effect, to find or to discover—not to create—the concepts in question.

From these two observations, Frege concluded that satisfaction of legitimate constraints on the use of concepts in mathematics leaves little room for their genuinely free use.

Hilbert rejected the first "observation." It went against the leading idea of his then-emerging proof-theoretic alternative to model-construction as a means of proving consistency.

His rejection did not, however, convince Frege. In part this was because Frege and Hilbert did not resolve, and scarcely even addressed, the differences between their respective conceptions of axiomatization. Partly too, though, it was because, at the time or their correspondence, Hilbert's comprehension of his proof-theoretic alternative was rudimentary and lacking in substantive detail. When Frege requested a more detailed description and an example, Hilbert was not in a position to provide them, and this essentially ended their correspondence.

5.1.1 Different Conceptions of Axiomatization

Frege maintained what was largely the traditional view of axiomatic reasoning. On this view, axioms are propositions with special epistemic qualifications.[2]

[1] Those concerned with conceptual freedom (i.e. freedom in the introduction and use of concepts in mathematics) included such otherwise diverse thinkers as Brouwer, Cantor, Dedekind, Frege, Hilbert, Kronecker, Peano and Weyl, to mention but some. Freedom was also a concern to some mathematicians (e.g. Gauss) who only occasionally showed interest in foundational matters.

[2] Traditionally, the list of such qualifications included such properties as self-evidence (or other forms of justificative immediacy), certainty, unprovability and rational undeniability.

These qualifications persuaded Frege and others (e.g. Peano) that there was not generally a need for, or, in a sense, even a genuine possibility of consistency proofs.

The correspondence with Hilbert didn't change this. In his letter of December 27, 1899 Frege stated the pivotal reasoning succinctly, for the case of geometry: "I call [geometrical, MD] axioms propositions that are true but are not proved because our knowledge of them flows from a source very different from the logical source, a source which might be called spatial intuition. From the truth of the axioms it follows that they do not contradict one another. There is therefore no need for a further proof." (Frege 1899).[3,4]

Hilbert could not have disagreed more. His disagreement was not, however, with Frege's description of the need for consistency proofs given acceptance of the traditional view of axiomatization. His disagreement was rather with the traditional conception itself and Frege's presumption of it.

In Hilbert's view, axioms were generally not to be seen as truths with special epistemic qualifications. They were not in fact to be seen as truths at all, or even as propositions. Rather, they were formulae, or sentence-schemata or, for some, propositional functions—in all events, items which have neither contents nor truth-values.[5]

Hilbert's conception of axiomatic method was thus very different from Frege's. He called attention to some of these differences in his letter of December 29, 1899, mentioning, in particular, what he described as "the cardinal point" of misunderstanding between the two of them. This concerned whether the axioms of his geometry were to be regarded as attempts to describe or capture contents that were in some sense given in advance of the axioms themselves (e.g. given antecedent understandings of such key terms as "point", "line", etc.), or whether, instead, it was the axioms themselves

[3] In the second volume of the *Grundgesetze* he would strengthen this to a statement of the impossibiity of giving a proof of consistency for a genuine set of axioms. "Axioms do not contradict one another because they are true; this admits of no proof." (Frege 1903, p. 321).

[4] Peano made similar though somewhat more restricted statements. He claimed, for example, that "A consistency proof for arithmetic, or for geometry, is ... not necessary. In fact, we do not create the axioms arbitrarily, but assume instead as axioms very simple propositions which appear explicitly or implicitly in every book of arithmetic or geometry. The axiom systems of arithmetic and geometry are satisfied by the ideas of number and point that every author in arithmetic or geometry knows. We think of numbers, and therefore numbers exist. A consistency proof can be useful if the axioms are intended as hypotheses which do not necessarily correspond to real facts." (Peano 1906, English trans. in Borga and Palladino (1992, p. 343)).

Peano's statement is curious because of the seeming tension between his claim that we do not create the axioms of arithmetic, but that numbers exist because we think of them. The standard for existence suggested by the latter claim seems, if anything, weaker than creativist standards. Creativists at least required that the "thinking" be consistent.

[5] Whitehead described the view this way, when applied to the axioms of geometry: "The points mentioned in the axioms are not a special determinate class of entities ... they are in fact any entities whatever, which happen to be inter-related in such a manner, that the axioms are true when considered as referring to those entities and their inter-relations. Accordingly—since the class of points is undetermined—the axioms are not propositions at all ... An axiom (in this sense) since it is not a proposition can neither be true or false." (Whitehead 1906, p. 1). Such descriptions were fairly common in the foundational literature of the early 20th century.

which, from the very start, determined whatever meanings the terms appearing in them ought rightly to be taken to have.

Frege, on the other hand, believed that the acceptance of at least some of Hilbert's axioms presumed prior understanding of the terms that occur in them. Hilbert flatly denied this. "I do not want to assume anything as known in advance" (Hilbert 1899), he wrote.

This absence of prior presumption concerning the contents of axioms was for Hilbert the distinguishing characteristic of modern axiomatic method. He took pains to make this clear in his most basic descriptions of the axiomatic approach—namely, those given for geometry and arithmetic in (Hilbert 1899) and (Hilbert 1900), respectively. Here's the core of the former:

> We think (denken) three different systems of things. The things of the first system we call *points* and designate them $A, B, C\ldots$ The things of the second system we call *lines* and designate them $a, b, c\ldots$ The things of the third system we call *planes* and designate them $\alpha, \beta, \gamma\ldots$
>
> We think (denken) the points, lines and planes in certain mutual relations ...
>
> The exact (genaue) and for mathematical purposes complete (vollständige) specification of these relationships is accomplished by the *axioms of geometry*.
>
> Hilbert (1899, Chap. 1, Sect. 1)[6]

The distinctive characteristic of abstract axiomatization was thus that it represented a "thinking." We "think" systems of things standing in certain relations. We do not observe or intuit them as doing so, and then express or describe the contents of these observations or intuitions in the axioms we give. Rather, we "think" the objects, and think them as standing in certain relations, with nothing being given prior to or in association with this thinking to serve as its intended contents.

Hilbertian axiomatization was thus intended to separate axioms (and therefore proofs and axiomatic systems) from contents. The very terminology of Hilbert's basic characterizations of axiomatization suggests this. The prominent use of 'denken' in particular suggests it.

It does so not least by recalling Kant's separation of thinking from knowing in the first critique: "I can think (denken) whatever I want", he wrote, "provided only that I do not contradict myself. This suffices for the possibility of the concept, even though I may not be able to answer for there being, in the sum of all possibilities, an object corresponding to it. Indeed, something more is required before I can ascribe to such a concept objective validity, that is, real possibility; the former possibility is merely logical." (Kant 1787, p. xxvi, note a).[7]

Hilbert thus conceived of axiomatization as a type of *denken*. Frege, by contrast, embraced the thoroughly contentual traditional conception of axiomatization. On

---

[6] Cf. Hilbert (1900, p. 181) for a closely parallel description of axiomatic method in arithmetic, with the same emphasis on its being a case of *denken*.

[7] The German of the core part of this remark is: "[D]enken kann ich, was ich will, wenn ich mir nur nicht selbst widerspreche, d. i. wenn mein Begriff nur ein möglicher Gedanke ist, ob ich zwar dafür nicht stehen kann, ob im Inbegriffe aller Möglichkeiten diesem auch ein Objekt korrespondiere oder nicht."

this conception, axioms were taken to be evidently true propositions, judged to be true by the axiomatic reasoner. Proofs were likewise finite sequences of judgments which began with axioms and proceeded to further propositions recognized as being true by dint of their perceived logical implication by the axioms. Theories, finally, were taken to be systems of true propositions bearing a variety of recognized logical relationships to one another.

5.1.2 Different Standards of Justification

Frege and Hilbert thus disagreed on both the essential characteristics and the aims or ideals of axiomatic method. It is hardly surprising, then, that they should also have disagreed on the requirements for the proper justification or foundation of axiomatic theories.

For Frege, this ultimately came down to making proper choices of axioms and methods of inference. The axioms were to be evidently true propositions, perhaps with additional qualifications. Similarly, the methods of inference were to be chosen for their clear validity and for their capacity (both individually and collectively) to eliminate logical gaps in our reasoning and thus to foster rigor in our proofs.

The standards for Hilbert's abstract axiomatization,[8] on the other hand, were quite different. A system of axioms was not to be associated with any particular contents. Accordingly, its adequacy was not to be judged by whether or not it captured such contents.

To put it roughly, in abstract or formal (formale) axiomatization, axioms were taken to determine contents, not contents axioms.[9]

> In formal axiomatization ...the basic relations are not taken as having already been determined contentually. Rather, they are determined implicitly by the axioms from the very start. And in all thinking with an axiomatic theory only those basic relations are used that are expressly formulated in the axioms Hilbert and Bernays (1934, p. 7)

On the abstract conception, then, axioms did not have contents. But if axioms do not have contents, by what, then, is the acceptability of an axiomatic system to be judged?

By and large, Hilbert's answer was "by its consistency." A little more accurately, it was by its consistency and its success in realizing the goals for the sake of whose realization it (i.e. the abstract axiomatic system in question) was wanted in the first place.[10]

---

[8] Or what he and Bernays generally termed *formal* (formale) axiomatization (cf. Hilbert and Bernays (1934, p. 1–2, 6–7).

[9] Somewhat more accurately, the axioms of an abstract axiomatization were not taken to themselves *have* contents, even though, as a distinct and separate matter, they were taken to admit of various interpretations, each of which associated contents with them.

[10] "[I]f the question of the justification (Berechtigung) of a procedure (Maßnahme) means anything more than proving its consistency, it can only mean determining whether the procedure fulfills

Consistency thus replaced evident truth as the basic measure of adequacy for axiomatic systems. Hilbert saw this replacement as making for greater freedom in mathematical thinking. As he put it: "The axiomatic method[11] … guarantees investigation (Forschung) the maximum freedom of movement (die vollste Bewegungsfreiheit) …" (Hilbert 1922, p. 201).

As noted at the beginning, this claim of the supposedly greater conceptual freedom afforded by the abstract axiomatic method was challenged by Frege. He believed that the abstractionist's commitment to consistency effectively neutralized any supposed increase in freedom that abstract axiomatization might otherwise support.

His reason for this was claim I., identified earlier—the claim that, practically speaking, the only way to establish the consistency of an abstract system is to find a "witnessing" interpretation, or what today we would call a "model" of its axioms.

Frege thus saw the abstractionist's reliance on such a proof as running contrary to his avowed aim to separate or disassociate axioms from contents. As a countermeasure, Hilbert launched his program to develop a so-called 'direct' approach to consistency proofs and, in particular, to a consistency proof for arithmetic. This was to be a proof that did not rely on the interpretation of arithmetic in other (i.e. non-arithmetical) terms. Rather, it was to be based directly on an analysis of what arithmetical reasoning, in its outwardly observable characteristics, revealed itself to be.

Frege sensed that Hilbert had some alternative to model-construction in mind as his envisioned means of proving the consistency of arithmetic.

> I believe I can discern, from some places in your lectures, that my arguments failed to convince you, which makes me all the more anxious to find out your counter-arguments. It seems to me that you believe yourself to be in possession of a principle for proving lack of contradiction (Widerspruchslosigkeit) which is essentially different from the one I formulated in my last letter and which, if I remember right, you alone apply in your Grundlagen der Geometrie.[12] If you were right in this, it could be of great significance; I do not yet believe in it, however, but suspect that such a principle could be reduced (wird zurückführen lassen) to the one I have formulated and that it cannot therefore have a wider application (grösseren Tragweite) than mine. It would help clear up matters if in your reply to my last letter—and I am still hoping for a reply—you could formulate such a principle precisely (genau formulieren) and perhaps elucidate its application by an example.
>
> Frege (1900b, p. 78)

Frege didn't specify the remarks that gave rise to this presentiment, but he was of course right to think that Hilbert was contemplating a radical alternative to model-construction as a means of proving the consistency of arithmetic.

Frege had raised this question in an earlier letter, where he challenged what he took to be Hilbert's belief in a principle to the effect that if a set of sentences is

(Footnote 10 continued)
its promised purpose. Indeed, success is necessary; here, too, it is the highest tribunal, to which everyone submits." (Hilbert 1926, p. 163).

[11] By which he meant "the abstract axiomatic method."

[12] Frege described this method as one of "point[ing] to an object" (Frege 1980, p. 47) that can be seen to satisfy the given set of axioms.

consistent then one can infer that there is an object (or domain of objects) satisfying those sentences. He said that such a principle was "not evident" (nicht einleuchtend) to him.

Neither, so far as I can see, did Hilbert think it was. Certainly, it was not generally regarded as evident at the time. In fact, it was widely regarded as problematic and even had a special name—the problem of *existence theorems*.[13,14]

I see little reason, then, to think that Hilbert regarded the inference from consistency to an existence theorem as unproblematic. His concern was not so much with the legitimacy of this inference, though, as with the nature of consistency proof itself. The common view was that one could only prove the consistency of an axiom system by proving an existence theorem for it. Hilbert disagreed. His having done so, however, ought not to be equated with his having regarded the inference from consistency to existence theorems as unproblematic.

After registering his concern with the legitimacy of the inference from consistency to existence theorems, Frege went on to what is my principal interest here—namely, his claim that such an inference would be useless even if were legitimate. His reasoning, as indicated above, was to invoke I. (i.e. the claim that there is no practical means of proving the consistency of a set of sentences except to identify an object which satisfies them)[15] and II. (i.e. the claim that if one could prove such an existence theorem, there would be "no need to demonstrate in a roundabout way that there is such an object by first demonstrating lack of contradiction" (Frege 1900a)).

It is this part of the exchange between Frege and Hilbert that concerns me here. Frege's argument is, in effect, that conceptual freedom in mathematics is largely an illusion. My aim is to get clearer on the role that conceptual freedom may have played as a motive for Hilbert's development of his proof theory. If I am right, it did play a role, though it was only one among a more complex scheme of motives. I hope to clarify both this larger scheme and the place that conceptual freedom had in it.

---

[13] Whitehead characterized an existence theorem for a system of axioms as a proposition to the effect that "there are entities so interrelated that the axioms become true propositions" when interpreted as applying to these entities and known relations between them [cf. Whitehead (1906, p. 2)]. He then described the problem as follows: "Some mathematicians solve the difficult problem of existence theorems by assuming the converse relation between existence theorems and consistency, namely that, if a set of axioms are consistent, there exists a set of entities satisfying them. Then consistency can only be guaranteed by a direct appeal to intuition, and by the fact that no contradiction has hitherto been deduced from the axioms. Such a procedure in the deduction of existence theorems seems to be founded on a rash reliance on a particular philosophical doctrine respecting the creative activity of the mind." (Whitehead 1906, p. 3–4).

[14] For completeness let me note that there have also been those who have questioned the inference from instantiation to consistency, describing it as nothing more than a dogma: "Is it possible that the only way we can determine whether a set is consistent is by seeing all the postulates actually exemplified in some one object? If so, we must arbitrarily assume that the object is self-consistent, so that the proof of consistency must ultimately rest on a dogma. As independence rests on consistency there are therefore no satisfactory proofs as yet of either independence and (sic!) consistency." (Weiss 1929, p. 468).

[15] "Is there some other means of proving non-contradictoriness than to identify (aufweist) an object (Gegenstand) which jointly has the properties (die Eigenschaften sämtlich hat)?" (see letter of January 76, 1900 to Hilbert, Frege (1900a), 70–71).

5.2 Background

Freedom and, in particular, what I will here refer to as conceptual freedom (i.e., freedom to justifiedly use concepts in mathematical reasoning without first abstracting their supposed contents from the contents of intuitions or experiences of their instances), were prominent concerns among 19th and 20th century mathematicians for whom foundational concerns were important.

For the most part, the views of these thinkers were of two broad types, one representing more conservative, the other more liberal views concerning the justified use of concepts/expressions in mathematics.

5.2.1 Conservative Views

More conservative thinkers favored security and (a certain type(s) of) objectivity in the use of concepts. Accordingly, they placed it under the control of what they and others regarded as relatively strict justificative requirements. At the core of their views, generally speaking, was some form of the following:

Conservative Standard (CS). Use of a concept in mathematical reasoning is justified if and only if (i) there is an intuition of an instance(s) of that concept and (ii) the contents of the concept that figure in its use can be obtained by available, reliable processes of abstraction from the contents of the intuition of said instance(s).[16,17]

Those who have advocated such a standard have generally believed that intuitive apprehension—or whatever more particular forms of cognition (e.g. finitary construction) might be favored—are epistemically privileged in the sense of being objective and especially reliable.

The idea that the knowledge mentioned should be objective was based on the idea that the contents of intuitions, as distinct from those of mere concepts, are in some sense "given to" or "impressed on" a knower, and that they therefore arise from sources that are not within the conceiving subject's voluntary control. For that reason and in that sense, then, they are considered to be objective rather than subjective.

In addition, since abstractive processes have commonly been taken to be of an essentially "subtractive" character (i.e. to only subtract from the intuitional contents to which they are properly applied), the abstractive component of conception keeps the contents of concepts within the bounds of contents supplied by objective sources.

This, roughly and briefly, is why the CS has commonly been taken to provide for the objectivity of conceptual contents.[18]

---

[16] It may be necessary to suppose that there is some mechanism for "sharing" or at least to facilitate some common access to intuitive apprehensions across the members of a (relevant) reasoning community.

[17] I will generally refer to (i) and (ii), taken together, as the *instantiation requirement*.

[18] In the history of mathematics, construction has been a prominent source of intuitional contents. Accordingly, the CS has often taken the form of a more specific requirement that the contents of

5.2.2 More Liberal Views

There has been considerable support for the CS as an appropriate standard for the justified use of concepts in mathematical reasoning. It has, in particular, been a point of emphasis for constructivists.

Others have rejected the CS as too restrictive and have advocated what they regard as satisfactory but less restrictive alternatives to it. The most common, or at least the most influential of these, is that which replaces the key requirement of the CS—namely, the requirement that conceptual contents be derivable from intuitive contents—with a simple requirement of consistency.

Liberal Standard (LS). Use of a concept in mathematical reasoning is justified if and only if it (i.e. the use in question) is consistent.[19,20]

Supporters of the LS have typically taken it to represent a more liberal standard of justified use of concepts because they have believed consistency to be less restrictive than the condition imposed by the CS—that is, the instantiation requirement. A basic belief underlying advocacy of the LS has thus been that

Reduced Restriction Thesis (RRT). The condition on justified use of concepts imposed by the LS is less restrictive than that imposed by the CS.

What "less restrictive" might or ought properly to mean in this context is of course a pivotal question and one that shall figure centrally in the discussion to follow. For the moment, though, I want simply to observe that there is supposed to be some difference in restrictiveness that has mattered to both advocates of the CS and advocates of the LS. The former, presumably, have seen the assumed greater restrictiveness of the CS as a point in its favor. The latter, on the other hand, have seen it as a point against it.

There have also been those who have rejected the RRT as false or unwarranted and who have therefore held that the LS's supposed support of conceptual freedom is illusory. These have argued that, at bottom, the LS is as restrictive of justified conceptual use as is the CS.[21]

(Footnote 18 continued)
concepts be abstractable from contents provided by construction (e.g. by the familiar constructions of classical synthetic geometry). Cf. Leslie (1821, p. 4) for a general statement of the common traditional view that quantities are *given* when they either have been or may readily be exhibited by construction.

[19] The common conception of consistency in this context has been not only that the uses of a given concept should be self-consistent, but that they should be consistent with such other uses of concepts as there might be reason to combine them with. Generally speaking, this second type of consistency is what Gödel and other writers have referred to as "outer" consistency.

[20] A second type of condition—one positing one or another form of fruitfulness—has often been added as well [cf. Cantor (1883, Sect. 8), Hilbert (1926, p. 163)]. I'll not comment on this further here because my focus will be the replacement of the abstraction-from-intuited-instance requirement with the consistency requirement, and whether such replacement may plausibly be seen as making for a comparative increase in conceptual freedom.

[21] The more direct among them have in fact maintained that the conditions placed by the two standards on the justified use of concepts are effectively *the same*.

This latter type of view will be my primary focus here. Of particular concern in this connection will be an argument of Frege's, which I call his *Futility Argument*. Before turning to it, though, I want to make some brief terminological remarks concerning my use of the term 'concept' and related expressions. My hope is that these will help to forestall objections based on what some may see as insufficient sensitivity to differences between concepts, on the one hand, and linguistic expressions, on the other.

Where I write 'use of a concept' I could as well write 'axiomatic use of a concept', since it is only axiomatic uses of concepts that concern me here. By an axiomatic use of a concept/expression, I mean, roughly, a use which consists in the application of an axiom and/or a rule of inference which application is supposed to constitute the use of the concept in question.[22]

What has just been said of concepts might equally well have been formulated with 'expression' in place of 'concept.' This is not to suggest that I take concepts and expressions to be identical, or that I see the use of a concept to be identical to the use of an expression. Rather, it is only to say that, for my purposes here, I regard the two as largely interchangeable. Accordingly, I'll often write 'concept' or 'use of concept' where I might more accurately, though also more cumbersomely, write 'concept or formal expression' or 'use of concept or use of formal expression.'

5.3 The RRT

According to the RRT, replacing the instantiation requirement with a consistency requirement is supposed to make for greater freedom in the justified use of concepts.[23]

Cantor took his famous claim that the essence of mathematics "lies precisely in its freedom" Cantor (Cantor 1883, Sect. 8) to be an expression of this view. In mathematics, as distinct from all other sciences, Cantor claimed, concepts do not have to have transient existence—that is, they do not have to apply to the objects of some realm of being independent of pure thought.

[22] To "apply" an axiom or rule of inference in a proof is essentially to incorporate it as an element of that proof.

There are of course different forms that such "incorporation" might take. These include incorporation as a premisory assertion, as a conclusory assertion, as an hypothesis, as an inferential instrument, and so on. Clarifying the different forms that incorporation might take is of course an important and unfinished task in our efforts to attain a better understanding of what proofs are and how they work. As I see it, though, neither the correctness nor the interest of what I have to say here depends in any substantial way on the particulars of such a clarification. I'll therefore not pursue the matter further here.

[23] A common phrase for free introduction and use of concepts among German writers was "freie Begriffsbildung" (often translated as "free concept formation"). Sometimes 'freie' or similar adjectives were use to modify such other phrases as "Einführung eines Begriffes" which seems better rendered as "concept-introduction" than as "concept-formation."

Rather, in mathematics, concepts need have only immanent existence—that is, their use needs only to be consistent. It need not be applicable to extra-subjective objects.

> This feature distinguishes mathematics from all other sciences and provides an explanation of the relatively easy and unconstrained manner with which one may operate with it. It is why it deserves the name 'free mathematics'. Mathematics is in its development entirely free and is bound only in the self-evident respect that its concepts must be consistent with each other ... Cantor (1883, Sect. 8)

Various other 19th and 20th century foundational writers made similar or related claims.[24]

The idea that replacing the instantiation condition by a consistency requirement should make for freer justified use of concepts in mathematics (i.e. the RRT) was thus common among foundational writers of the late 19th and early 20th centuries.

There were also those who disagreed, of course, and it is their views that I now want to consider. For the most part, those who have challenged the idea that replacing instantiation with consistency makes for greater freedom in the justified use of concepts have done so by arguing that there is no significant *practical* difference between proving consistency and proving satisfaction of the instantiation condition (i.e. proving the existence of what is evidently a model). They have not argued that there is no theoretical difference or difference in principle between establishing consistency and establishing satisfaction of the instantiation condition.

This was a widely held view in the late 19th and early 20th centuries, though it was seldom supported by argument. Frege *did* give an argument for it, though, and it is to his argument that I now turn.

5.4 Frege's Futility Argument

Frege stated this argument in a number of places (cf. Frege (1884, Sect. 95) and Frege's letter to Hilbert of January 6, 1900 [(Frege 1976, p.71, 75)]. Perhaps the clearest statement was the following, from the second volume of the *Grundgesetze*.

> [T]he power to create (*schöpferische Macht*) is ... restricted by the proviso that the properties must not be mutually inconsistent.[25] But how does one know that the properties do not contradict one another? There seems to be no criterion for this except the occurrence of the

---

[24] Cf. Peacocke (1830, Sect. 78), Hankel (1869, p. 20), Dedekind (1888, p. 335, 338), Durège (1896, p. 8–10), Burkhardt (1897, p. 2), Young (1911, p. 52–53), Keyser (1915, p. 679–680), Hilbert (1920, p. 18–20), Menger (1937, p. 333–334), Courant and Robbins (1981, p. 88–89).

[25] Frege seems to have seen a difference between the conditions of legitimate introduction of concepts and the conditions of legitimate general use of concepts. He suggested that consistency is a legitimate condition on the latter though not on the former.

> A concept is still admissible (*zulässig*) even though its defining characteristics (*Merkmale*) do contain a contradiction: all that we are forbidden to do, is to presuppose that something falls under it. From the assumption that a concept contains no contradiction, however, we cannot

> properties in question in one and the same object. But the creative power with which many mathematicians credit themselves then becomes practically worthless. For as it is they must certainly prove, before they perform a creative act, that there is no inconsistency between the properties they want to assign to the objects that are to be, or have already been, constructed; and apparently they can do this only by proving that there is an object with all these properties together. But if they can do that, they need not first create (*schaffen*) such an object.
>
> (Frege 1903, Sect. 143)

The objection to the RRT I find suggested here centers on the following two claims: (I.a.) the only practicable way to prove the consistency of a theory is to construct a model for it, and (I.b.) to construct a model for a theory is, in effect, to satisfy the instantiation constraint with respect to it. Both claims are worth considering more carefully. Since, however, it is (I.a.) rather than (I.b.) that Hilbert's "direct" approach to consistency was intended to challenge, it will be my focus here.[26,27]

As Frege rightly sensed, Hilbert did have an alternative to model-construction in mind as a means of proving consistency. Less clear, though, is what reason he may have had for thinking that application of such an alternative might succeed where application of a model-theoretic approach to consistency would not—namely, in securing the freedom taken to be represented by general adoption of an abstract conception of axiomatization. It is this question that I now want to consider.

(Footnote 25 continued)

> infer that for that reason something falls under it. If such concepts were not admissible, how could we generally prove that a concept does not contain a contradiction?
>
> (Frege 1884, Sect. 94)

In his final question, Frege suggests that the need to prove the consistency of the characteristic properties of a concept would not be a sensible requirement were it not possible for inconsistent concepts (or sets of concepts) to be in some meaningful sense legitimately "introduced." He thus seems to have relied on a distinction between a concept's being in some sense "admissible" into our thinking and its doing what concepts ordinarily do, namely, comprehend instances.

[26] I see I.a. and I.b. as separable components of thesis I., presented in the introductory section.

[27] I say that the objection is *suggested by* Frege's argument, not that it was, in all respects, his actual objection. I make this qualification because what I describe as Frege's objection is a reconstruction of what he says in his argument. This reconstruction, being formulated in terms of the contemporary notion of a model as it is, makes use of ideas that were not strictly speaking part of Frege's conceptual arsenal.

Specifically, it renders his claim that "There seems to be no criterion for this [i.e. for consistency, MD] except the occurrence of the properties in question in one and the same object" as "there seems to be no criterion for consistency except the construction of a model for the properties (axioms) in question." For my purposes, this appeal to the contemporary notion of a model is satisfactory. It is unlikely, though, that it would have figured in any explicit way in Frege's thinking.

This said, I would also note, though, that Frege described the method he had in mind as the one Hilbert used in (Hilbert 1899). That method, as I understand it, is one of model-construction. It may therefore be that the method of proving consistency that Frege referred to was, as a matter of fact, a method which may with some right be described as a method of model-construction—and this despite the fact that he himself might not have described it, or even thought of it in that way.

5.5 Hilbert's 'Direct' Approach to Consistency

Perhaps the first thing to note in this connection is that Hibert believed that establishing consistency always required proof.

> [P]rior to our specifically proving the consistency of our axioms, we can never be certain of it.
> Hilbert (1922, p. 161)

He thus rejected the idea of Peano, Frege and others that, when axioms truly are axioms, their evidentness will obviate the need for a proof of their consistency, and may, in fact, make genuine proof of consistency in some sense impossible. As Hilbert saw it, there is a general

> *Need for Proof*: The full resolution of a consistency problem for an axiomatic system generally requires development of a proof that does not depend on appeal to the evidentness of the axioms of that system.

Why Hilbert should have committed himself to Need for Proof is perhaps less clear. One likely reason was his support of abstract axiomatization as the general model for a scientific mathematics. Another reason may have been a general concern like that described by Dedekind in (Dedekind 1888)—namely, that the sheer complexity of the set of proofs which belongs to an axiomatic system makes the risk of failing to discern an inconsistency that is actually there non-negligible.[28]

The Need for Proof thesis does not in itself, of course, make a case for taking a direct approach to consistency, not generally, and not even in the case of arithmetic. More is thus required to motivate Hilbert's call for a direct proof in the case of arithmetic.

One reason was the use that he and others had made of interpretation in arithmetic as a means of solving the consistency problems for various non-arithmetical theories. To anchor these proofs (i.e. to provide for the eventual "detachment" of their conclusions), a non-relativized proof of the consistency of arithmetic was needed.[29]

Hilbert thus intended that his 'direct' proof of the consistency of arithmetic be direct in the sense of not relying on an interpretation of arithmetic in other terms.

---

[28] Dedekind made this observation in his well-known letter to Keferstein:

> After the essential nature of the simply infinite system, whose abstract type is the number sequence N, had been recognized in my analysis (articles 71 and 73), the question arose: does such a system exist at all in the realm of our ideas? Without a logical proof of existence it would always remain doubtful whether the notion of such a system might not perhaps contain internal contradictions. Hence the need for such proofs (articles 66 and 72 of my essay).
>
> Heijenoort (1967, p. 101)

Dedekind limited his remark to his theory of simply infinite systems. The point he raises, though, seems to generalize. It would not be unreasonable to think that such a concern may have been part of what led Hilbert to his Need for Proof thesis too.

[29] By a non-relativized consistency proof I do not mean of course a proof which does not rely on premises. Such a proof would be impossible.

> In geometry, the proof of the consistency of the axioms can be effected by constructing a suitable field of numbers, such that analogous relations between the numbers of this field correspond to the geometrical axioms. ... In this way the desired proof for the consistency of the geometrical axioms is made to depend upon the theorem of the consistency of the arithmetical axioms.
>
> On the other hand a direct method is needed for the proof of the consistency of the arithmetical axioms.
>
> Hilbert (1900, pp. 264–265)[30]

Hilbert's call for a direct proof of the consistency of arithmetic was often met with skepticism. He had not articulated the direct approach to a point where it could be seen to offer a workable alternative to model-construction as a practical means of proving consistency. In the view of some, then, Hilbert's announcement of his so-called direct approach to the consistency of arithmetic was largely the raising of a theoretical possibility. It did not change the fact that the only known, generally workable means of proving consistency at that time was model-construction. J.W. Young put the point this way:

> The only test for the consistency of a body of propositions is that which connects with the abstract theory a concrete representation of it. We are dealing here with a collection of symbols. If we can give them a concrete interpretation which satisfies, or appears to satisfy, all our assumptions, then every conclusion that we derive formally from those assumptions will have to be a true statement concerning this concrete interpretation.
>
> (Young 1911, p. 43)[31]

Hilbert was aware of the popularity of such views, and he seems to have seen logicism as representing a kind of final hope for them. One could perhaps use model-construction to solve even the consistency problem for arithmetic if one could find an interpretation of arithmetic in logic.

Hilbert took this possibility seriously, but, in the end, rejected it. As he saw it, even logical laws ultimately require appeal to certain rudimentary forms of arithmetical or proto-arithmetical evidence for their justification [cf. Hilbert (1905)]. This being so, the evidence needed to establish a logical interpretation of arithmetic as a model of arithmetic would lack the type of independence from arithmetical evidence that a proof of the consistency of arithmetic ought properly to have.

---

[30] Some twenty years later, Hilbert mentioned that the task of securing the unsecured relative consistency proofs for various non-arithemetical areas of mathematics remained unresolved: "The proof of the consistency of axioms succeeds in many cases, for example, in geometry, in thermodynamics, in the theory of radiation, and in other physical disciplines, by taking it back to the question of the consistency of the axioms of analysis; this question is in its turn an as yet unsolved problem." [Hilbert (1922, p. 16)]. Others made similar observations earlier [cf. Moore (1903, p. 405)].

[31] Veblen and Young gave a similar statement in their well-known text on projective geometry. They wrote: "[I]n general, *a set of assumptions is said to be consistent if a single concrete representation of the assumptions can be given.* Veblen and Young (1910, p. 3), emphasis in text]. See Brown (1906, p. 530), (1908, p. 629) and Coolidge (1909, 72) for similar statements. Veblen and Young went on to note what they took to be a concerning feature of such proofs—namely, that "they merely shift the difficulty from one domain to another" (*loc. cit.*). Others expressed similar and, in some cases, even stronger reservations [cf. Weiss (1929, p. 468), Nagel (1929, pp. 484–485), Emch (1936, pp. 185–186)].

As Hilbert saw it, logic and arithmetic were intertwined in such a way as to make a "simultaneous" development of their foundations necessary. He granted that arithmetical evidence may often depend upon logical evidence for its acceptance. He believed as well, though, that logical evidence sometimes depends for its acceptance on arithmetical evidence.

> One often designates arithmetic to be a part of logic, and what have traditionally been regarded as logical notions are usually presupposed (voraussetzen) when it comes to establishing a foundation for arithmetic. Careful attention reveals, though, that in the traditional representation (Darstellung) of the laws of logic certain fundamental concepts of arithmetic are already used, for example, the notion of set and, to some extent, also that of number, especially as cardinal number (Anzahl). We therefore find ourselves turning in a circle (Zwickmühle), and that is why a partly simultaneous development (teilweise gleichzeitige Entwicklung) of the laws of logic and of arithmetic is required if paradoxes are to be avoided.
>
> Hilbert (1905, pp. 245–246)[32]

A purely logical foundation for arithmetic was thus not, in Hilbert's view, a genuine possibility. What had been thought by some to be an independent and appropriately more basic theory to which to turn for a model of arithmetic—namely, logic—in fact was not. Hilbert thus concluded that

> the usual method of suitable specialization or construction of examples (übliche Methode der geeigneten Spezialisierung oder Bildung von Beispielen),[33] which is otherwise customary for such proofs [i.e. proofs of consistency]—in geometry, in particular—necessarily fails here [that is, in the case of arithmetic].
>
> Hilbert (1905, 252) (square brackets and their contents added)

Later, Bernays and he developed further reasons for rejecting model-construction as a means of proving the consistency of arithmetic. This was perhaps most clearly spelled out in what Bernays described as an updated description of Hilbert's "new methodological approach to the foundation of arithmetic" [Bernays (Bernays 1922, 10)].

In this chapter, Bernays set out to sharpen and clarify the "quite indistinct" (recht dunkeln, loc. cit.) description of the "direct" approach given in Hilbert's 1900 Paris address and in his 1904 Heidelberg lecture [upon which (Hilbert 1905) was based]. One particular point of emphasis concerned the supposedly rudimentary intuitive character of the evidence upon which a direct proof of the consistency of arithmetic was to be based.

Bernays remarked that in pursuing this type of cognitive elementarity in the knowledge upon which a direct proof of consistency was to be based, Hilbert was following a broader trend in the foundations of the exact sciences—namely, to

[32] That logic is in some sense properly seen as depending on what is in effect arithmetical evidence for its justification was an enduring theme of Hilbert's foundational thinking. For a relatively late statement to this effect, see Hilbert (1928, pp. 1–2).

[33] 'Method of suitable specialization' and 'construction of examples' were Hilbert's terms for what I have been referring to as 'instantiation' or 'model-construction.'

> exclude, to the fullest extent possible, the finer organs of knowledge (die feineren Organe der Erkenntnis), and to use only the most primitive means of knowledge (die primitivsten Erkenntnismittel)
>
> Bernays (1922, 11)

in what is counted as basic evidence.

If Bernays was right, then, Hilbert accepted some such principle as this:

> *Foundational Primitivity*: The evidence on which a foundation for arithmetic ought properly to be based is evidence of the epistemologically most primitive type.

Together with Hilbert's belief that

> *Centrality of Consistency*: A consistency proof for arithmetic is the central element of a proper foundation for it,

Foundational Primitivity implies

> *Primitive Proof of Consistency*: The evidence on which a proof of the consistency of arithmetic ought properly to be based is evidence of the epistemologically most primitive type.

If to this we further add the substantive claim that

> *Intuitive Character*: The epistemologically most primitive evidence is evidence of a rudimentary intuitive character,[34]

we may then conclude that

> *Intuitive Proof of Consistency*: The evidence on which a proof of the consistency of arithmetic ought properly to be based is evidence of a rudimentary intuitive character.

I have taken this plodding route to Intuitive Proof of Consistency for two reasons. The first is that it reveals the deep influence of a normative element—Foundational Primitivity—in Hilbert's finitary outlook. This seems not to have been widely recognized.

The centrality of this normative element to the argument for Intuitive Proof of Consistency indicates that the search for a finitary proof of consistency was not motivated solely or even primarily out of consideration for the reliability of finitary evidence. Rather, it was motivated more basically out of consideration of its (i.e.

[34] Bernays stated that the foundational challenge for arithmetic is to determine "whether it is possible to ground (begründen) …[the] transcendent acceptances (transzendenten Annahmen)" (Bernays 1922, p. 11) of arithmetic in such a way that "only *primitive intuitive cognitions are appealed to* (*primitive anschauliche Erkenntnisse zur Anwendung kommen*)" (loc. cit., emphases in text; square brackets and their contents added). If this is an accurate statement of Hilbert's views, and I think it is, then he took the epistemologically most primitive evidence to be evidence of a rudimentary intuitive type.

finitary evidence's) presumed epistemological primitivity. Even more basically, it was motivated by a general foundational norm which requires that a purported foundation for a body of knowledge trace that knowledge back to primitive epistemological sources.

The second reason for developing Intuitive Proof of Consistency in the deliberate way I have is to prepare the way for what Hilbert and Bernays evidently saw as a key difference between a direct approach to the consistency problem for arithmetic and a model-construction approach to it—namely, that the latter lacks the primitive intuitive character required by the combination of Foundational Primitivity and Intuitive Character.

Bernays put the point this way:

> Appeal to an intuitive grasp of the number series or of a manifold of magnitudes ... can not be a question of an intuition in the primitive sense, for certainly no infinite manifolds are given to us in the primitive intuitive mode of representation (*in der primitiven anschaulichen Vorstellungsweise ... gegeben*).
>
> Bernays (1922, p. 11)

For our purposes, the central point of relevance here is the suggestion that

Non-Primitivity of Model Recognition: For any would-be model $\mathcal{M}$ of arithmetic, knowledge that $\mathcal{M}$ is a model of arithmetic can not properly be based on evidence of a primitive intuitive type.

This seems to be significant. In particular it makes possible the elimination of model-construction as a normatively appropriate means of establishing the consistency of arithmetic. More precisely, from Non-Primitivity of Model Recognition and Intuitive Proof of Consistency it follows that a proof of the consistency of arithmetic *ought not* to be based on the construction of a model for it.[35] To the extent that Non-Primitivity of Model Recognition and Intuitive Proof of Consistency are accepted, then, an alternative to model-construction as a means of proving the consistency of arithmetic becomes urgent.

Before passing to our final concern—the possible role of freedom as a motive for the direct approach—I want briefly to consider another motive which seems generally to have been overlooked. This concerns the linkage between the consistency or inconsistency of a body of reasoning and the rules or norms which regulate its conduct.

The line of thinking here begins with something like the following premise: (i) the consistency or inconsistency of a body of reasoning is determined by (a) the rules which are taken to govern its conduct and by (b) whether these rules are adhered to in its actual conduct. From this it is then inferred that (ii) the consistency or inconsistency of a body of reasoning ought to be discernible from a sufficiently thorough examination and analysis of the rules taken to govern its conduct and sufficiently careful observation concerning whether these rules have been adhered

---

[35] This may hold for other cases too—perhaps, for example for, any theories all of whose models are infinite. Bernays did not elaborate on why he believed intuition of the number series or of a manifold of magnitudes not to qualify as primitive intuition in his sense.

to. In my view, this is a basic line of motivational thinking not only for Hilbert's direct approach to consistency problems, but for proof theory generally.

Noticing this, various foundational writers of the early 20th century remarked the relative unsatisfactoriness of adopting an intepretational rather than a direct approach to consistency. Young (1911, p. 43f), for example, remarked the unsatisfactoriness of taking an interpretational approach to the consistency and independence problems for abstract geometries. In his view, it amounted to answering the fundamental question (namely, the consistency question)

> only by reference to a concrete representation of the abstract ideas involved, and it is such concrete representations that we wished especially to avoid.[36] At the present time, however, no absolute test for consistency is known.

Young (Young 1911, pp. 43–44)

Interpretational approaches to consistency problems for abstract theories thus relied on the very interpretations which advocates of abstract axiomatization tried to separate axiomatic reasoning from.

The philosopher Ralph Eaton raised the more general type of concern I'm trying to highlight here—namely, the connection of methods of solving the consistency problem for a body of reasoning with the observation and analysis of the conduct of that reasoning.

> The proofs of independence, consistency and completeness usually employed in connection with sets of postulates are theoretically unsatisfactory; they involve an appeal to interpretations. ... There should be some analytical way—purely in the realm of the abstract, without interpretation—of establishing these properties of a set of postulates. This is an important problem that awaits solution.

Eaton (1931, p. 474 (footnote 2))

Eaton seems right to me. The consistency or inconsistency of a body of reasoning ought ultimately to depend on how that reasoning is conducted. That is, it ought ultimately to depend on the principles or rules in accordance with which we try to conduct it, and on how successful our attempts to conduct it in accordance with those principles/rules really are. This applies as much to abstract axiomatic proving as to any other type of reasoning.

The demands of rigor—generally speaking, the demands of full disclosure of all judgments used to establish that something is a proof in a given axiomatic system—imply a separation of the criteria of proof from considerations of contents. We specify axioms and rules by *exhibiting* them, not by *expressing* them (i.e. by giving their semantical contents). To put it another way, we specify them according to their external appearances (i.e., their broadly syntactical characteristics), not their semantical contents. We likewise develop similarly external or non-contentual standards for their use or application.

Reasoning that is conducted according to such standards can be observed. Once such reasoning is embedded in a suitably idealized framework, it should be possible to address such questions as its syntactical consistency.

---

[36] Wished to especially avoid, that is, according to the abstract conception of axiomatization that Young was contrasting to the traditional conception of axiomatization.

What has changed since Eaton's writing, of course, is our estimation of what is required for such analysis and our estimation of how significant the variation of what is required may be in moving from one theory to another. In this connection, it is worth noting that even in his later writings, Hilbert emphasized what he saw as an important "uniformity" of the direct approach which he took to set it apart from interpretational approaches to consistency.

> Formal axiomatization ... depends on evidence for the conduct of deductions and for the proof of consistency. There is, however, an essential difference—namely, that the type of evidence in this case does not depend on a special epistemological relationship to any specific field. Rather, it is one and the same for every axiomatization. Specifically, it is that primitive kind of knowledge that is the precondition for every exact theoretical investigation whatever.
>
> Hilbert and Bernays (1934, Sect. 1, 2)[37]

5.6 Freedom and the Direct Approach

We come finally to the question of freedom, and its possible role as a motive for proof theory. Here we should first recall a point made ealier—namely, that freedom, and in particular what I am calling conceptual freedom, was a clear motive for advocating the abstract conception of axiomatization.

Hilbert took abstract axiomatization to offer "the maximum freedom of movement (die vollste Bewegungsfreiheit) ..." Hilbert (1922, p. 201) in our thinking. That it should do so was in his view largely due to the fact that it required only consistency rather than truth in its theories.

That this was so was in turn due to the fact that it (i.e. abstract axiomatization) represented what Hilbert and Bernays described as a modern tendency in science—a tendency, specifically, towards idealization and away from sheer description or the capturing of (or confirmation by) contents presumed to in some sense be given "prior to" a given theory.

> In science we are predominantly if not always concerned with theories that are not completely given to representing reality, but whose significance (Bedeutung) consists in the *simplifying idealization* (*vereinfachende Idealisierung*) they offer of reality. This idealization results from the extrapolation by which the concept formations (Begriffsbildungen) and basic laws (Grundsätze) of the theory go beyond (überschreitet) the realm of experiential data (Erfahrungsdaten) and intuitive evidence (anschauliche Evidenz).
>
> Hilbert and Bernays (1934, pp. 2–3), emphases in original

Abstract axiomatization, though, represented more than just a tendency towards idealization and away from descriptive adequacy as the controlling condition on the acceptability of theories. It represented an active attempt to *separate* axiomatic reasoning from contents.

---

[37] How significant the supposed uniformity mentioned by Hilbert and Bernays may be is unclear to me. Is there generally less variation or less significant variation between proof-theoretic proofs of consistency for different theories than for model-construction proofs of consistency for different theories? This seems to be a difficult question and one about which not much is known.

Such separation was judged to be necessary for the attainment of rigor in the logical or purely deductive reasoning that is seemingly so important to proof. Hilbert remarked that on his conception of axiomatization, contentual inference was to be replaced by "externally guided manipulation of signs according to rules (äußeres Handeln nach Regeln)" (Hilbert 1928, p. 4). In this, of course, he was following a line of development begun by Pasch. In Pasch's view, if geometrical proof was to become truly deductive or rigorous, it needed to be separated from the senses and referents of geometrical terms. That is, it was to be conducted according to a standard which made it possible to confirm the validity of individual steps of reasoning in geometrical proofs in complete abstraction from the meanings of the geometrical terms involved. Were this not possible, Pasch said, the rigor (or genuinely deductive character) of the reasoning ought to be held in doubt.[38]

If I am not mistaken, Hilbert also took separation from contents to be necessary for another type of rigor—what we might call declarative or specificational rigor. As I am thinking of it, this is rigor (in the form of full disclosure) in the specification or identification of axioms and rules of inference of a system. Since this is a place where, at least potentially, undisclosed use of information might enter a proof process, it is a place where the development of reasonable counter-measures is appropriate.

This is the type of concern I take Hilbert to have had in mind when, in the remark cited above from Hilbert (1928, p. 4), he said his standards of proof required that proofs should be distinguishable from each other and from non-proofs by externally evident (hence, non-contentual) characteristics. I think it may also be what he had in mind when he emphasized that proper axiomatization should provide an "exact" (genaue) and "complete" (vollständige) specification of the axioms and rules of inference of a given system.[39]

Here there seems to be at least tacit use of some distinction between specification of axioms (or rules) by *expression* (i.e. expression of contents) versus specification of axioms by something like *exhibition*. The idea, if I am right, is that exact and complete specification cannot rely on interpretion or association of contents. An axiom cannot be exactly specified by giving a sentence that expresses it—that is, by giving a sentence which, interpreted in a certain way, semantically signifies it. Rather, it can be exactly and completely specified only by giving a sentence which *is* it, or more accurately, a sentence which *exhibits* or externally exemplifies it.[40]

The thinking is that if the axioms of an abstract science are to admit of exact and complete specification in a sense like that just described, they cannot be propositions. Propositions can, after all, only be expressed by—that is, can only be interpretationally associated with—sentences. They are not exhibited by them. Or, to put it another way, they cannot be obtained by processes of "copying" (i.e. simulating certain exter-

---

[38] Cf. Pasch (1882, p. 98).

[39] Cf. Hilbert (1899, Chap. 1, Sect. 1), Hilbert (1900, p. 181).

[40] Here by exhibition or external exemplification, I mean the giving of a concrete expression which serves as a type of exemplar for other expressions—an exemplar such that concrete expressions whose external features are sufficiently similar to those of the exemplar qualify as tokens of the same axiom it betokens.

nal characteristics of) them. This being so, to express an axiom as a content is not to give it in a way that allows it to be identified by external characteristics. Rather, it is to rely on the application of interpretive processes for its identification. If, therefore, the axioms of an axiomatic theory are to be exactly and completely specifiable, they must be sentences or formulae, not propositions or other semantical contents.

If this is essentially right, then, abstract axiomatization is to be seen as aiming at the separation of contents from proof not only for purposes of deduction, but also for purposes of specification. The question we must now consider is whether separation from contents is likewise necessary for purposes of establishing the consistency of abstract axiomatic theories.

If the reasoning set out in Sect. 5.5 is correct, then the answer would seem to be an at least qualified "yes." By Intuitive Proof of Consistency, Non-Primitivity of Model Recognition, Centrality of Consistency and Foundational Primitivity, a proof of the consistency of arithmetic cannot contribute to the foundation of arithmetic in the way(s) expected or desired if it depends upon the recognition of a model for it. There is thus a need for at least this type of separation of a consistency proof for arithmetic from contents. This may not be so compelling or so radical a call for separation as the call for inferential or specificational separation, but it is roughly as compelling as the combination of the principles noted above is.

Preservation of whatever conceptual freedom may be natural to abstract axiomatization may also favor separation from contents for purposes of proving consistency. On the abstract conception, an axiomatization represents a *denken*, and the freedom represented by a *denken* may be expected to decrease if it is required to be associatable with specific contents for purposes of proving its consistency. *Denken* constrained in this way would thus seem, at least generally, to be more constrained, hence less free, than abstract Hilbertian *denken*.

Frege was therefore right to emphasize that to make provision of the usual sort of consistency proof (i.e. proof by model-construction) a condition for the legitimate use of an axiomatic theory was in effect to compromise any increase in conceptual freedom that might otherwise be represented by the shift to abstract axiomatization. In the worst case, it might transform what would otherwise be an abstract or formal (formale) axiomatization into what is effectively a contentual (inhaltlich) axiomatization.

By offering an alternative route to consistency—a route not involving the association of specific contents with the axioms of an axiomatic theory—Hilbert's direct approach to consistency thus offered a possible hedge against compromising whatever gains in conceptual freedom might generally be represented by the shift from the traditional to the abstract conception of axiomatization.

5.7 Conclusion

A possible hedge, yes. A certain hedge, no. The need for any proof of consistency, whether direct or by model-construction, promises to constrain the legitimate freedom represented by abstract axiomatization. Whether solving the consistency problem for arithmetic by means of model-construction would be overall more confining to axiomatic thinking in arithmetic than would be solving it by direct proof is much more difficult to say.

What I hope to have made clearer here is that the elements of Hilbert's foundational program which seem most to have favored a direct approach to the consistency problem for arithmetic are the normative principle Foundational Primitivity and the empirical or quasi-empirical claims Intuitive Character and Non-Primitivity of Model Recognition.

This is not at all to suggest that concern to maximize conceptual freedom was not among Hilbert's reasons for pursuing a direct approach to the consistency of arithmetic. I believe it was. As such a reason, though, it does not seem to have been either so logically or so strategically central as the considerations just mentioned.

The possibility of a direct approach to the consistency problem for arithmetic did offer a pleasingly direct and substantial response to Frege's Futility Argument. This, however, was due to the exposure to which Frege's advocacy of I. (or I.a.) opened him. It was not because Hilbert offered convincing reasons to think that a direct consistency proof for arithmetic would be less confining to axiomatic thinking in arithmetic than would a model-theoretic proof of its consistency.

Neither, in retrospect, does the exposure which results from commitment to I. (or I.a.) seem to be so great as it would likely have seemed to Hilbert in the early years of the 20th century. With the benefit of hindsight, we can see that the direct approach to consistency has not been developed in such ways as would make it a major practical competitor with model-construction as a means of proving consistency.[41]

More important than any of this, I think, are the questions which the exchange between Frege and Hilbert raise for us today. These include such larger questions as the continuing significance of the distinction between abstract and traditional conceptions of axiomatization.

They include as well such more particular problems as what the conceptual freedom represented by an abstract axiomatic theory might plausibly and illuminatingly be taken to consist in, under what conditions its exercise may be justified and how significant a factor its legitimate exercise has been as a force shaping the historical development of mathematics.

My hope is to have provided a starting point for renewed thinking about such questions.

Acknowledgments Thanks to the *Agence nationale de la recherche* (ANR) of France for their generous support of my research under their *chaires d'excellence* program. Thanks too to the HPS group at the Université de Paris 7–Diderot, the Philosophy Department and the Archives Henri Poincaré at the Université de Lorraine, the chairs for epistemology and the philosophy of language at the Collège de France and the University of Notre Dame for their support as well.

---

[41] The reasons for this, though, are complicated and may in part be due to consistency problems having become an overall less prominent part of the larger landscape of mathematical concern than they were in Hilbert's day.

For useful discussions of various parts of the material in this chapter, I am grateful to audiences at the Ohio State University (Foundational Adventures Conference), SPHERE (Paris-Diderot), the Université de Lorraine, the IHPST (Paris 1), the ENS, the Collège de France, the Universidade de Coimbra (Gentzen Centenary Conference), Cambridge University, the IREM (Paris-Diderot), Carnegie Mellon University, the Georg-August-Universität Göttingen (Lichtenberg-Kolleg), McGill University, the University of Montreal, the Université de Toulouse Paul Sabatier and the University of Virginia.

References

Bernays, P. (1922). *Über Hilbert's Gedanken zur Grundlegung der Arithmetik* (pp. 10–19). Jahresberichte der deutschen Mathematiker-Verienigung 30: . English translation in [46].

Borga, M., & Palladino, D. (eds. and trans.) (1992). Logic and foundations of mathematics in Peano's school. *Modern Logic*, *3*, 18–44.

Brown, H. C. (1906). Review of [54] Journal of Philosophy. *Psychology and Scientific Method*, *3*, 530–531.

Brown, H. C. (1908). Infinity and the Generalization of the Concept of Number. *Journal of Philosophy, Psychology and Scientific Methods*, *5*, 628–634.

Burkhardt, H. (1897). *Einführung in die Theorie der analytischen Functionen einer complexen Veränderlichen*. Leipzig: Verlag Veit & Comp.

Cantor, G. (1883). Ueber unendliche, lineare Punktmannichfaltigkeiten. 5. Fortsetzung. *Mathematische Annalen21*, 545–591. Page references are to the reprinting in [7].

Cantor, G. (1883). *Grundlagen einer allgemeinen Mannigfalltigkeitslehre: ein mathematisch-philosophischer Versuch in der Lehre des Unendlichen*. Leipzig: Teubner.

Coolidge, J. (1909). *The Elements of Non-Euclidean Geometry*. Oxford: Clarendon.

Courant, R., & H, Robbins. (1981). *What is Mathematics?*. Oxford: Oxford University Press.

Dedekind, R. (1888). *Was sind und was sollen die Zahlen?*, Braunschweig: Vieweg, Reprinted in [11]. Page references are to this reprinting.

Dedekind, R. (1932). In Fricke, R., Noether, E., & Ore, O. (Eds.), *Gesammelte mathematische Werke*. Braunschweig: F. Vieweg & Sohn.

Detlefsen, M. (2005). Oxford Handbook of Philosophy of Mathematics and Logic. In S. Shapiro (Ed.), *Formalism* Chap. 8. (pp. 236–317). Oxford: Oxford University.

Durège, H. (1896). *Elements of the theory of functions of a complex variable, with especial reference to the methods of Riemann*. Authorized English trans. from the fourth German edition. G. Fisher & I. Schwatt, Philadelphia.

Eaton, R. (1931). *General logic: an introductory survey*. New York: C. Scribner's Sons.

Emch, A. (1936). Consistency and Independence in Postulational Technique. *Philosophy of Science*, *3*, 185–196.

Ewald, W. (Ed.). (1996). *From kant to hilbert: a source book in the foundations of mathematics* (Vol. 2). Oxford: Oxford University Press.

Frege, G. (1884). *Die Grundlagen der Arithmetik. Eine logisch mathematische Untersuchung über den Begriff der Zahl*. In W. Koebner, Breslau. English translation in [18].

Frege, G. (1953). *The foundations of arithmetic; a logico-mathematical enquiry into the concept of number*. In G. L. Austin (trans.), Blackwell, Oxford.

Frege, G. (1899). Letter to Hilbert, December 27. In [24]. English trans. in [25].

Frege, G. (1900a). Letter to Hilbert of January 6, (1900). In [24]. English trans. in [25].

Frege, G. (1990b). Letter to Hilbert, September 16. In [24]. English trans. in [25].

Frege, G. (1893). *Grundgesetze der Arithmetik: begriffsschriftlich abgeleitet* (Vol. I). Jena: H. Pohle.

Frege, G. (1903). *Grundgesetze der Arithmetik: begriffsschriftlich abgeleitet* (Vol. II). Jena: H. Pohle.

Frege, G. (1976). In Gabriel, G., et al. (Eds.), *Wissenschaftlicher briefwechsel*. Hamburg: Felix Meiner Verlag.
Frege, G. (1980). In Gabriel, G., et al. (Eds.), *Gottlob Frege: Philosophical and Mathematical Correspondence*. Chicago: University of Chicago Press.
Hankel, H. (1869). *Die Entwicklung der Mathematik in den letzten Jahrhunderten*. Tübingen: L. Fr. Fues'sche Sortimentsbuchhandlung.
van Heijenoort, J. (Ed.). (1967). *From Frege to Gödel: A Source Book in Mathematical Logic, 1879–1931*. Cambridge: Harvard University Press.
Hilbert, D. (1899). *Die Grundlagen der Geometrie, Published in Festschrift zur Feier der Enthullung des Gauss-Weber Denkmals in Göttingen*. Leipzig: Teubner.
Hilbert, D. (1899). Letter to Frege, December 29, 1899. In [24]. English translation in [25].
Hilbert, D. (1900). Mathematische probleme, *Göttinger Nachrichten*. 253–297.
Hilbert, D. (1900). Über den Zahlbegriff. *Jahresbericht der Deutschen Mathematiker-Vereinigung*, *8*, 180–194.
Hilbert, D. (1905). Über die Grundlagen der Logik und der Arithmetik, *Verhandlungen des Dritten Internationalen Mathematiker-Kongresses in Heidelberg vom 8. bis 13.* August 1904, Teubner, Leipzig. Page references are to the reprinting as appendix VII in [33].
Hilbert, D. (1913). *Grundlagen der Geometrie* (4th ed.). Leipzig: Teubner.
Hilbert, D. (1920). Probleme der mathematischen Logik. Lecture notes (as recorded by Schönfinkel and Bernays), summer semester 1920. Page reference is to partial translation of this in [16].
Hilbert, D. (1922). Neubegründung der Mathematik. Erste Mitteilung, *Abhandlungen aus dem Mathematischen Seminar der Hamburgischen Universität 1*. 157–77. Page references are to the reprinting in [41], vol. III.
Hilbert, D. (1922). *The New Grounding of Mathematics-Part I*. English trans. by P. Mancosu of [35], in [46].
Hilbert, D. (1926). Über das Unendliche. *Mathematische Annalen*, *95*, 161–190.
Hilbert, D. (1928). Die Grundlagen der mathematik, Abhandlungen aus dem mathematischen Seminar der Hamburgischen Universität, *6*, 5–85. Reprinted in Hamburger Einzelschriften 5, pp. 1–21. Teubner, Leipzig. Page references are to the latter.
Hilbert, D. (1930). *Naturerkennen, und Logik: Königsberg*, Springer Verlag, Berlin,. (1971). *Printed in [41]*, vol. III: Page references are to this printing.
Hilbert, D., & Bernays, P. (1934). *Grundlagen der Mathematik*. Berlin: Springer Verlag. Page references are to [42], the second edition of this text.
Hilbert, D. (1932–1935). *Gesammelte Abhandlungen*, three volumes, Chelsea Pub. Co., Bronx, NY.
Hilbert, D., & Bernays, P. (1968). *Grundlagen der Mathematik* (2nd ed.). Berlin: Springer.
Kant, I. (1787). *Kritik der reinen Vernunft* (2nd ed.). Riga: J. F. Hartknoch.
Keyser, C. (1915). *Science (new series)*, *42*(1089), 663–680.
Leslie, J. (1821). *Geometrical analysis, and geometry of curve lines*. Longman, Hurst, Rees, Orme, & Brown, London.
Mancosu, P. (ed. and trans.) (1998). *From Brouwer to Hilbert: The debate on the foundations of mathematics in the 1920s*, Oxford: Oxford University Press.
Menger, K. (1933). Fünf Wiener Vorträge. In H. F. Mark, et al. (Eds.), *Die neue Logik", in Krise und Neuaufbau in den Exakten Wissenschaften*. Leipzig: F. Dueticke.
Menger, K. (1937). *The New Logic*, Philosophy of Science, *4*, 299–336. English trans. of [47].
Moore, E. H. (1903). On the Foundations of Mathematics. *Bulletin of the American Mathematical Society*, *9*, 402–424.
Nagel, E. (1929). Intuition, Consistency & the Excluded Middle. *Journal of Philosophy*, *26*, 477–489.
Pasch, M. (1882). *Vorlesungen über neuere Geometrie (1882)*. Leipzig: Teubner.
Peacocke, G. (1830). *A Treatise on Algebra*. Cambridge, England: J. & J.J. Deighton.
Peano, G. (1906). Super theorema de Cantor-Bernstein, *Revista de mathematica*, *8*, 136–143. Reprinted in Opere scelte, a cura dell'Unione matematica italiana e col contributo del Consiglio nazionale della ricerche, Edizioni cremonese, Rome, 1957–59. English translation in [2].

Pieri, M. (1906). Sur la compatibilité des axioms de l'arithmetique. *Revue de Métaphysique et de Morale*, *14*, 196–207.
van Heijenoort, Jean. (1967). *From Frege to Gödel: A Sourcebook in Mathematical Logic, 1879–1931*. Cambridge: Harvard University Press.
Veblen, O., & Young, J. W. (1910). *Projective Geometry* (Vol. I). Boston: Ginn & Co.
Weiss, P. (1929). The Nature of Systems. II. *The Monist*, *39*, 440–472.
Whitehead, A. N. (1906). *The Axioms of Projective Geometry*. Cambridge: University Press.
Young, J.W., (1911). In W.W. Denton & U.G. Mitchell. *Lectures on Fundamental Concepts of Algebra and Geometry*. Macmillan Co., New York.

Chapter 6
Inferential Semantics

Kosta Došen

Abstract Prawitz's views concerning inferences and their validity are examined in the light of opinions about general proof theory and proof-theoretic semantics inspired by categorial proof theory. The frame for these opinions, and for the ensuing examination of those of Prawitz, is provided by what has been qualified as a dogmatic position that proof-theoretic semantics shares with model-theoretic semantics.

Keywords Inference · Deduction · Proposition · Name · Consequence · Truth · Provability · Validity, ground · Proof-theoretic semantics · General proof theory · Categorial proof theory

6.1 Introduction

The expression that makes the title of this paper, "inferential semantics", may be understood either as a synonym of "proof-theoretic semantics", the explanation of meaning through inference (see Sect. 6.1 for references), or as the semantics of inferences, i.e. the explanation of the meaning of inferences, which for Prawitz consists primarily in defining the notion of validity of an inference. Both of these readings cover the subject-matter of the paper.

The goal is to examine Prawitz's views concerning inferences and their validity, views that are based on the notions of ground and operation. I will concentrate on a section of a rather recent paper of Prawitz (2006), which I chose because it is self-

Work on this paper was supported by the Ministry of Education, Science and Technological Development of Serbia, and the Alexander von Humboldt Foundation has supported its presentation at the workshop "Dag Prawitz on Proofs and Meaning", in Bochum, in September 2012. I am indebted in particular to Heinrich Wansing, who invited me to write the paper and enabled me to present it at this workshop, which he organized.

K. Došen (✉)
Faculty of Philosophy, University of Belgrade, and Mathematical Institute,
Knez Mihailova 36 P.f. 367, 11001 Belgrade SANU, Serbia
e-mail: kosta@mi.sanu.ac.rs

H. Wansing (ed.), *Dag Prawitz on Proofs and Meaning*, Outstanding Contributions to Logic 7, DOI 10.1007/978-3-319-11041-7_6

contained and, I believe, fairly representative of his current views on proof-theoretic semantics and the semantics of inference.

My examination will be made in the light of some opinions I have concerning general proof theory and proof-theoretic semantics, which are inspired partly by categorial proof theory. I will spend more time in summarizing these opinions than in exposing those of Prawitz (much better known than mine). His opinions will not be examined before the last section. I have exposed the philosophical side of my opinions in a paper I have published rather recently (Došen 2011), and I will repeat much of what is in that short paper. This is done in order not to oblige the reader to keep returning to that previous paper. Concerning more technical matters I cannot do better than give a number of references.

The exposition of these opinions I give here differs however from the preceding one in being organized with respect to what Schroeder-Heister in Schroeder-Heister and Contu (2005) and Schroeder-Heister (2008) qualified as a dogmatic position that proof-theoretic semantics shares with model-theoretic semantics (see Sect. 6.3). My opinions are shown to go against these dogmas. I did not know of this idea of Schroeder-Heister when I wrote (Došen 2011), and I was happy to find how well what I said may be framed within the context provided by this idea. Besides two dogmas that I derive from Schroeder-Heister and Contu (2005) and Schroeder-Heister (2008), I add a third one, of the same kind, and with that I obtain a position from which, armed with a poignant terminology, I can examine Prawitz's opinions. Though he propounds a new proof-theoretic semantics, and though some of his ideas may point towards something else, his opinions are found to be in accordance with the dogmas.

6.2 Propositions, Names and Inferences

As a legacy of the last century, and to a certain extent of the nineteenth century too, the primary notion in the philosophy of language is nowadays the notion of proposition. With propositions we perform what is considered, as part of the same legacy, the main act of language, the main act of speech: namely, we assert something. Concurrently, the key to a theory of meaning would be the quality propositions have when it is legitimate to assert them; namely, their correctness, which may be either their truth, as in classical semantics, or something of the same kind like their provability, as in the constructivist understanding of mathematics. From a classical semantical perspective, inspired by logical model theory, the central notion is the notion of truth. The motto that may be put over this legacy is Frege's principle from the introduction of the *Grundlagen der Arithmetik* "never to ask for the meaning of a word in isolation, but only in the context of a proposition" (Frege 1974). The notion of proposition is primary in what Dummett calls the order of explanation of the function of language (see Dummett 1973, Chap. 1).

In the previous tradition, that should be traced back to Aristotle, the main act of speech and the key to a theory of meaning was in the act of naming. The primary notion in the philosophy of language was the notion of name. This old tradition is the one so forcibly called into question at the beginning of Wittgenstein's second book (see Wittgenstein 1953, Sect. 1). At the same place however Wittgenstein calls into question the primacy of asserting over other speech acts. It is usually taken that in his first book (Wittgenstein 1921) Wittgenstein took asserting as the main speech act, but the *Tractatus* allows many, if not contradictory, readings, and it has been found in Marion (2001) that the young Wittgenstein may even be understood as a precursor of proof-theoretic semantics. The views of the later Wittgenstein are however closer to the ideas of this semantics, and it seems right to say that the young Wittgenstein is on the trace of these ideas to the extent that in his first book he anticipates the second.

Proof-theoretic semantics (the term is due to Schroeder-Heister; see Kahle and Schroeder-Heister 2006), which is inspired by Gentzen's work on natural deduction (and to a lesser degree sequent systems; see Gentzen (1935), in particular, Sect. 2.5.13), covers ideas about meaning like those propounded by Prawitz and Dummett since the 1970s (see Prawitz 1971 and Dummett 1973, Appendix to Chap. 12). This is primarily an approach to logical semantics that looks for the meaning of logical constants in the role that they play in inference. (Although I prefer the term *deduction*, following Prawitz's usage I will speak of *inference* in this text; the two terms should mean the same.) This logical semantics may lead to analogous ideas about meaning outside logic. The proof-theoretic semantic conceptions of Prawitz and Dummett are constructivist; they stem from natural deduction for intuitionistic logic.

With proof-theoretic semantics, one could perhaps assume that the notion of inference should replace the notion of proposition as the primary notion of the philosophy of language. In the order of explanation of language, in the sense of Dummett, the notion of inference should now presumably precede that of proposition. The main act of language should not be any more the act of asserting, but the act of inferring (or deducing). This should perhaps be so, but it is not exactly so in the texts of Prawitz and Dummett.

6.3 Two Dogmas of Semantics

Categorical notions are those that are not hypothetical. (This use of *categorical* should not be confused with *categorial*, which, according to the *Oxford English Dictionary* (Simpson and Weiner 1989), means "relating to, or involving, categories"; unfortunately, in mathematical category theory *categorical* dominates in the sense of *categorial*.) The best example we may give for the distinction, an example taken from our subject-matter, is with categorical and hypothetical proofs. The latter is a proof under hypotheses, while the former depends on no hypothesis.

Schroeder-Heister (together with Contu in Schroeder-Heister and Contu 2005, Sect. 4, in Schroeder-Heister 2008, Sect. 3, and in Schroeder-Heister 2012) found that constructivist proof-theoretic semantics shares with classical semantics based on model theory two *dogmas*, which one may formulate succinctly as follows:

(1) Categorical notions have primacy over hypothetical notions.
(2) The correctness of the hypothetical notions reduces to the preservation of the correctness of the categorical ones.

The formulations of these dogmas in English in Schroeder-Heister (2008) and Schroeder-Heister (2012) is not the same as ours: Schroeder-Heister speaks of *consequence* in both dogmas, and in the second dogma he has *transmission* of correctness instead of *preservation* of correctness (which is not an essential difference); he does not speak about the correctness of consequence, but just about consequence. Our formulations are however close enough.

The second dogma may be understood as a corollary of the first one, and Schroeder-Heister spoke first of a single dogma. In Schroeder-Heister (2012) he speaks however of two dogmas, and adds a third one, which we will mention in Sect. 6.6 (where another third dogma is discussed). It is preferable (as Quine suggests) to have two dogmas, which makes referring to them easier.

The first dogma (1) is accepted if we take the notion of proposition, a categorical notion, to have primacy over the notion of inference, a hypothetical notion. Let us also illustrate the second, preservation, dogma (2). In the classical, model-theoretical, case the correctness, i.e. validity, of a consequence relation, which is something hypothetical, is defined in terms of the preservation of the correctness, i.e. truth, of propositions, which are categorical. In the constructivist case, a correct hypothetical proof should preserve categorical provability when one passes from the hypotheses to the conclusion.

6.4 Concerning the First Dogma

In Došen (2011) I have argued myself against the first dogma (1) in general proof theory, and I will repeat here what I said on that matter. The first dogma is present in general proof theory in the Curry-Howard correspondence, where we find typed lambda terms t as codes of natural deduction derivations. If t codes the derivation that ends with the formula B as the last conclusion, then this may be written $t : B$, and we say that t is of *type* B. (Formulae are of course of the grammatical category of propositions.) Our derivation may have uncancelled hypotheses. That will be seen by t's having possibly a free variable x, which codes an occurrence of a formula A as hypothesis; i.e. we have $x : A$, an x of type A.

All this makes conclusions prominent, while hypotheses are veiled. Conclusions are clearly there to be seen as types of terms, while hypotheses are hidden as types of free variables, which are cumbersome to write always explicitly when the variables occur as proper subterms of terms. The desirable terms are closed terms, which code derivations where all the hypotheses have been cancelled. These closed terms are supposed to play a key role in understanding intuitionistic logic. The categorical has precedence over the hypothetical.

An alternative to this coding would be a coding of derivations that would allow hypotheses to be as visible as conclusions, and such an alternative coding exists in categorial proof theory. There one writes $f: A \vdash B$ as a code for a derivation that starts with premise A and ends with conclusion B. The arrow term f is an arrow term of a category (a *cartesian closed category* if we want to cover the conjunction-implication fragment of intuitionistic logic; see Lambek and Scott (1986), Part I). The *type* of f is now not a single formula, but an ordered pair of formulae (A, B). The notation $A \vdash B$ serves just to have something more suggestive than (A, B) (In categories one usually writes $f: A \to B$ instead of $f: A \vdash B$, but $\to$ is sometimes used in logic for implication, and we should not be led to confuse inference with implication just because of notation).

If B happens to be derived with all hypotheses cancelled, then we will have $f: \top \vdash B$, with the constant formula $\top$ standing for the empty collection of hypotheses. If we happen to have more than one hypothesis, but as usual a finite number of them, then we will assume that with the help of conjunction all these hypotheses have been joined into a single hypothesis. So the categorial notation $f: A \vdash B$ with a single premise does not introduce a cramping limitation; at least not for the things intended to be said here.

The typed lambda coding of the Curry-Howard correspondence, involving finite product types and functional types, and the categorial coding in cartesian closed categories are equivalent in a very precise sense. This has been first shown by Lambek (see Lambek 1974; Lambek and Scott 1986; Došen 1996; Došen 2001). The import of the two formalisms is however not exactly the same. The typed lambda calculus suggests something different about the subject matter than category theory. It makes prominent the *proofs* $t: B$—and we think immediately of the categorical ones, without hypotheses—while category theory is about the *inferences* $f: A \vdash B$.

Another asymmetry is brought by the usual format of natural deduction, where there can be more than one premise, but there is always a single conclusion. This format favours intuitionistic logic, and in this logic the coding with typed lambda terms works really well with implication and conjunction, while with disjunction there are problems.

The categorial coding of derivations allows hypotheses to be treated on an equal footing with conclusions also with respect to multiplicity. With such a coding we could hope to deal too with classical logic, with all its Boolean symmetries, and with disjunction as well as with conjunction.

As an aside, let us note that the asymmetry of natural deduction with respect to premises and conclusions is most unfortunate when one has to formulate precisely what Prawitz calls the Inversion Principle (see Prawitz 1965, Chap. II), which, following Gentzen's suggestion of Gentzen (1935) (Sect. 2.5.13), relates the introduction and elimination rules for the logical constants. Dummett encounters analogous problems with his principle of harmony (see Dummett 1991b).

With Gentzen's plural (or multiple-conclusion) sequents we overcome this asymmetry, and we may formulate rules for the logical constants as *double-line rules*, i.e. invertible rules, going both ways, from the premises to the conclusion and back. The inversion of the Inversion Principle is now really inversion. I believe

Gentzen was aware of double-line rules, because they are implicit in Ketonen (1944), a doctoral thesis he supervised. (Older references concerning these rules are listed in Došen 1980, Sect. 25, and Došen 1997, end of Sect. 7; see also Schroeder-Heister 1984.)

What one will not realize by passing simply to sequents, without introducing the categorial coding mentioned above, is that with double-line rules we have just a superficial aspect of adjoint situations. This matter, which explains how the Inversion Principle is tied to a most important mathematical phenomenon, would however lead us too far afield, and I will here just list some references: Lawvere (1969), Došen (1999), Došen (2001) (see in particular Sect. 9), Došen (2006) (Sect. 6.4), Došen and Petrić (2008) (Sect. 6.7) and Došen and Petrić (2009) (Sect. 1.4).

Let us return now to the dogmas of semantics. The first dogma (1) is manifested also in the tendency to answer the question what is an inference by relying on the notion of proposition as more fundamental. It is as if Frege's recommendation from the *Grundlagen der Arithmetik* to look after meaning in the context of a proposition (see Sect. 6.2) was understood to apply not only to bits of language narrower than propositions, which should be placed in the broader propositional context, but also to something broader than propositions, as inferences, in which propositions partake, which should be explained in terms of the narrower notion of proposition.

An inference is usually taken as something involving propositions. Restricting ourselves to inferences with single premises, as we agreed to do above, for ease of exposition, we may venture to say that an inference consists in *passing* from a proposition called premise to a proposition called conclusion.

What could "passing" mean here? Another principle of Frege from the introduction of the *Grundlagen der Arithmetik* would not let us understand this passing as something happening in our head. Such an understanding would expose us to being accused of psychologism. No, this passing should be something objective, something done or happening independently of any particular thinking subject, something sanctioned by language and the meaning it has.

The temptation of psychologism is particularly strong here, but as a proposition is not something mental that comes into being when one asserts a sentence, so an inference should not be taken as a mental activity of passing from sentences to sentences or from propositions to propositions. Such a mental activity exists, as well as the accompanying verbal and graphical activities, but the inference we are interested in is none of these activities. It is rather something in virtue of which these activities are judged to be correctly performed or not. It is something tied to rules governing the use of language, something based on these rules, which are derived from the meaning of language, or which confer meaning to it.

When inspired by categorial proof theory we reject the first dogma (1) we do not take that categorical and hypothetical notions are on an equal footing, but we give priority to the hypothetical notions. This is related to the priority that category theory gives to arrows over objects. When in the category with sets as objects and functions as arrows, which is the paradigmatic example of a category, one has to explain what is an ordered pair, i.e. one has to characterize the operation of cartesian product on sets, one does not look *inside* the cartesian product of two sets, but one characterizes cartesian product from the *outside*. This may be achieved in terms of some special functions—namely, the projection functions—next, in terms of an operation on functions—namely, the operation of pairing that applied to the functions $f : C \to A$ and $g : C \to B$ gives the function $\langle f, g \rangle : C \to A \times B$—and, finally, in terms of equations between functions concerning these special functions and operation on functions (see Lambek and Scott 1986, Sect. 1.3, and Došen and Petrić 2004, Sect. 9.1).

In categorial proof theory inferences are taken as arrows and propositions as objects, and inferences have priority over propositions. When one has to characterize the connective of conjunction, one may do it in terms of the inferences of conjunction elimination, which correspond to projection functions, in terms of the rule of conjunction introduction, which corresponds to the pairing operation on functions, and in terms of equations between inferences concerning implication elimination and implication introduction. These equations, the same as the equations of cartesian product mentioned above, correspond to β and η reduction rules, which one encounters in reductions to normal form in natural deduction and sequent systems.

6.5 Concerning the Second Dogma

We shall now consider matters, taken again from Došen (2011), that lead to dissent with the second dogma of semantics (2).

Can inferences be reduced to consequence relations? So that having an inference from A to B means just that B is a consequence of A. This would square well with the objective character of inferences we have just talked about, because B's being a consequence of A is something objective. "Consequence" here can be understood as semantical consequence, and the objectivity of consequence would have semantical grounds.

Since B is a consequence of A whenever the implication $A \to B$ is true or correct, there would be no essential difference between the theory of inference and the theory of implication. An inference is often written vertically, with the premise above the conclusion,

$$\frac{A}{B}$$

and an implication is written horizontally $A \to B$, but besides that, and purely grammatical matters, there would not be much difference.

This reduction of inference to implication, which squares well with the second dogma of semantics, is indeed the point of view of practically all of the philosophy of logic and language in the twentieth century. This applies not only to classically minded theories where the essential, and desirable, quality of propositions, their correctness, is taken to be truth, but also to other theories, like constructivism in mathematics, or verificationism in science, where this quality is something different.

It may be deemed strange that even in constructivism, where the quality is often described as *provability*, inferences are not more prominent. Rather than speak about inferences, constructivists, such as intuitionists, tend to speak about something more general covered by the portmanteau word “construction”. (Constructions produce mathematical objects as well as proofs of mathematical propositions, which are about these objects.) Where above we spoke about *passing*, a constructivist would presumably speak about *constructions*.

By reducing inferences from A to B to ordered pairs (A, B) in a consequence relation we would loose the need for the categorial point of view. The f in $f : A \vdash B$ would become superfluous. There would be at most one arrow with a given source and target, which means that our categories would be preordering relations (i.e. reflexive and transitive relations). These preorderings are consequence relations.

With that we would achieve something akin to what has been achieved for the notion of function. This notion has been *extensionalized*. It has been reduced to a set of ordered pairs. If before one imagined functions as something like a *passing* from an argument to the value, now a function is just a set of ordered pairs made of arguments and values. Analogously, inferences would be the ordered pairs made of premises and conclusions.

The extensionalizing of the notion of inference which consists in its reduction to the notion of consequence relation can be called into question if we are able to produce examples of two different inferences with the same premise and the same conclusion. Here is such an example of formal, logical, inferences, which involve conjunction, the simplest and most basic of all logical connectives.

From $p \wedge p$ to p there are two inferences, one obtained by applying the first rule of conjunction elimination, the first projection rule,

$$\frac{A \wedge B}{A}$$

and the other obtained by applying the second projection rule

$$\frac{A \wedge B}{B}$$

This and other such examples from logic redeem the categorial point of view (see Došen and Petrić 2004, 2007). In this example we have $\pi^1 : p \wedge p \vdash p$ and $\pi^2 : p \wedge p \vdash p$ with $\pi^1 \neq \pi^2$.

A category where these arrows are exemplified is $\mathcal{C}$, which is the category with binary product freely generated by a set of objects. The category $\mathcal{C}$ models inferences involving only conjunction. It does so for both classical and intuitionistic conjunction, because the inferences involving this connective do not differ in the two alternative logics. This is a common ground of these two logics.

The notion of binary product codified in $\mathcal{C}$ is one of the biggest successes of category theory. The explanation of the extremely important notion of ordered pair in terms of this notion is the most convincing corroboration of the point of view that mathematical objects should be characterized only up to isomorphism. It is remarkable that the same matter should appear at the very beginning of what category theory has to say about inferences in logic, in connection with the connective of conjunction (see the end of the preceding section).

For the category $\mathcal{C}$ there exists a kind of completeness theorem, which categorists call a *coherence* result. There is namely a faithful functor from $\mathcal{C}$ to the model category that is the opposite of the category of finite ordinals with functions as arrows. With this functor, π^1 and π^2 above correspond respectively to

Another example of two different formal inferences with the same premise and the same conclusion, which involves graphs of a slightly more complicated kind, is given by

The first inference is made by conjunction elimination, while the second by modus ponens.

Coherence is one of the main inspirations of categorial proof theory (see Došen and Petrić 2004). The other, older, inspiration, which works for inferences in intuitionistic logic, comes ultimately from the notion of adjunction (for which we gave references in Sect. 6.4).

In model categories such as we find in coherence results we have models of equational theories axiomatizing identity of inferences. These are not models of the theorems of logic. The arrows of the model categories are however hardly what inferences really are. It is not at all clear that these categories provide a real semantics of inferences (cf. Došen 2006).

It is not clear what, from the point of view of proof-theoretical semantics, should be the semantics of inferences, i.e. the explanation of meaning of inferences. If inferences provide meaning, how can their meaning be reduced to something more primitive, in the style of the dogmas?

Invoking now another principle of Frege's *Grundlagen der Arithmetik*, we might look for an answer to the question "What is an inference?" by looking for a criterion of identity of inferences. Prawitz introduced in Prawitz (1971) the field of general proof theory with that question. (It is remarkable that the same question was raised by Hilbert in his discarded 24th problem; see Thiele (2005)).

We would strive to define a significant and plausible equivalence relation on derivations as coded by arrow terms of our syntactical categories, and equivalence classes of derivations, or equivalence classes of arrow terms, which are the arrows of our syntactical categories, would stand for inferences.

An inference $f: A \vdash B$ would be something sui generis, that does not reduce to its type, the ordered pair (A, B). It would be represented by an arrow in a category, to which is tied a criterion of identity given by the system of equations that hold in the category. The category should not be a preorder.

In the arrow term f of $f: A \vdash B$, the inference rules involved in building an inference are made manifest as operations for building this term. The theory of inference is as a matter of fact the theory of such operations (usually partial). It is an algebraic theory codifying with equations the properties of these operations.

It is rather to be expected that the theory of inference should be the theory of inference rules, as arithmetic, the theory of natural numbers, is the theory of arithmetic operations (addition, multiplication etc.). Extensionalizing the notion of inference by reducing it to consequence (as in the preceding section) makes us forget the inference rules, which are prominent in categorial proof theory.

From a classical point of view, the desirable quality of propositions, their correctness, is their truth. If the notion of inference is something sui generis, not reducible to the notion of proposition, why should the desirable quality of inferences, their correctness, be reducible to the desirable quality of propositions?

If we abandon the second dogma of semantics, a correct inference would not be just one that preserves truth—a correct consequence relation could be that. An inference $f: A \vdash B$ is not just $A \vdash B$; we also have the f. As a matter of fact, the inference is f. It may be a necessary condition for a correct $f: A \vdash B$ that B be a consequence of A, but this is not sufficient for the correctness of f. This is not what the correctness of f consists in. The correctness of an inference would be, as the notion of inference itself, something sui generis, not to be explained in terms of the correctness of propositions.

We might perhaps even try to turn over the positions, and consider that the correctness of propositions should be explained in terms of the correctness of inferences. This is presumably congenial to a point of view like that found in intuitionism, where the correctness of propositions is taken to be provability, i.e. deducibility from an empty collection of premises. We could however take an analogous position with a classical point of view, where the *truth* of analytic propositions would be guaranteed by the correctness of some inferences (cf. Dummett 1991a, p. 26). The correctness of the inferences underlies the truth, and not the other way round.

6.6 A Third Dogma of Semantics

In (Schroeder-Heister 2012), Schroeder-Heister considers a third dogma of semantics, which he formulates as follows:

> Consequence means the same as correctness of inference.

To understand properly this and the alternative view of proof-theoretic semantics Schroeder-Heister proposes would require entering into the technical notions of Schroeder-Heister (2012), where the matter is presented with respect to an understanding of consequence within the framework of a specific theory of definition. (Among these notions are definitional clause, definitional closure, definitional reflection, and others.) We will not do that here, and will not try to determine how much Schroeder-Heister's views and terminology accord with ours. Our purpose is not that, but to examine Prawitz's views in the light of the two dogmas previously formulated, and a third one, which, though maybe related, is not Schroeder-Heister's. Before I try to formulate this third dogma, which I think constructivist proof-theoretic semantics shares with classical model-theoretic semantics, some explanations are needed.

In classical semantics, it is not the categorical notion of proposition that is really central. Its place is taken by the notion of correctness of a proposition, namely *truth*. Similarly, the hypothetical notion of consequence is really less central than the notion of correctness of a consequence, i.e. its *validity*, in terms of which consequence is defined. The notion of consequence reduces to its validity.

In proof-theoretic semantics the situation is not much different. The notion of proposition seems to be less central than its correctness, namely *provability*, and for consequences, or inferences (if these two notions are distinguished), one concentrates again on their validity.

I will formulate the third dogma of semantics as follows:

(3) The notions of correctness of the notions mentioned in (2) are more important than these notions themselves.

The third dogma stems probably from the old, venerable and general, denotational perspective on semantics, which Wittgenstein has criticized at the beginning of the *Philosophical Investigations* (mentioned in Sect. 6.2). As names stand for objects, as they refer to them, so propositions stand for truth-values, and the correct ones stand for truth, i.e. the value *true*. They need not exactly behave as names (as the later Frege thought), and they do not exactly refer to the truth-values, but still they somehow stand for them, in a manner not quite foreign to the manner in which a name stands for the object it denotes. The insistence on validity when speaking of consequence should have the same denotational roots.

I think that the influence of the third dogma on proof-theoretic semantics is more pernicious when we talk of inferences than when we talk of propositions. The notion of inference should be here more important than the notion of proposition, and the notion of inference itself should be more important than the notion of correctness of an inference, which Prawitz calls validity, and on whose definition he has worked for a long time.

It is questionable whether in proof-theoretic semantics we need at all the notion of validity of an inference. What I have just said may be surprising. How come—I will be told—that valid inferences are not our subject matter? I would reply that invalid inferences do not exist in a certain sense.

Some people may pass from A to B when this is not sanctioned by an inference $f : A \vdash B$, and we might say that we have here an invalid inference, but this is only a manner of speaking, and not a good one, because it leads us astray. Invalid inferences may exist as interesting psychological entities, but in logic they cannot have this status. Instead of saying that we have an invalid inference from A to B, we should say that for the type $A \vdash B$ we have no inference. It is as if all inferences are valid, and since they are all valid, their validity need not be mentioned expressly.

To make an analogy, illegal chess moves may exist in the physical world, but they do not exist in the world of chess. To characterize all the legal chess moves is to characterize all the possible moves. Impossible moves are excluded.

The situation is similar with the notion of formula. There used to be a time when logicians spoke much of *well-formed* formulae, as if there were formulae that are not well formed. This bad usage has, fortunately, died out, and nowadays the attribute "well-formed" is much less often applied to formulae. A formula that is not well formed is better characterized as a word in our alphabet that is not a formula. All formulae are well formed. There are no other formulae in the perspective of logic. There might be a badly formed word an individual would take wrongly for a formula, and call so. This is something that might perhaps be interesting for psychology, but need hardly concern us as logicians. The notion of *well-formedness* of a formula is hardly something separate from a formula, which could serve to make an important semantical point. A semantical theory based on the abstract notion of well-formedness seems as suspect as a physiological theory concerning opiates based on the notion of *virtus dormitiva*.

I surmise that the notion of validity of an inference is very much like the notion of well-formedness of a formula. A valid inference is like a well-formed syntactical object. In logic, derivations, like other syntactical objects, are defined inductively, and inferences are equivalence classes of derivations.

A term f, of type $A \vdash B$, is defined inductively as well, and usually it stands only for valid inferences, which yields that the implication $A \to B$ is correct. We may envisage, for technical reasons, into which we cannot go here (see Došen 2003, Sect. 7, Došen and Petrić 2004, Sect. 1.6, and Chaps. 12 and 13), arrow terms $f : A \vdash B$, where $A \to B$ is not correct, but this is not what is usually done.

Even if we assume that the notion of validity of an inference is legitimate and, by following the second dogma, assume that this validity is to be defined in terms of the correctness of propositions, which amounts to the correctness of implications, why would this oblige us to follow the third dogma too, and take that the notion of validity of an inference should be more central in our proof-theoretic semantical theory than the notion of inference itself? It seems that the second dogma need not entail the third dogma, but since they probably have similar roots, those following one would be inclined to follow the other.

6.7 Inferences as Operations on Grounds

The title of the present section is taken from the title of Sect. 7 of Prawitz's paper Prawitz (2006). Our goal is to examine Prawitz's views expressed at that particular place in the light of our critique of the three dogmas of semantics. As I said in the introduction, I chose that particular paper, and that particular section, because it is a fairly recent and self-contained piece, which is I believe representative of his current views (he may have written similar things elsewhere).

Prawitz's goal in that section is to reconsider the concept of inference "to get a fresh approach to the concept of valid inference". In the second paragraph Prawitz mentions the "intuitive understanding of an inference as consisting of something more than just a conclusion and some premises". This is something that accords well with the critique from Sect. 6.5 above of the extensionalizing of inference by its reduction to consequence.

Prawitz says that his "main idea is thus to take an inference as given not only by its premises and conclusion (...), but also by an operation defined for grounds for the premises". Prawitz does not define the term "operation" more precisely, and it is not clear at the beginning that he has something very mathematical in mind here, but it is natural to suppose that these operations correspond to our arrows f from Sect. 6.4. Prawitz's first example of such an operation is mathematical induction. The other examples, which we will mention below, involve rules for conjunction.

The term *ground for a sentence* is more difficult to understand (*sentence* here should mean what we meant by *proposition* throughout this text). It is something very wide (approaching in the width of its scope the term *construction* of the intuitionists). Prawitz says it is "what a person needs to be in possession of in order to be justified in holding the sentence true". Although at the beginning of the section he suggests that grounds are premises—hence sentences, i.e. propositions as we would say—he concludes that "the premises from which a conclusion is inferred do not constitute grounds for the conclusion—rather the premises have their grounds, and it is by operating on them that we get a ground for the conclusion". Grounds are perhaps not linguistic at all. Could they be, for example, sense perceptions? But how does one operate with inferences on sense perceptions?

Prawitz does not mention the Curry-Howard correspondence in the present context, but his exposition accords rather well with its assumptions, including the first dogma of semantics (1). The categorical is primary, and a ground is like a typed lambda term coding a natural deduction derivation. Even if the term is not closed, it is of a different kind from the operation on typed lambda terms that stands for an inference. Examples of such operations are surjective pairing and projections, which Prawitz mentions on pp. 21–22, without calling them so. (In that context he gives two equations for inferences—he says grounds—which may be understood as the equations of the lambda calculus with product types derived from β-reduction; there are corresponding equations of categories with products.)

The alternative point of view of categorial proof theory would take grounds to be of the same kind as the operations that correspond to inferences (intuitionists would use the word "construction" for both). Instead of Prawitz's ground $t: A$, i.e. ground t for A, to which we apply the inference, i.e. operation, $o: A \vdash B$ in order to get the ground $o(t): B$, we would have the inference $f: \top \vdash A$, which we compose with the inference $o: A \vdash B$ in order to get the inference $o \circ f: \top \vdash B$. We would understand grounds as something inferential too. They may be distinguished from the hypothetical inferences by their type—they have a $\top$ on the left—but otherwise they are of the same nature. (One may treat axioms of logical systems just as rules with no premises.) Operations on grounds correspond to inferences, but we have also to deal with rules of inference that correspond to operations on inferences (see the ends of Sects. 6.4 and 6.5).

Speaking about *individual inferences* Prawitz expresses some quite psychologistic views, but he soon moves to *inference forms*, in which psychologistic ingredients disappear. His *inference figures*, or *schemata*, or *schemes*, which "abstract from the operation left in an inference form", are something like consequence; we would say they are the *type* $A \vdash B$ of an inference obtained from $o: A \vdash B$ by forgetting the o.

Then Prawitz gives (on p. 19) a characterization of the validity of an inference in which the basic notion is validity of an individual inference:

> An individual inference is valid if and only if the given grounds for the premises are grounds for them and the result of applying the given operation to these grounds is in fact a ground for the conclusion.

The notion of individual inference is psychologistic, but the definition of its validity does not seem to depend very much on that, and the characterization of the validity of an inference form is essentially the same, except that instances are mentioned. An inference figure $A \vdash B$ is taken to be valid when there is an o such that $o: A \vdash B$ is a valid inference form.

Prawitz's discussion is intermingled with epistemological considerations, which are very important for his concerns. (A paper contemporaneous with Prawitz 2006 where this is even more clear is Prawitz 2008.) I think this has to do with the psychologistic ingredients in his position.

Prawitz's opinions accord well with the second dogma of semantics (2). His characterization of validity of an inference, sketched above, as well as his older views upon that matter, clearly give precedence to something tied to propositions, here called grounds. The validity of an inference consists in the preservation of *groundedness*.

I think that Prawitz's opinions accord pretty well with the third dogma of semantics (3), too. From a semantical perspective, the notion of validity of an inference seems to be for him more important than the notion of inference. The main task of semantics seems to be for him the definition of this validity.

He makes what may be interpreted as a move away from the dogmas by taking that an inference is an operation, understood non-extensionally. Take a typical operation like addition of natural numbers. One may speak about the correctness of an individual application of addition, an individual performance of it by a human being (or perhaps a machine), but one does not usually speak about the correctness of the abstract notion of addition, independently of its applications. How would correct addition differ from addition tout court? One does not speak either about the validity of addition. What would that be? Were it not for his psychologistic inclinations, we could then surmise that the understanding of inferences as operations might have led Prawitz to question the third, and the other dogmas, too.

References

Došen, K. (1980). *Logical constants: An essay in proof theory*, Doctoral thesis. University of Oxford. http://www.mi.sanu.ac.rs/~kosta/publications.htm

Došen, K. (1996) Deductive completeness. *The Bulletin of Symbolic Logic*, 2 (pp. 243–283, 523). (For corrections see Došen 2001, Sect. 5.1.7, and Došen 2003).

Došen, K. (1997). Logical consequence: A turn in style. In M. L. Dalla Chiara et al. (Eds.), *Logic and Scientific Methods, Volume I of the 10th International Congress of Logic, Methodology and Philosophy of Science, Florence 1995* (pp. 289–311). Dordrecht: Kluwer. http://www.mi.sanu.ac.rs/~kosta/publications.htm

Došen, K. (2001). Abstraction and application in adjunction. In Z. Kadelburg (Ed.), *Proceedings of the Tenth Congress of Yugoslav Mathematicians, Faculty of Mathematics, University of Belgrade, Belgrade* (pp. 33–46). http://arXiv.org

Došen, K. (2003). Identity of proofs based on normalization and generality. *The Bulletin of Symbolic Logic, 9*, 477–503. (Version with corrected remark on difunctionality available at: http://arXiv.org)

Došen, K. (2006). Models of deduction, proof-theoretic semantics. In R. Kahle, & P. Schroeder-Heister (Eds.), *Proceedings of the Conference "Proof-Theoretic Semantics, Tübingen 1999", Synthese*, vol. 148 (pp. 639–657). http://www.mi.sanu.ac.rs/~kosta/publications.htm

Došen, K. (2011). A prologue to the theory of deduction. In M. Peliš, V. Punčochář (Eds.), *The logica yearbook 2010* (pp. 65–80). London: College Publications (http://www.mi.sanu.ac.rs/~kosta/publications.htm)

Došen, & K., Petrić, Z. (2004). *Proof-theoretical coherence*. London: KCL Publications (College Publications). (Revised version of 2007 available at: http://www.mi.sanu.ac.rs/~kosta/publications.htm)

Došen, & K., Petrić, Z. (2007). *Proof-net categories*. Monza: Polimetrica. (preprint of 2005 available at: http://www.mi.sanu.ac.rs/~kosta/publications.htm)

Došen, & K., Petrić, Z. (2008). Equality of proofs for linear equality. *Archive for Mathematical Logic, 47*, 549–565. (http://arXiv.org)

Došen, & K., Petrić, Z. (2009). Coherence in linear predicate logic. *Annals of Pure and Applied Logic, 158*, 125–153. (http://arXiv.org)

Došen, K. (1999). *Cut elimination in categories*. Dordrecht: Kluwer.

Dummett, M. A. E. (1973). *Frege: Philosophy of language*. London: Duckworth.

Dummett, M. A. E. (1991a). *Frege: Philosophy of mathematics*. London: Duckworth.

Dummett, M. A. E. (1991b). *The logical basis of metaphysics*. London: Duckworth.

Frege, G., *Die Grundlagen der Arithmetik: Eine logisch mathematische Untersuchung über den Begriff der Zahl*, Verlag von Wilhelm Koebner, Breslau, 1884 (English translation by J. L. Austin: *The Foundations of Arithmetic: A Logico-Mathematical Enquiry into the Concept of Number, second revised edition*, Blackwell, Oxford, 1974).

Gentzen, G. (1935). Untersuchungen über das logische Schließen, Mathematische Zeitschrift, 39 (1935) 176–210, 405–431. (English translation: Investigations into logical deduction (1969), In M.E. Szabo, (Ed.), *The collected papers of Gerhard Gentzen* (pp. 68–131, 312–317) North-Holland: Amsterdam).

Kahle, R., & Schroeder-Heister, P. (2006) Introduction: Proof-theoretic semantics. In R. Kahle, P. Schroeder-Heister, (Eds.), *Proof-Theoretic Semantics, Proceedings of the Conference "Proof-Theoretic Semantics, Tübingen 1999"*, *Synthese* (Vol. 148, pp. 503–506).
Ketonen, O. (1944). Untersuchungen zum Prädikatenkalkül. *Annales Academiae Scientiarum Fennicae, series A, I Mathematica-physica*, *23*, 71.
Lambek, J. (1974). Functional completeness of cartesian categories. *Annals of Mathematical Logic*, *6*, 259–292.
Lambek, J., & Scott, P. J. (1986). *Introduction to higher order categorical logic*. Cambridge: Cambridge University Press.
Lawvere, F. W. (1969). Adjointness in foundations. *Dialectica*, *23*, 281–296.
Marion, M. (2001). Qu'est-ce que l'inférence? Une relecture du Tractatus logico-philosophicus. *Archives de Philosophie*, *64*, 545–567.
Prawitz, D. (2006) Validity of inferences, to appear in the *Proceedings of the 2nd Lauener Symposium on Analytical Philosophy, Bern 2006* (http://people.su.se/~prawd/Bern2006.pdf)
Prawitz, D. (1965). *Natural deduction: a proof-theoretical study*. Stockholm: Almqvist & Wiksell.
Prawitz, D. (1971). Ideas and results in proof theory. In J. E. Fenstad (Ed.), *Proceedings of the Second Scandinavian Logic Symposium* (pp. 235–307). Amsterdam: North-Holland.
Prawitz, D. (2008). Inference and knowledge. In M. Peliš (Ed.), *The logica yearbook, 2009* (pp. 175–192). London: College Publications.
Schroeder-Heister, P. (2008). Proof-theoretic versus model-theoretic consequence. In Peliš, M. (Ed.), *The logica yearbook, 2007* (pp. 187–200). Prague: Filosofia.
Schroeder-Heister, P. (2012) The categorical and the hypothetical: A critique of some fundamental assumptions of standard semantics. In Lindström, S. et al., (Eds.), *The Philosophy of Logical Consequence and Inference, Proceedings of the Workshop "The Philosophy of Logical Consequence, Uppsala, 2008"*, *Synthese* (Vol. 187, pp. 925–942).
Schroeder-Heister, P. (1984). Popper's theory of deductive inference and the concept of a logical constant. *History and Philosophy of Logic*, *5*, 79–110.
Schroeder-Heister, P., & Contu, P. (2005). In W. Spohn, P. Schroeder-Heister, & E. J. Olsson (Eds.), *Folgerung, Logik in der philosophie* (pp. 247–276). Heidelberg: Synchron.
Simpson, J., & Weiner, E. (Eds.). (1989). *The Oxford english dictionary* (2nd ed.). Oxford: Oxford University Press.
Thiele, R., (2005), Hilbert and his twenty-four problems. In Van Brummelen, G., Kinyon, M. (Eds.), *Mathematics and the Historian's Craft: The Kenneth O. May Lectures*, Canadian Mathematical Society (pp. 243–295). New York: Springer.
Wittgenstein, L., (1921). Logisch-philosophische Abhandlung, *Annalen der Naturphilosophie*, 14 (pp. 185–262) (English translation by C. K. Ogden: *Tractatus logico-philosophicus*, Routledge, London, 1922, new translation by D. F. Pears and B. F. McGuinness, Routledge, London, 1961).
Wittgenstein, L., (1953). *Philosophische Untersuchungen*. Blackwell: Oxford (English translation by G. E. M. Anscombe: *Philosophical investigations, fourth edition with revisions* by P. M. S. Hacker and J. Schulte, Wiley-Blackwell, Oxford, 2009).

Chapter 7
Cut Elimination, Substitution and Normalisation

Roy Dyckhoff

Abstract We present a proof (of the main parts of which there is a formal version, checked with the Isabelle proof assistant) that, for a **G3**-style calculus covering all of intuitionistic zero-order logic, with an associated term calculus, and with a particular strongly normalising and confluent system of cut-reduction rules, every reduction step has, as its natural deduction translation, a sequence of zero or more reduction steps (detour reductions, permutation reductions or simplifications). This complements and (we believe) clarifies earlier work by (e.g.) Zucker and Pottinger on a question raised in 1971 by Kreisel.

Keywords Intuitionistic logic · Minimal logic · Sequent calculus · Natural deduction · Cut-elimination · Substitution · Normalisation

7.1 Introduction

It is well-known that, in intuitionistic logic, sequent calculus derivations (with or without *Cut*) are recipes for constructing natural deductions, and that, by the Curry-Howard correspondence, with care about variable discharge conventions, one can represent both the former and the latter using typed lambda terms. Natural deduction terms may, by various standard reductions, be reduced; but there are many sequent calculi S, reduction systems R for S and reduction strategies for R, including but not limited to those given by Gentzen. We present here, for a complete single-succedent sequent calculus **G3ip** (roughly that in Troelstra and Schwichtenberg (2000) and Negri and von Plato (2001)), a reduction system (of 32 rules) for cut-elimination, with the virtues that (a) it is strongly normalising; (b) it is confluent; (c) it is explicitly given using a term notation; and (d) it allows a homomorphism (as described below) from cut-elimination to normalisation.

R. Dyckhoff (✉)
School of Computer Science, St Andrews University,
Scotland, St Andrews, Fife KY16 9SX, UK
e-mail: rd@st-andrews.ac.uk

H. Wansing (ed.), *Dag Prawitz on Proofs and Meaning*, Outstanding Contributions to Logic 7, DOI 10.1007/978-3-319-11041-7_7

Kreisel (1971) asked about the relation between cut-elimination and normalisation. Troelstra and van Dalen in (2000, p. 565) comment that "The combinatorial relationship between normalization and cut-elimination has been investigated by Zucker (1974) and Pottinger (1977). Normalization corresponds to cut-elimination under a homomorphic mapping provided the basic cut-elimination steps are suitably chosen. Recently a still better correspondence has been achieved by Diller and Unterhalt". It is however not clear what the suitable choices of the "basic cut-elimination steps" should be. Diller has written (19 December 2012) that "I am very sorry that I cannot point at a publication of Unterhalt's thesis or at a paper published by the two of us. After quarter of a century since Unterhalt's PhD thesis (1986), I cannot even give a concise explanation of what the progress of Unterhalt's work was in comparison to the work or methods of Zucker and Pottinger. I think that the central points are contained in Troelstra and Dalen (2000). Unterhalt mainly studies ... semantics of E-logic (...), but he also establishes a transfer to cut-elimination and normalization".

Zucker (1974), using Gentzen's cut-elimination steps and an innermost-first strategy, gave a partial answer, but had difficulties with disjunction, including a failure of strong normalisation. His calculus has explicit rules of *Weakening* and *Contraction* and context-splitting inference rules. Pottinger (1977) gave a positive answer covering disjunction; but, as pointed out by Urban (2014), the notion of normality for the sequent calculus proof terms does not coincide with cut-freedom, and this renders Pottinger's answer "subtly" defective—in our view overtly defective, despite Pottinger's claim that the difference is "trivial". (Moreover, the closest system in Pottinger (1977) to a conventional sequent calculus is his H_L; but, although it is complete for derivability of formulae, it does not admit *Contraction*; it does not derive, for example, the sequent $p \Rightarrow p \wedge p$. Nor does it admit *Weakening* or derive the sequent $p \wedge p \Rightarrow p \wedge p$. There is a section explaining what one might do if *Contraction* is added as a primitive rule, with no explanation of how the cut-reduction rules might change—one is reminded of Gentzen's difficulties with *Contraction* and his avoidance thereof with his *Mix* rule.)

von Plato (2001) and (with Negri) (Negri and von Plato 2001) consider related issues, with a focus not on cut reduction and normalisation steps but on obtaining an isomorphism between sequent calculus and natural deduction, achieved by using generalised elimination rules in the latter. We see this as a solution to a different problem.

The purpose of this paper is thus to clarify matters in our preferred context, namely the sequent calculus **G3ip** (with context-sharing inference rules and without explicit rules of *Weakening* and *Contraction*) and a standard natural deduction calculus from (Gentzen 1935; Prawitz 1965). Such sequent calculi (widely studied in Negri and von Plato (2001)) correspond better than others (such as those of (Gentzen 1935; Pottinger 1977; Zucker 1974)) to calculi used for root-first proof search, either as sequent calculi or (inverted) as tableau calculi. We also choose to use a term notation (with an appropriate binding mechanism) to allow the concise presentation of reduction rules: this has the extra virtue of simplifying automation and verification.

Urban's solution also uses, for representing sequent derivations, a term notation, deriving from his work in classical logic (Urban and Bierman 2001), with names and co-names. His notation improves on Pottinger's and Zucker's, but his result (that there is such a homomorphism) applies only to the ($\supset, \wedge, \forall$)-fragment of the logic.

Borisavljević (2004) gives a detailed explanation of Zucker's difficulties and proposes a solution (for intuitionistic predicate logic) using generalised elimination rules (from (Negri and von Plato 2001; von Plato 2001)), concluding that "the problem in connections of conversions from the full systems (with $\vee$ and $\exists$) δ and $\mathcal{N}$ is the consequence of the different forms of elimination rules for $\wedge$, $\supset$ and $\forall$ on the one side, and $\vee$ and $\exists$ on the other side, in the system $\mathcal{N}$" (where $\mathcal{N}$ is Zucker's natural deduction system). We prefer not to adopt a natural deduction system with generalised elimination rules but to use the original systems of Gentzen (1935) and Prawitz (1965).

Kikuchi (2006) treats an aspect of the relationship between normalisation and cut-elimination, but only for implicational logic, and with a very different goal: the simulation of normalisation by cut-elimination, i.e. the lifting of reduction steps in natural deduction back to cut reduction steps (from which one can infer strong normalisation of natural deduction from the corresponding property of the sequent calculus).

Thus, we present here a treatment of this issue for **all** of intuitionistic zero-order logic, using a standard natural deduction calculus, a standard sequent calculus and a standard notation for terms, and without Pottinger-style defects, allowing a clear understanding of exactly what reductions are required in the natural deduction calculus and in the sequent calculus for the following to hold: let L and L' be sequent derivations so that $L \rightsquigarrow^* L'$ by some sequence of cut-elimination steps; then $N \rightsquigarrow^* N'$, where $N \equiv \phi(L)$ and $N' \equiv \phi(L')$ are the natural deductions constructed from the recipes L and L' by means of the Gentzen-Prawitz translation ϕ. Like others, we consider this to give a *homomorphism* from one reduction system to another.

No claim is made about the converse; ensuring that a cut-reduction system can simulate beta-reduction is tricky. See Kikuchi (2006), and also Dyckhoff and Urban (2003) for a solution involving a restricted sequent calculus, Herbelin's **LJT**.

Our result is for **all** the connectives of intuitionistic zero-order logic, including disjunction. Given that there are examples in (for example) Urban (2014) illustrating the difficulty with disjunction, this may be surprising. The solution is given by the complex reduction rules such as (7.6.18) and (7.6.20). A computer-checked verification of the results is available (Dyckhoff and Chapman 2009), using Nominal Isabelle (Urban 2008). Work extending both the theory and the formal verification to first-order logic is not yet undertaken: no major difficulties are anticipated.

An extended abstract of an earlier version of this paper appeared as (Dyckhoff 2011).

7.2 Technical Background

Where distinct meta-variables x, y, etc. are used, they stand for distinct variables except where explicitly stated or indicated otherwise. We will use x and x', for example, to indicate two variables which can but need not be distinct from each other. The symbol i will be understood as ranging over $\{1, 2\}$. *Atoms* are proposition variables $p, q, \ldots$ or $\bot$; *formulae* $A, B, C, D, E, \ldots$ are built up from atoms using implication, conjunction and disjunction. *Contexts* Γ are (as usual) sets of expressions $x : A$ where each x is associated with at most one formula A.

7.2.1 Natural Deduction, in Logistic Style

We present the typing rules for typed lambda calculus in natural deduction format, but using Gentzen's logistic style.

$$\frac{\Gamma \Rightarrow N : \bot}{\Gamma \Rightarrow ef(N) : C}\ \bot E \qquad \frac{}{x : A, \Gamma \Rightarrow x : A}\ Ax$$

$$\frac{\Gamma \Rightarrow N : A \supset B \quad \Gamma \Rightarrow N' : A}{\Gamma \Rightarrow ap(N, N') : B}\ \supset E \qquad \frac{x : A, \Gamma \Rightarrow N : B}{\Gamma \Rightarrow \lambda x.N : A \supset B}\ \supset I$$

$$\frac{\Gamma \Rightarrow N : A_1 \wedge A_2}{\Gamma \Rightarrow pr_i(N) : A_i}\ \wedge E_i \qquad \frac{\Gamma \Rightarrow N_1 : A_1 \quad \Gamma \Rightarrow N_2 : A_2}{\Gamma \Rightarrow (N_1, N_2) : A_1 \wedge A_2}\ \wedge I$$

$$\frac{\Gamma \Rightarrow N : A \vee A' \quad x : A, \Gamma \Rightarrow N' : C \quad x' : A', \Gamma \Rightarrow N'' : C}{\Gamma \Rightarrow D(N, x.N', x'.N'') : C}\ \vee E \qquad \frac{\Gamma \Rightarrow N : A_i}{\Gamma \Rightarrow in_i(N) : A_1 \vee A_2}\ \vee I_i$$

We write $ap(N, N')$ just as NN', or maybe as (NN') to avoid ambiguity; sometimes however we use the original form for emphasis. D is short for "decide". When (for some context Γ, term N and formula A) one can infer that $\Gamma \Rightarrow N : A$, we also say that (in the context Γ) N has *type* A.

7.2.2 Reductions of Lambda Terms

We use $[N/x]N'$ to indicate the result of *substituting* the term N *for* free occurrences of the variable x *in* the term N'. It is, as usual, capture-avoiding: bound variables are, if necessary, renamed to avoid capture. The order N, x, N' of the sub-expressions in this notation is deliberately chosen to match the order in which they appear in the premisses of the (admissible, by induction on the structure of N') typing rule for the operation:

$$\frac{\Gamma \Rightarrow N : C \quad \Gamma, x : C \Rightarrow N' : B}{\Gamma \Rightarrow [N/x]N' : B}\ Subs.$$

Lemma 7.2.1 *Let N, N' and N'' be terms and let x and y be distinct variables, with x not free in N''. Then*

$$[N''/y]\,[N'/x]\,N \equiv [[N''/y]N'/x]\,[N''/y]\,N.$$

Proof See Barendregt (1984, p. 27); the proof extends without difficulty to cover all the connectives. □

For reference, we give some standard reductions ("detour reductions" (7.2.1)–(7.2.3), "permutation reductions" (7.2.4)–(7.2.7), $\bot$-reductions (7.2.8)–(7.2.10) and a "simplification" (7.2.11)) of lambda terms:

$$(\lambda x.N)N' \rightsquigarrow [N'/x]N \tag{7.2.1}$$

$$pr_i((N_1, N_2)) \rightsquigarrow N_i \tag{7.2.2}$$

$$D(in_i(N), x_1.N_1, x_2.N_2) \rightsquigarrow [N/x_i]N_i \tag{7.2.3}$$

$$D(N, x_1.N_1, x_2.N_2)N' \rightsquigarrow D(N, x_1.N_1N', x_2.N_2N') \tag{7.2.4}$$

$$pr_i(D(N, x_1.N_1, x_2.N_2)) \rightsquigarrow D(N, x_1.pr_i(N_1), x_2.pr_i(N_2)) \tag{7.2.5}$$

$$D(D(N, x_1.N_1, x_2.N_2), y'.N', y''.N'') \rightsquigarrow D(N, x_1.D(N_1, y'.N', y''.N''), x_2.D(N_2, y'.N', y''.N'')) \tag{7.2.6}$$

$$ef(D(N, x_1.N_1, x_2.N_2)) \rightsquigarrow D(N, x_1.ef(N_1), x_2.ef(N_2)) \tag{7.2.7}$$

$$ef(N)N' \rightsquigarrow ef(N) \tag{7.2.8}$$

$$pr_i(ef(N)) \rightsquigarrow ef(N) \tag{7.2.9}$$

$$D(ef(N), x_1.N_1, x_2.N_2)) \rightsquigarrow ef(N) \tag{7.2.10}$$

$$ef(ef(N)) \rightsquigarrow ef(N) \tag{7.2.11}$$

Other reductions might be considered; but, these suffice for our purposes. These include all those given in Pottinger (1977), with, in addition, those ((7.2.8), (7.2.9) and (7.2.10)) required by our avoidance of the restriction in Pottinger (1977) of $\bot E$ to an atomic conclusion C.

Various freshness conditions are required: in (7.2.4), neither x_i is free in N'; in (7.2.6), neither x_i is free in N' or N''.

Confluence and strong normalisation of this system is well-known; see Prawitz (1971) (where the restriction of $\bot E$ to atomic conclusions is inessential). A term-based strong normalisation proof can also be built on the basis of the techniques in von Raamsdonk and Severi (1995), Joachimski and Matthes (2003); see for example Schwichtenberg and Wainer (2012).

7.2.3 Sequent Calculus

First, for clarity, we present a sequent calculus without proof terms. We choose, for reasons discussed elsewhere (Vestergaard 1999) by Vestergaard, a **G3i**-style calculus with principal formulae in the antecedent of the conclusion duplicated into the premisses. All two-premiss rules are context-sharing. Antecedents Γ are (temporarily) multisets of formulae:

$$\frac{}{\bot,\Gamma \Rightarrow C}\ L\bot \qquad \frac{}{A,\Gamma \Rightarrow A}\ Ax$$

$$\frac{A \supset B,\Gamma \Rightarrow A \quad B, A \supset B,\Gamma \Rightarrow C}{A \supset B,\Gamma \Rightarrow C}\ L\supset \qquad \frac{A,\Gamma \Rightarrow B}{\Gamma \Rightarrow A \supset B}\ R\supset$$

$$\frac{A_i, A_1 \wedge A_2,\Gamma \Rightarrow C}{A_1 \wedge A_2,\Gamma \Rightarrow C}\ L\wedge_i \qquad \frac{\Gamma \Rightarrow A \quad \Gamma \Rightarrow B}{\Gamma \Rightarrow A \wedge B}\ R\wedge$$

$$\frac{A_1, A_1 \vee A_2,\Gamma \Rightarrow C \quad A_2, A_1 \vee A_2,\Gamma \Rightarrow C}{A_1 \vee A_2,\Gamma \Rightarrow C}\ L\vee \qquad \frac{\Gamma \Rightarrow A_i}{\Gamma \Rightarrow A_1 \vee A_2}\ R\vee_i$$

$$\frac{\Gamma \Rightarrow A \quad A,\Gamma \Rightarrow C}{\Gamma \Rightarrow C}\ Cut$$

It is convenient to represent derivations by terms, in such a way that a (unique) derivation can be recovered from a term; these appear in the following typing rules. We use short names like X for constructors for typographic reasons; note that A and C are used both as formula meta-variables and (with A for "apply" and C for "cut") as such constructors. The variable Γ now ranges over "contexts", i.e. assignments of formulae (aka "types") to variables, as announced at the start of this Section. Terms are terms of a simple lambda calculus, with a variable binding mechanism, as in for example $C(L, x.L')$ where occurrences of x in L' are bound.

Now for the typing rules:

$$\frac{}{x:\bot,\Gamma \Rightarrow X(x):C}\ L\bot \qquad \frac{}{x:A,\Gamma \Rightarrow x:A}\ Ax$$

$$\frac{z:A \supset B,\Gamma \Rightarrow L:A \quad y:B,z:A \supset B,\Gamma \Rightarrow L':C}{z:A \supset B,\Gamma \Rightarrow A(z,L,y.L'):C}\ L\supset \qquad \frac{x:A,\Gamma \Rightarrow L:B}{\Gamma \Rightarrow \lambda x.L:A \supset B}\ R\supset$$

$$\frac{x:A_i,z:A_1 \wedge A_2,\Gamma \Rightarrow L:C}{z:A_1 \wedge A_2,\Gamma \Rightarrow E_i(z,x.L):C}\ L\wedge_i \qquad \frac{\Gamma \Rightarrow L:A \quad \Gamma \Rightarrow L':B}{\Gamma \Rightarrow (L,L'):A \wedge B}\ R\wedge$$

$$\frac{x:A_1,z:A_1 \vee A_2,\Gamma \Rightarrow L:C \quad x':A_2,z:A_1 \vee A_2,\Gamma \Rightarrow L':C}{z:A_1 \vee A_2,\Gamma \Rightarrow W(z,x.L,x'.L'):C}\ L\vee \qquad \frac{\Gamma \Rightarrow L:A_i}{\Gamma \Rightarrow in_i(L):A_1 \vee A_2}\ R\vee_i$$

$$\frac{\Gamma \Rightarrow L:A \quad x:A,\Gamma \Rightarrow L':C}{\Gamma \Rightarrow C(L,x.L'):C}\ Cut$$

Freshness constraints:

1. in $L\bot$, x is fresh for Γ;
2. in Ax, x is fresh for Γ;
3. in $L\supset$, z is fresh for Γ and y is fresh for $\Gamma, z : A \supset B$;
4. in $R\supset$, x is fresh for Γ;
5. in $L\wedge_i$, z is fresh for Γ and x is fresh for $\Gamma, z : A_1 \wedge A_2$.
6. in $L\vee$, z is fresh for Γ, and both x and x' are fresh for $\Gamma, z : A_1 \vee A_2$;

7. in Cut, x is fresh for Γ.

Various easy consequences follow, e.g. that, in the rule $L\supset$, the variable y is not free in L.

7.3 Translation from Sequent Calculus into Natural Deduction

The following translation (Gentzen 1935; Prawitz 1965; Zucker 1974) is standard: details are only given for complete clarity.

$$\phi(X(x)) \equiv ef(x) \tag{7.3.1}$$
$$\phi(x) \equiv x \tag{7.3.2}$$
$$\phi(A(z, L, y.L')) \equiv [z\phi(L)/y]\,\phi(L') \tag{7.3.3}$$
$$\phi(\lambda x.L) \equiv \lambda x.\phi(L) \tag{7.3.4}$$
$$\phi(E_i(z, x.L)) \equiv [pr_i(z)/x]\,\phi(L) \tag{7.3.5}$$
$$\phi((L, L')) \equiv (\phi(L), \phi(L')) \tag{7.3.6}$$
$$\phi(W(z, x'.L', x''.L'')) \equiv D(z, x'.\phi(L'), x''.\phi(L'')) \tag{7.3.7}$$
$$\phi(in_i(L)) \equiv in_i(\phi(L)) \tag{7.3.8}$$
$$\phi(C(L, x.L')) \equiv [\phi(L)/x]\,\phi(L') \tag{7.3.9}$$

7.4 Translation from Natural Deduction to Sequent Calculus

The following translation, due to Gentzen (1935), is only used here as evidence that the translation ϕ given above is surjective.

$$\psi(ef(x)) \equiv X(x) \tag{7.4.1}$$
$$\psi(x) \equiv x \tag{7.4.2}$$
$$\psi(ap(N, N')) \equiv C(\psi(N), x.A(x, \psi(N'), y.y)) \tag{7.4.3}$$
$$\psi(\lambda x.N) \equiv \lambda x.\psi(N) \tag{7.4.4}$$
$$\psi(pr_i(N)) \equiv C(\psi(N), x.E_i(x, y.y)) \tag{7.4.5}$$
$$\psi((N, N')) \equiv (\psi(N), \psi(N')) \tag{7.4.6}$$
$$\psi(D(N, x'.N', x''.N'')) \equiv C(\psi(N), z.W(z, x'.\psi(N'), x''.\psi(N''))) \tag{7.4.7}$$
$$\psi(in_i(N)) \equiv in_i(\psi(N)) \tag{7.4.8}$$

It is routine to observe that the composite of the two translations translates every natural deduction term to itself; thus ϕ is surjective (indeed, a retraction) and ψ is injective. In other words, for every N, $\phi(\psi(N)) = N$. The argument is, exploiting freshness of variables, by induction and case analysis:

1. $\phi(\psi(X(x))) \equiv \phi(ef(x)) \equiv X(x)$
2. $\phi(\psi(x)) \equiv \phi(x) \equiv x$
3. $\phi(\psi(ap(N, N')))$
   $\equiv \phi(C(\psi(N), x.A(x, \psi(N'), y.y)))$
   $\equiv [\phi(\psi(N))/x]\phi(A(x, \psi(N'), y.y))$
   $\equiv [N/x]\phi(A(x, \psi(N'), y.y))$
   $\equiv [N/x][ap(x, \phi(\psi(N')))/y]y$
   $\equiv [N/x]ap(x, \phi(\psi(N')))$
   $\equiv [N/x]ap(x, N')$
   $\equiv ap(N, N')$
4. $\phi(\psi(\lambda x.N)) \equiv \phi(\lambda x.\psi(N)) \equiv \lambda x.\phi(\psi(N)) \equiv \lambda x.N$
5. $\phi(\psi(pr_i(N)))$
   $\equiv \phi(C(\psi(N), x.E_i(x, y.y)))$
   $\equiv [\phi(\psi(N))/x]\phi(E_i(x, y, y))$
   $\equiv [\phi(\psi(N))/x][pr_i(x)/y]y$
   $\equiv [\phi(\psi(N))/x]pr_i(x)$
   $\equiv [N/x]pr_i(x)$
   $\equiv pr_i(N)$
6. $\phi(\psi((N, N'))) \equiv \phi((\psi(N), \psi(N'))) \equiv (\phi(\psi(N)), \phi(\psi(N'))) \equiv (N, N')$
7. $\phi(\psi(D(N, x'.N', x''.N'')))$
   $\equiv \phi(C(\psi(N), z.W(z, x'.\psi(N'), x''.\psi(N''))))$
   $\equiv [\phi(\psi(N))/z]\phi(W(z, x'.\psi(N'), x''.\psi(N'')))$
   $\equiv [N/z]\phi(W(z, x'.\psi(N'), x''.\psi(N'')))$
   $\equiv [N/z]D(z, x'.\phi(\psi(N')), x''.\phi(\psi(N'')))$
   $\equiv D(N, x'.N', x''.N'')$
8. $\phi(\psi(in_i(N))) \equiv \phi(in_i(\psi(N))) \equiv in_i(\phi(\psi(n))) \equiv N$

7.5 Substitution

We now present 32 lemmas about substitution; their only interest is that they are exactly what is required to show that cut-reduction steps translate to sequences of reduction steps. It is however of interest to see exactly what properties of substitution are required for the main result to hold.

Lemma 7.5.1 *Let N be a term and x and y be distinct variables. Then*

$$[N/x]y \equiv y.$$

Proof By definition of substitution. □

Lemma 7.5.2 *Let N be a term and x be a variable. Then*

$$[N/x]x \equiv N.$$

Proof By definition of substitution. □

Lemma 7.5.3 *Let N be terms and x and y be distinct variables, with y not free in N. Then*

$$[N/x](\lambda y.N') \equiv \lambda y.[N/x]N'.$$

Proof By definition of substitution. □

Lemma 7.5.4 *Let N, N' and N'' be terms and x, y and z be distinct variables, with x not free in N and z not free in N or N'. Then*

$$[N/x][yN'/z]N'' \equiv [y[N/x]N'/z][N/x]N''.$$

Proof By Lemma 7.2.1, since z is not free in N, the LHS $\equiv$

$$[[N/x](yN')/z][N/x]N''$$

and, since, $x \not\equiv y$, $[N/x](yN') \equiv y[N/x]N'$. The result now follows. □

Lemma 7.5.5 *Let N' and N'' be terms and w, x and z be distinct variables, with z not free in N'. Then*

$$[w/x][xN'/z]N'' \equiv [w[w/x]N'/z][w/x]N''.$$

Proof By Lemma 7.2.1, since $z \neq w$, the LHS $\equiv [[w/x](xN')/z][w/x]N''$ and, by definition of substitution, $[w/x](xN') \equiv w[w/x]N'$. The result now follows. (The freshness hypothesis about z is not used.) □

Lemma 7.5.6 *Let N, N^*, N' and N'' be terms and w, x, y and z be distinct variables, with x not free in N or N^*, y not free in N, N^* or N' and z not free in N, N' or N''. Then*

$$[[wN/z]N^*/x][xN'/y]N'' \equiv [wN/z][N^*/x][xN'/y]N''.$$

Proof Letting $M \equiv [xN'/y]N''$, observe that the RHS is just $[wN/z][N^*/x]M$; by Lemma 7.2.1, since x is not free in wN, this $\equiv [[wN/z]N^*/x][wN/z]M$. But, $z \neq x$ and z not free in N or N'' imply that $[wN/z]M \equiv M$. By symmetry of $\equiv$, the result now follows. (The freshness hypotheses about x and y and that about z w.r.t. N are not used.) □

Lemma 7.5.7 *Let N, N' and N'' be terms and w, x and y be distinct variables, with w not free in N' or N'', x not free in N and y not free in N or N'. Then*

$$[\lambda w.N/x][xN'/y]N'' \quad \rightsquigarrow^* \quad [[\lambda w.N/x]N'/w][N/y][\lambda w.N/x]N''.$$

Proof Let $M' \equiv [\lambda w.N/x]N'$ and $M'' \equiv [\lambda w.N/x]N''$. By definition of substitution, $M' \equiv (\lambda w.N)M'$. By Lemma 7.2.1, since y is not free in $\lambda w.N$, the LHS (of the present lemma) $\equiv [[\lambda w.N/x](xN')/y]M''$, which (by definition of substitution) $\equiv [(\lambda w.N)M'/y]M''$. This reduces (in 0 or more steps) by Rule 7.2.1 to $[[M'/w]N/y]M''$; there will be 0 steps if, for example, y is not free in N'', but in general there may be several such steps. But, by Lemma 7.2.1, since y is not free in M' (because not free in N or N'), we have $[M'/w][N/y]M'' \equiv [[M'/w]N/y]M''$; by symmetry of $\equiv$ the result follows. (The freshness hypotheses about w and x are not used.) □

Lemma 7.5.8 *Let N, N' and N'' be terms and x be a variable, not free in N. Then*

$$[N/x](N', N'') \equiv ([N/x]N', [N/x]N'').$$

Proof By definition of substitution. □

Lemma 7.5.9 *Let N and N' be terms and x, y and z be distinct variables, with x not free in N and z not free in N. Then*

$$[N/x][pr_i(y)/z]N' \equiv [pr_i(y)/z][N/x]N'.$$

Proof By Lemma 7.2.1, since z is not free in N and, from $x \neq y$, $[N/x]pr_i(y) \equiv pr_i(y)$. □

Lemma 7.5.10 *Let N' be a term and w, x and z be distinct variables. Then*

$$[w/x][pr_i(x)/z]N' \equiv [pr_i(w)/z][w/x]N'.$$

Proof By Lemma 7.2.1, since $z \neq w$ and $[w/x]pr_i(x) \equiv pr_i(w)$. □

Lemma 7.5.11 *Let N' and N'' be terms and w, x, y and z be distinct variables, with x not free in N'', y not free in N' and z not free in N''. Then*

$$[[pr_i(w)/y]N''/x]\,[pr_j(x)/z]\,N' \equiv [pr_i(w)/y]\,[N''/x]\,[pr_j(x)/z]\,N'.$$

Proof By Lemma 7.2.1, since $x \not\equiv w$, we obtain, with $N''' \equiv [pr_j(x)/z]N'$,

$$[pr_i(w)/y]\,[N''/x]\,N''' \equiv [[pr_i(w)/y]N''/x]\,[pr_i(w)/y]\,N'''.$$

But, $y \neq x$ and y is not free in N', so, y is not free in N'''. By definition of substitution, we obtain

$$[[pr_i(w)/y]N''/x]\,[pr_i(w)/y]\,N''' \equiv [[pr_i(w)/y]N''/x]\,N'''.$$

The result now follows by symmetry and transitivity of $\equiv$. □

Lemma 7.5.12 *Let N, N' and N'' be terms and w, x, y and z be distinct variables, with x not free in N, y not free in N or N' and z not free in N' or N''. Then*

$$[[pr_i(w)/z]N/x]\,[xN'/y]N'' \equiv [pr_i(w)/z]\,[N/x]\,[xN'/y]N''.$$

Proof By Lemma 7.2.1 and symmetry of $\equiv$, since w and x are distinct and since z is not free in $[xN'/y]N''$. □

Lemma 7.5.13 *Let N, N' and N'' be terms and w, x, y and z be distinct variables, with x not free in N or N'', y not free in N or N' and z not free in N or N''. Then*

$$[[wN/y]N''/x]\,[pr_i(x)/z]N' \equiv [wN/y]\,[N''/x]\,[pr_i(x)/z]N'.$$

Proof Let $M \equiv [pr_i(x)/z]N'$; by symmetry of $\equiv$, we have to show that

$$[wN/y]\,[N''/x]M \equiv [[wN/y]N''/x]M.$$

Since x is not free in wN and y is not free in M, this follows by Lemma 7.2.1. □

Lemma 7.5.14 *Let N_1, N_2 and N' be terms and x and z be distinct variables, with x not free in N_1 or N_2 and z not free in N_1 or N_2. Then*

$$[(N_1, N_2)/x][(pr_i(x)/z]N' \quad \rightsquigarrow^* \quad [N_i/z][(N_1, N_2)/x]N'.$$

Proof By Lemma 7.2.1, since z is not free in (N_1, N_2), we obtain

$$[(N_1, N_2)/x][pr_i(x)/z]N' \equiv [pr_i([(N_1, N_2)/x]x)/z][(N_1, N_2)/x]N'$$

which, by definition of substitution for x in x, is just

$$[(N_1, N_2)/x][(pr_i(x)/z]N' \equiv [pr_i((N_1, N_2))/z][(N_1, N_2)/x]N'$$

But also

$$[pr_i((N_1, N_2))/z][(N_1, N_2)/x]N' \quad \rightsquigarrow^* \quad [N_i/z][(N_1, N_2)/x]N'$$

whence the result. □

Lemma 7.5.15 *Let N and N' be terms and x be a variable, with x not free in N. Then*

$$[N/x][in_i(N') \equiv in_i([N/x]N'.$$

Proof By definition of substitution. □

Lemma 7.5.16 *Let N, N' and N'' be terms and x, y, z' and z'' be distinct variables, with x not free in N, z' not free in N and z'' not free in N. Then*

$$[N/x]D(y, z'.N', z''.N'') \equiv D(y, z'.[N/x]N', z''.[N/x]N'').$$

Proof By definition of substitution. □

Lemma 7.5.17 *Let N' and N'' be terms and w, x, z' and z'' be distinct variables. Then*

$$[w/x]D(x, z'.N', z''.N'') \equiv D(w, z'.[w/x]N', z''.[w/x]N'').$$

Proof By definition of substitution. □

Lemma 7.5.18 *Let N_1, N_2, N' and N'' be terms and w, w_1, w_2, x, x' and x'' be distinct variables, with w_1 not free in N_2 or N' or N'', w_2 not free in N_1 or N' or N'', x not free in N_1 or N_2, x' not free in N_1 or N_2 or N'' and x'' not free in N_1 or N_2 or N'. Then, with $M \equiv D(w, w_1.N_1, w_2.N_2)$,*

$$\begin{array}{c}[D(w, w_1.N_1, w_2.N_2)/x]\ D(x, x'.N', x''.N'') \\ \rightsquigarrow^* \\ D(w,\ w_1.[N_1/x]D(x, x'.[M/x]N', x''.[M/x]N''), \\ \quad w_2.[N_2/x]D(x, x'.[M/x]N', x''.[M/x]N'')).\end{array}$$

Proof Note that neither x nor x' nor x'' is free in M. By definition of substitution,

$$[M/x]\ D(x, x'.N', x''.N'') \equiv D(M, x'.[M/x]N', x''.[M/x]N'').$$

By permutation reduction rule 7.2.6, the RHS of this reduces to

$$D(w,\ w_1.D(N_1, x'.[M/x]N', x''.[M/x]N''),\ w_2.D(N_2, x'.[M/x]N', x''.[M/x]N'')).$$

But, since x is not free in M, this is $\equiv$ to

$$\begin{array}{l}D(w,\ w_1.[N_1/x]D(x, x'.[M/x]N', x''.[M/x]N''), \\ \quad w_2.[N_2/x]D(x, x'.[M/x]N', x''.[M/x]N'')),\end{array}$$

as required. □

Lemma 7.5.19 *Let N_1, N_2 and N be terms and x, y_1 and y_2 be distinct variables, with x not free in N, y_1 not free in N and y_2 not free in N. Then*

$$[in_i(N)/x]D(x, y_1.N_1, y_2.N_2)) \quad \rightsquigarrow^* \quad [N/y_i][in_i(N)/x]N_i.$$

Proof The LHS is $\equiv$ to $D(in_i(N), y_1.[in_i(N)/x]N_1, y_2.[in_i(N)/x]N_2))$, by definition of substitution; this reduces by 7.2.3 to the RHS. □

Lemma 7.5.20 *Let N_1, N_2, N' and N'' be terms and x, y, z, w_1 and w_2 be distinct variables, with x not free in N_1 or N_2, y not free in N_1, N_2 or N', w_1 not free in*

N' or N'' and w_2 not free in N' or N''. Then (writing $M \equiv D(z, w_1.N_1, w_2.N_2)$ for brevity)

$$\begin{array}{c}[M/x]\,[xN'/y]N'' \\ \rightsquigarrow^* \\ [D(z, w_1.[N_1/x](x[M/x]N'), w_2.[N_2/x](x[M/x]N'))/y]\,[M/x]\,N''\end{array} \tag{7.5.1}$$

Proof By Lemma 7.2.1, and since y is free in M,

$$[M/x]\,[xN'/y]N'' \equiv [M[M/x]N'/y]\,[M/x]\,N'';$$

By reduction rule 7.2.4, $M[M/x]N' \rightsquigarrow^* D(z, w_1.N_1[M/x]N', w_2.N_2[M/x]N')$, whence

$$[M[M/x]N'/y]\,[M/x]\,N'' \quad \rightsquigarrow^*$$

$$[D(z, w_1.N_1[M/x]N', w_2.N_2[M/x]N')/y]\,[M/x]\,N''.$$

Since x is not free in M, and thus not free in $[M/x]N'$, this is $\equiv$ to

$$[D(z, w_1.[N_1/x](x[M/x]N'), w_2.[N_2/x](x[M/x]N'))/y]\,[M/x]\,N'',$$

as required. □

Lemma 7.5.21 *Let N_1, N_2, N and N'' be terms and x, w and z be distinct variables, distinct from w_1 and w_2, with x not free in N or N'' and z not free in N, N_1 or N_2. Then*

$$[[wN/z]N''/x]D(x, w_1.N_1, w_2.N_2) \equiv [wN/z][N''/x]D(x, w_1.N_1, w_2.N_2).$$

Proof Essentially the same as (7.5.6). In other words, letting $M \equiv D(x, w_1.N_1, w_2.N_2)$, observe that the RHS is just $[wN/z][N''/x]M$; by Lemma 7.2.1, since x is not free in wN, this $\equiv [[wN/z]N''/x][wN/z]M$. But, $z \neq x$ and z not free in N_1 or N_2 imply that $[wN/z]M \equiv M$. By symmetry of $\equiv$, the result now follows. □

Lemma 7.5.22 *Let N_1, N_2 and N be terms and x, y and z be distinct variables, distinct from w_1 and w_2, with x not free in N_1 or N_2 and y not free in N_1 or N_2 and w_1 not free in N and w_2 not free in N. Then*

$$\begin{array}{c}[D(z, w_1.N_1, w_2.N_2)/x]\,[(pr_i(x)/y]\,N \\ \rightsquigarrow^* \\ [D(z, w_1.pr_i(N_1), w_2.pr_i(N_2))/y]\,[D(z, w_1.N_1, w_2.N_2)/x]\,N\ .\end{array} \tag{7.5.2}$$

Proof By Lemma 7.2.1 and the conditions on y, and with $N' \equiv [D(z, w_1.N_1, w_2.N_2)/x]N$, we have

$$[D(z, w_1.N_1, w_2.N_2)/x]\,[(pr_i(x)/y]\,N \quad \equiv \quad [pr_i(D(z, w_1.N_1, w_2.N_2))/y]N'.$$

By the reduction 7.2.5, this $\equiv$

$$[D(z, w_1.pr_i(N_1), w_2.pr_i(N_2))/y]N'$$

as required. □

Lemma 7.5.23 *Let N, N_1, N_2 be terms and let x, y and z be distinct variables, distinct from w_1 and w_2, with x not free in N and y not free in N_1 or N_2. Then*

$$[[pr_i(z)/y]N/x]\;D(x, w_1.N_1, w_2.N_2) \;\equiv\; [pr_i(z)/y]\;[N/x]\;D(x, w_1.N_1, w_2.N_2).$$

Proof Again, essentially the same as Lemma 7.5.6. Let $M \equiv D(x, w_1.N_1, w_2.N_2)$. By Lemma 7.2.1, the RHS $\equiv$

$$[[pr_i(z)/y]N]/x][pr_i(z)/y]M.$$

But, since y is not free in M, this is $\equiv$ the LHS. □

Lemma 7.5.24 *Let N be a term and x and y be distinct variables, with x not free in N. Then*

$$[N/x](ef(y)) \equiv ef(y).$$

Proof By definition of substitution. □

Lemma 7.5.25 *Let w and x be distinct variables. Then*

$$[w/x](ef(x)) \equiv ef(w).$$

Proof By definition of substitution. □

Lemma 7.5.26 *Let w and x be distinct variables. Then*

$$[ef(w)/x](ef(x)) \quad \leadsto^* \quad ef(w).$$

Proof By definition of substitution and Rule 7.2.11. □

Lemma 7.5.27 *Let N and N' be terms and x, y and z be distinct variables, with y not free in N. Then*

$$[ef(z)/x][xN/y]N' \quad \leadsto^* \quad [ef(z)/y][ef(z)/x]N'.$$

Proof By Lemma 7.2.1, the LHS $\equiv [ef(z)[ef(z)/x]N/y][ef(z)/x]N'$. It now suffices to show that

$$ef(z)[ef(z)/x]N \leadsto^* ef(z);$$

this follows by Rule 7.2.8. □

Lemma 7.5.28 *Let N and N' be terms and w, x and z be distinct variables, with x not free in N or N' and z not free in N. Then*

$$[[wN/z]N'/x]\, ef(x) \equiv [wN/z]\, [N'/x]\, ef(x).$$

Proof Since x is not free in wN, by Lemma 7.2.1 the RHS is $[[wN/z]N'/x]\, [wN/z]\, ef(x)$. Since $z \not\equiv x$, and so is not free in $ef(x)$, this is $\equiv$ to the LHS. □

Lemma 7.5.29 *Let N be a term and x, y and z be distinct variables. Then*

$$[ef(z)/x]\, [pr_i(x)/y]\, N \quad \rightsquigarrow^* \quad [ef(z)/y]\, [ef(z)/x]\, N.$$

Proof By Lemma 7.2.1, the LHS $\equiv [pr_i(ef(z))/y]\, [ef(z)/x]\, N$. By Rule 7.2.9, this reduces to

$$[ef(z)/y]\, [ef(z)/x]\, N.$$

□

Lemma 7.5.30 *Let N be a term and w, x and z be distinct variables, with x not free in N. Then*

$$[[pr_i(w)/z]N/x]ef(x) \equiv [pr_i(w)/z][N/x]ef(x)$$

Proof Since $x \not\equiv w$ and z is not free in $ef(x)$, by Lemma 7.2.1 the RHS simplifies to the LHS. □

Lemma 7.5.31 *Let N' and N'' be terms and x, z, y' and y'' be distinct variables. Then*

$$[ef(z)/x]D(x, y'.N', y''.N'') \quad \rightsquigarrow^* \quad ef(z)$$

Proof The LHS is just $D(ef(z), y'.[ef(z)/x]N', y''.[ef(z)/x]N'')$; by Rule 7.2.10 this reduces to the RHS. □

Lemma 7.5.32 *Let N and N' be terms and x, w, y and y' be distinct variables. Then*

$$[D(w, y.N, y'.N')/x]ef(x) \quad \rightsquigarrow^* \quad D(w, y.[N/x]ef(x), y'.[N'/x]ef(x)).$$

Proof The LHS is, by definition of substitution, $ef(D(w, y.N, y'.N'))$; this reduces by Rule 7.2.7 to $D(w, y.ef(N), y'.ef(N'))$, which, by definition of substitution, is just the RHS. □

7.6 Cut Reduction

Consider the following rules for reducing *Cut*:

$$C(L, x.y) \rightsquigarrow y \tag{7.6.1}$$

$$C(L, x.x) \rightsquigarrow L \tag{7.6.2}$$

$$C(L, x.\lambda y.L') \rightsquigarrow \lambda y.C(L, x.L') \tag{7.6.3}$$

$$C(L, x.A(y, L', z.L'')) \rightsquigarrow A(y, C(L, x.L'), z.C(L, x.L'')) \tag{7.6.4}$$

$$C(w, x.A(x, L', z.L'')) \rightsquigarrow A(w, C(w, x.L'), z.C(w, x.L'')) \tag{7.6.5}$$

$$C(A(w, L, z.L^*), x.A(x, L', y.L'')) \rightsquigarrow A(w, L, z.C(L^*, x.A(x, L', y.L''))) \tag{7.6.6}$$

$$C(\lambda w.L, x.A(x, L', y.L'')) \rightsquigarrow C(C(\lambda w.L, x.L'), w.C(L, y.C(\lambda w.L, x.L''))) \tag{7.6.7}$$

By Lemmata (7.5.1)–(7.5.7) respectively, these rules are *semantically sound*, i.e. the translations of these rules into the language of natural deduction are either identities or (for the last rule) 0 or more reductions. The various hypotheses about distinctness and freeness in those lemmata come direct from the conditions implicit in the notation of these rules.

We now add further rules, for dealing with conjunction (and the interaction between conjunction and implication):

$$C(L, x.(L', L'')) \rightsquigarrow (C(L, x.L'), C(L, x.L'')) \tag{7.6.8}$$

$$C(L, x.E_i(y, z.L')) \rightsquigarrow E_i(y, z.C(L, x.L')) \tag{7.6.9}$$

$$C(w, x.E_i(x, z.L')) \rightsquigarrow E_i(w, z.C(w, x.L')) \tag{7.6.10}$$

$$C(E_i(w, y.L''), x.E_j(x, z.L')) \rightsquigarrow E_i(w, y.C(L'', x.E_j(x, z.L'))) \tag{7.6.11}$$

$$C(E_i(w, z.L), x.A(x, L', y.L'')) \rightsquigarrow E_i(w, z.C(L, x.A(x, L', y.L''))) \tag{7.6.12}$$

$$C(A(w, L, y.L''), x.E_i(x, z.L')) \rightsquigarrow A(w, L, y.C(L'', x.E_i(x, z.L'))) \tag{7.6.13}$$

$$C((L_1, L_2), x.E_i(x, z.L')) \rightsquigarrow C(L_i, z.C((L_1, L_2), x.L')) \tag{7.6.14}$$

Their semantic soundness follows from Lemmata (7.5.8)–(7.5.14) respectively.

Now we add the rules for disjunction (and its interactions with conjunction and implication):

$$C(L, x.in_i(L')) \rightsquigarrow in_i(C(L, x.L')) \tag{7.6.15}$$

$$C(L, x.W(y, z'.L', z''.L'')) \rightsquigarrow W(y, z'.C(L, x.L'), z''.C(L, x.L'')) \tag{7.6.16}$$

$$C(w, x.W(x, z'.L', z''.L'')) \rightsquigarrow W(w, z'.C(w, x.L'), z''.C(w, x.L'')) \tag{7.6.17}$$

$$\begin{aligned} C(W(w, w_1.L_1, w_2.L_2), x.W(x, x'.L', x''.L'')) \rightsquigarrow \\ W(w, w_1.C(L_1, x.W(x, x'.C(W(w, w_1.L_1, w_2.L_2), x.L'), \\ x''.C(W(w, w_1.L_1, w_2.L_2), x.L''))), \\ w_2.C(L_2, x.W(x, x'.C(W(w, w_1.L_1, w_2.L_2), x.L'), \\ x''.C(W(w, w_1.L_1, w_2.L_2), x.L'')))) \end{aligned} \tag{7.6.18}$$

$$C(in_i(L), x.W(x, y_1.L_1, y_2.L_2)) \rightsquigarrow C(L, y_i.C(in_i(L), x.L_i)) \quad (7.6.19)$$

$$\begin{aligned}&C(W(z, w_1.L_1, w_2.L_2), x.A(x, L', y.L'')) \rightsquigarrow \\ &\quad C(W(z, w_1.C(L_1, x.A(x, C(W(z, w_1.L_1, w_2.L_2), x.L'), y.y)), \\ &\quad\quad w_2.C(L_2, x.A(x, C(W(z, w_1.L_1, w_2.L_2), x.L'), y.y))), \\ &\quad y.C(W(z, w_1.L_1, w_2.L_2), x.L''))\end{aligned} \quad (7.6.20)$$

$$\begin{aligned}&C(A(w, L, z.L''), x.W(x, w_1.L_1, w_2.L_2)) \rightsquigarrow \\ &\quad A(w, L, z.C(L'', x.W(x, w_1.L_1, w_2.L_2)))\end{aligned} \quad (7.6.21)$$

$$\begin{aligned}&C(W(z, w_1.L_1, w_2.L_2), x.E_i(x, y.L)) \rightsquigarrow \\ &\quad C(W(z, w_1.C(L_1, x.E_i(x, y.y)), w_2.C(L_2, x.E_i(x, y.y))), \\ &\quad y.C(W(z, w_1.L_1, w_2.L_2), x.L))\end{aligned} \quad (7.6.22)$$

$$C(E_i(z, y.L), x.W(x, w_1.L_1, w_2.L_2)) \rightsquigarrow E_i(z, y.C(L, x, W(x, w_1.L_1, w_2.L_2))) \quad (7.6.23)$$

Their semantic soundness follows from Lemmata (7.5.15)–(7.5.23) respectively. Finally the rules for absurdity (and its interactions with the other logical constants):

$$C(L, x.X(y)) \rightsquigarrow X(y) \quad (7.6.24)$$

$$C(w, x.X(x)) \rightsquigarrow X(w) \quad (7.6.25)$$

$$C(X(w), x.X(x)) \rightsquigarrow X(w) \quad (7.6.26)$$

$$C(X(z), x.A(x, L, y.L')) \rightsquigarrow C(X(z), y.C(X(z), x.L')) \quad (7.6.27)$$

$$C(A(w, L, z.L'), x.X(x)) \rightsquigarrow A(w, L, z.C(L', x.X(x))) \quad (7.6.28)$$

$$C(X(z), x.E_i(x, y.L)) \rightsquigarrow C(X(z), y.C(X(z), x.L)) \quad (7.6.29)$$

$$C(E_i(w, z.L), x.X(x)) \rightsquigarrow E_i(w, z.C(L, x.X(x))) \quad (7.6.30)$$

$$C(X(z), x.W(x, y.L, y'.L')) \rightsquigarrow X(z) \quad (7.6.31)$$

$$C(W(w, y.L, y'.L'), x.X(x)) \rightsquigarrow W(w, y.C(L, x.X(x)), y'.C(L', x.X(x)))) \quad (7.6.32)$$

Their semantic soundness follows from Lemmata (7.5.24)–(7.5.32) respectively.

7.7 Completeness

Where the cut formula is non-principal in the second premiss, we use one of the reductions (7.6.1), (7.6.3), (7.6.4), (7.6.8), (7.6.9), (7.6.15), (7.6.16) and (7.6.24), according to the form of the last step of the second premiss.

Otherwise, we consider the cases where the cut formula is principal in the second premiss, presented in tabular form (where the first column indicates the last step of the first premiss and the top row indicates the last step of the second premiss). Note that some pairs, e.g. $R\supset/L\wedge$, can never arise.

It follows that every cut (except possibly where one of the premisses ends with a cut) can be matched to the LHS of one of the 32 reduction rules.

Table 7.1 ...

	Ax	$L\supset$	$L\wedge$	$L\vee$	$L\bot$
Ax	7.6.2	7.6.5	7.6.10	7.6.17	7.6.25
$L\supset$	↓	7.6.6	7.6.13	7.6.21	7.6.28
$L\wedge$	↓	7.6.12	7.6.11	7.6.23	7.6.30
$L\vee$	↓	7.6.20	7.6.22	7.6.18	7.6.32
$L\bot$	↓	7.6.27	7.6.29	7.6.31	7.6.26
$R\supset$	↓	7.6.7	–	–	–
$R\wedge$	↓	–	7.6.14	–	–
$R\vee$	↓	–	–	7.6.19	–

7.8 Counterexamples

Now that all the reduction rules have been presented, we are able to present some counterexamples (suggested by Graham-Lengrand). An alternative reduction for the antecedent of (7.6.18) is just to permute the cut into the first premiss without any adjustments. Thus, with $R \equiv W(x, x'.L', x''.L'')$,

$$C(W(w, w_1.L_1, w_2.L_2), x.R)$$

would be transformed to

$$W(w, w_1.C(L_1, x.R), w_2.C(L_2, x.R)).$$

Consider, with $w : A_1 \vee A_2$ and $L_i \equiv z_i : B_1 \vee B_2$, the term (of type $B_1 \vee B_2$)

$$L \equiv C(W(w, w_1.z_1, w_2.z_2), x.W(x, x_1.x, x_2.x)),$$

which reduces by this rule to

$$L' \equiv W(w, w_1.C(z_1, x.W(x, x_1.x, x_2.x)), w_2.C(z_2, x.W(x, x_1.x, x_2.x))).$$

The natural deduction image $\phi(L)$ of L is

$$[D(w, w_1.z_1, w_2.z_2)/x]D(x, x_1.x, x_2.x)$$

i.e.

$$N \equiv D(D(w, w_1.z_1, w_2.z_2), x_1.D(w, w_1.z_1, w_2.z_2), x_2.D(w, w_1.z_1, w_2.z_2)).$$

The only reduction applicable to N is the permutative reduction (7.2.6), which reduces it to the normal term

$$N_1 \equiv D(w, w_1.D(z_1, x_1.S, x_2.S), w_2.D(z_2, x_1.S, x_2.S))$$

where

$$S \equiv D(w, w_1.z_1, w_2.z_2).$$

But, the natural deduction image $\phi(L')$ of L' is

$$D(w, w_1.[z_1/x]D(x, x_1.x, x_2.x)), w_2.[z_2/x]D(x, x_1.x, x_2.x)))$$

i.e.

$$N' \equiv D(w, w_1.D(z_1, x_1.z_1, x_2.z_1)), w_2.D(z_2, x_1.z_2, x_2.z_2))).$$

But N' is neither N nor N_1; so it is not the case that

$$N \rightsquigarrow^* N'.$$

It is also not the case that $N' \rightsquigarrow^* N_1$; since N_1 is normal (and the reduction system is confluent), we conclude that it is not even the case that $N = N'$ (in the equational theory generated by $\rightsquigarrow^*$).

Let us consider adding "immediate simplification" to the reduction system for **NJ**, i.e. adding the reduction of an $\vee E$ step, when a minor premiss does not use the extra assumption, to the derivation given by the minor premiss. Such immediate simplifications destroy confluence, so we prefer to avoid them; in their presence, however, it is now the case that $N_1 \rightsquigarrow^* N'$ (and hence $N \rightsquigarrow^* N'$), as follows:

$$N_1 \equiv D(w, w_1.D(z_1, x_1.S, x_2.S), w_2.D(z_2, x_1.S, x_2.S))$$

$$\rightsquigarrow^*$$

$$N' \equiv D(w, w_1.D(z_1, x_1.z_1, x_2.z_1), w_2.D(z_2, x_1.z_2, x_2.z_2))$$

by using immediate simplifications $S \rightsquigarrow z_i$ on the four different copies of S.

But we can block these immediate simplifications. A temporary abbreviation will be useful: we let xM abbreviate $A(x, M, u.u)$.

Recall that

$$L \equiv C(W(w, w_1.z_1, w_2.z_2), x.W(x, x_1.x, x_2.x)) : B_1 \vee B_2.$$

Each term $z_i : B_1 \vee B_2$ in L is replaced by $y_i w_i$ where $y_i : A_i \supset (B_1 \vee B_2)$; unlike z_i, this depends on w_i. The term $R \equiv W(x, x_1.x, x_2.x)$ in L is replaced by

$$T \equiv W(x, x_1.A(u_1, x_1, z.zx), x_2.A(u_2, x_2, z.zx)),$$

where $u_i : B_i \supset ((B_1 \vee B_2) \supset E)$. The fresh bound variable z herein is of type $(B_1 \vee B_2) \supset E$; so zx is of type E. Note that $A(u_i, x_i, z.zx)$ depends on x_i.

So the new term $L_0 : E$ whose cut, in the context

$w : A_1 \vee A_2, y_1 : A_1 \supset (B_1 \vee B_2), y_2 : A_2 \supset (B_1 \vee B_2), u_1 : B_1 \supset ((B_1 \vee B_2) \supset E), u_2 : B_2 \supset ((B_1 \vee B_2) \supset E)$

we reduce by our questionable reduction, is

$$L_0 \equiv C(W(w, w_1.y_1w_1, w_2.y_2w_2), x.T) : E$$

which reduces to

$$L_0' \equiv W(w, w_1.C(y_1w_1, x.T), w_2.C(y_2w_2, x.T)) : E.$$

We now consider their natural deduction images $N_0 \equiv \phi(L_0)$ and $N_0' \equiv \phi(L_0')$. Carrying out the substitutions, we find that

$$N_0 \equiv [D(w, w_1.y_1w_1, w_2.y_2w_2)/x]D(x, x_1.u_1x_1x, x_2.u_2x_2x)$$

i.e.

$$N_0 \equiv D(D(w, w_1.y_1w_1, w_2.y_2w_2), x_1.u_1x_1D(w, w_1.y_1w_1, w_2.y_2w_2),\\ x_2.u_2x_2D(w, w_1.y_1w_1, w_2.y_2w_2))$$

and similarly

$$N_0' \equiv D(w, w_1.[y_1w_1/x]D(x, x_1.u_1x_1x, x_2.u_2x_2x)),\\ w_2.[y_2w_2/x]D(x, x_1.u_1x_1x, x_2.u_2x_2x))$$

i.e.

$$N_0' \equiv D(w, w_1.D(y_1w_1, x_1.u_1x_1(y_1w_1), x_2.u_2x_2(y_1w_1)),\\ w_2.D(y_2w_2, x_1.u_1x_1(y_2w_2), x_2.u_2x_2(y_2w_2))).$$

The immediate simplifications are now blocked: there is no sequence of reductions from N_0 to N_0'. All we can do is apply the permutative conversion (7.2.6) and we obtain (with some renaming to avoid confusion of variables) the normal term

$$D(w, w_1.D(y_1w_1, x_1.u_1x_1D(w, w_1'.y_1w_1', w_2'.y_2w_2'),\\ x_2.u_2x_2D(w, w_1'.y_1w_1', w_2'.y_2w_2')), w_2.\ldots).$$

(But, since we no longer have confluence, the normality of this term is unhelpful.)

7.9 Strong Normalisation of Cut Reduction

Weak normalisation (in the typed system) is shown by observing that the rank of the cut is always reduced, where the *rank* of a cut is the triple comprising the cut formula, the height of the first premiss and the height of the second premiss, with ranks lexicographically ordered and formulae ordered by the "is a subformula of" relation.

Strong normalisation (in the typed system) is shown by a lexicographic path order argument (as in Baader and Nipkow (1998)). First one removes the variable binders, so we are dealing with a system of first-order terms (i.e. terms with no binders). (An infinite reduction sequence with binders would translate to an infinite reduction sequence without binders.) We then order the constructors as follows: every cut constructor C exceeds every other constructor (such as W), and cut constructors C are ordered according to the size of the cut formula (which is not made explicit in the term; this could be done, but less readably). The details are tedious, but routine, since one merely has to check, for each rule, that $LHS > RHS$; they have been checked using a Prolog program (Dyckhoff 2013) that implements the LPO method.

The rules (for LPO ordering $>$ of terms $s, t, \ldots$) are as follows (where $\rhd$ indicates the relation between a term and each of its immediate subterms, $\lhd$ is the converse, and $>^{lex}$ is the lexicographic extension of $>$ to tuples, with associated rule $>_{lex}$):

$$\frac{\exists u \lhd s.\ u \geq t}{s > t}\ >_i \qquad\qquad \frac{s \gg t \quad \forall u \lhd t.\ s > u}{s > t}\ >_{ii}$$

$$\frac{f > g}{f(s_1,\ldots,s_m) \gg g(t_1,\ldots,t_n)}\ \gg_i \qquad \frac{(s_1,\ldots,s_n) >^{lex} (t_1,\ldots,t_n)}{f(s_1,\ldots,s_n) \gg f(t_1,\ldots,t_n)}\ \gg_{ii}$$

An illustrative example (7.6.8) with the constructor p for 'pair', is thus:

$$\dfrac{\dfrac{C > p}{C(L,p(L',L'')) \gg p(C(L,L'),C(L,L''))}\ \gg_i \qquad \dfrac{\dfrac{\dfrac{\dfrac{L' \geq L'}{p(L',L'') > L'}\ >_i}{(L,p(L',L'')) >^{lex} (L,L')}\ >_{lex}}{C(L,p(L',L'')) \gg C(L,L')}\ \gg_{ii} \quad \ldots \quad \ldots}{C(L,p(L',L'')) > C(L,L')}\ >_{ii} \quad \ldots}{C(L,p(L',L'')) > p(C(L,L'),C(L,L''))}\ >_{ii}$$

and neither space nor their intrinsic interest permits the inclusion here of the 31 other cases. (The missing bits indicated here by... are similar, concluding that $C(L, p(L', L'')) > L$, that $C(L, p(L', L'')) > L'$ and that $C(L, p(L', L'')) > C(L, L'')$ respectively.)

7.10 Subject Reduction

As always, one has to show that if a term L has (in the context Γ) type A, and $L \rightsquigarrow L'$, then also L' has type A. We could consider the 32 separate cases; but we only show two in detail.

One of these is routine. Using D for $A \supset B$ for brevity and omitting the context Γ, Rule 7.6.7 transforms

$$\dfrac{\dfrac{\dfrac{\cdots}{w:A \Rightarrow L:B}}{\Rightarrow \lambda w.L:D}\, R\supset \qquad \dfrac{\dfrac{\cdots}{x:D \Rightarrow L':A} \qquad \dfrac{\cdots}{x:D, y:B \Rightarrow L'':E}}{x:A\supset B \Rightarrow A(x,L',y.L''):E}\, L\supset}{\Rightarrow C(\lambda w.L, x.A(x,L',y.L'')):E}\, \mathit{Cut}$$

into (using Wk for some weakening steps)

$$\dfrac{\dfrac{\dfrac{\cdots}{\Rightarrow \lambda w.L:D} \qquad \dfrac{\cdots}{x:D \Rightarrow L':A}}{\Rightarrow C(\lambda w.L,x.L'):A}\, \mathit{Cut} \qquad \dfrac{\dfrac{\cdots}{w:A \Rightarrow L:B} \qquad \dfrac{\dfrac{\cdots}{y:B,w:A \Rightarrow \lambda w.L:D}\, \mathit{Wk} \qquad \dfrac{\cdots}{x:D,y:B,w:A \Rightarrow L'':E}\, \mathit{Wk}}{y:B,w:A \Rightarrow C(\lambda w.L,x.L''):E}\, \mathit{Cut}}{w:A \Rightarrow C(L,y.C(\lambda w.L,x.L'')):E}\, \mathit{Cut}}{\Rightarrow C(C(\lambda w.L,x.L'),w.C(L,y.C(\lambda w.L,x.L''))):E}\, \mathit{Cut}$$

The other example, however, is more complicated. We recall Rule 7.6.18:

$$\begin{aligned}
&C(W(w, w_1.L_1, w_2.L_2), x.W(x, x'.L', x''.L'')) \rightsquigarrow \\
&\qquad W(w, w_1.C(L_1, x.W(x, x'.C(W(w, w_1.L_1, w_2.L_2), x.L'), \\
&\qquad\qquad x''.C(W(w, w_1.L_1, w_2.L_2), x.L''))), \\
&\qquad\quad w_2.C(L_2, x.W(x, x'.C(W(w, w_1.L_1, w_2.L_2), x.L'), \\
&\qquad\qquad x''.C(W(w, w_1.L_1, w_2.L_2), x.L''))))
\end{aligned}$$

We need some abbreviations. Let

$$M \equiv W(w, w_1.L_1, w_2.L_2),$$

$$M^* \equiv W(x, x'.C(M, x.L'), x''.C(M, x.L'')),$$

$$M_i \equiv C(L_i, x.M^*).$$

So we can simplify the rule to

$$\begin{aligned}
&C(W(w, w_1.L_1, w_2.L_2), x.W(x, x'.L', x''.L'')) \\
&\qquad \rightsquigarrow\ W(w, w_1.C(L_1, x.M^*), w_2.C(L_2, x.M^*))).
\end{aligned}$$

With $A \equiv A_1 \vee A_2$ and $B \equiv B' \vee B''$, the *LHS* of this type-checks, we suppose, as follows, in the presence of some context Γ left implicit:

$$\dfrac{\dfrac{\dfrac{\cdots}{w:A,\ w_1:A_1 \Rightarrow L_1:B} \quad \cdots}{w:A \Rightarrow M:B}\,L\vee \qquad \dfrac{\dfrac{\cdots}{x:B,\ w:A,\ x':B' \Rightarrow L':E} \quad \cdots}{x:B,\ w:A \Rightarrow W(x,x'.L',x''.L''):E}\,L\vee}{w:A \Rightarrow C(M,x.W(x,x'.L',x''.L'')):E}\,Cut \, .$$

It is transformed into a derivation too large to be conveniently displayed. We begin by considering the derivation of the first upper right premiss: $x : B, w : A, x' : B' \Rightarrow L' : E$; we rename, for some fresh variable x^*, the variable x to x^*, obtaining a derivation of $x^* : B, w : A, x' : B' \Rightarrow L'^* : E$; this can now be weakened with $x : B$ to a derivation of $x : B, x^* : B, w : A, x' : B' \Rightarrow L'^* : E$. The left premiss $w : A \Rightarrow M : B$ can, after a weakening, be cut with this, giving

$$\dfrac{\dfrac{\cdots}{x:B,\ w:A,\ x':B' \Rightarrow M:B} \qquad \dfrac{\cdots}{x:B, x^*:B,\ w:A,\ x':B' \Rightarrow L'^*:E}}{x:B, w:A, x':B' \Rightarrow C(M,x^*.L'^*):E}\,Cut$$

and the same is done to obtain a derivation of $x : B, w : A,\ x'' : B'' \Rightarrow C(M, x^*.L''^*) : E$. Since $B \equiv B' \vee B''$, we use $L\vee$ to obtain a derivation of

$$x : B, w : A \Rightarrow W(x, x'.C(M, x^*.L'^*), x''.C(M, x^*.L''^*)) : E,$$

i.e. of

$$x : B, w : A \Rightarrow M^* : E.$$

We now (for each $i = 1, 2$) perform, after some weakenings, another cut:

$$\dfrac{\dfrac{\cdots}{x:B, w:A,\ w_i:A_i \Rightarrow L_i:B} \qquad \dfrac{\cdots}{x:B,\ w:A,\ w_i:A_i \Rightarrow M^*:E}}{x:B,\ w:A,\ w_i:A_i \Rightarrow C(L_i,x.M^*):E}\,Cut$$

and conclude as follows:

$$\dfrac{\dfrac{\cdots}{w:A,\ w_1:A_1 \Rightarrow C(L_1,x.M^*):E} \qquad \dfrac{\cdots}{w:A,\ w_2:A_2 \Rightarrow C(L_2,x.M^*):E}}{w:A \Rightarrow W(w,w_1.C(L_1,x.M^*),w_2.C(L_2,x.M^*)):E}\,L\vee.$$

All the renamings are, by use of alpha-conversion, omitted in the earlier presentation (rule 7.6.18). Note that we have used the height-preserving admissible rule of *Weakening*.

7.11 Confluence of Cut Reduction

The system of cut-reduction rules (7.6.1–7.6.32) is a left-linear orthogonal pattern-rewrite system, without critical pairs; by the results of (Mayr and Nipkow 1998), confluence is immediate. That there are no critical pairs is simply the observation that every term matches at most one LHS from the set of rules (7.6.1–7.6.32), and any LHS from one of these rules fails to match any non-variable proper subterm of any of these rules.

7.12 Conclusion

Putting the various results together, we have the following:

Theorem 7.12.1 *The system of cut reduction rules* (7.6.1–7.6.32) *is complete (for reducing cuts), confluent, strongly normalising (on typed terms) and satisfies the subject reduction property; moreover, every cut reduction rule translates via ϕ to a sequence of zero or more reductions in the natural deduction setting.*

Acknowledgments Thanks are due to Jan von Plato, Peter Chapman, Jacob Howe, Stéphane Graham-Lengrand and Christian Urban for helpful comments and (to the last of these) for a copy of (Urban 2014) prior to its publication, albeit many years after its 2001 presentation in Rio. Chapman's work (Dyckhoff and Chapman 2009) (incorporating also ideas by Urban) was invaluable in checking the correctness of all the lemmata about substitution. The work was motivated by requirements for some not yet published work (joint with James Caldwell) supported by EPSRC grant EP/F031114/1.

References

Baader, F., & Nipkow, T. (1998). *Term rewriting and all that*. Cambridge: Cambridge University Press.

Barendregt, H. (1984). *The lambda calculus*. Amsterdam: Elsevier Science. revised ed.

Borisavljević, M. (2004). Extended natural-deduction images of conversions from the system of sequents. *Journal of Logic & Computation*, *14*, 769–799.

Dyckhoff, R., & Urban, C. (2003). Strong normalization of Herbelin's explicit substitution calculus with substitution propagation. *Journal of Logic & Computation*, *13*, 689–706.

Dyckhoff, R., & Chapman, P. (2009). Isabelle proof script, available from its first author.

Dyckhoff, R. (2011). Cut-elimination, substitution and normalisation, Section A3. Logic and Computation, *14th Congress of Logic, Methodology and Philosophy of Science*, Nancy.

Dyckhoff, R. (2013). LPO checker in Prolog, available from the author.

Gentzen, G. (1935). Untersuchungen uber das logische Schließen. *Mathematische Zeitschrift*, *39*, 176–210 and 405–431.

Joachimski, F., & Matthes, R. (2003). Short proofs of normalization for the simply-typed lambda-calculus, permutative conversions and Gödel's T. *Archive for Mathematical Logic*, *42*, 59–87.

Kikuchi, K. (2006). On a local-step cut-elimination procedure for the intuitionistic sequent calculus. In *Proceedings of the 13th International Conference on Logic for Programming Artificial Intelligence and Reasoning (LPAR'06)*, 4246 of LNCS (pp. 120–134). Springer.

Kreisel, G. (1971). A survey of proof theory II, In J. E. Fenstad (Ed.), *Proceedings of the 2nd Scandinavian Logic Symposium*, North-Holland (pp. 109–170).

Mayr, R., & Nipkow, T. (1998). Higher-order rewrite systems and their confluence. *Theoretical Computer Science*, *192*, 3–29.

Negri, S., & von Plato, J. (2001). *Structural proof theory*. Cambridge: Cambridge University Press.

Pottinger, G. (1977). Normalisation as a homomorphic image of cut-elimination. *Annals of Mathematical Logic 12*, 323–357. New York: North-Holland.

Prawitz, D. (1965). *Natural deduction: A proof-theoretical study*. Stockholm: Almqvist & Wiksell.

Prawitz, D. (1971). Ideas and results in proof theory. In J. E. Fenstad (Ed.), *Proceedings of the 2nd Scandinavian Logic Symposium*, North-Holland (pp. 235–307).

Schwichtenberg, H., & Wainer, S. (2012). *Proofs and computations*, Perspectives in Logic Series, Association for Symbolic Logic. Cambridge: Cambridge University Press.

Troelstra, A., & van Dalen, D. (2000). *Constructivism in mathematics: An introduction, Vol II* (Vol. 123)., Studies in logic and the foundations of mathematics Amsterdam: North-Holland.

Troelstra, A., & Schwichtenberg, H. (2000). *Basic proof theory* (2nd ed.). Cambridge: Cambridge University Press.

Urban, C. (2014). Revisiting Zucker's work on the correspondence between cut-elimination and normalisation. In L. C. Pereira, E. H. Haeusler, & V. de Paiva (Eds.), *Advances in natural deduction. A celebration of Dag Prawitz's work* (pp. 31–50). Dordrecht: Springer.

Urban, C., & Bierman, G. (2001). Strong normalisation of cut-elimination in classical logic. *Fundamenta Informaticae*, *45*, 123–155.

Urban, C. (2008). Nominal techniques in Isabelle/HOL. *Journal of Automated Reasoning*, *40*, 327–356.

Vestergaard, R. (1999). Revisiting Kreisel: A computational anomaly in the Troelstra-Schwichtenberg G3i system, Unpublished.

von Plato, J. (2001). Natural deduction with general elimination rules. *Archive for Mathematical Logic*, *40*, 541–567.

von Raamsdonk, F. & Severi, P. (1995). On normalisation, Centrum voor Wiskunde en Informatica, Amsterdam, CS-R9545.

Zucker, J. (1974). The correspondence between cut-elimination and normalisation. *Annals of Mathematical Logic*, *7*, 1–112.

Chapter 8
Inversion Principles and Introduction Rules

Peter Milne

Abstract Following Gentzen's practice, borrowed from intuitionist logic, Prawitz takes the introduction rule(s) for a connective to show how to prove a formula with the connective dominant. He proposes an inversion principle to make more exact Gentzen's talk of deriving elimination rules from introduction rules. Here I look at some recent work pairing Gentzen's introduction rules with general elimination rules. After outlining a way to derive Gentzen's own elimination rules from his introduction rules, I give a very different account of introduction rules in order to pair them with general elimination rules in such a way that elimination rules can be read off introduction rules, introduction rules can be read off elimination rules, and both sets of rules can be read off classical truth-tables. Extending to include quantifiers, we obtain a formulation of classical first-order logic with the subformula property.

Keywords Introduction rules · Elimination rules · General elimination rules · Inversion principle · Sequent calculus

Famously, in Sect. 5.13 of his article 'Untersuchungen über das logische Schließen', Gentzen said

> The introductions represent, as it were, the "definitions" of the symbols concerned, and the eliminations are no more, in the final analysis, than the consequences of these definitions. This fact may be expressed as follows: In eliminating a symbol, we may use the formula with whose terminal symbol we are dealing only "in the sense afforded it by the introduction of that symbol". (Gentzen 1935, p. 80)

Gentzen went on to say,

> By making these ideas more precise it should be possible to display the E-inferences as unique functions of their corresponding I-inferences, on the basis of certain requirements. (Gentzen 1935, p. 81)

I thank audiences at the Universities of Tübingen and St. Andrews for comments on some of the material presented here.

P. Milne (✉)
School of Arts and Humanities - Law and Philosophy, University of Stirling,
Scotland FK9 4LA, UK
e-mail: peter.milne@stir.ac.uk

H. Wansing (ed.), *Dag Prawitz on Proofs and Meaning*, Outstanding Contributions to Logic 7, DOI 10.1007/978-3-319-11041-7_8

Surprisingly, in view of this claim, Gentzen says nothing about how any such display might go. Indeed, in the 1934 paper he says nothing about the character of introduction rules in general, nor about the provenance of his own introduction rules for &, ∨ ⊃, ¬, ∀ and ∃, save for very quickly sketching examples of reasoning in which &-introduction, &-elimination, ∨-introduction, ∨-elimination, ∀-introduction, ∀-elimination, ∃-introduction, ∃-elimination and *reductio ad absurdum* may be discerend to be in play (Gentzen 1935, Sect. 2.1).[1] In 'Die Widerspruchsfreiheit der reinen Zahlentheorie', published in 1936, he is more forthcoming. Here there is an extended analysis of Euclid's proof of the infinity of prime numbers (Gentzen 1936, Sect. 2.4). Moreover, after presenting the introduction and elimination rules in sequent guise (but these are the sequent formulations of the natural deduction rules, not the left and right introduction rules for sequent calculus from Gentzen (1935)) in Gentzen (1936), he goes on, in Sect. 10, to discuss 'the finitist interpretation of ∀, &, ∃, and ∨ in transfinite propositions', and explains how the rules given are in harmony with this interpretation: "the '*finitist* interpretation' of the logical connectives ∀, &, ∃ and ∨ in transfinite propositions described in Sect. 10 agrees essentially with the interpretation of the *intuitionists*" (Gentzen 1936, p. 169). He then points to difficulties in giving an intuitionist account of the meanings of the conditional and negation. Nonetheless we have some reason to suppose the specific form of his introduction rules owes something to Heyting's formalisation of intuitionistic logic in Heyting (1930), the more so as this is one of the few articles Gentzen cites in either his 1934–1935 or his 1936 and this one he cites in both.

Certainly, Gentzen's practice fits the intuitionist pattern summed up neatly by the late Michael Dummett:

> The meaning of each constant is to be given by specifying, for any sentence in which that constant is the main operator, what is to count as a proof of that sentence, it being assumed that we already know what is to count as a proof of any of the constituents.
>
> (Dummett 2000, p. 8)

This is the form of rule Dag Prawitz codified in *Natural Deduction*:

> An introduction rule for a logical constant γ allows the inference *to* a formula A that has γ as principal sign from formulae that are subformulae of A; i.e. an instance of such a rule has a consequence A with γ as principal sign and has one or two premisses which are subformulae of A. (Prawitz 1965, p. 32)[2]

In the effort to invest Gentzen's idea of deriving elimination rules from introduction rules with some precision, Prawitz said to the reader,

---

[1] Strictly, the negation rule in play is weaker than *reductio ad absurdum*. It's the weak principle which, added to positive logic, yields what Allen Hazen calls *subminimal logic*: from $\neg\psi$ and a proof of ψ with ϕ as assumption, infer $\neg\phi$ and discharge the assumption ϕ (Hazen 1995).

[2] The restriction to one or two premisses is clearly inessential. What is surprising, though, is the lack of mention of subformulae possibly occurring as assumptions discharged in the application of the rule. As this characterisation stands, it would seem to preclude Gentzen's rule for ⊃-introduction (conditional proof). Prawitz gives a much more general form at (Prawitz 1978, p. 35) which corrects this oversight.

> Observe that an elimination rule is, in a sense, the inverse of the corresponding introduction rule: by an application of an elimination rule one essentially only restores what had already been established if the major premiss of the application was inferred by an application of the introduction rule.

and then proposed this *inversion principle*:

> Let α be an application of an elimination rule that has B as consequence. Then, deductions that satisfy the sufficient conditions [given in the introduction rule(s)] for deriving the major premiss of α when combined with deductions of the minor premisses of α (if any), already "contain" a deduction of B; the deduction of B is thus obtainable directly from the given deductions without the addition of α. (Prawitz 1965, p. 33)

The inversion principle suggests, he says, this *inversion theorem*

> If $\Gamma \vdash A$ then there is a deduction of A from Γ in which no formula occurrence is both the consequence of an application of an I-rule and major premiss of an application of an E-rule.
>
> (Prawitz 1965, p. 34)

This, the key idea behind Prawitz's normalisation theorems for intuitionist and classical logics in natural deduction, is very close to Gentzen's informal account of his *Hauptsatz*:

> The *Hauptsatz* says that every purely logical proof can be reduced to a definite, though not unique, normal form. Perhaps we may express the essential properties of such a normal proof by saying: it is not roundabout. No concepts enter into the proof other than those contained in its final result, and their use was therefore essential to the achievement of that result.
>
> (Gentzen 1935, pp. 68–69)

Despite the fact that, strictly speaking, he proved the *Hauptsatz* for sequent calculi, Gentzen has natural deduction in mind here—at this point in the article he has made no mention and given no hint of sequent calculi. (We know from Jan von Plato's researches that he had obtained something very much like Prawitz's normalisation theorem for intuitionist logic—see Plato (2008)—but that is perhaps not quite to the point, for Gentzen goes on immediately to say, 'In order to be able to ennunciate and prove the *Hauptsatz* in a convenient form, I had to provide a logical calculus especially suited to the purpose'.)

In an article published in 1978, Prawitz sets about making good Gentzen's claims that elimination rules are consequences and functions of the introduction rules. He presents a general form for introduction rules, then outlines a general procedure for obtaining the elimination rules:

> Gentzen suggested that 'it should be possible to display the elimination rules as unique functions of the corresponding introduction rules on the basis of certain requirements'. One has indeed a strong feeling that the elimination rules are obtained from the corresponding introduction in a uniform way. For the introduction rules of the schematic type, we can easily describe how the elimination rules are obtained uniformly from the corresponding introduction rule. (Prawitz 1978, pp. 36–37)

This proposal gave rise to two very closely intertwined lines of research and tied in closely with a third to which Prawitz' (1978) contributed: formulations of natural deduction employing *general elimination rules*, work on inversion principles, and the delineation and proof of proof-theoretic conceptions of expressive adequacy.[3] General elimination rules refashion Gentzen's elimination rules on the model of his elimination rule for disjunction (proof by cases). They are familiar from the work of Schroeder-Heister (1984, 1984b), Tennant (1992), Read (2000, 2004, 2010), Negri and Plato (2001), Plato (2001), Francez and Dyckhoff (2012).[4] Progress on inversion principles is conveniently surveyed by Moriconi and Tesconi (2008); they credit Schroeder-Heister et al. with raising new perspectives on the inversion principle.

Despite these developments, recent authors have seen fit to say

> Gentzen does not give the function which maps introduction rules to elimination rules and to my knowledge neither does anyone else. This is remarkable, given there seems to exist a certain consensus amongst workers in the field of how to 'read off' elimination rules from introduction rules. (Kurbis 2008, p. 89)

and

> [T]he general method (hoped for by many) for generating an elimination rule from an arbitrary set (possibly empty) of introduction rules for a logical constant leads to a mess. (Dyckhoff 2009, p. 2)

These authors exaggerate, I suggest. But what it is to the point to observe is that what one takes the role of introduction rules to be and what one takes to be the right way to obtain elimination rules from introduction rules matters. I shall first discuss some recent work pairing Gentzen's introduction rules with general elimination rules, next outline a way to derive Gentzen's elimination rules from his introduction rules, then, lastly, present different introduction rules which pair elegantly with general elimination rules and admit a formulation of classical first-order logic with the subformula property.

8.1 Inversion Principles

There is a reading of the inversion principle that has been employed in recent work to which Negri and Plato give succinct expression:

> Whatever follows from the direct grounds for asserting a proposition must follow from the proposition Negri. (2001, p. 6)[5]

[3] On the last, *cf.* Zucker and Tragesser (1978).

[4] Schroeder-Heister (2004, p. 33, n. 10), (2014, n. 2) says that he developed and investigated general elimination rules following not just Prawitz (1978) but also earlier work by Kutschera (1968).

[5] Francez and Dyckhoff (2012, Sect. 3.1) give the same formulation as 'another formulation of the idea behind the inversion-principle'. In her (2002, pp. 571–572), Negri says, 'Whatever follows from the grounds for deriving a formula must follow from that formula'. I thank Jan von Plato (personal communication) for bringing (Negri 2002) to my attention.

In the Introduction to their book, Negri and von Plato call what they give a generalisation of the inversion principle (2001, p. xiv). They see it as a generalisation precisely because it does not require that 'elimination rules conclude the immediate grounds for deriving a proposition instead of arbitrary consequences of these grounds' (Negri and von Plato 2001). Moriconi and Tesconi (2008) point out, though, that Negri and von Plato's principle is, rather, the opposite of Prawitz's principle, 'one requiring that the consequences of a proposition be no more than the consequences of its immediate grounds, the other prescribing that they be no less'. In Prawitz's formulation of the inversion principle, the conclusion of an elimination rule must follow from each set of grounds provided by an introduction rule. In the only sane way to read Negri and von Plato's formulation, given connectives like $\vee$ which has two introduction rules, each consequence common to every set of grounds provided by introduction rules must follow from some elimination rule.

Prawitz's formulation leads straightforwardly to what Dummett (1991, p. 248) calls "the levelling of local peaks"; as Prawitz himself puts it, "The inversion principle says in effect that nothing is 'gained' by inferring a formula through introduction for use as a major premiss in an elimination" (Prawitz 1965, pp. 33–34). As Prawitz is aware, though, this leaves open the possibility of a gap between what is sufficient according to the introduction rule(s) and what is necessary according to the elimination rule(s):

> [A] solution to the problem suggested by Gentzen should also establish that these elimination rules are the strongest possible elimination rules corresponding to the given introduction rules, i.e. the strongest possible elimination rules that are correct in virtue of the meaning of the logical constant stated by the condition for asserting sentences of the form in question.
>
> (Prawitz 1978, p. 37)

Negri and von Plato's inversion principle admits a problem which one might think of a different order entirely: there is incoherence of a certain kind if what is necessary according to the elimination rule(s) is logically stronger than what is sufficient according to the introduction rule(s), but their statement of their inversion principle does nothing to preclude such cases. What their inversion principle does secure is *uniqueness*: if one considers the introduction rules to apply to one connective, the elimination rules to another, then a proposition with the second dominant is logically at least as strong as the homologous formula with the first dominant.

Recalling Nuel Belnap's (1962) requirements on definitions of logical constants, Prawitz's inversion principle speaks to Belnap's demand for conservativeness (and normalisation theorems answer it fully), Negri and von Plato's speaks to Belnap's requirement for uniqueness. As stated, the two inversion principles aim for different proof-theoretical virtues and allow for different proof-theoretical infelicities. Both principles outlaw Prior's *tonk*, however only Prawitz's does so if one modifies *tonk* so as to give it the two introduction rules of $\vee$ and the two elimination rules of &; conversely, a connective governed by the introduction rule of & and the elimination rule of $\vee$ is permitted by Prawitz's principle but not by Negri and von Plato's.

Dummett was aware that levelling of local peaks might allow—as I have put it above—a gap. Read writes:

He [Dummett] wrote:

> We may thus provisionally identify harmony between the introduction and elimination rules for a given logical constant with the possibility of carrying out this procedure, which we have called the levelling of local peaks. (Dummett 1991, p. 250)

He later abandons that provisional identification in favour of "total harmony", namely, conservativeness. His reason is a fear that intrinsic harmony may be too weak a requirement. Perhaps the E-rule does not permit all that the I-rule justifies. General-elimination harmony (ge-harmony for short) is designed to exclude that possibility.

The idea of ge-harmony is that we may infer from an assertion *all and only* what follows from the various grounds for that assertion. (Read 2010, p. 563, my emphasis)

Schroeder-Heister's Condition 4.2 (1984, p. 1293) is a more formal statement of this idea. Ge-harmony combines Prawitz's and Negri and von Plato's inversion principles. As a first approximation, we say:

> Each consequence common to every set of grounds provided by introduction rules must follow from some elimination rule *and* whatever follows from some elimination rule must follow from each set of grounds provided by an introduction rule.

At first sight, this is, in practice, the result of what we might call Negri and von Plato's *inversion procedure*. For example, from the standard introduction rule for &,

$$\frac{\phi \qquad \psi}{\phi \,\&\, \psi}$$

they obtain this general elimination rule

$$\frac{\phi \,\&\, \psi \qquad \begin{array}{c}\underbrace{[\phi]^m, [\psi]^m}\\ \vdots \\ \chi\end{array}}{\chi}\,m$$

and here it looks as though it is exactly those propositions that follow from the grounds presented in the & -introduction rule that are licensed as inferable from the conjunction—see Negri and von Plato (2001, p. 6) and Schroeder-Heister (1984, p. 1294).

As is only too familiar, Gentzen's introduction rules for (inclusive) disjunction are

$$\frac{\phi}{\phi \vee \psi} \quad \text{and} \quad \frac{\phi}{\phi \vee \psi}.$$

The two grounds explicitly given for assertion of $\phi \vee \psi$ are ϕ and ψ. Now, one way to follow to the letter our inversion principle, and, indeed, Negri and von Plato's, is to say that our single elimination rule is this

$$\frac{\phi \vee \psi \quad \begin{matrix} [\![\phi]\!]^m \\ \vdots \\ \chi \end{matrix} \quad \begin{matrix} [\![\psi]\!]^m \\ \vdots \\ \chi \end{matrix}}{\chi}\, m\,,$$

where the bracketing indicates that in the subproofs at most ϕ and at most ψ may occur as assumptions, for exactly they are the grounds given by the introduction rules. That is, the elimination rule we obtain is the weak $\vee$-elimination rule of quantum logic. This rule suffices for the levelling of local peaks and proofs of uniqueness. What it doesn't allow is the sort of permutation of applications of rules that is needed for proof of normalisation: as is well known, adding Gentzen's rules for $\supset$ leads to a non-conservative extension of quantum logic.

With this in mind, we see, too, that in the elimination rule for &, *at most* ϕ and ψ should occur as assumptions on which χ depends in the subproof.

There are other signs that all is not well. This inversion principle does not get us the general elimination recasting of *modus ponendo ponens* as the elimination rule for $\supset$. Gentzen's introduction rule is

$$\frac{\begin{matrix} [\phi]^m \\ \vdots \\ \psi \end{matrix}}{\phi \supset \psi}\, m\,.$$

What we should get from this is what Schroeder-Heister does obtain by recourse to his "natural extension of natural deduction", namely

$$\frac{\phi \supset \psi \quad \begin{matrix} [\phi \Rightarrow \psi]^m \\ \vdots \\ \chi \end{matrix}}{\chi}\, m\,.$$

where $\phi \Rightarrow \psi$ stands for the existence of a proof with conclusion ψ in which ϕ occurs as an assumption.

Among the consequences of a proof with conclusion ψ in which ϕ occurs as an assumption are, *in the presence of* ϕ *and assuming transitivity*, the consequences of ψ, so we can derive Negri and von Plato's (2001, p. 8), Read's (2010), and Francez and Dyckhoff's (2012) general $\supset$-elimination rule,[6]

$$\frac{\phi \supset \psi \quad \phi \quad \begin{matrix} [\psi]^m \\ \vdots \\ \chi \end{matrix}}{\chi}\, m.$$

---

[6] In his (2010), Read doesn't discuss $\supset$ explicitly but we can extrapolate from his discussion of negation and the general case. The derivation here is exactly in accord with how he arrives at the standard general elimination rule.

On the face of it, we cannot immediately say that this rule exhausts the content of the elimination rule above—and even if it does, it is *not* the rule obtained by means of the inversion procedure. Following Schroeder-Heister, we might offer a proof that it does in fact exhaust the content of the elimination rule proper:

> With ψ in place of χ we get, *assuming reflexivity and transitivity*, the familiar *modus ponendo ponens* argument-form: we have then a proof of ψ in which ϕ and $\phi \supset \psi$ occur as premisses. Thus anything that follows from the assumption that there is a proof of ψ in which ϕ occurs as assumption, can now be drawn down as a conclusion dependent on the assumptions $\phi \supset \psi$ and ϕ; and hence may be drawn down *tout court* in the presence of $\phi \supset \psi$ and ϕ.
>
> (*Cf.* Schroeder-Heister 1984, pp. 1294–1295, Lemma 4.4, 2014, pp. 5–6)

But so what? The point at issue is that the general elimination version of *modus ponendo ponens* is *not* what one reads off the $\supset$-introduction rule *via* Read's ge-harmony principle/Schroeder-Heister's Condition 4.2 and the more fully articulated principle above. Francez and Dyckhoff (2012) have the decency at this point to reformulate the inversion principle they employ. Negri and von Plato (2001, p. 8) brazen it out. They say

> The direct ground for a deriving $A \supset B$ is the existence of a hypothetical derivation of B from the assumption A. The fact that C can be derived from the existence of such a derivation can be expressed by:
>
> If C follows from B it already follows from A.
>
> This is achieved precisely by the elimination rule
>
> $$\dfrac{A \supset B \qquad A \qquad \begin{matrix}[B] \\ \vdots \\ C\end{matrix}}{C} \supset\text{-}E\ .$$
>
> (Negri and von Plato 2001, p. 8)

We know, thanks to Schroeder-Heister's proof, that we lose nothing by adopting this rule, but Negri and von Plato seem not to recognise that, on the one hand, this is a point that needs argument and, on the other, 'if C follows from B it already follows from A' is not literally what 'C can be derived from the existence of a hypothetical derivation of B from the assumption A' *says*, for Schroeder-Heister's proof requires appeal to structural rules. In the light of the structural rules of reflexivity and transitivity, the general elimination version of *modus ponendo ponens* can do the work of the rule the inversion procedure/Schroeder-Heister's Condition 4.2 delivers.[7]

[7] Read (2010) emphasises the role of structural rules in proving the equivalence of different forms of elimination rules.

8.2 From Gentzen's Introduction Rules to Gentzen's Elimination Rules (Approximately)

Gentzen says that 'it should be possible to display the E-inferences as unique functions of their corresponding I-inferences, on the basis of certain requirements'. Nothing he says implies that there should be a general form for introduction rules and a general method, based on that general form, for obtaining corresponding elimination rules. Nevertheless, introduction rules have a common function and we may exploit that function in order to go some way towards showing that elimination rules are, in the last analysis, no more than the consequences of the corresponding introduction rules. In the light of what we saw above regarding $\vee$-elimination, however, it would be naïve to think that we can straightforwardly obtain the standard $\vee$-elimination rule. This was, I think, Gentzen's considered view. The question is how to get side premisses into place in the $\vee$-elimination rule. Gentzen (1936, p. 153), himself says, 'The formulation using additional hypotheses … may appear rather artificial in the case of the $\vee$ and $\exists$-elimination, if these rules are compared with the corresponding examples of inferences (4.5)', (4.5) being the subsection of the article in which we find a 'classification of the individual forms of inference by reference to examples from Euclid's proof'; he goes on to say, 'However, the formulation is smoothest if in the *distinction of cases* ($\vee$-elimination) the two possibilities that result are simply regarded as *assumptions* which become redundant as soon as the same result has been obtained from each', where, tacitly, he is allowing, in accordance with a common pattern of reasoning, the use of other assumptions in arriving at that common result. I think that in the appeal to smoothness one can hear, if not a retreat from then at least, a qualification to the bold claims regarding elimination rules as consequences and functions of the introduction rules in Gentzen (1935).

We might guess that Gentzen realised the importance of having the standard, rather than the restricted, rule for proving the *Hauptsatz*. Instead of the familiar example of adding a conditional governed by Gentzen's rules to quantum logic, here's another example showing that lack of the standard rule threatens the *Hauptsatz*. If $\Sigma_1, \phi \vdash \neg\chi$ and $\Sigma_2, \psi \vdash \neg\chi$ then, using only the rules for $\&$ and $\neg$, $\Sigma_1, \Sigma_2, \neg(\neg\phi \mathbin{\&} \neg\psi) \vdash \neg\chi$. We need only the restricted $\vee$-elimination rule of quantum logic and the weak form of *reductio ad absurdum* mentioned in footnote 1 to show that $\phi \vee \psi \vdash \neg(\neg\phi \mathbin{\&} \neg\psi)$. But now, how to avoid passage through $\neg(\neg\phi \mathbin{\&} \neg\psi)$ in the absence of the standard $\vee$-elimination rule?

Allowing for a little bit of a fudge in the case of $\vee$-elimination and $\exists$-elimination, how do we get from Gentzen's introduction rules to Gentzen's elimination rules? Introduction rules place upper bounds on the logical strength of the introduced proposition. For example, $\supset$-introduction tells us that for any set of formulae Σ, $\Sigma \vdash \phi \supset \psi$ if $\Sigma, \phi \vdash \psi$. At first sight we might think that what we want the elimination rule(s) to do is to turn that 'if' into an 'if and only if' but that cannot be right in general: $\vee$-introduction tells us that $\Sigma \vdash \phi \vee \psi$ if $\Sigma \vdash \phi$ or $\Sigma \vdash \psi$ but it is contrary to how we use 'or' to suppose that for, arbitrary ϕ and ψ, $\phi \vee \psi \vdash \phi$ or $\phi \vee \psi \vdash \psi$. Rather, what we want the elimination rule or rules to do is to show that the introduced proposition represents an attained least upper bound. Put another way, the elimination rule(s) must show that the introduced proposition is strong enough to allow for its own introduction in the circumstances set out in the introduction rule. We read off from the introduction rule what is required.

- &: If $\phi \,\&\, \psi$ is to allow for its own introduction, we must, at the very least (but this is all that we need), be permitted to infer both ϕ and ψ from it, which is exactly what Gentzen's two elimination rules for & do.
- $\supset$: A proof in which ϕ occurs as assumption and ψ occurs as conclusion is a proof which, in the presence of ϕ, yields ψ. If $\phi \supset \psi$ is to allow for its own introduction, we must, at the very least (but this is all that we need), obtain from ϕ and $\phi \supset \psi$ a proof with ψ as conclusion, which is exactly what *modus ponendo ponens*, Gentzen's elimination rule for $\supset$, provides.
- $\neg$: Gentzen's rules of $\neg$-introduction and $\neg$-elimination are, respectively, the special cases of $\supset$-introduction and $\supset$-elimination in which the consequent is $\curlywedge$ so negation does not require separate treatment. (Note that for Gentzen $\curlywedge$, which has no introduction rule, is *not* a logical constant. He says it is a "definite proposition, 'the false proposition'" (Gentzen 1935, p. 70) and what we nowadays call the $\curlywedge$-elimination rule 'occupies a special place among the schemata: It does not belong to a logical symbol, but to the propositional symbol $\curlywedge$' (Gentzen 1935, p. 81).[8])
- $\vee$: From the introduction rules for $\vee$, ϕ and ψ separately entail $\phi \vee \psi$. Now, suppose χ is a common consequence of assumptions including ϕ and of assumptions including ψ. $\phi \vee \psi$ is also a common consequence. In order for $\phi \vee \psi$ to be an upper bound, we must say that χ is a consequence of $\phi \vee \psi$ together with the totality of assumptions in play. But, $\phi \vee \psi$ being a common consequence, this is sufficient for $\phi \vee \psi$ to allow for its own introduction.[9]
- $\forall$: Gentzen's introduction rule is:

$$\frac{\phi(a)}{\forall x \phi(x)}$$

  where the eigenvariable/free object variable a does not occur in $\forall x \phi(x)$, nor in any assumption on which it depends (Gentzen 1935, p. 77).[10] Gentzen's rule is doubly schematic: it is schematic in ϕ but it also schematic with respect to the eigenvariable a. Gentzen says of the restriction on a that it gives more precise

[8] If $\curlywedge$ isn't a logical constant then, strictly speaking, Gentzen's rules for negation are not *impure* in the sense of Dummett (1991, p. 257), for then only one logical constant figures in them.

[9] One may reasonably object that the standard $\vee$-elimination rule does not provide the *minimal* conditions that have $\phi \vee \psi$ allow for its own introduction; the restricted $\vee$-introduction rule of quantum logic also does that. But the pursuit of minimality is not an end in itself and since the standard rule allows for the levelling of local peaks, (i) we have no proof-theoretic motivation to restrict the elimination rule and (ii) the standard rule accords with the use of proof by cases in mathematics (and elsewhere).

[10] This is how Gentzen puts the restriction. He does not say, as we would nowadays expect, that a does not occur in any assumption on which $\phi(a)$ depends.

expression to the 'presupposition' that a is "completely arbitrary" (Gentzen 1935, p. 78). Now, if one has a proof of $\phi(a)$ from assumptions in which a does not occur, one can (although perhaps only with judicious changing of bound variables) turn any such proof into a proof of $\phi(t)$, for *any* term t, in the language notionally under consideration *and* in extensions of it. Thus the grounds on which $\forall x\phi(x)$ is derived are akin to those on which an indefinitely large conjunction of its instances would hold. This indefinite collection of instances provides the upper bound, the *attained* upper bound, and so the elimination rule for $\forall$ takes its familiar form:

$$\frac{\forall x\phi(x)}{\phi(t)},$$

for any term t. This is almost Gentzen's rule in his (1935). There he has the eigenvariable a in place of t, which is surely an unintended restriction as he has no rule for substituting other terms for eigenvariables and, in any case, he lifts the restriction in Gentzen (1936, p. 152).

- $\exists$: Gentzen's introduction rule for $\exists$, as amended in Gentzen (1936), is:

$$\frac{\phi(t)}{\exists x\phi(x)}$$

where t is any term substituting for the occurrences of x in $\phi(x)$. Thus $\exists x\phi(x)$ is an upper bound on *all* substitution instances. The $\exists$-elimination rule must show that $\exists x\phi(x)$ behaves as an attained least upper bound. First, then, we must think how to represent χ's being an upper bound on all substitution instances of $\exists x\phi(x)$ *in the presence of side formulae*. Now, if proofs are to remain surveyable, finite objects, although the substitution instances range across all terms in the language notionally under consideration and extensions of it, the side formulae may not. Hence, for indefinitely many terms t, there are side formulae Σ not involving the term t from which χ may be derived in the presence of $\phi(t)$. But then t occurs "parametrically" in the derivation, and so can be replaced by an eigenvariable foreign to Σ and χ. With such a derivation in hand, though, all cases now fall out of this one (possibly with judicious changes of bound variables in play). The restrictions on the eigenvariable a in the $\exists$-elimination rule thus fall out naturally from the least-upper-bound conception of the $\exists$-introduction rule together with the requirement that proofs be finite, surveyable objects.

We obtain elimination rules in general elimination format in the case of the two logical constants for which there are, at least implicitly, more than a single introduction rule. Gentzen, I suggest, would have seen no particular merit in rewriting the elimination rules for &, $\supset$, $\neg$, and $\forall$ in the general elimination format, for his own forms conform more closely to the way these logical constants are used in mathematical reasoning.

8.3 General Elimination Rules and a Rethinking of the Role of Introduction Rules

In a preliminary remark on his calculus *NJ* "for 'natural' intuitionist derivations of true formulae [richtigen Formeln]", Gentzen characterizes what is the foundation stone of the natural deduction formulation of a logic:

> Externally, the essential difference between '*NJ*-derivations' and derivations on the systems of Russell, Hilbert, and Heyting is the following: In the latter systems true formulae are derived from a sequence of 'basic logical formulae' by means of a few forms of inference. Natural deduction, however, does not, in general, start from basic logical propositions, but rather from *assumptions* ... to which logical deductions are applied. By means of a later inference the result is then again made independent of the assumption. (Gentzen 1935, p. 75)

With that in mind, and on the assumptions (1) that logically complex formulae may occur as assumptions and (2) that when they so occur they do not feature only as input to introduction rules for formulae of greater complexity (whether or not their internal structure is pertinent to such an application of an introduction rule), we need elimination rules to tell us "what to make of" a so-introduced formula, how to progress. But presumably not only this, for if a logically complex formula of the same form may occur as an intermediate conclusion in a deduction from assumptions, we need too, granted the assumption that it not so occur only as input to an introduction rule for a formula of greater complexity, to know "what to make of" the formula, to know how to progress, in this context as well, even though the formula does not occur as an assumption. In principle there could, then, be two kinds of elimination rule, one for when a logically complex formula occurs as an assumption, one for when it occurs as an intermediate conclusion in a derivation. But if at this point we hold to Gentzen's aim 'to set up a formalism that reflects as accurately as possible the actual logical reasoning involved in mathematical proofs' (Gentzen 1935, p. 74) and note the practices, on the one hand, of proving lemmata whose conclusions then stand as premisses in subsequent proofs and, on the other, of drawing conclusions on the basis of conjectures and hypotheses often subsequently proved, we see that we lack motivation to draw any such distinction among elimination rules. There is, then, a single role for elimination rules. In *Natural Deduction* Prawitz sums it up thus:

> An elimination rule for a logical constant γ allows an inference *from* a formula that has γ as principal sign, i.e. an instance of such a rule has a major premiss A, which has γ as principal sign. In an application of such a rule, subformulae of the major premiss occur either as consequence and as minor premisses (if any) or as assumptions discharged by the application. (Prawitz 1965, pp. 32–33)

In keeping with the refashioning of general elimination rules, we strike out the occurrence of 'as consequence' here and arrive at this general form for an elimination rule[11]:

[11] I do not claim that this is the most general form possible. It is, I suggest, the most straightforward general form matching Prawitz's characterisation and consonant with the refashioning employed in general elimination rules.

$$\dfrac{\star(\phi_1,\phi_2,\ldots,\phi_n)\quad \begin{matrix}[\phi_{i_1}]^m\\ \vdots\\ \chi\end{matrix}\quad \begin{matrix}[\phi_{i_2}]^m\\ \vdots\\ \chi\end{matrix}\quad \begin{matrix}\cdots\\ \cdots\\ \cdots\end{matrix}\quad \begin{matrix}[\phi_{i_k}]^m\\ \vdots\\ \chi\end{matrix}\qquad \phi_{j_1}\ \phi_{j_2}\ \cdots\ \phi_{j_l}}{\chi}\ m\ \star\text{-elimination}$$

where $k + l \leq n$ and $i_p \neq j_q, 1 \leq p \leq k, 1 \leq q \leq l$.

Let us call the formulae $\phi_{i_1}, \phi_{i_2}, \ldots\phi_{i_k}, \phi_{j_1}, \phi_{j_2}, \ldots\ \phi_{j_l}$ *side formulae* and let us say that the ϕ_{i_p}'s occur *hypothetically*, the ϕ_{j_q}'s *categorically* in the rule (*cf.* Francez and Dyckhoff 2012, Sect. 2); we shall apply the same terminology to ntroduction rules below.

A given rule is weakened by adding side formulae—not necessarily from among $\phi_1, \phi_2, \ldots, \phi_n$ —, whether categorically or hypothetically. We insist that $i_p \neq j_q, 1 \leq p \leq k, 1 \leq q \leq l$, for a rule that breached this requirement would be a weakening of the rule

$$\dfrac{\begin{matrix}[\phi]^m\\ \vdots\\ \chi\end{matrix}\qquad \phi}{\chi}\ m$$

which, in the context of most systems of natural deduction, does no work, and otherwise, in the context of a system in which it were not redundant, would (presumably) be an unwelcome addition.[12]

I note in passing that the general form allows the two &-elimination rules

$$\dfrac{\phi \,\&\, \psi \qquad \begin{matrix}[\phi]^m\\ \vdots\\ \chi\end{matrix}}{\chi}\ m \quad \text{and} \quad \dfrac{\phi \,\&\, \psi \qquad \begin{matrix}[\psi]^m\\ \vdots\\ \chi\end{matrix}}{\chi}\ m\,.$$

I prefer these over the single (general) elimination rule

$$\dfrac{\phi \,\&\, \psi \qquad \begin{matrix}\underbrace{[\phi]^m, [\psi]^m}\\ \vdots\\ \chi\end{matrix}}{\chi}\ m$$

favoured by most of our authors—Schroeder-Heister (1984; 1984b), Tennant (1992), Read (2000; 2004), Negri (2001) and von Plato (2001), Francez and Dyckhoff (2012); Read (2010, pp. 565–566) is an exception.[13]

[12] Prawitz's general procedure for obtaining elimination rules from introduction rules in his (1978) yields a rule of this form as elimination rule for $\supset$. This anomaly led Schroeder-Heister to his higher-order rule—see Schroeder-Heister (2014, n. 2).

[13] Negri (2002, pp. 573, 574) has the single elimination rule for the multiplicative conjunction of linear logic and the pair of elimination rules for the additive conjunction.

What of introduction rules? As we have seen, in *Natural Deduction* Prawitz characterizes them so:

> An introduction rule for a logical constant γ allows the inference *to* a formula A that has γ as principal sign from formulae that are subformulae of A; i.e. an instance of such a rule has a consequence A with γ as principal sign and has one or two premisses which are subformulae of A. (Prawitz 1965, p. 32)

Thinking of the role of assumptions as what is characteristic of natural deduction, the point to remark here is that it is not immediately obvious how one arrives at this conception of introduction rules.

Elimination rules allow us to progress from assumptions. Now, of course, by making appropriate assumptions, one can prove anything (from appropriate assumptions). What is important in mathematical practice is the choice of assumptions, and, perhaps most importantly, the whittling away of assumptions, the elimination of axioms, conjectures and hypotheses, either as redundant, or, more commonly, as following from more basic assumptions. What is important, then, is to know when a logically complex assumption is unnecessary. *One* way to do this, the route Gentzen takes, is to lay down conditions under which one may draw logically complex formulae as conclusions. What is of the utmost importance is to recognise that it is just *one* way; it is *not* the only way. What *is* essential to an introduction rule is that it characterises conditions under which a logically complex assumption is unnecessary and hence may be discharged without loss. I propose, then, this general form for introduction rules:

$$\dfrac{\begin{array}{c}[\star(\phi_1,\phi_2,\ldots,\phi_n)]^m\\ \vdots\\ \chi\end{array}\quad \begin{array}{c}[\phi_{r_1}]^m\\ \vdots\\ \chi\end{array}\quad \begin{array}{c}[\phi_{r_2}]^m\\ \vdots\\ \chi\end{array}\quad \begin{array}{c}\ldots\\ \vdots\\ \ldots\end{array}\quad \begin{array}{c}[\phi_{r_t}]^m\\ \vdots\\ \chi\end{array}\quad \phi_{s_1}\ \phi_{s_2}\ \ldots\ \phi_{s_u}}{\chi}\ m\ \star\text{-introduction}$$

where $t + u \leq n$ and $r_p \neq s_q$, $1 \leq p \leq t$, $1 \leq q \leq u$. I propose, too, to call such rules *general introduction rules*.[14]

The standard introduction rules for conjunction and disjunction that we have from Gentzen can obviously be recast in this mould, but at first sight this template may seem unduly restrictive. The insistence that all subproofs have a uniform conclusion prohibits this inessential reformulation of Gentzen's $\supset$-introduction:

$$\dfrac{\begin{array}{c}[\phi \supset \psi]^m\\ \vdots\\ \chi\end{array}\qquad \begin{array}{c}[\phi]^m\\ \vdots\\ \psi\end{array}}{\chi}\ m\ .$$

[14] The term 'general introduction rule' has been used already by Negri and von Plato. My general introduction rules are similar in form but not identical to those of Negri and von Plato's (2001, p. 217) natural deduction system NG for intuitionist propositional logic, and again similar in style but not identical to the rules of Negri (2002). For example, Negri and von Plato adopt what I call immediately below an inessential reformulation of Gentzen's $\supset$-introduction rule, a rule which does not comply with my conception of general introduction rules.

Far from being restrictive, however, this format for introduction rules admits strictly classical rules for negation and the conditional. For negation, we adopt the *rule of dilemma*, suggested by Gentzen himself as a rule governing classical negation (Gentzen 1936, p. 154):

$$\dfrac{\begin{matrix}[\neg\phi]^m\\ \vdots\\ \chi\end{matrix}\qquad\qquad\begin{matrix}[\phi]^m\\ \vdots\\ \chi\end{matrix}}{\chi}\,m\,.$$

For the conditional we have these two general introduction rules:

$$\dfrac{\begin{matrix}[\phi\supset\psi]^m\\ \vdots\\ \chi\end{matrix}\qquad\qquad\begin{matrix}[\phi]^m\\ \vdots\\ \chi\end{matrix}}{\chi}\,m \qquad\text{and}\qquad \dfrac{\begin{matrix}[\phi\supset\psi]^m\\ \vdots\\ \chi\end{matrix}\qquad\qquad\psi}{\chi}\,m\,,$$

the first of which I called 'Tarski's Rule' in Milne (2008, 2010) as it bears the same relation to the tautology sometimes called 'Tarski's Law' as the better known Peirce's Rule bears to Peirce's Law; from these we readily obtain both Gentzen's introduction rule and the inessential variant as derived rules:

$$\dfrac{[\phi\supset\psi]^m \qquad \dfrac{[\phi\supset\psi]^n \qquad \begin{matrix}[\phi]^m\\ \vdots\\ \psi\end{matrix}}{\phi\supset\psi}\,n}{\begin{matrix}\phi\supset\psi\\ \vdots\\ \chi\end{matrix}}\,m\,.$$

My favoured formulation of classical propositional logic with all of conjunction, disjunction, the conditional, and negation treated as primitive, has, then, the introduction rules just noted, the general elimination reformulations of standard elimination rules for conjunction and the conditional, *ex falso quodlibet*, which I take already to be in general elimination format, and Gentzen's $\vee$-elimination as elimination rules.[15]

We should note that, quite generally, two rules that differ only in having a side formula occur hypothetically in one and categorically in the other can be "merged" by the removal of this side formula. Our negation rules establish the legitimacy of this procedure. Given the rules

$$\dfrac{\mathit{stuff}\qquad\begin{matrix}[\phi]^m\\ \vdots\\ \chi\end{matrix}}{\chi}\,m \qquad\text{and}\qquad \dfrac{\mathit{stuff}\qquad\phi}{\chi}\,,$$

[15] *Two comments:* Firstly, the use of general elimination rules is new here, it is not present in Milne (2008, 2010). Secondly, the system of Milne (2008, 2010) and its extension to first order employing the existential quantifier only as primitive were discovered independently by Tor Sandqvist.

we proceed as follows:

$$\dfrac{\dfrac{\mathit{stuff} \qquad \dfrac{[\phi]^m \qquad [\neg\phi]^n}{\chi}\,\neg\text{-e}}{\chi}\,m \qquad \dfrac{\mathit{stuff} \qquad [\phi]^n}{\chi}}{\chi}\,n\ \neg\text{-i}\,,$$

which gives us the template for a derivation of χ from the resources labelled 'stuff'. (Our $\supset$-elimination rule and Tarski's Rule could be used to the same effect in place of the negation rules.)[16]

8.4 General Introduction Rules and General Elimination Rules

In either Gentzen's original form or in the style I have adopted for them, if a formula may be introduced in the circumstances indicated in its introduction rule(s), so too may any logically weaker formula. (Of course, we need an elimination rule or rules to draw out consequences and so help make manifest just which formulae are weaker, but the thought stands nonetheless.) Now, if we were to treat the introduced formula as weaker than need be, its consequences would be too weak to permit its own introduction; or, turning that around, given what we could infer from it, it would have an unnecessarily restrictive introduction rule. Thus it is the job of an elimination rule to ensure that the introduced formula is taken to be as strong as its introduction rule allows or rules allow: in particular, it should be strong enough to permit its own introduction. As we saw, applied to Gentzen's rule for the introduction of the conditional, this line of thinking tells us that ϕ and $\phi \supset \psi$ must jointly supply a proof with ϕ as assumption and ψ as conclusion and this is exactly what the standard $\supset$-elimination rule does tell us they do (*cf.* Gentzen 1935, pp. 80–81). Applied to my *two* introduction rules, $\phi \supset \psi$ must provide a proof of χ in which it occurs as an assumption to be discharged (on grounds of redundancy) in the presence of either (i) a proof of χ from ϕ or (ii) a proof of ψ; the minimal condition that ensures this is that in the presence of $\phi \supset \psi$, ϕ and a proof of χ in which ψ occurs as assumption, we obtain a proof of χ in which the assumption ψ may be discharged. That is, we obtain the standard rule of $\supset$-elimination recast in general elimination form.

More generally, to derive the elimination rule or rules governing the logical connective $\star$ from the introduction rule(s) governing it, we proceed as follows:

- $\star(\phi_1, \phi_2, \ldots, \phi_n)$ must occur as an assumption in a proof of χ *and occur in such a way as to be able to be to be discharged in the circumstances set out in the introduction rule(s)* for $\star(\phi_1, \phi_2, \ldots, \phi_n)$;
- it does this only if, for *each* introduction rule, it provides a means to obtain a proof of χ not dependent on $\star(\phi_1, \phi_2, \ldots, \phi_n)$;

[16] Merging is a special case of the operation on rules called *splicing* in Milne (Milne 2012).

- this is achieved if, for one side formula in each introduction rule, the elimination rule provides a means to obtain a proof of χ employing that side formula;
- if the side-formula occurs hypothetically in the introduction rule, having the same formula occur categorically in the elimination rule gives us a proof of χ from the formula in question in the circumstances set out in that introduction rule; if the side-formula occurs categorically in the introduction rule, having the same formula occur hypothetically in the elimination rule as an assumption in a proof of χ lets us obtain a proof of χ from the formula in question in the circumstances set out in that introduction rule;
- we maintain our standing restriction on introduction and elimination rules, namely that no formula occurs both hypothetically and categorically in the same rule.

Examples may make this clearer. Consider first the binary connective with this pair of introduction rules:

$$\dfrac{\begin{array}{c}[\phi + \psi]^m \\ \vdots \\ \chi\end{array} \qquad \begin{array}{c}[\phi]^m \\ \vdots \\ \chi\end{array} \qquad \psi}{\chi}\, m \quad \text{and} \quad \dfrac{\begin{array}{c}[\phi + \psi]^m \\ \vdots \\ \chi\end{array} \qquad \phi \qquad \begin{array}{c}[\psi]^m \\ \vdots \\ \chi\end{array}}{\chi}\, m\,.$$

Selecting ϕ from the first rule we must select ψ from the second to obtain the first of the following rules; selecting ψ from the first, we obtain the second:

$$\dfrac{\phi + \psi \qquad \phi \qquad \psi}{\chi}\, m \quad \text{and} \quad \dfrac{\phi + \psi \qquad \begin{array}{c}[\phi]^m \\ \vdots \\ \chi\end{array} \qquad \begin{array}{c}[\psi]^m \\ \vdots \\ \chi\end{array}}{\chi}\, m\,.$$

The introduction and elimination rules are well matched, for the introduction rules for $+$ allow us to derive $\phi + \psi$ from $(\phi \vee \psi)\ \&\ \neg(\phi\ \&\ \psi)$ and the elimination rules we have just obtained permit the converse derivation. '$+$' stands for exclusive disjunction.

Consider next the ternary connective with these introduction rules:

$$\dfrac{\begin{array}{c}[\looparrowright(\phi, \psi, \chi)]^m \\ \vdots \\ \upsilon\end{array} \qquad \begin{array}{c}[\phi]^m \\ \vdots \\ \upsilon\end{array} \qquad \chi}{\upsilon}\, m \quad \text{and}$$

$$\dfrac{\begin{array}{c}[\looparrowright(\phi, \psi, \chi)]^m \\ \vdots \\ \upsilon\end{array} \qquad \phi \qquad \psi}{\upsilon}\, m\,.$$

The procedure outlined above produces *three* elimination rules:

$$\frac{\looparrowright(\phi, \psi, \chi) \qquad \phi \qquad \begin{array}{c}[\psi]^m\\ \vdots\\ \upsilon\end{array}}{\upsilon}\,m\,,$$

$$\frac{\looparrowright(\phi, \psi, \chi) \qquad \begin{array}{c}[\phi]^m\\ \vdots\\ \upsilon\end{array} \qquad \begin{array}{c}[\chi]^m\\ \vdots\\ \upsilon\end{array}}{\upsilon}\,m\,, \quad \text{and}$$

$$\frac{\looparrowright(\phi, \psi, \chi) \qquad \begin{array}{c}[\psi]^m\\ \vdots\\ \upsilon\end{array} \qquad \begin{array}{c}[\chi]^m\\ \vdots\\ \upsilon\end{array}}{\upsilon}\,m\,.$$

However, weakening the last by adding ϕ categorically gives us a weakening of the first rule, weakening it by adding ϕ hypothetically gives us a weakening of the second, showing the third to be redundant. '$\looparrowright(\phi, \psi, \chi)$' is the computer programmer's 'if ϕ then ψ, else χ'.[17]

We shall call this method of obtaining elimination rules from introduction rules *the classical inversion procedure*. Let us call a general elimination rule or set of general elimination rules obtained by this method from the general introduction rule or set of general introduction rules for a connective *harmonious*.

The form of general introduction and elimination rules allows us to establish:

- derivability of (harmonious) general introduction rules from general elimination rules;
- levelling of local peaks in something like Dummett's sense;
- derivability of classical truth-functional characterisations of propositional connectives governed by harmonious general introduction and elimination rules;
- derivability of harmonious general introduction and elimination rules for a classical truth-functional connective from its truth-table;
- completeness of harmonious introduction and elimination rules with respect to the classical semantics for the connectives;
- the obtaining of the subformula property for any fragment of classical logic containing only connectives governed by harmonious general introduction and elimination rules;
- uniqueness in Belnap's sense and harmony in the sense of Neil Tennant's *Natural Logic*, for connectives governed by harmonious general introduction and elimination rules.

We shall go through this list in order.

[17] The rules here are obtained by the splicing technique when $\looparrowright(\phi, \psi, \chi)$ is identified either with $(\phi \supset \psi) \,\&\, (\neg\phi \supset \chi)$ or with $(\phi \,\&\, \psi) \vee (\neg\phi \,\&\, \chi)$ — see Milne (2012). Compare the rules given here for 'if ...then ...else ...' with those in Francez and Dyckhoff (2012) which mention negation explicitly.

8.4.1 Deriving Introduction Rules From Elimination Rules

We can use *exactly* the same procedure to read off introduction rules from elimination rules. Why so? Each elimination rule in general elimination format gives us the form of a proof of arbitrary χ in which the eliminated complex formula $\star(\phi_1, \phi_2, \ldots, \phi_n)$ stands as an assumption. By attending to the "side conditions" given in an elimination rule we can then read off circumstances that would render the proof in which the eliminated formula occurs as an assumption redundant; we must ensure, though, that this holds good for *every* elimination rule. In other words, given any one elimination rule, there is a separate introduction rule containing each side formula that occurs in the elimination rule; and each introduction rule must contain one formula that occurs in each elimination rule. It suffices that a formula occur categorically in an introduction rule if it occurs hypothetically in an elimination rule and conversely; by our standing restriction on introduction and elimination rules, no formula occurs redundantly, i.e both hypothetically and categorically, in any introduction rule.[18]

It is readily seen that we may obtain the introduction rules for '$+$' and '$\looparrowright$' given above from the elimination rules previously derived. Quite generally, our classical inversion procedure is reversible.

8.4.2 The Levelling of Local Peaks

Adapting to the form general introduction rules take, we make one change to Prawitz's previously quoted characterisation of the inversion principle: substituting 'discharging' for 'deriving', we obtain

> Let α be an application of an elimination rule that has B as consequence. Then, deductions that satisfy the sufficient conditions [given in the introduction rule(s)] for discharging the major premiss of α when combined with deductions of the minor premisses of α (if any), already "contain" a deduction of B; the deduction of B is thus obtainable directly from the given deductions without the addition of α.

Of course, we have a different conception of the details of how the deductions are combined but the spirit is the same.

The classical inversion procedure gives rise to harmonious sets of introduction and elimination rules. Satisfaction of Prawitz' inversion principle is what Dummett called the levelling of local peaks. As an example of a local peak involving standard rules, we might have

---

[18] The procedure here and in the previous section is close to the *two* procedures outlined in Kurbis (2008), but Kurbis's method for deriving elimination from introduction rule applies only to constants with a single introduction rule and, likewise, his method for deriving introduction rules from elimination rule applies only to constants with a single elimination rule.

Negri (2002, pp. 571, 575) uses a 'dual inversion principle' to obtain her general introduction rules from general elimination rules. The dual inversion principle states, 'Whatever follows from a formula follows from the sufficient grounds for deriving the formula'. Cf. Negri (2001, p. 217).

$$\begin{array}{c}\dfrac{\begin{array}{c}\vdots\\ \phi\end{array} \qquad \dfrac{\begin{array}{c}[\phi]^m\\ \vdots\\ \psi\end{array}}{\phi \supset \psi}\,{}_m \supset\text{-i}}{\psi}\supset\text{-e}\,.\\ \vdots\\ \chi\end{array}$$

The maximum formula/local peak/hillock[19] can be "levelled". The deduction of ψ with ϕ as assumption, which, as the application of the introduction rule spells out, is sufficient for deriving the major premiss of the application of the elimination rule, namely $\phi \supset \psi$, when combined with a deduction of the minor premiss of that application, does indeed already contain, quite literally, a deduction of ψ, the conclusion of the application of the elimination rule. The proof simplifies:

$$\begin{array}{c}\vdots\\ \phi\\ \vdots\\ \psi\\ \vdots\\ \chi\end{array}\;.$$

When we transcribe a local peak for conjunction using general introduction and general elimination rules, we see what form a local peak takes in the present setting:

$$\begin{array}{c}\dfrac{\dfrac{\begin{array}{c}\vdots\\ \phi\end{array}\quad\begin{array}{c}\vdots\\ \psi\end{array}}{\phi \mathbin{\&} \psi}\,\&\text{-i}}{\phi}\,\&\text{-e}\\ \vdots\\ \chi\end{array} \quad \text{becomes} \quad \dfrac{\begin{array}{c}\vdots\\ \phi\end{array}\quad\begin{array}{c}\vdots\\ \psi\end{array}\quad \dfrac{[\phi \mathbin{\&} \psi]^n \qquad \begin{array}{c}[\phi]^m\\ \vdots\\ \chi\end{array}}{\chi}\,{}_m\,\&\text{-e}\,.}{\chi}\,{}_n\,\&\text{-i}$$

In the general case we have the major premiss $\star(\phi_1, \phi_2, \dots, \phi_n)$ in an application of a $\star$-elimination rule with conclusion χ standing as assumption discharged in an application of $\star$-introduction rule, the conclusion χ of the elimination subproof being brought down *immediately* as conclusion of the application of the introduction rule.

Whether one starts from elimination rules or from introduction rules, our classical inversion procedure *guarantees* satisfaction of the modified inversion principle: one of the side formulae in the elimination rule must occur in the introduction rule, moreover the formula occurs hypothetically in one, categorically in the other, and so we obtain a proof of χ from the same or fewer assumptions with the maximum formula excised, the peak/hillock levelled.

[19] We have 'maximum formula' from Prawitz (1965, p. 34), 'local peak' from Dummett (1991, p. 248), and 'hillock' from Gentzen (von Plato 2008, p. 243).

For example,

$$\dfrac{\dfrac{[\phi \supset \psi]^n \quad \phi \quad \begin{matrix}[\psi]^m \\ \vdots \\ \chi\end{matrix}}{\chi}\,{\scriptstyle m}\ \supset\text{-e} \qquad \begin{matrix}[\phi]^n \\ \vdots \\ \chi\end{matrix}}{\chi}\,{\scriptstyle n}\ \supset\text{-i}$$

reduces to

$$\begin{matrix}\phi \\ \vdots \\ \chi\end{matrix} \quad .$$

8.4.3 From Syntax to Semantics and Back

From an introduction rule together with *ex falso quodlibet*, our elimination rule for negation, we get

$$\neg\phi_{i_1}, \neg\phi_{i_2}, \ldots, \neg\phi_{i_k}, \phi_{j_1}, \phi_{j_2}, \ldots, \phi_{j_l} \vdash \star(\phi_1, \phi_2, \ldots, \phi_n);$$

from an elimination rule together with *both ex falso quodlibet and* the rule of dilemma, we get

$$\neg\phi_{i_1}, \neg\phi_{i_2}, \ldots, \neg\phi_{i_k}, \phi_{j_1}, \phi_{j_2}, \ldots, \phi_{j_l} \vdash \neg \star (\phi_1, \phi_2, \ldots, \phi_n).$$

We may proceed as follows. For an n-place connective $\star(\phi_1, \phi_2, \ldots, \phi_n)$, first draw up a table of all 2^n truth-value assignments to $\phi_1, \phi_2, \ldots, \phi_n$; next, corresponding to each introduction rule, treat categorical occurrences as *true*, hypothetical as *false*, and mark each row thus matching the pattern of occurrence in the introduction rule as *true*; when this has been done for each introduction rule, mark each remaining row as *false*. This is the truth-table for $\star$.

We can now read elimination rules off the resulting truth-table. Corresponding to each row marked false we construct an elimination rule by having ϕ_i occur categorically if marked *true* in the row and hypothetically if marked *false*. We then proceed to prune the collection of elimination rules by merging rules: if two rules differ only in that in one a formula occurs categorically, in the other hypothetically, we replace the two by a single rule in which that formula does not occur.

For example, starting with the previously encountered pair of introduction rules for the three-place connective $\looparrowright$:

Table 8.1 Truth-table for $\looparrowright(\phi, \psi, \chi)$

ϕ	ψ	χ	$\looparrowright(\phi, \psi, \chi)$
t	t	t	t
t	t	f	t
t	f	t	f
t	f	f	f
f	t	t	t
f	t	f	f
f	f	t	t
f	f	f	f

$$\dfrac{\begin{matrix}[\looparrowright(\phi, \psi, \chi)]^m \\ \vdots \\ \upsilon\end{matrix} \qquad \begin{matrix}[\phi]^m \\ \vdots \\ \upsilon\end{matrix} \qquad \chi}{\upsilon}\, m \quad \text{and}$$

$$\dfrac{\begin{matrix}[\looparrowright(\phi, \psi, \chi)]^m \\ \vdots \\ \upsilon\end{matrix} \qquad \phi \qquad \psi}{\upsilon}\, m\,.$$

we obtain this truth-table from which we read off the *four* rules (Table 8.1)

$$\dfrac{\looparrowright(\phi, \psi, \chi) \qquad \phi \qquad \chi \qquad \begin{matrix}[\psi]^m \\ \vdots \\ \upsilon\end{matrix}}{\upsilon}\, m\,,$$

$$\dfrac{\looparrowright(\phi, \psi, \chi) \qquad \phi \qquad \begin{matrix}[\psi]^m \\ \vdots \\ \upsilon\end{matrix} \qquad \begin{matrix}[\chi]^m \\ \vdots \\ \upsilon\end{matrix}}{\upsilon}\, m\,,$$

$$\dfrac{\looparrowright(\phi, \psi, \chi) \qquad \psi \qquad \begin{matrix}[\phi]^m \\ \vdots \\ \upsilon\end{matrix} \qquad \begin{matrix}[\chi]^m \\ \vdots \\ \upsilon\end{matrix}}{\upsilon}\, m\,, \quad \text{and}$$

$$\dfrac{\looparrowright(\phi, \psi, \chi) \quad \begin{matrix}[\phi]^m \\ \vdots \\ \upsilon\end{matrix} \qquad \begin{matrix}[\psi]^m \\ \vdots \\ \upsilon\end{matrix} \qquad \begin{matrix}[\chi]^m \\ \vdots \\ \upsilon\end{matrix}}{\upsilon}\, m\,,$$

which reduce by merging to our previous pair of elimination rules.

Conversely, again draw up a table of all 2^n truth-value assignments to $\phi_1, \phi_2, \ldots, \phi_n$, then, corresponding to each elimination rule, treat categorical occurrences as *true*, hypothetical as *false*, and mark each row thus matching the pattern of occurrence in the elimination rule as *false*; when this has been done for each elimination rule, mark each remaining row as *true*. We have the truth-table for $\star$ again. — And from it, we can now read off the introduction rules. Corresponding to each row marked true we construct an introduction rule by having ϕ_i occur categorically if marked *true* in the row and hypothetically if marked *false*; we then prune the collection of rules by merging.

In this setting, elimination rules are, very immediately, "unique functions" of their corresponding introduction rules, as Gentzen said, *and vice versa*. Moreover, we see how to read both introduction and elimination rules off truth-tables. The rules produced in this way are guaranteed to be harmonious because each row in the truth-table not matching an introduction rule differs from *every* row matching an introduction rule (and conversely).

8.4.4 The Restricted Lindenbaum Construction

General introduction and elimination rules are *ideally* fitted to the Lindenbaum's Lemma construction standardly used in proving (model theoretic) completeness. The model-theoretic technique of Milne (2008) and Milne (2010) can be used to prove an especially strong generic completeness result: employing harmonious general introduction and general elimination rules for each of *any* set of connectives, we have a deductive system complete with respect to classically valid inferences involving those connectives (where the classical truth conditions for each connective can be read off the truth-tables derived from the rules as in the previous section); moreover, the system of rules has the subformula property, so that any valid inference can be derived using rules for only the connectives occurring in the premisses and conclusion *and* these rules need be applied only to subformulae of the premisses and/or conclusion.

Let $\Sigma \nvdash_{\Sigma\cup\{\chi\}} \chi$, where $\vdash_{\Sigma\cup\{\chi\}}$ represents proof employing only rules for connectives occurring in χ and the formulae in Σ, restricted in application to subformulae of χ and the formulae in Σ. Let $\psi_1, \psi_1, \ldots, \psi_n \ldots$ be an enumeration of the subformulae of χ together with the subformulae of the formulae in Σ.

- $\Gamma_0 = \Sigma$;
- $\Gamma_{n+1} = \Gamma_n \cup \{\psi_n\}$, if $\Gamma_n, \psi_n \nvdash_{\Sigma\cup\{\chi\}} \chi$;
- $\Gamma_{n+1} = \Gamma_n$ otherwise;
- $\Gamma_I = \bigcup_{n\in I} \Gamma_n$, where I indexes $\psi_1, \psi_1, \ldots, \psi_n, \ldots$.

For any ψ a subformula of χ or a subformula of a formula in Σ, we have that $\psi \notin \Gamma_I$ if, and only if, $\Gamma_I, \psi \vdash_{\Sigma\cup\{\chi\}} \chi$.

Suppose that $\star(\phi_1, \phi_2, \ldots, \phi_n)$ is one of the subformulae of interest. Let

$$\dfrac{\begin{array}{cccccccccc}[\star(\phi_1,\phi_2,\ldots,\phi_n)]^m & [\phi_{r_1}]^m & [\phi_{r_2}]^m & \ldots & [\phi_{r_t}]^m & & & & \\ \vdots & \vdots & \vdots & \vdots & \vdots & & & & \\ \chi & \chi & \chi & \ldots & \chi & \phi_{s_1} & \phi_{s_2} & \ldots & \phi_{s_u}\end{array}}{\chi}\, m$$

be an introduction rule for $\star$; suppose that $\phi_{r_1} \notin \Gamma_I$, that $\phi_{r_2} \notin \Gamma_I$, ..., that $\phi_{r_t} \notin \Gamma_I$, that $\phi_{s_1} \in \Gamma_I$, that $\phi_{s_2} \in \Gamma_I$, ..., and that $\phi_{s_u} \in \Gamma_I$. If $\star(\phi_1, \phi_2, \ldots, \phi_n) \notin \Gamma_I$, then we have $\Gamma_I, \phi_{r_1} \vdash_{\Sigma\cup\{\chi\}} \chi, \Gamma_I, \phi_{r_2} \vdash_{\Sigma\cup\{\chi\}} \chi, \ldots, \Gamma_I, \phi_{r_t} \vdash_{\Sigma\cup\{\chi\}} \chi, \Gamma_I \vdash_{\Sigma\cup\{\chi\}} \phi_{s_1}$, $\Gamma_I \vdash_{\Sigma\cup\{\chi\}} \phi_{s_2}, \ldots, \Gamma_I \vdash_{\Sigma\cup\{\chi\}} \phi_{s_u}$, and $\Gamma_I, \star(\phi_1, \phi_2, \ldots, \phi_n) \vdash_{\Sigma\cup\{\chi\}} \chi$, hence, by the $\star$ introduction rule, $\Gamma_I \vdash_{\Sigma\cup\{\chi\}} \chi$, contrary to hypothesis. Thus associated with the introduction rule there is a closure (or completeness) condition: if $\phi_{r_1} \notin \Gamma_I, \phi_{r_2} \notin \Gamma_I$, ..., $\phi_{r_t} \notin \Gamma_I, \phi_{s_1} \in \Gamma_I, \phi_{s_2} \in \Gamma_I$, ..., and $\phi_{s_u} \in \Gamma_I$ then $\star(\phi_1, \phi_2, \ldots, \phi_n) \in \Gamma_I$.

Next, let

$$\dfrac{\begin{array}{cccccccccc} & [\phi_{i_1}]^m & [\phi_{i_2}]^m & \ldots & [\phi_{i_k}]^m & & & & \\ & \vdots & \vdots & \vdots & \vdots & & & & \\ \star(\phi_1,\phi_2,\ldots,\phi_n) & \chi & \chi & \ldots & \chi & \phi_{j_1} & \phi_{j_2} & \ldots & \phi_{j_l}\end{array}}{\chi}\, m$$

be an elimination rule for $\star$; suppose that $\phi_{i_1} \notin \Gamma_I$, that $\phi_{i_2} \notin \Gamma_I$, ..., that $\phi_{i_k} \notin \Gamma_I$, that $\phi_{j_1} \in \Gamma_I$, that $\phi_{j_2} \in \Gamma_I$, ..., and that that $\phi_{j_l} \in \Gamma_I$. If $\star(\phi_1, \phi_2, \ldots, \phi_n) \in \Gamma_I$, then we have $\Gamma_I, \phi_{i_1} \vdash_{\Sigma\cup\{\chi\}} \chi, \Gamma_I, \phi_{i_2} \vdash_{\Sigma\cup\{\chi\}} \chi, \ldots, \Gamma_I, \phi_{i_k} \vdash_{\Sigma\cup\{\chi\}} \chi, \Gamma_I \vdash_{\Sigma\cup\{\chi\}}$ $\phi_{j_1}, \Gamma_I \vdash_{\Sigma\cup\{\chi\}} \phi_{j_2}, \ldots, \Gamma_I \vdash_{\Sigma\cup\{\chi\}} \phi_{j_l}$, and $\Gamma_I \vdash_{\Sigma\cup\{\chi\}} \star(\phi_1, \phi_2, \ldots, \phi_n)$, hence, by the $\star$ elimination rule, $\Gamma_I \vdash_{\Sigma\cup\{\chi\}} \chi$, contrary to hypothesis. Thus associated with the elimination rule there is an exclusion (or soundness) condition: if $\phi_{i_1} \notin \Gamma_I, \phi_{i_2} \notin \Gamma_I$, ..., $\phi_{i_k} \notin \Gamma_I, \phi_{j_1} \in \Gamma_I, \phi_{j_2} \in \Gamma_I$, ..., and $\phi_{j_l} \in \Gamma_I$ then $\star(\phi_1, \phi_2, \ldots, \phi_n) \notin \Gamma_I$.

Consequently, when we build up the classical truth-valuation inductively, starting with those atomic subformulae of χ and the formulae in Σ belonging to Γ_I as true and those not belonging as false, the introduction and elimination rules for $\star$ ensure that the valuation matches $\star$'s truth-table. — In the standard way, this establishes completeness of the system of rules with respect to the truth-conditions codified in the truth-tables. We have, moreover, restricted the means of proof so that the subformula property is guaranteed to obtain when we use only introduction and elimination rules harmonious in the sense introduced above.

We can now easily establish uniqueness in the sense of Belnap (1962) and harmony in the sense of Tennant (1978). Let $\star$ and $*$ be two connectives governed by homologous rules. Let $\phi_1, \phi_2, \ldots, \phi_n$ be atomic and suppose that $\star(\phi_1, \phi_2, \ldots, \phi_n)$ does not entail $*(\phi_1, \phi_2, \ldots, \phi_n)$ using only the rules for $\star$ and $*$ restricted in application to $\phi_1, \phi_2, \ldots, \phi_n$, $\star(\phi_1, \phi_2, \ldots, \phi_n)$, and $*(\phi_1, \phi_2, \ldots, \phi_n)$. Apply the Lindenbaum construction to this set of formulae. We obtain a set Γ containing $\star(\phi_1, \phi_2, \ldots, \phi_n)$ but not $*(\phi_1, \phi_2, \ldots, \phi_n)$. For each introduction rule

$$\dfrac{\begin{array}{cccccccccc}[*(\phi_1,\phi_2,\ldots,\phi_n)]^m & [\phi_{r_1}]^m & [\phi_{r_2}]^m & \ldots & [\phi_{r_t}]^m & & & & \\ \vdots & \vdots & \vdots & \vdots & \vdots & & & & \\ \chi & \chi & \chi & \ldots & \chi & \phi_{s_1} & \phi_{s_2} & \ldots & \phi_{s_u}\end{array}}{\chi}\, m$$

for $*$, we have that at least one of the ϕ_{r_p}'s, $1 \leq r \leq t$, is in Γ or at least one of the ϕ_{s_q}'s, $1 \leq s \leq u$, is not. But by the classical inversion procedure, this means that we have an elimination rule

$$\dfrac{\star(\phi_1, \phi_2, \ldots, \phi_n) \quad \begin{matrix}[\phi_{i_1}]^m \\ \vdots \\ \chi\end{matrix} \quad \begin{matrix}[\phi_{i_2}]^m \\ \vdots \\ \chi\end{matrix} \quad \begin{matrix}\ldots \\ \vdots \\ \ldots\end{matrix} \quad \begin{matrix}[\phi_{i_k}]^m \\ \vdots \\ \chi\end{matrix} \quad \phi_{j_1} \quad \phi_{j_2} \quad \ldots \quad \phi_{j_l}}{\chi}\,m$$

for $\star$ for which *none* of the ϕ_{i_p}'s belong to Γ and *all* of the ϕ_{j_q}'s belong to Γ, yet $\star(\phi_1, \phi_2, \ldots, \phi_n)$ belongs to Γ, contradicting the exclusion condition associated with the rule! Hence $\star(\phi_1, \phi_2, \ldots, \phi_n) \vdash_{\{\phi_1, \phi_2, \ldots, \phi_n, \star(\phi_1, \phi_2, \ldots, \phi_n), *(\phi_1, \phi_2, \ldots, \phi_n)\}} *(\phi_1, \phi_2, \ldots, \phi_n)$. — The rules governing $\star$ alone establish uniqueness in Belnap's sense. Furthermore, we see that we have appealed only to the introduction rule(s) for $*$ and the elimination rule(s) for $\star$ so we have, in effect shown, as Tennant's notion of harmony requires (Tennant 1978, pp. 74–76), that the elimination rules for $\star$ show that any formula playing the role set out in the introduction rules for $\star(\phi_1, \phi_2, \ldots, \phi_n)$ is entailed by $\star(\phi_1, \phi_2, \ldots, \phi_n)$, hence $\star(\phi_1, \phi_2, \ldots, \phi_n)$ is, up to interderivability, the strongest that can be introduced, and that the introduction rules for $\star$ show that any formula playing the role set out in the elimination rules for $\star(\phi_1, \phi_2, \ldots, \phi_n)$ entails $\star(\phi_1, \phi_2, \ldots, \phi_n)$, hence $\star(\phi_1, \phi_2, \ldots, \phi_n)$ is, up to interderivability, the weakest that can be eliminated.

8.4.5 Falsum, Verum, and Bullet

falsum

For Gentzen ⅄, *falsum*, was not a logical constant. Subsequently, it has been taken to be such, a zero-place connective with no introduction rule and this elimination rule (which is in general elimination format):

$$\frac{⅄}{\chi}\,.$$

If we start with the elimination rule then, following the classical inversion procedure, because there are no side conditions in the elimination rule, we can read off *no* circumstances that would render the proof in which the eliminated formula occurs as an assumption redundant. Consequently, there is *no* introduction rule—just as tradition requires!

Starting from the absence of an introduction rule, we are looking for side conditions to put in place of '?' in the elimination rule

$$\frac{⅄ \quad ?}{\chi}\,.$$

And exactly because there is no introduction rule, there is nothing to put in its place: so we obtain the standard elimination rule.

This account of $\curlywedge$ ties in, as we should expect it would, with the connection developed above between harmonious general introduction and elimination rules and truth-tables, for $\curlywedge$ has this one-row, one-column truth-table:

$$\begin{array}{c} \curlywedge \\ \hline\hline f \end{array}$$

Applying the analysis of Sect. 8.4.3, it is no surprise we obtained the rules just given.

verum

The orthodox rule for a *verum* constant, $\curlyvee$, would have

$$\frac{\phi}{\curlyvee}\;;$$

in general introduction format, this becomes

$$\dfrac{\phi \qquad \begin{array}{c} [\curlyvee]^m \\ \vdots \\ \chi \end{array}}{\chi}\, m$$

which simplifies to

$$\dfrac{\begin{array}{c} [\curlyvee]^m \\ \vdots \\ \chi \end{array}}{\chi}\, m$$

since ϕ is arbitrary.

If we start with the introduction rule then, following the classical inversion procedure, because there are no side conditions in the introduction rule, we cannot read off any way in which $\curlyvee$ may occur as a premise in a proof of χ so that it can be discharged in the circumstances set out in the introduction rule. In short, there is no elimination rule.

Starting from the absence of an elimination rule, we are looking for side conditions to put in place of '?' in the introduction rule

$$\dfrac{\begin{array}{c} [\curlyvee]^m \\ \vdots \\ \chi \end{array} \qquad ?}{\chi}\, m\;.$$

Exactly because there is no elimination rule, there is nothing to put in its place: so we obtain the introduction rule above.

This analysis is confirmed by $\curlyvee$'s uninteresting one-row, one-column truth-table:

$$\frac{\curlyvee}{\overline{t}}$$

bullet Stephen Read has written about a further, paradoxical, one-place connective which, symbolised by '$\bullet$', is sometimes called 'bullet': *bullet* is interderivable with its negation. In his (2000), Read gives it these (impure) introduction and elimination rules:

$$\frac{\neg\bullet}{\bullet}\ \bullet\text{-i} \qquad \text{and} \qquad \frac{\bullet \qquad \begin{array}{c}[\neg\bullet]^m\\ \vdots \\ \chi\end{array}}{\chi}\ \bullet\text{-e}$$

(Read 2000, p. 141).[20]

In his (Read 2010) it has these (impure) rules:

$$\frac{\begin{array}{c}[\bullet]^m\\ \vdots \\ \curlywedge\end{array}}{\bullet}\ {}_m\ \bullet\text{-i} \qquad \text{and} \qquad \frac{\bullet}{\curlywedge}\ \bullet\text{-e}$$

(Read 2010, p. 571).[21]

According to Read, these rules satisfy the requirements of general elimination harmony. So much the worse for ge-harmony. I don't for a moment consider the rules governing $\bullet$ to be harmonious. In view of the elimination rule for $\curlywedge$ and *provided we do not call in doubt the transitivity of derivability*, we lose nothing by taking the elimination rule for $\bullet$ to be

$$\frac{\bullet}{\chi}\ ,$$

and now we see that, from the present perspective (in which we are not calling in doubt the usual structural rules)—but surely also from Read's perspective—, there is something very badly wrong. If the elimination rule for *falsum* is genuinely harmoniously paired with no introduction rule (as we have seen it is), it should not also be harmoniously paired with the introduction rule for $\bullet$.

It gets worse—or, to be strictly accurate, *in a classical setting* it gets worse. Classically, $\bullet$'s introduction rule simplifies. Using Gentzen's $\neg$-elimination rule and our $\neg$-introduction rule, we obtain this derivation

$$\frac{\dfrac{\dfrac{[\bullet]^m \qquad [\neg\bullet]^n}{\curlywedge}\ \neg\text{-e}}{\bullet}\ {}_m\ \bullet\text{-i} \qquad \begin{array}{c}[\bullet]^n\\ \vdots \\ \chi\end{array}}{\chi}\ {}_n\ \neg\text{-i}\ ,$$

[20] *Cf.* Schroeder-Heister (2004, pp. 36–37)

[21] Strictly, because he is being very careful regarding structural rules, Read has two occurrences of $\bullet$ "above the line" in the elimination rule.

which shows that we could, in the present setting, just as well adopt this introduction rule for $\bullet$:

$$\dfrac{\begin{matrix}[\bullet]^m\\ \vdots\\ \chi\end{matrix}}{\chi}\,m\,.$$

And again we see that, from the present perspective, there is something very badly wrong. If the introduction rule for *verum* is genuinely harmoniously paired with no elimination rule (as we have seen it is), it cannot also be harmoniously paired with the elimination rule for $\bullet$ (unless we have lost all sense in which elimination rules are consequences and unique functions of introduction rules).

Put another way, in a classical setting, $\bullet$ can be thought of as having these standard rules:

$$\frac{\phi}{\bullet}\,\bullet\text{-i} \qquad \text{and} \qquad \frac{\bullet}{\psi}\,\bullet\text{-e}\,,$$

making it a very *tonkish* connective.[22] $\bullet$, of course, has no classical truth-table, confirming the lack of harmony in any set of general introduction and general elimination rules proposed for it.

8.5 The Extension to First-Order

We can apply our classical inversion procedure to first-order rules. The existential quantifier has this general introduction rule:

$$\dfrac{\phi(t) \qquad \begin{matrix}[\exists x\phi(x)]^m\\ \vdots\\ \chi\end{matrix}}{\chi}\,m\,.$$

Now, just as Gentzen's introduction rule is doubly schematic, so too this rule stands for an indefinite number of introduction rules varying with the term t employed, indefinite for the rule applies in all extensions of the language one starts off from. With that in mind, our classical inversion procedure produces an equally indefinite, single elimination rule

$$\dfrac{\exists x\phi(x) \qquad \begin{matrix}[\phi(t_1)]^m\\ \vdots\\ \chi\end{matrix} \quad \begin{matrix}[\phi(t_2)]^m\\ \vdots\\ \chi\end{matrix} \quad \begin{matrix}\ldots\\ \vdots\\ \ldots\end{matrix} \quad \begin{matrix}[\phi(t_n)]^m,\\ \vdots\\ \chi\end{matrix} \quad \begin{matrix}\ldots\\ \vdots\\ \ldots\end{matrix}}{\chi}\,m\,.$$

[22] In fact, it *is* Heinrich Wansing's `*tonk` (Wansing 2006, p. 657). Read takes $\bullet$ to show that (ge-)harmony does not entail consistency, let alone conservativeness—see (Read 2000, pp. 141–142, Read 2010, pp. 570–573); Schroeder-Heister (2004, p. 37) takes it to show failure of cut/transitivity of deduction (thus preserving consistency).

If proofs are to be finitary objects, the resources called on in such a rule must be finite, and hence the side premisses involved cannot vary indefinitely. In particular, we cannot have a different set of side premisses for each term. Implicitly, then, there are cases in which the side premisses do not mention the term t. But if that is so, the term occurs parametrically—*any* other would serve as well. And so we arrive at the standard elimination rule:

$$\dfrac{\exists x \phi(x) \qquad \begin{matrix} [\phi(a)]^m \\ \vdots \\ \chi \end{matrix}}{\chi}\ m$$

where a occurs neither in χ nor in any side premiss upon which χ depends.

Deriving the introduction rule from the elimination rule by means of our classical inversion procedure proceeds similarly. If we take the elimination rule to have the extended form above, then, for each term t, we see exactly how it may feature in a proof of χ in which the assumption $\exists x \phi(x)$ is redundant, namely exactly as in the general introduction reformulation of the standard introduction rule given above.

In similar fashion, we may obtain the standard introduction rule for the universal quantifier from the general elimination reformulation of the standard elimination rule and *vice versa*.

It is at this point that the analogy with the propositional case weakens and matters start to unravel a little. Adding the existential quantifier governed by what are, in effect, the standard rules to our favoured propositional logic gives us a classically complete system *with the subformula property* (Milne 2010, Sect. 3.3, Sandqvist 2012).[23] Adding anything like standard rules for the universal quantifier, however, scuppers any prospect for holding on to the subformula property: there is no hope of demonstrating the classical theorem

$$\forall x Fx \vee \exists x \neg Fx$$

without violating the subformula property. In Milne (2010) I freed up the standard restrictions on $\forall$-introduction with a view to restoring the subformula property. But, on the one hand, the result is extremely inelegant (if, in fact, it works) Tor Sandqvist (2012, p. 719) says that 'he is inclined to think that the formula-token-marking machinery they call for lies beyond the bounds of what may reasonably be labeled "natural deduction"' and I'm not going to disagree and, on the other, the standard rules (modified inessentially to fit the general introduction and elimination style) are surely the right rules.[24] The better course must then be, to treat $\forall x \phi$ simply as an abbreviation of $\neg\exists x \neg\phi$ and forego taking the universal quantifier as primitive, as Sandqvist (2012) does.

Gentzen did not bring identity into the realm of the logical constants. If we choose to, clearly we should start with the elimination rule:

---

[23] The proof in Milne (2010) is not constructive. The proof in Sandqvist (2012) is constructive.

[24] They are the rules obtained *via* the classically valid equivalence of $\forall x \phi$ and $\neg\exists x \neg\phi$ from the rules for negation and the existential quantifier by splicing (Milne 2012).

$$\frac{\phi(t) \qquad t = u \qquad \begin{matrix}[\phi(u)]^m \\ \vdots \\ \chi\end{matrix}}{\chi}\ m\ \text{=-e}\,,$$

where $\phi(u)$ results from $\phi(t)$ by substitution of the closed term u for some among the occurrences of the closed terms t in $\phi(t)$

In something like the same way as ∃-introduction and ∀-elimination, this rule is doubly schematic but here it is the formula ϕ that we must think of varying indefinitely. Applying the classical inversion procedure, we obtain indefintely many introduction rules of the form

$$\frac{\begin{matrix}[t=u]^m \\ \vdots \\ \chi\end{matrix} \quad \begin{matrix}[\phi_{i_1}(t)]^m \\ \vdots \\ \chi\end{matrix} \quad \begin{matrix}[\phi_{i_2}(t)]^m \\ \vdots \\ \chi\end{matrix} \quad \ldots \quad \begin{matrix}[\phi_{i_l}(t)]^m \\ \vdots \\ \chi\end{matrix} \quad \ldots \quad \phi_{j_1}(u) \quad \phi_{j_2}(u) \quad \ldots \quad \phi_{j_n}(u) \quad \ldots}{\chi}\ m$$

where, notionally, *every* one-place predicate occurs either hypothetically as one of the $\phi(t)$'s or categorically as one of the $\phi(u)$'s (and not both). What we have is not quite unmanageable, for the totality of these rules does the same job as this indefinite *single* rule:

$$\frac{\begin{matrix}[t=u]^m \\ \vdots \\ \chi\end{matrix} \qquad \phi_{i_1}(t) \supset \phi_{i_1}(u) \quad \phi_{i_2}(t) \supset \phi_{i_2}(u) \quad \ldots \quad \phi_{i_k}(t) \supset \phi_{i_k}(u) \quad \ldots}{\chi}\ m\,.$$

If that seems odd, notice that the =-elimination rule says, in effect, that $t = u$ has the same elimination rule as an indefinite conjunction of conditionals of the form $\phi(t) \supset \phi(u)$.

We want now to reduce the introduction rule to more primitive terms and to "finitize" it in something like the way we have done with ∀-introduction and ∃-elimination. Here our having two introduction rules for the conditional is a handicap. We need a uniform introduction rule: for this we return to Gentzen's introduction rule for the conditional. What we end up with as an appropriate introduction rule for '=' is this:

$$\frac{\begin{matrix}[t=u]^m \\ \vdots \\ \chi\end{matrix} \qquad \begin{matrix}[\phi(t)]^m \\ \vdots \\ \phi(u)\end{matrix}}{\chi}\ m\ \text{=-e}\,.$$

where the predicate ϕ does not occur in any assumption upon which $\phi(u)$ depends other than $\phi(t)$. Essentially, what we have as general introduction and general elimination rules for identity are the rules of Read (2004).

Read's introduction rule adds nothing in deductive power over the standard introduction rule which, recast in general introduction style, is

$$\dfrac{\begin{array}{c}[t = t]^m \\ \vdots \\ \chi \end{array}}{\chi}\, m\,.$$

What Read's introduction rule allows is levelling of local peaks (Read 2004, pp. 116–117). It also allows proof of uniqueness in Belnap's sense and harmony in the sense of Tennant's *Natural Logic* (Milne 2007, pp. 37–38).

8.6 Sequent Rules

There are three ways we can translate general introduction and elimination rules into sequent-calculus form. Firstly, we might follow von Plato (2001) and turn the general elimination rules of natural deduction into left-introduction rules: the natural deduction rule

$$\dfrac{\star(\phi_1, \phi_2, \ldots, \phi_n) \quad \begin{array}{c}[\phi_{i_1}]^m \\ \vdots \\ \chi\end{array} \quad \begin{array}{c}[\phi_{i_2}]^m \\ \vdots \\ \chi\end{array} \quad \begin{array}{c}\ldots \\ \vdots \\ \ldots\end{array} \quad \begin{array}{c}[\phi_{i_k}]^m \\ \vdots \\ \chi\end{array} \quad \phi_{j_1} \quad \phi_{j_2} \quad \ldots \quad \phi_{j_l}}{\chi}\; m \;\star\text{-elimination}$$

translates as the sequent rule

$$\dfrac{\Sigma_{i_1}, \phi_{i_1} \vdash \chi \quad \Sigma_{i_2}, \phi_{i_2} \vdash \chi \quad \ldots \quad \Sigma_{i_k}, \phi_{i_k} \vdash \chi \quad \Sigma_{j_1} \vdash \phi_{j_1} \quad \Sigma_{j_2} \vdash \phi_{j_2} \quad \ldots \quad \Sigma_{j_l} \vdash \phi_{j_l}}{\Sigma_{i_1}, \Sigma_{i_2}, \ldots, \Sigma_{i_k}, \Sigma_{j_1}, \Sigma_{j_2}, \ldots, \Sigma_{j_l}, \star(\phi_1, \phi_2, \ldots, \phi_n) \vdash \chi}\;\text{left } \star\text{-i}\,.$$

But whereas standard natural deduction introduction rules go over nicely into right-introduction rules in sequent calculus, this is not such a natural way to translate *general* introduction rules. They are more naturally read as left *elimination* rules (as Peter Schroeder-Heister pointed out to me). The natural deduction general introduction rule

$$\dfrac{\begin{array}{c}[\star(\phi_1, \phi_2, \ldots, \phi_n)]^m \\ \vdots \\ \chi\end{array} \quad \begin{array}{c}[\phi_{r_1}]^m \\ \vdots \\ \chi\end{array} \quad \begin{array}{c}[\phi_{r_2}]^m \\ \vdots \\ \chi\end{array} \quad \begin{array}{c}\ldots \\ \vdots \\ \ldots\end{array} \quad \begin{array}{c}[\phi_{r_t}]^m \\ \vdots \\ \chi\end{array} \quad \phi_{s_1} \quad \phi_{s_2} \quad \ldots \quad \phi_{s_u}}{\chi}\;\text{left } \star\text{-introduction}$$

translates as the sequent calculus left elimination rule

$$\dfrac{\Sigma, \star(\phi_1, \phi_2, \ldots, \phi_n) \vdash \chi \; \Sigma_{r_1}, \phi_{r_1} \vdash \chi \; \Sigma_{r_2}, \phi_{r_2} \vdash \chi \; \ldots \; \Sigma_{r_t}, \phi_{r_t} \vdash \chi \; \Sigma_{r_t} \vdash \phi_{s_1} \; \Sigma_{s_2} \vdash \phi_{s_2} \; \ldots \; \Sigma_{s_u} \vdash \phi_{s_u}}{\Sigma, \Sigma_{r_1}, \Sigma_{r_2}, \ldots, \Sigma_{r_t}, \Sigma_{s_1}, \Sigma_{s_2}, \ldots, \Sigma_{s_u} \vdash \chi}\;\text{left } \star\text{-e}\,.$$

This puts most of the action on the left hand side. And that is one way to translate from general introduction and general elimination rules into sequent calculus.

Another way is to stick with translating natural deduction general introduction rules as left elimination rules but to change our treatment of natural deduction general elimination rules. Instead of left introduction rules, we translate them as right elimination rules. The general form is

$$\frac{\Sigma \vdash \star(\phi_1, \phi_2, \ldots, \phi_n) \quad \Sigma_{i_1}, \phi_{i_1} \vdash \chi \quad \Sigma_{i_2}, \phi_{i_2} \vdash \chi \ \ldots \ \Sigma_{i_k}, \phi_{i_k} \vdash \chi \quad \Sigma_{j_1} \vdash \phi_{j_1} \quad \Sigma_{j_2} \vdash \phi_{j_2} \ \ldots \ \Sigma_{j_l} \vdash \phi_{j_l}}{\Sigma, \Sigma_{i_1}, \Sigma_{i_2}, \ldots, \Sigma_{i_k}, \Sigma_{j_1}, \Sigma_{j_2}, \ldots, \Sigma_{j_l} \vdash \chi} \text{right } \star\text{-e}.$$

What we get from the most direct reading of sequent rules off general natural deduction rules is a system with left and right elimination rules (in contradistinction to Gentzen, who obtained left and right introduction rules). In this form, we directly transcribe natural deduction derivations line by line. The difference is purely notational: in the sequent calculus form, each formula on the right-hand side of the turnstile drags the assumptions it depends on along with it on the left-hand side of the turnstile. We can retain Gentzen's restriction according to which all initial sequents are of the form $\phi \vdash \phi$, but cannot go further, as some have done, and restrict ϕ to being atomic.[25]

The third way involves an idea of Gentzen's and, unlike the previous two, admits multiple formulae in the succeedent. Each (general) introduction rule and elimination rule is turned into a basic sequent—an axiom of the logistic system, to Gentzen's way of thinking. The natural deduction introduction rule

$$\frac{\begin{array}{ccccccccc} [\star(\phi_1, \phi_2, \ldots, \phi_n)]^m & [\phi_{r_1}]^m & [\phi_{r_2}]^m & \ldots & [\phi_{r_t}]^m & & & & \\ \vdots & \vdots & \vdots & \vdots & \vdots & & & & \\ \chi & \chi & \chi & \ldots & \chi & \phi_{s_1} & \phi_{s_2} & \ldots & \phi_{s_u} \end{array}}{\chi} \; m \ \star\text{-introduction}$$

is transmuted into the sequent calculus axiom

$$\phi_{s_1}, \phi_{s_2}, \ldots, \phi_{s_u} \vdash \star(\phi_1, \phi_2, \ldots, \phi_n), \phi_{r_1}, \phi_{r_2}, \ldots, \phi_{r_t};$$

the elimination rule

$$\frac{\begin{array}{ccccccccc} & [\phi_{i_1}]^m & [\phi_{i_2}]^m & \ldots & [\phi_{i_k}]^m & & & & \\ & \vdots & \vdots & \vdots & \vdots & & & & \\ \star(\phi_1, \phi_2, \ldots, \phi_n) & \chi & \chi & \ldots & \chi & \phi_{j_1} & \phi_{j_2} & \ldots & \phi_{j_l} \end{array}}{\chi} \; m \ \star\text{-elimination}$$

becomes the axiom

$$\phi_{j_1}, \phi_{j_2}, \ldots, \phi_{j_l}, \star(\phi_1, \phi_2, \ldots, \phi_n) \vdash \phi_{i_1}, \phi_{i_2}, \ldots, \phi_{i_k}.$$

The only rules are structural rules. As Gentzen says, this leads to a simplification of the sequent calculus, but to find the simplification congenial we would have to 'attach no importance to the *Hauptsatz*' (Gentzen 1935, pp. 85–86), for the Cut rule now does nearly all the work.

[25] That we obtain a sequent calculus with only elimination rules from a direct transcription of our system of natural deduction rules is in marked contrast with the standard case (on which see Prawitz 1971, p. 243).

8.7 Conclusion

Prawitz sees in Gentzen's way of thinking of introduction rules in natural deduction a manifestation of verificationism:

> The basic idea of verificationism as construed here is thus that the meaning of a sentence is given by what counts as a direct verification of it. Gentzen's suggestion that the meanings of the logical constants are determined by their introduction rules can be seen as a special case of this verificationist idea. (Prawitz 2006, p. 520)[26]

On another occasion he went further, saying 'there is a close correspondence between the *constructive* meaning of a logical constant and its introduction rule'. He went on to note that in Gentzen's approach, 'the classical deductive operations are analysed as consisting of the constructive ones plus a principle of indirect proof for atomic sentences' and remarked that 'one may doubt this is the proper way of analysing classical inferences' (Prawitz 1971, p. 244-245). Given the close connection, developed above in Sect. 8.4.3, between general introduction and general elimination rules for propositional connectives and their truth-tables, one might hold that what I have presented above is closer to being "the proper way of analysing classical inferences" (in propositional logic, if perhaps not first-order). At least, that might be so if one sees this task in the way Marco Mondadori did:

> [M]y aim is the construction of a calculus characterizing *classical* first order proofs in the same sense in which, according to Prawitz, [Gentzen's] NJ characterizes *constructive* first order proofs. The relevant sense is that the inference rules of the calculus must capture the essential logical content of *classical* logical operations in the languages considered. [...]
>
> (C1) The meaning of a logical constant, $*$, is completely determined when the conditions under which a sentence of the form $(A * B)$ is respectively true and false are specified in terms of the truth or falsity of A or/and B.
>
> (C2) For all sentences A, A is either true or false. (Mondadori 1988, p. 211)

Since its truth-table can be read off either the set of general introduction rules governing a connective or the set of general elimination rules governing it when the introduction and elimination rules are harmoniously related, (C1) is satisfied. (C2) is implicit in the way we read harmonious general introduction and general elimination rules off truth-tables, thus implicit in the classical inversion procedure.

By taking as starting point the role of logically complex assumptions in proofs, I have taken a step away from what Peter Schroeder-Heister sees as the assertion-drivenness of the standard notion of proof (Schroeders-Heister 2004, p. 29). And perhaps I have taken a further step away in providing a system of introduction and elimination rules that in effect *is* a sequent calculus (albeit a deviant sequent calculus

[26] Ruy de Queiroz (2008) decries the verificationist tendency in proof-theoretic semantics and seeks to oppose it by finding in Wittgenstein early and late a focus on the consequences of assertions, thus playing up the significance of elimination rules.

with left and right elimination rules), thus narrowing the gap between Gentzen's N and L formulations.[27]

I have employed the same inversion procedure, namely that one reads off from an introduction rule or rules an elimination rule or rules just strong enough to show that the proposition introduced by an application of any one of the introduction rules allows for its own introduction in the circumstances set out in that or those introduction rule or rules, to obtain Gentzen's elimination rules from Gentzen's introduction rules and to obtain general elimination rules from what I have called general introduction rules. In the latter case, we obtain a system for classical logic with many of the virtues sought by the proof-theoretic semanticist. Thus I take myself to have shown that the revisionist challenge to classical logic does not and cannot have its roots solely within logic: what sort of introduction rules are permissible cannot be decided on considerations internal to our proof-theoretic practices. On the other hand, in giving the starring role to the use of complex propositions as assumptions, I may seem to have severed connection with any aspect of the use of sentences rightly considered central to our mastery of language, such as, to pick one that features prominently in the work of Dummett and Prawitz, knowledge of the conditions under which it correct to assert a sentence. That matter is hardly clear cut, though, as, on the one hand, Gentzen's introduction rules are derived rules in the system elaborated here and, on the other, the content of the sentences assumed in instances of use of $\forall$-introduction and $\exists$-elimination seems to outrun the contents of sentences we might assert, for in assuming something of the form 'a is F' in the course of a proof of 'It's not the case that an F exists' from 'Everything is not-F', 'a is F' seems not to have the semantics of any ordinary sentence of English—see (Milne 2007, pp. 24–33).

To return to our starting point, Gentzen himself indicated that what is distinctive about natural deduction is the role of assumptions; I have taken that observation at face value in providing a novel account of introduction rules and therewith harmony and inversion. What is missing, not just in the philosophy of logic but in the philosophy of language more generally, is an account of the making of assumptions "for the sake of argument" as a distinctive linguistic act, but in the making of assumptions containing parametric occurrences of proper-name-like items for which no reference has been secured, as we do in employing Gentzen's rule of $\exists$-elimination, it does seem that a distinctive kind of proposition-like entity is employed. In the Quine-inspired tradition that uses in place of Gentzen's rule a rule of existential instantiation—see Prawitz (1965, Appendix C, Sect. 3) and Prawitz (1967)—some have sought to explain the role of the introduced names in terms of Hilbert's ϵ-operator—e.g. Suppes (1957, p. iv), Lemmon (1965).[28] But if Hilbert's ϵ-operator does underlie the

---

[27] Pavel Tichý maintained that, despite what Gentzen said, natural deduction only makes sense when viewed in the same way as sequent calculus: as a calculus of logically true statements concerning entailments (Tichy 1988, Chap. 13).

[28] The manipulation of general introduction and general elimination rules by splicing shows that the Gentzen and Quine approaches are more closely related than may first appear (Milne 2012, Sect. 6).

(natural language) semantics of the role of seeming proper names in assumptions, that is most certainly a story for another occasion.

References

Belnap, N., Jr. (1962). Tonk, plonk and plink. *Analysis*, *22*, 130–134 (Reprinted in Philosophical Logic by P. F. Strawson (ed.) (1967). Oxford University Press, Oxford, 132–137).

Dummett, M. A. E. (2000). *Elements of intuitionism* (2nd ed.). Oxford: Oxford University Press (1st ed. 1977).

Dummett, M. A. E. (1991). *The logical basis of metaphysics*. London: Duckworth.

Dyckhoff, R. (2009). Generalised elimination rules and harmony, talk given at St Andrews. Retrieved May 26, 2009, http://www.cs.st-andrews.ac.uk/rd/talks/2009/GE.pdf

Francez, N., & Dyckhoff, R. (2012). A note on harmony. *Journal of Philosophical Logic*, *41*, 613–628.

Gentzen, G. (1934–1935). Untersuchungen über das logische schließen. *Mathematische Zeitschrift*, *39*, 176–210, 405–431. (English translation by Manfred Szabo as 'Investigations into Logical Deduction', American Philosophical Quarterly, 1, 1964, 288–306, 2, 1965, 204–218, also in Szabo, M.E. (ed.) (1969). *The Collected Papers of Gerhard Gentzen* (pp. 68–131). North-Holland, Amsterdam).

Gentzen, G. (1936). Die Widerspruchfreiheit der reinen Zahlentheorie. *Mathematische Annalen*, *112*, 493–565. (English translation by Manfred Szabo as The Consistency of Elementary Number Theory in Szabo, M. E. (ed.) (1969). *The Collected Papers of Gerhard Gentzen* (pp. 132–201). North-Holland, Amsterdam).

Hazen, A. (1995). Is even minimal negation constructive? *Analysis*, *55*, 105–107.

Heyting, A. (1930). Die formalen Regeln der intuitionistischen logik und mathematik. In P. Mancosu (ed.). *Sitzungsberichte der Preußischen Akademie der Wissenschaften, Physikalisch-mathematische Klasse* (pp. 42–65). (English translation, (1998) 'The formal rules of intuitionist logic', In P. Mancosu (ed.). From Brouwer to Hilbert: the debate in the foundations of mathematics in the 1920s (pp. 311–328). Oxford: Oxford University Press).

Kurbis, N. (2008). Stable harmony. In M. Peliš (Ed.), *The logica yearbook 2007* (pp. 87–96). Institue of Philosophy, Czech Academy of Sciences, Prague: Filosofia.

von Kutschera, F. (1968). Die vollständigkeit des operatorensystems { $\neg$, $\wedge$, $\vee$, $\rightarrow$ } für die intuitionistische aussagenlogik im rahmen der gentzensemantik. *Archiv für Math Logik Grundlagenforschung*, *11*, 3–16.

Lemmon, E. (1965). A further note on natural deduction. *Mind*, *74*, 594–597.

Milne, P. (2007). Existence, freedom, identity, and the logic of abstractionist realism. *Mind*, *116*, 23–53.

Milne, P. (2008). A formulation of first-order classical logic in natural deduction with the subformula property. In M. Peliš (Ed.), *The logica yearbook 2007* (pp. 97–110). Institue of Philosophy, Czech Academy of Sciences, Prague: Filosofia.

Milne, P. (2010). Subformula and separation properties in natural deduction via small Kripke models. *Review of Symbolic Logic*, *3*, 175–227.

Milne, P. (2012). Inferring, splicing, and the Stoic analysis of argument. In O. T. Hjortland & C. Dutilh Novaes (Ed.), *Insolubles and consequences: Essays in honour of Stephen Read* (pp. 135–154). College Publications: London.

Mondadori, M. (1988). On the notion of classical proof. In C. Cellucci & G. Sambin (Eds.), *Atti del congresso temi e prospettive della logica e della filosofial della scienza contemporanee, Cesana 7–10 gennaio 1987* (Vol. 1, pp. 211–214). Bologna: Cooperativa Libraria Universitaria Editrice Bologna.

Moriconi, E., & Tesconi, L. (2008). On inversion principles. *History and Philosophy of Logic*, *29*, 103–113.
Negri, S., & von Plato, J. (2001). *Structural proof theory*. Cambridge: Cambridge University Press (with an appendix by Arne Ranta).
Negri, S. (2002). Varieties of linear calculi. *Jounral of Philosophical Logic*, *31*, 569–590.
von Plato, J. (2001). Natural deduction with general elimination rules. *Archive for Mathematical Logic*, *40*, 541–567.
von Plato, J. (2008). Gentzen's proof of normalization for natural deduction. *Bulletin of Symbolic Logic*, *14*, 240–257.
Prawitz, D. (1965). *Natural deduction: A proof-theoretical study*. Stockholm: Almqvist & Wiksell (reprinted with new preface, Mineola, NY: Dover Publications, 2006).
Prawitz, D. (1971). Ideas and results in proof theory. In J. E. Fenstad (Ed.), *Proceedings of the Second Scandinavian Logic Symposium, Studies in Logic and the Foundations of Mathematics* (Vol. 63). Amsterdam: North-Holland.
Prawitz, D. (1978). Proofs and the meaning and completeness of the logical constants. In J. Hintikka, I. Niiniluoto, & E. Saarinen (Eds.), *Essays on Mathematical and Philosophical Logic: Proceedings of the Fourth Scandinavian Logic Symposium and of the First Soviet-Finnish Logic Conference, Jyväskylä, Finland, June 29-July 6, 1976, Synthese Library* (Vol. 122, pp. 25–40). Dordrecht: Reidel.
Prawitz, D. (1967). A note on existential instantiation. *Journal of Symbolic Logic*, *32*, 81–82.
Prawitz, D. (2006). Meaning approached via proofs. *Synthese*, *148*, 507–524.
de Queiroz, R. J. (2008). On reduction rules, meaning-as-use, and proof-theoretic semantics. *Studia Logica*, *90*, 211–247.
Read, S. (2000). Harmony and autonomy in classical logic. *Journal of Philosophical Logic*, *29*, 123–154.
Read, S. (2004). Identity and harmony. *Analysis*, *64*, 113–119.
Read, S. (2010). General-elimination harmony and the meaning of the logical constants. *Journal of Philosophical Logic*, *39*, 557–576.
Sandqvist, T. (2012). The subformula property in natural deduction established constructively.*Review of Symbolic Logic*, *5*, 710–719.
Schroeder-Heister, P. (1984). A natural extension of natural deduction. *Journal of Symbolic Logic*, *49*, 1284–1300.
Schroeder-Heister, P. (1984). Generalized rules for quantifiers and the completeness of the intuitionistic operators & $\vee$, $\rightarrow$, $\curlywedge$, $\forall$, $\exists$. In M. Richter., E. Börger., W. Oberschelp., B. Schinzel., W. Thomas (Eds.), *Computation and Proof Theory. Proceedings of the Logic Colloquium held in Aachen*, July 18–23, 1983, Part II, Lecture Notes in Mathematics (pp. 399–426, vol 1104). Berlin, Heidelberg, New York, Tokyo: Springer.
Schroeder-Heister, P. (2004). On the notion of assumption in logical systems. In: R. Bluhm & C. Nimtz (Eds.). *Selected Papers Contributed to the Sections of GAP5, Fifth International Congress of the Society for Analytical Philosophy* (pp. 27–48), Bielefeld. Paderborn: Mentis Verlag. Retrieved 22–26 Sept 2003, fromhttp://www.gap5.de/proceedings/.
Schroeder-Heister, P. (2014). Generalized elimination inferences, higher-level rules, and the implications-as-rules interpretation of the sequent calculus. In L. C. Pereira, & E. H. Haeusler, & V. de Paiva (Eds.), *Advances in natural deduction: A Celebration of Dag Prawitz's Work* (pp. 1–29). Dordrecht, Heidelberg, New York, London.
Suppes, P. (1957). *Introduction to logic*. New York: Van Nostrand Rienhold.
Tennant, N. (1978). *Natural logic* (1st ed.). Edinburgh: Edinburgh University Press (Reprinted in paperback with corrections, 1991).
Tennant, N. (1992). *Autologic*. Edinburgh: Edinburgh University Press.
Tichý, P. (1988). *The foundations of Frege's logic*. Berlin, New York: Walter de Gruyter.
Wansing, H. (2006). Connectives stranger than tonk. *Journal of Philosophical Logic*, *35*, 653–660.
Zucker, J. I., & Tragesser, R. S. (1978). The adequacy problem for inferential logic. *Journal of Philosophical Logic*, *7*, 501–516.

Chapter 9
Intuitionistic Existential Instantiation and Epsilon Symbol

Grigori Mints

Abstract A natural deduction system for intuitionistic predicate logic with existential instantiation rule presented here uses Hilbert's ϵ-symbol. It is conservative over intuitionistic predicate logic. We provide a completeness proof for a suitable Kripke semantics, sketch an approach to a normalization proof, survey related work and state some open problems. Our system extends intuitionistic systems with ϵ-symbol due to Dragalin and Maehara.

Keywords Hilbert's ϵ-symbol · Intuitionistic predicate logic · Existential instantiation · Natural deduction · Sequent calculus

9.1 Introduction

In natural deduction formulations of classical and intuitionistic logic, the existence-elimination rule is usually taken in the form

$$\dfrac{\exists x A(x) \qquad \begin{array}{c} A(a) \\ \vdots \\ C \end{array}}{C}\ \exists^-$$

where a is a fresh variable. Existential instantiation is the rule

$$\dfrac{\exists x A(x)}{A(a)}\ \exists i$$

Prof. Mints passed away May 29, 2014

G. Mints (✉)
Department of Philosophy, Stanford University, Building 90, Stanford, CA 94305-2155, USA

H. Wansing (ed.), *Dag Prawitz on Proofs and Meaning*, Outstanding Contributions to Logic 7, DOI 10.1007/978-3-319-11041-7_9

where a is a fresh constant. It is sound and complete (with suitable restrictions) in the role of existence-elimination rule in classical predicate logic but is not sound intuitionistically, since it makes possible for example the following derivation:

$$\dfrac{\dfrac{\dfrac{\dfrac{C \to \exists x A(x), C \Rightarrow \exists x A(x)}{C \to \exists x A(x), C \Rightarrow A(a)}\ \exists i}{C \to \exists x A(x) \Rightarrow C \to A(a)}}{C \to \exists x A(x) \Rightarrow \exists x (C \to A(x))}}{\Rightarrow (C \to \exists x A(x)) \to \exists x (C \to A(x))}$$

There are several approaches in the literature to introduction of restrictions making this rule conservative over intuitionistic predicate calculus.

We present an approach using intuitionistic version of Hilbert's epsilon-symbol and strengthening works by Dragalin Dragalin (1974) and Maehara (1970) where ϵ-terms are treated as partially defined. Then a survey of extensions and related approaches including the important paper by Shirai (1971) is given and some problems are stated. There is an obvious connection with the problem of Skolemization of quantifiers. The role of existence conditions in that connection is prominent in the work by Baaz and Iemhoff (2006).

We do not include equality since in this case adding of ϵ-symbol with natural axioms is not conservative over intutionistic logic (Mints 1974; Osswald 1975). A simple counterexample due (in other terms) to Smorynski (1977) is

$$\forall x \exists y P(x, y) \to \forall x x' \exists y y' (Pxy \& Px'y' \& (x = x' \to y = y'))$$

In our natural deduction system NJϵ axioms and propositional inference rules are the same as in ordinary intuitionistic natural deduction, the same holds for $\forall$-introduction. The remaining rules are as follows:

$$\dfrac{\Gamma \Rightarrow \exists x F(x)}{\Gamma \Rightarrow F(\epsilon x F(x))}\ \exists i \tag{9.1}$$

existential instantiation,

$$\dfrac{\Gamma \Rightarrow t\downarrow \quad \Delta \Rightarrow \forall z F(z)}{\Gamma, \Delta \Rightarrow F(t)} \qquad \dfrac{\Gamma \Rightarrow t\downarrow \quad \Delta \Rightarrow F(t)}{\Gamma, \Delta \Rightarrow \exists z F(z)} \tag{9.2}$$

where

$$\epsilon x A(x) \downarrow := \exists y (\exists x A(x) \to A(y)), \tag{9.3}$$

and $t\downarrow := \top$ (the constant "true") if t is a variable or constant.

Two semantics are given for NJϵ, or more precisely to an equivalent Gentzen-style system IPCϵ (Sect. 9.2). The first semantics, which is incomplete but convenient for a proof of conservative extension property over IPC is defined in Sect. 9.3.

The second semantics with a completeness proof for IPCϵ is given in Sect. 9.4.

Section 9.4.1 presents a sketch of a possible proof of a normal form theorem. Section 9.5 surveys some of the previous work and Sect. 9.6 outlines some open problems.

9.2 Gentzen-Style System IPCϵ

Let us state our Gentzen-style rules for the intuitionistic predicate calculus IPCϵ with ϵ-symbol. For simplicity we assume that the language does not have function symbols except constants. Formulas and terms are defined by familiar inductive definition plus one additional clause:

If $A(x)$ is a formula then $\epsilon x A(x)$ is a term.

Derivable objects of IPCϵ are *sequents* $\Gamma \Rightarrow A$ where Γ is a finite set of formulas, A is a formula. This means in particular that structural rules are implicitly included below.

First, let's list the rules of the intuitionistic predicate calculus IPC without ϵ-symbol.

Axioms:

$$\Gamma, A \Rightarrow A, \quad \Gamma, \bot \Rightarrow A\,.$$

Inference rules:

$$\frac{\Gamma \Rightarrow A \quad \Gamma \Rightarrow B}{\Gamma \Rightarrow A \& B} \Rightarrow \& \qquad \frac{A, B, \Gamma \Rightarrow \Delta}{A \& B, \Gamma \Rightarrow \Delta} \& \Rightarrow$$

$$\frac{A, \Gamma \Rightarrow \Delta \quad B, \Gamma \Rightarrow \Delta}{A \vee B, \Gamma \Rightarrow \Delta} \vee \Rightarrow \qquad \frac{\Gamma \Rightarrow A}{\Gamma \Rightarrow A \vee B} \Rightarrow \vee \frac{\Gamma \Rightarrow B}{\Gamma \Rightarrow A \vee B}$$

$$\frac{\Gamma \Rightarrow A \quad B, \Gamma \Rightarrow \Delta}{A \to B, \Gamma \Rightarrow \Delta} \to\Rightarrow \qquad \frac{A, \Gamma \Rightarrow B}{\Gamma \Rightarrow A \to B} \Rightarrow\to$$

$$\frac{\Gamma \Rightarrow A(t)}{\Gamma \Rightarrow \exists x A(x)} \Rightarrow \exists \qquad \frac{\Gamma \Rightarrow A(b)}{\Gamma \Rightarrow \forall x A(x)} \Rightarrow \forall$$

$$\frac{A(b), \Gamma \Rightarrow G}{\exists x A(x), \Gamma \Rightarrow G} \exists \Rightarrow \qquad \frac{A(t), \Gamma \Rightarrow G}{\forall x A(x), \Gamma \Rightarrow G} \forall \Rightarrow$$

$$\frac{\Gamma \Rightarrow C \quad C, \Gamma \Rightarrow G}{\Gamma \Rightarrow G} \, Cut$$

For IPCϵ quantifier-inferences $\Rightarrow \exists$, $\forall \Rightarrow$ are modified by requirement that the term t substituted in the rule should be "defined" (cf. (9.3)).

$$\frac{\Gamma \Rightarrow t{\downarrow} \quad F(t), \Delta \Rightarrow \Delta}{\forall z F(z), \Gamma, \Delta \Rightarrow \Delta} \forall \Rightarrow \qquad \frac{\Gamma \Rightarrow \Delta, t{\downarrow} \quad \Gamma \Rightarrow \Delta, F(t)}{\Gamma \Rightarrow \Delta, \exists z F(z)} \Rightarrow \exists \tag{9.4}$$

$\exists \Rightarrow$-rule is also changed for IPCϵ:

$$\frac{A(\epsilon x A(x))\Gamma \Rightarrow G}{\Gamma, \exists x A(x) \Rightarrow G} \exists_\epsilon \Rightarrow \tag{9.5}$$

A routine proof shows that IPCϵ is equivalent to a Hilbert-style system obtained by weakening familiar axioms for quantifiers to

$$(\epsilon Q1)\ t{\downarrow}\&\forall x A(x) \rightarrow A(t)$$

$$(\epsilon Q2)\ t{\downarrow}\&A(t) \rightarrow \exists x A(x)$$

and adding the axiom

$$\exists x A(x) \rightarrow A(\epsilon x A(x))$$

9.2.1 Equivalence of IPCϵ and NJϵ

Let us recall that in natural deduction a sequent

$$A_1, \ldots, A_n \Rightarrow A$$

is used to indicate that A is deducible from assumptions $A_1, \ldots, A_n$.

Theorem 9.1 *A sequent is provable in NJϵ iff it is provable in IPCϵ.*

Proof The proof is routine: every rule of one of these systems is directly derivable in the other system. Let's show derivations of the rules $\exists i$ and $\exists \Rightarrow$ from each other using abbreviation $e := \epsilon x F(x)$.

$$\frac{\Gamma \Rightarrow \exists x F(x) \quad \dfrac{F(e) \Rightarrow F(e)}{\exists x F(x) \Rightarrow F(e)} \exists \Rightarrow}{\Gamma \Rightarrow F(e)} \qquad \frac{\dfrac{\exists x F(x) \Rightarrow \exists x F(x)}{\exists x F(x) \Rightarrow F(e)} \exists i \quad \dfrac{F(e), \Gamma \Rightarrow G}{\Gamma \Rightarrow F(e) \rightarrow G}}{\exists x F(x), \Gamma \Rightarrow G}$$

9.3 A Kripke Semantics for Intuitionistic ϵ-symbol

To prove that IPCϵ is conservative over IPC we present an incomplete semantics modifying a semantics from Dragalin (1974). The main modification is in the definition of $t{\downarrow}$ and treatment of atomic formulas containing ϵ-terms $\epsilon x A$.

Definition 9.1 Let w be a world in a Kripke model. Denote

$$\epsilon x A(x) \downarrow w :\equiv \ w \models \epsilon x A(x) \downarrow .$$

We say that a term $\epsilon x A$ is *defined* in w iff $\epsilon x A(x) \downarrow w$.

The symbol $\perp$ in the next definition indicates the condition (9.6) below.

Definition 9.2 An intuitionistic Kripke $\epsilon\perp$-model (or simply model in this section)

$$\mathcal{M} = (W, <, D, \models, V)$$

has to satisfy the following conditions:

$(W, <)$ is a Kripke frame with a strict partial ordering $<$,

D is a domain function assigning to every $w \in W$ a non-empty set $D(w)$ monotone with respect to $<$,

$w \models A$ is a relation between worlds $w \in W$ and atomic formulas A with constants from

$$D := \cup_{w \in W} D(w)$$

monotonic with respect to $\leq$ and such that

$$w \not\models A \text{ if } A \text{ contains at least one constant in } D - D(w). \tag{9.6}$$

V is a valuation function assigning a constant $V(e, w) \in D$ to any ϵ-term e (possibly containing constants from D) and $w \in W$.

The relation $\models$ is extended to composite formulas in the familiar way. The components of an ϵ-model have to satisfy the following conditions.

$$V(\epsilon x B(x, \epsilon y C), w) = V(\epsilon x B(x, V(\epsilon y C, w)), w), \tag{9.7}$$

$$w \models A(\epsilon y C) \leftrightarrow w \models A(x, V(\epsilon y C)), \tag{9.8}$$

where substitution of $\epsilon y C$ is safe, that is no free variable of $\epsilon y C$ becomes bound. Also, if $e \downarrow w$ for a term $e := \ \epsilon x A(x)$, then

$$V(e, w) \in D(w) \text{ and } V(e, w') = V(e, w) \text{ for every } w' \geq w.$$

Note. The requirement (9.6) is sound as we prove at the rest of this section, but it leads to incompleteness. For example the formula

$$P(\epsilon x P(x)) \to \exists x P(x)$$

is valid: if $\epsilon x P(x)$ is undefined in a world w then the premise is false in w, otherwise the conclusion is true. However this formula is not derivable, since its instance

$$(C \to P(e)) \to \exists x(C \to P(x)) \tag{9.9}$$

where C is a propositional variable and $e = \epsilon x(C \to P(x))$, implies

$$(C \to \exists x P(x)) \to \exists x(C \to P(x)). \tag{9.10}$$

Indeed, the following figure (where $e_0 = \epsilon x P(x)$) is a derivation, and one application of (9.9) yields (9.10).

$$\frac{C \Rightarrow C \qquad \dfrac{\dfrac{\exists x P(x) \Rightarrow e_0\downarrow \quad \dfrac{P(e_0) \Rightarrow C \to P(e_0)}{\exists x P(x) \Rightarrow C \to P(e_0)}}{\exists x P(x) \Rightarrow \exists x(C \to P(x))} \qquad \dfrac{C, C \to P(e) \Rightarrow P(e)}{C, \exists x(C \to P(x)) \Rightarrow P(e)}}{C, \exists x P(x) \Rightarrow P(e)}}{\dfrac{C \to \exists x P(x), C \Rightarrow P(e)}{C \to \exists x P(x) \Rightarrow C \to P(e)}} \tag{9.11}$$

We continue the proof of soundness. The proofs of the next lemmata are routine.

Lemma 9.1 *Let t be a closed term, A a closed formula with constants from D. Then*

$$w \leq w' \to \ (t\downarrow w \to t\downarrow w' \& \ (w \models A \to w' \models A))$$

Proof Simultaneous induction on t, A.

Lemma 9.2 *If Γ is a set of formulas, G a formula then $\Gamma \vdash G$ in* IPCϵ *implies $\Gamma \models G$.*

Proof Induction on derivations. Checking the rule $\exists \Rightarrow$ uses the fact that $\exists x A(x)$ implies $\epsilon x A(x) \downarrow$. It may be interesting to check whether any other properties of the formula $t\downarrow$ are used. □

Theorem 9.2 *If A, B are formulas without ϵ-symbol then $A \vdash B$ in* IPCϵ *implies $A \vdash B$ in intuitionistic predicate logic IPC.*

Proof We need to prove that for every Kripke model

$$\mathcal{M}_0 = (W, <, D, \models_0)$$

for intutionistic predicate logic refuting $A \to B$ there is an IPCϵ-model refuting $A \to B$. Before applying the construction from Dragalin (1974), let us recall a refinement of a completeness theorem for intuitionistic predicate logic IPC.

Lemma 9.3 *The following additional requirements to the definition of a Kripke frame $(W, <, D)$ for IPC are still complete:*

1. *W is a countable tree with a root* **0** *such that each $w \in W$ except* **0** *has unique immediate $<$-predecessor and the number of predecessors of w is finite.*

2. *domains D(w) are strictly increasing: if $w < w'$ then $D(w)$ is a proper subset of $D(w')$.*

Proof The requirement 1 is satisfied by the canonical proof search tree for a given sequent, see for example Mints (2000). To satisfy the second requirement, note that an infinite branch of the canonical proof search tree does not have "leaf worlds": for every $w \in W$ there exists a $w' > w$. Now take a fixed element $e \in D(w_0)$ and duplicate it by a fresh element, say e_w in every world w. More precisely for the new domain function D' define

$$e_w \in D'(w) - D'(w^-),$$

where w^- is the immediate predecessor of w. Let's extend the relation $\models$ by identifying e_w and e, more precisely define for atomic formulas $P(c_1, \ldots, c_n)$ with constants $c_i \in D'(w)$

$$w \models P(c_1, \ldots, c_n) := w \models P(c_1^-, \ldots, c_n^-)$$

where $c_i^- = e$, if $c_i = e_w$ and $c_i^- = c_i$ otherwise. It is easily proved by induction on formulas that this property extends to all formulas:

$$w \models A(c_1, \ldots, c_n) \text{ implies } w \models A(c_1^-, \ldots, c_n^-)$$

so that the new model verifies (and refutes) the same formulas. ⊣

Proof of the Theorem 9.2. We extend the model for IPC satisfying the previous Lemma by the definition of values for ϵ-terms without changing domains $D(w)$, which is done by induction on construction of the term. Assume that the elements of D are well-ordered by a relation $\prec$ in some arbitrary way. In view of the condition (9.7) it enough to define $V(\epsilon x A, w)$ when $\epsilon x A$ does not have proper non-closed ϵ-subterms. In that case,

if $\epsilon x A(x) \downarrow w$, take the <-minimal element $v \leq w$ such that $\epsilon x A \downarrow v$, then define

$$V(\epsilon x A(x), w) := \text{ the } \prec -\text{first}\, d \in D(v)(v \models (\exists x A(x) \to A(d)))$$

If not $\epsilon x A(x) \downarrow w$, define $V(\epsilon x A(x), w)$ as the $\prec$-first $d \in D - D(w)$. □

9.4 Completeness Proof for IPCϵ

We prove that removing condition (9.6) but preserving familiar monotonicity requirement

$$w \leq w' \to (w \models A \to w' \models A) \tag{9.12}$$

leads to a complete semantics for IPCϵ.

For simplicity consider term models where individual domain $D(w)$ for every world w consists of terms, and the evaluation function for terms is identity: value of a term t is t. In particular the value of $\epsilon x A$ is $\epsilon x A$.

Definition 9.3 An intuitionistic Kripke (term) ϵ-model (or simply ϵ-model)

$$\mathcal{M} = (W, <, D, \models, V)$$

has to satisfy the following conditions.

$(W, <)$ is a Kripke frame with a strict partial ordering $<$,

D is a domain function assigning to every $w \in W$ a non-empty set $D(w)$ (of terms) monotone with respect to $<$,

$w \models A$ is a relation between worlds w and atomic formulas A with constants from

$$D := \cup_{w \in W} D(w)$$

monotonic with respect to $\leq$.

V is a valuation function assigning a constant $V(e, w) \in D$ to any ϵ-term e (possibly containing constants from D) and $w \in W$. (In a term model $V(e, w) = e$.)

The relation $\models$ is extended to composite formulas in the familiar way. The components of an ϵ-model have to satisfy the following conditions.

$$V(\epsilon x B(x, \epsilon y C), w) = V(\epsilon x B(x, V(\epsilon y C, w)), w) \tag{9.13}$$

$$w \models A(\epsilon y C) \leftrightarrow w \models A(x, V(\epsilon y C)) \tag{9.14}$$

where substitution of $\epsilon y C$ is safe, that is no free variable of $\epsilon y C$ becomes bound. Also if $e \downarrow w$ for a term $e := \epsilon x A(x)$, then

$$V(e, w) \in D(w) \text{ and } V(e, w') = V(e, w) \text{ for every } w' \geq w.$$

Let's present a completeness proof along familiar lines.

Definition 9.4 An infinite sequent is a pair of sets Γ, Δ of formulas such that there is an infinite number of variables not in $\Gamma \cup \Delta$. An infinite sequent w is written as $\Gamma \Rightarrow \Delta$ and notation

$$w_a := \Gamma, \; w_s := \Delta$$

is used for its antecedent and succedent.

L_w denotes the set of all terms and formulas with free variables and constants occurring in formulas of w.

$D(w)$ is the set of all terms $t \in L_w$ such that $(t\downarrow) \in w_a$. In other worlds $D(w)$ consists of all free variables and constants in w plus all ϵ-terms $\epsilon x A(x)$ such that $\exists y(\exists x A(x) \to A(y)) \in w_a$.

An infinite sequent w is *consistent*, if it is underivable, that is if no finite sequent $\Gamma \Rightarrow \Delta$ with $\Gamma \subset w_a,\ \Delta \subset w_s$ is derivable in IPCϵ.

A consistent infinite sequent w is *maximal consistent* if $w_a \cup w_s$ is the whole set of formulas in L_w.

Lemma 9.4 *Every consistent infinite sequent w_0 can be extended to a maximal consistent sequent.*

Proof Enumerate all formulas containing only free variables and constants in L_{w_0}, then add them one by one to w_a or w_s preserving consistency. At the n-th stage of this process a sequent w_n, an extension of w_0 by a finite number of formulas is generated.

It cannot happen that at some stage n of this process a formula A fits none of w_a^n, w_s^n, i.e., both of

$$w_a^n \Rightarrow w_s^n, A_n \qquad A_n, w_a^n \Rightarrow w_s^n$$

are inconsistent, since in that case $w_a^n \Rightarrow w_s^n$ is inconsistent by a cut rule. □

Important example. If w is $\forall x P(x) \Rightarrow P(\epsilon x Q(x))$ with $P \neq Q$, and the first "undecided" formula is $\exists y(\exists x Q(x) \rightarrow Q(y))$ then this formula is added to the succedent, since adding it to the antecedent results in an inconsistent sequent.

Lemma 9.5 *Every maximal consistent infinite sequent w is closed under invertible rules of multiple-succedent version of* IPCϵ, *that is under all rules except* $\Rightarrow \forall, \Rightarrow\rightarrow$. *More precisely*

$$(A\&B) \in w_a \textit{ implies } A \in w_a \textit{ and } B \in w_a,$$

$$(A \rightarrow B) \in w_a \textit{ implies } A \in w_s \textit{ or } B \in w_a,$$

$$(A \vee B) \in w_a \textit{ implies } A \in w_a \textit{ or } B \in w_a,$$

$$(\forall x A(x)) \in w_a \textit{ implies } (\forall t \in D(w))(A(t) \in w_a)$$

$$(\exists x A(x)) \in w_a \textit{ implies } A(\epsilon x A(x)) \in w_a$$

$$(A \vee B) \in w_s \textit{ implies } A \in w_a \textit{ and } B \in w_a,$$

$$(A\&B) \in w_s \textit{ implies } A \in w_a \textit{ or } B \in w_a,$$

$$(\exists x A(x)) \in w_s \textit{ implies } (\forall t \in D(w))(A(t) \in w_s)$$

Proof Suppose $(A\&B) \in w_a$. If $A \notin w_a$ then by maximality $A \in w_s$. Therefore w is inconsistent.

Suppose $\forall x A \in w_a$. If $A(t) \notin w_a$ for some $t \in D(w)$ then by maximality $A(t) \in w_s$. Therefore $\forall x A \Rightarrow A(t)$ is derived by one application of the $\forall \Rightarrow$-rule, and hence w is inconsistent. Note that additional premise $t\downarrow$ of this rule is available by $t \in D(w)$.

Other cases are similar. □

Definition 9.5 For infinite sequents w, w' define

$$w < w' \text{ iff } w_a \subseteq w'_a \text{ and } D(w) \subseteq D(w')$$

Lemma 9.6 *The set of maximal consistent sequents is closed under non-invertible rules* $\Rightarrow\rightarrow$, $\Rightarrow\forall$. *More precisely,*

For every maximal consistent sequent w, *if* $(A \rightarrow B) \in w_s$ *then there exists a maximal consistent sequent* $w' > w$ *with* $A \in w'_a$, $B \in w'_s$.

For every maximal consistent sequent w, *if* $\forall x A(x) \in w_s$ *then there exists a maximal consistent sequent* $w' > w$ *with* $A(a) \in w'_s$ *for some variable* a, $a \in D(w')$.

Proof If $(A \rightarrow B) \in w_s$ then the sequent $A, w_a \Rightarrow B$ is consistent, since otherwise one application of the rule $\Rightarrow\rightarrow$ leads to inconsistency of w. Now extend $A, w_a \Rightarrow B$ to a complete consistent sequent.

If $\forall x A(x) \in w_s$ then the sequent $w_a \Rightarrow A(a)$ for a fresh variable a is consistent, since otherwise one application of the rule $\Rightarrow \forall$ leads to inconsistency of w. Now extend $A, w_a \Rightarrow B$ to a complete consistent sequent. □

Definition 9.6 (Canonical model) Consider the following model

$$M = (W, <, V, \models).$$

W is the set of all maximal complete sequents, $<$, V are as above,

$w \models A$ iff $A \in w_a$ for atomic formulas A.

This definition implies that $w \not\models A$ for atomic $A \in w_s$, since otherwise w is inconsistent.

Lemma 9.7 *The relation* $\models$ *for atomic formulas and the function* D *is monotonic.*

Proof Consider only $D(w)$. Let $w < w'$. All variables and constants in $D(w)$ are in $D(w')$ by the definition of $<$. Assume $\epsilon x A(x) \in D(w)$, that is $\epsilon x A(x) \downarrow \in w_a$. Then $\epsilon x A(x) \downarrow \in w'_a$ by $w < w'$, and hence $\epsilon x A(x) \in D(w')$. □

Lemma 9.8 *For every formula* $A \in L_w$

1. $A \in w_a$ *implies* $w \models A$,
2. $A \in w_s$ *implies* $w \not\models A$,

Proof Induction on formulas using Lemmata 9.5 and 9.6. For example, if $A \& B \in w_a$ then $A, B \in w_a$, therefore $w \models A, w \models B$ by induction hypothesis, and hence $w \models A \& B$.

If $\forall x A \in w_s$ then there exists $w' > w$ such that $A(a) \in w'_s$ for some variable $a \in D(w')$. Therefore $w' \not\models A(a)$ and hence $w \not\models \forall x A(x)$. □

Theorem 9.3 *The system IPCϵ is sound and complete.*

Proof Soundness is checked as before. For completeness take an arbitrary underivable formula A, then extend sequent $\Rightarrow A$ to a maximal consistent set w. By the previous Lemma $w \not\models A$. □

9.4.1 About a Cut-Free Formulation

It is plausible that the completeness proof for the rules with cut given above can be modified to provide completeness of a cut-free formulation. As our example (9.11) shows, complete cut-elimination is impossible. One may need to admit cuts for formulas of the form $\epsilon x A(x) \downarrow$ where $\epsilon x A(x)$ occurs in the conclusion, and subformulas of such formulas. The following proof where $e := \epsilon x P(x)$ is another example.

$$\dfrac{\dfrac{\dfrac{\dfrac{\exists x P(x) \Rightarrow e\downarrow \quad \neg P(e), P(e), \exists x P(x) \Rightarrow P(0)}{\forall x \neg P(x), P(e), \exists x P(x) \Rightarrow P(0)}}{\forall x \neg P(x), P(e) \Rightarrow \exists x P(x) \to P(0)}}{\forall x \neg P(x), P(e) \Rightarrow \exists y(\exists x P(x) \to P(y))} \quad \dfrac{e\downarrow \Rightarrow e\downarrow \quad \neg P(e), P(e) \Rightarrow}{e\downarrow, \forall x \neg P(x), P(e) \Rightarrow}}{\forall x \neg P(x), P(e) \Rightarrow}$$

9.5 Comparison with Previous Work

9.5.1 System $IPC\Omega\epsilon$

Let $\exists\epsilon x A(x) := \exists x A(x)$.

A. Dragalin's system $IPC\Omega\epsilon$ from Dragalin (1974) for a given language $\Omega\epsilon$ is obtained by weakening familiar axioms for quantifiers

$$(\epsilon Q1)\ \exists t \& \forall x A(x) \to A(t)$$

$$(\epsilon Q2)\ \exists t \& A(t) \to \exists x A(x)$$

and adding the axiom

$$\exists x A(x) \to A(\epsilon x A(x))$$

A. Dragalin (Dragalin 1974) tried to avoid as much as possible dealing with a value of an ϵ-term in a world w where the term is not defined. Values (in a given world w) are assigned only to ϵ-terms defined in w, and many intermediate results are proved only for the case when all relevant ϵ-terms are defined. Nevertheless soundness is established for all formulas, without any restrictions. As pointed out earlier, this system is not complete.

In Sect. 9.3 we changed the definition of a model from Dragalin (1974) to a more uniform version: ϵ-term e which is not defined at the world w is assigned a value at w, but this value does not belong to the individual domain $D(w)$. To make this possible, the Kripke frame underlying the model and the domain function should satisfy additional conditions that still guarantee completeness.

Let us consider other systems in the literature.

9.5.2 Systems with $\exists y(\exists x A(x) \to A(y))$ as Existence Condition

In systems due to Maehara (1970) and Shirai (1971), instead of using $\exists x A(x)$ as a discriminating criterion, a weaker formula $\exists y(\exists x A(x) \to A(y))$ is employed. This still allows to anticipate a correct future value of the term $\epsilon x A(x)$ in a world w even if $\exists x A(x)$ fails in w.

Sh. Maehara treats a weaker language than ours: $\epsilon x A(x)$ is a syntactically correct term only if it is closed. He proves (using partial cut-elimination and other syntactic transformations) conservativity over IPC of the rules

$$\frac{\Gamma, \exists x A(x) \quad A(\epsilon x A(x)), \Delta \Rightarrow G}{\Gamma, \Delta \Rightarrow G}\ \exists_\epsilon$$

$$\frac{\Gamma \Rightarrow t\downarrow \quad F(t), \Delta \Rightarrow G}{\forall z F(z), \Gamma, \Delta \Rightarrow G} \qquad \frac{\Gamma \Rightarrow t\downarrow \quad \Delta \to F(t)}{\Gamma, \Delta \Rightarrow \exists z F(z)} \tag{9.15}$$

where

$$\epsilon x A(x)\downarrow := \exists y(\exists x A(x) \to A(y)); \ a\downarrow := \top \tag{9.16}$$

Here $\top$ is the constant true, a is an arbitrary variable.

Note that the first of these rules contains a hidden cut. This conservativity result is used to establish a kind of completeness theorem for IPC over a modification of Kripke semantics, although this modification is not stated explicitly. More precisely, Sh. Maehara proves Kripke-style soundness and completeness results for the relation $A \in \alpha$ between formulas A and complete consistent (in his sense) subsets α of the set of formulas. Only his condition for $\forall$ is not standard:

$\forall x A(x) \in \alpha \leftrightarrow (\exists B)(B \in \alpha \& \forall \beta \forall t[B \in \beta \to (t \in D_\beta \to A(t) \in \beta)])$

To establish this condition he uses admissibility of the following rule in his system:

$$\frac{\exists y(\exists x \neg A(x) \to \neg A(y)) \to A(\underline{x} \neg A(x))}{\forall x A(x)}$$

This rules approximates equivalence

$$\forall x A(x) \leftrightarrow A(\epsilon x \neg A(x))$$

which is valid only classically.

Shirai (1971) removes the restriction to closed ϵ-terms. He considers a language with the existence predicate denote by D. Instead of the rules used by Maehara he considers the following axioms:

$$D(t), \exists y(\exists x A(x, t) \rightarrow A(y, t) \Rightarrow D(\epsilon x A(x)) \tag{9.17}$$

$$D(t), \exists x A(x, t) \Rightarrow A(\epsilon x A(x, t), t)$$

plus standard modifications of quantifier rules for the system with existence predicate D.

He proves conservativity of his system over IPC by a combination of a partial cut-elimination and Maehara's argument.

Leivant (1973) and Smirnov (1979) define logical systems with ϵ-symbol conservative over IPC by requiring that assumptions discharged in natural deduction rules contain no ϵ-symbol. These systems are probably much weaker than IPCϵ. The system introduced by the author (Mints 1974) is certainly weaker than IPCϵ: a sequent containing subterm $\epsilon x A(x, y)$ with a bound variable y is syntactically correct only provided $\forall y \exists x A(x, y)$ is a member of the antecedent.

9.6 Further Work

Complete the proof of cut-elimination for IPCϵ and of the normal form theorem for NJϵ.

Give a syntactic proof of cut-elimination for IPCϵ and of normalization for NJϵ.

Provide a semantics for the systems by Maehara (1970) and Shirai (1971) and find out whether these systems admit cut-elimination. It seems that the system by Shirai provides the most general formulation of the idea that ϵ-terms are partially defined in some arbitrary way. The restriction $D(t)$ allowing to quantify over value of t can be an arbitrary predicate with the only condition (9.17).

References

Baaz, M., & Iemhoff, R. (2006). On the skolemization of existential quantifiers in intuitionistic logic. *Annals of Pure and Applied Logic*, *142*(1–3), 26–295.

Dragalin, A. (1974). Intuitionistic logic and Hilbert's ϵ-symbol, (Russian) Istoriia i metodologiia estestvennykh nauk, Moscow, MGU, pp. 78–84, republished in: Albert Grigorevich Dragalin,

Konstruktivnaia Teoriia Dokazatelstv I Nestandartnyi Analiz, s. 255–263, 2003, Moscow, Editorial Publisher.

Fitting, M. (1969). *Intuitionistic logic, model theory and forcing*. Amsterdam: North-Holland.

Kripke, S. (1965). Semantical analysis of intuitionistic logic I'. In J. N. Crossley & M. A. E. Dummett (Eds.), *Formal systems and recursive functions* (pp. 92–130). Amsterdam: North-Holland.

Leivant, D. (1973). 'Existential instantiation in a system of natural deduction for intuitionistic arithmetics', Stichtung Mathematisch Centrum, Technical Report ZW 13/73, Amsterdam

Maehara, Sh. (1970). A general theory of completeness proofs. *Annals of the Japan Association for Philosophy of Science*, *3*, 242–256.

Mints, G. (1966). Skolem method of elimination of positive quantifiers in sequential calculi. *Soviet Math. Dokl.*, *7*(4), 861–864.

Mints, G. (1974). The Skolem method in intuitionistic calculi. *Proc. Inst. Steklov*, *121*, AMS, 73–109.

Mints, G., Heyting Predicate Calculus with Epsilon Symbol (Russian), Zapiski Nauchnykh Seminarov Leningradskogo Otdeleniya Matematicheskogo Instituta im. V.A. Steklova AN SSSR, Vol. 40, pp. 110–118, 1974. English translation in (Mints 1992, pp. 97–104).

Mints, G. (1992). *Selected papers in proof theory*. Napoli/Amsterdam: Bibliopolis/North-Holland.

Mints, G. (1998). Linear lambda-terms and natural deduction. *Studia Logica*, *60*, 209–231.

Mints, G. (2000). *A short introduction to intuitionistic logic*. New York: Kluwer Academic-Plenum.

Osswald, H. (1975). *Über Skolemerweiterungen in der Intuitionistischen Logik mit Gleichheit*, Lecture Notes in Mathematics Heidelberg: Springer.

Shirai, K. (1971). Intuitionistic Predicate Calculus with ϵ-symbol. *Annals of the Japan Association for Philosophy of Science*, *4*, 49–67.

Smorynski, C. (1977). On axiomatizing fragments. *Journal of Symbolic Logic*, *42*(4), 530–544.

Smirnov, V. (1979). Theory of quantification and E-calculi. In J. Hintikka, I. Niiniluoto, & E. Saarinen (Eds.), *Essays on mathematical and philosophical logic* (pp. 41–49). Dordrecht: Kluwer Academic. Proc. 4th Scand. Logic Symp.

Chapter 10
Meaning in Use

Sara Negri and Jan von Plato

Abstract The historical origins of provability semantics are illustrated by so far unexplored manuscript passages of Gentzen and Gödel. Next the determination of elimination rules in natural deduction through a generalized inversion principle is treated, proposed earlier by the authors as a pedagogical device. The notion of validity in intuitionistic logic is related to the notion of formal provability through a direct translation. Finally, it is shown how the correspondence between rules and meaning can be used for setting up complete labelled sequent calculi, first for intuitionistic logic with the remarkable property of invertibility of all the logical rules, and then for modal and related logics.

Keywords Meaning explanation · Inversion principles · General elimination rules · Natural deduction · Sequent calculus · Labelled deduction.

10.1 Meaning Explanations and Provability Conditions

The discussions about proof theory and meaning in the past few decades date back to the early years of intuitionistic logic. The very name "BHK-explanations" reminds us of this fact. The *locus classicus*, however, was not penned down by Brouwer, Heyting, or Kolmogorov, but by Gentzen. He writes in his published thesis that the introduction rules of natural deduction are definitions of sorts of the logical connectives, and that the elimination rules are consequences of these definitions (Gentzen 1934–1935, III.5.12). He suggests further that it should be possible to actually determine the elimination rules, "as unique functions of the I-rules." Unfortunately the topic is not pursued further.

S. Negri (✉) · J. von Plato
Department of Philosophy, University of Helsinki, P.O.Box 24, 00014 Helsinki, Finland
e-mail: sara.negri@helsinki.fi

J. von Plato
e-mail: jan.vonplato@helsinki.fi

H. Wansing (ed.), *Dag Prawitz on Proofs and Meaning*, Outstanding Contributions to Logic 7, DOI 10.1007/978-3-319-11041-7_10

It turned out in 2005, when the second author found a handwritten manuscript version of the thesis, that the passage in the printed thesis was taken directly from a longer discourse that begins with:

> The "introductions" present, so to say, the "definitions" of the signs in question, and the "eliminations" are actually just consequences thereof, expressed more or less as follows: In the elimination of a sign, the proposition the outermost sign of which is in question, must "be used only as what it means on the basis of the introduction of this sign." An example will clarify what is meant hereby: The proposition $A \rightarrow B$ could be introduced when a derivation of B from the assumption A was at hand. If one now wants to use $A \rightarrow B$ further with the elimination of the sign $\rightarrow$ (uses for the construction of *longer* propositions, such as $A \rightarrow B. \vee C$ (OI), are naturally also possible), one can do it straightaway so that one concludes at once B from A that has been proved (FE). For $A \rightarrow B$ documents the existence of a derivation of B from A. Note well: It is not necessary to rely on a "contentful sense" of the sign $\rightarrow$.
>
> I think one could show, by making precise this idea, that the E-inferences are, through certain conditions, *unique* consequences of the respective I-inferences.

Save for a few stylistic changes and the horseshoe implication in the published version in place of the arrow, the two texts are the same to this point, but the manuscript version has the following continuation:

> I shall limit myself to the indication of a consequence of this connection, one that can be established purely formally. It will form the basis of later investigations into decidability and consistency. It goes as follows:
>
> If in an NII-proof an introduction (I) of a sign is followed immediately by its elimination (E), the proposition with the sign in question (as its outermost sign) can be eliminated through a simple "reduction" of the proof.
>
> These reductions proceed after the following schemes: ($\alpha, \beta, \ldots, \varepsilon, \zeta$ denote the further lines of the proof, in a way that can be easily seen. Square brackets mean that the respective part of the proof is to be written as many times as there occurred the respective assumption before the reduction.)

& ∨

$$\begin{matrix}\dfrac{\dfrac{\begin{matrix}\alpha & \beta\\ A & B\end{matrix}}{A\&B}\,AI}{A}\,AE\\ \varepsilon\end{matrix} \quad \text{becomes:} \quad \begin{matrix}\alpha\\ A\\ \varepsilon\end{matrix} \qquad\qquad \begin{matrix}\dfrac{\dfrac{\begin{matrix}\alpha\\ A\end{matrix}}{A \vee B}\,OI \quad \begin{matrix}[A]\\ \gamma\\ C\end{matrix} \quad \begin{matrix}[B]\\ \delta\\ C\end{matrix}}{C}\,OE\\ \varepsilon\end{matrix} \quad \text{into:} \quad \begin{matrix}\begin{bmatrix}\alpha\\ A\end{bmatrix}\\ \gamma\\ C\\ \varepsilon\end{matrix}$$

> (it is quite analogous with the other form of AE resp. OI.)

These "simple reductions" of a proof are nothing but steps of conversion to normal form. At the time of writing the above, Gentzen had not yet proved the normalization theorem of intuitionistic natural deduction, but just conjectured it. The passage continues with the conversion schemes for the quantifiers and implication and ends with:

> It requires some considerations to see to it that a correct proof is in fact produced in each case. I shall refrain from the exact realization, because I will not make any use of these facts, but rather present them for the sake of intuitiveness. —

The manuscript does contain an "exact realization," though, for it has a chapter added later with a detailed proof of the normalization theorem, published for the first time seventy-five years after it was written (cf. von Plato 2008). Gentzen's idea at this time was to extend the normal form of derivations from pure logic to arithmetic, in a proof of the consistency of arithmetic; These are the "later investigations into decidability and consistency" that he mentions.

Gentzen had finalized his set of logical principles of proof, the system of natural deduction, by September 1932. His analysis of "actual proofs" in mathematics led to intuitionistic logic, a topic well-defined after Arend Heyting's article of 1930 that was based on standard axiomatic logic in the tradition of Frege, Peano, Russell, Hilbert, and Bernays. There is in Heyting's subsequent article (1931) a brief explanation of negation through "a proof procedure that leads to a contradiction." Next, it is stated that a proof of a disjunction consists in a proof of one of the disjuncts.

Heyting's explanations evolved later into the well-known proof conditions: $A \& B$ is proved whenever A and B have been proved separately, $A \vee B$ is proved whenever one of A and B has been proved, $A \supset B$ is proved whenever any proof of A turns into some proof of B. For the quantifiers, $\forall x A(x)$ is proved whenever $A(y)$ is proved for an arbitrary y, and $\exists x A(x)$ is proved whenever $A(a)$ is proved for some object a. It was realized soon that the explanation of implication need not reduce a proof of $A \supset B$ into something simpler, for A could have been obtained by any proof. The difficulty is mentioned by Gentzen in a manuscript from the fall of 1932 with no reference to Heyting, and by Gödel repeatedly in the late 1930s. Heyting's short article of 1931 suggestive of the BHK-explanations was in a volume that contained some of the proceedings of a conference held in Königsberg in September 1930, the very occasion in which Gödel made his incompleteness result public. So we can trust that Heyting's paper had been read by those involved.

Gentzen's stenographic notes contain an item from the fall of 1932, some twenty-five dense pages, with a few pages added in the next spring and ten more in October 1934. The title is "Formal conception of the notion of contentful correctness in pure number theory, relation to proof of consistency" (*Die formale Erfassung des Begriffs der inhaltlichen Richtigkeit in der reinen Zahlentheorie, Verhältnis zum Widerspuchsfreiheitsbeweis*). Most of it was written within a month in October-November, and it was meant to be a groundwork for systematic formal studies, after the basic structure of mathematical reasoning had been cleared in September. We abbreviate the manuscript in the same way he did, as **INH**. The first task in it is to explain the notion of *correctness* for intuitionistic logic and arithmetic, quite similarly to Heyting's explanations:

14.X. *Contentful correctness in intuitionistic proofs*

One defines contentful correctness like this: The mathematical axioms are correct. $A\&B$ is correct when A is correct and B is correct, $A \vee B$ when at least one of them is correct, Ax when for each number substitution for x this correct, the same with [the universally quantified] $x\ Ax$, Aa when a number can be given so that Aa holds, the same for $Ex\ Ax$, $A \rightarrow B$ when from the correctness of A that of B can be concluded, $\neg A$ when from A a contradiction can be concluded.

It is to be shown now that the result of a proof is correct.

In the case of $A \,\&\, B$ and $A \vee B$, a well-founded notion is achieved, but $A \supset B$ remained problematic. A few weeks later, Gentzen writes:

> 3.XI *The* $\rightarrow$ plays a special role in the *definition of correctness*, because correctness is always reduced with the other signs to the correctness of smaller propositions. This does not happen with $\rightarrow$. The correctness of $A \rightarrow B$ can be conceived as the existence of a proof of B from A. However, there is a circle in this conception once the proof operates in its turn with $\rightarrow$. Maybe one has to do a recursion of a theory to one closest below (of which the former is the metatheory).

As can be seen, Gentzen is requesting that if $A \supset B$ is provable, it should have a proof that is somehow made up from the components of $A \supset B$. The correctness of a notion of proof with this property would not be circular.

Doubts about the explanation of implication through hypothetical proof were raised from early on also by others. Here is a passage from Bernays (Hilbert and Bernays 1934):

> The methodological point of "intuitionism" that is at the basis of Brouwer, is formed by a certain *extension of the finitary position*, namely, an extension in so far as Brouwer allows the introduction of an assumption about the presence of a consequence, resp. of a proof, even if such a consequence, resp. proof, is not determined in respect of its visualizable constitution. For example, from Brouwer's point of view, propositions of the following forms are permitted: "If proposition B holds under assumption A, also C holds," and also: "The assumption that A is refutable leads to a contradiction," or in Brouwer's mode of expression, "the absurdity of A is absurd."

It is hard to believe that the idea of a hypothetical proof, so common today, was taken to be *the* new methodological idea of intuitionism. The passage calls for a revision of the view of the tradition of axiomatic logic, from Frege to Hilbert, to the effect that it was exclusively based on a categorical notion of truth as in the logicist thesis.

Here is another discussion of the intuitionistic meaning of implication:

> By far the most important and interesting of these notions here is $P \rightarrow Q$. Now to explain the meaning of a proposition in a constructive system means to state under which circumstances one is entitled to assert it. And the answer in this case is: If one is able to deduce Q from the assumption P. But one has to be careful: the assumption P in a constructive logic means the assumption, that a proof for P is given, since truth in itself without proof makes no sense in a constructive logic. So $P \rightarrow Q$ means: Given a proof for P one can construct a proof for Q or in other words: One has a method to continue any given proof of P to a proof of Q. It is quite essential that $\rightarrow$ is not interpreted as meaning Q is deducible from the assumption that P is true, because certain theorems of intuitionistic logic don't hold for it.

This is not Dummett or Prawitz, but Gödel himself in the lectures on intuitionism he gave in Princeton in 1941. The influence of Gentzen in the passage seems clear.

Hypothetical reasoning has its pitfalls, as indicated by Gödel. His warning in the passage goes equally well for classical logic: If from the truth of P the truth of Q follows, $P \rightarrow Q$ need not be derivable. By the completeness of propositional logic, substitute truth by derivability and you have: If from $\vdash P$ it follows that $\vdash Q$, it need not follow that $\vdash P \rightarrow Q$. Thus, the former condition is that Q is derivable whenever P is, the latter the stronger condition that Q be derivable from P. After eighty years, the erroneous conclusion can still be found in books and articles written by otherwise competent logicians, even dubbed "failure of the deduction theorem" by those who commit the error of mixing an assumption about provability with an assumption (see Hakli and Negri 2012, for a detailed treatment).

It was a real pity that Gentzen did not present his normalization theorem for natural deduction in the published thesis or explain it to Heyting and Gödel in correspondence. Bernays seems not to have realized that Gentzen had the result (see von Plato 2012, p. 330). The normalization theorem would have cut short the talk about the possible circularity of Heyting's explanation of implication, at least in first-order logic: First assertions without open assumptions are covered by the fact that their normal derivations end with an introduction rule, as in the BHK-explanations. Gentzen calls these "direct proofs" in his 1936 paper on the consistency of arithmetic (end of §10.3). Then hypothetical assertions are covered in the sense that whenever their hypotheses receive direct proofs, a direct proof of the assertion can be obtained through normalization. This explanation is found very clearly stated in Gentzen's 1936 paper. The central point of that work was to extend such a meaning explanation to cover also the rule of induction: The inductive step consists in a derivation of $A(x+1)$ from the hypothesis $A(x)$ that may be "transfinite," and the conclusion is $\forall x A(x)$. Whenever a numerical instance $A(n)$ is concluded, the hypothesis can be made disappear through a composition of the sequence $A(0), A(0) \supset A(1), A(1) \supset A(2), \ldots, A(n-1) \supset A(n)$ (ibid., §10.5).

10.2 Determination of the Elimination Rules

A minimum condition for the "unique determination" Gentzen is calling for is given by the Gentzen-Prawitz *inversion principle*[1]: The elimination rule of a connective or quantifier should bring back that which is *included in the sufficient conditions* for introducing that connective or quantifier. The detour conversion schemes, as in the above quote from Gentzen, have been seen as a formal manifestation of this idea: They justify the elimination rules in terms of the introduction rules, by showing how the immediate grounds for introducing a formula are recovered in the conversions.

The Gentzen-Prawitz inversion principle does not meet Gentzen's requirement of actually determining the elimination rules from the introduction rules, instead of only justifying them. Thus, the possibility remains that the elimination rules are in some way too weak. The principle can be generalized, as in Negri and von Plato (2001, p. 6), into one in which one looks at the *arbitrary consequences of the sufficient grounds* for introducing a formula, instead of just those grounds. For conjunction, the

[1] We do not enter into the discussion of the background of this principle beyond Prawitz, but refer to Moriconi and Tesconi (2008) for that.

grounds are A and B separately, and their arbitrary consequences give the following *general* elimination rule:

$$\frac{A\&B \quad \begin{matrix}[A,B] \\ \vdots \\ C\end{matrix}}{C}\,\&E$$

For implication, the sufficient ground for introducing $A \supset B$ is, in Gentzen's words, "the existence of a derivation of B from A" (1934–1935, II.5.23). First-order logic is not able to represent formally within its language the existence of a derivation. Therefore (Schroeder-Heister 1984b) considered a system of higher-order rules. In 1980, with publication in his (1984), Martin-Löf formulated a scheme for elimination rules in his constructive type theory in which the existence of a derivation can be expressed. The general lesson from his discourse is that introduction rules correspond to "constructor" functions in an inductive definition, and a general elimination scheme for any such functions is a principle of recursive definition of functions over the inductively defined class.

In this light, the Gentzen-Prawitz inversion principle covers the base case of the recursive definition of functions over proofs of a compound formula, the one in which the arbitrary consequences of the sufficient grounds for introducing the formula are just these sufficient grounds. In the case of conjunction elimination, the way the elimination scheme computes a proof of the consequence C from a proof of $A \& B$ and a proof of A from assumed proofs of A and B separately has the base case that the proof of A is the proof of A, and the second base case that the proof of C is the proof of B. Thus, Martin-Löf's general elimination scheme gives us for these base cases the two rule instances:

$$\frac{A\&B \quad [A]}{A}\,\&E \qquad \frac{A\&B \quad [B]}{B}\,\&E$$

To recover the Gentzen-Prawitz elimination rules, it is sufficient to leave unwritten the degenerate derivations of the minor premisses in these two rule instances.

For implication, the sufficient ground for concluding $A \supset B$ is that there is a derivation of B from the assumption A. Such an existence can be indicated only schematically, and no way has been found to express in first-order logic the idea that C is the consequence of the existence of a derivation. In Negri and von Plato (2001, p. 8), the following is suggested: If there is a derivation of B from A, then, whatever follows from B follows already from A. Thus, the rule scheme becomes:

$$\frac{A \supset B \quad A \quad \begin{matrix}[B] \\ \vdots \\ C\end{matrix}}{C}\,{\supset}E$$

As with conjunction elimination, the standard rule comes out as the base case when the derivation of the minor premiss C is degenerate.

What has been said of implication goes also for universal quantification: The sufficient ground for concluding $\forall x A$ is the existence of a derivation of $A(y/x)$ for an *arbitrary* y. Type theory can hypothesize the existence of a higher-order function that produces, for any value of y, a proof of $A(y/x)$, and express that a formula C follows from the existence of such a function. In first-order logic, an elimination rule can be written, with t an arbitrary term, as:

$$\dfrac{\forall x A \qquad \begin{matrix}[A(t/y)] \\ \vdots \\ C\end{matrix}}{C}\ \forall E$$

The type-theoretical version of this rule is presented in Martin-Löf (1984, preface), and the first-order rule in Schroeder-Heister (1984a). The full set of general elimination rules is found in Dyckhoff (1988), then in Tennant (1992), Lopez-Escobar (1999), and von Plato (2000, 2001).

Natural deduction with general elimination rules can be brought into a direct correspondence with the left rules of sequent calculus, with the following result, as established in von Plato (2001):

Isomorphism between natural deduction and sequent calculus. *A cut-free derivation in sequent calculus translates isomorphically into a derivation in natural deduction with general elimination rules with the property that all major premisses of elimination rules are assumptions.*

The correspondence between left rules and elimination rules and right rules and introduction rules, as well as the order of the logical rules, is maintained by the translation.

The translation goes also in the other direction, from natural deduction to sequent calculus, and the property singled out by the isomorphism gives a simple notion of normal derivability:

Normal derivations. *A derivation in natural deduction with general elimination rules is normal if all major premisses of elimination rules are assumptions.*

Further results include that instances of the structural rules of weakening and contraction in sequent calculus correspond to vacuous and multiple discharges, respectively, of assumptions in natural deduction. These results come out directly from the isomorphic translation between derivations in natural deduction and sequent calculus. The normalization of derivations is carried through so that cases with major premisses of elimination rules derived by other elimination rules are first removed in what are known as *permutative conversions.* Such conversions for disjunction and existence elimination were first published in Prawitz (1965) but actually known and used already by Gentzen (von Plato 2008). With general elimination rules, there are permutative conversions for all the elimination rules. After permutative conversions have been exhausted, there come the cases of major premisses of elimination rules that are derived by the corresponding introduction rules, i.e., the *detour convertibilities.* A direct proof of normalization for natural deduction with general elimination rules was given in Negri and von Plato (2001, pp. 199–201), (see also Negri and von Plato 2011, pp. 27–28). The related result of strong normalization was proved in Joachimski and Matthes (2003).

The last rule in a normal derivation of a theorem, i.e., a derivation without open assumptions, must be an introduction rule, because an elimination rule would leave its major premiss as an open assumption. Results that were earlier proved through sequent calculus, such as the disjunction and existence properties of intuitionistic logic, can now be carried through in natural deduction. There are many later applications of the very strong property of normal derivability that is made possible by general elimination rules, such as the existence property of Heyting arithmetic (von Plato 2006).

The point with the inversion principle of *Structural Proof Theory* was mainly a pedagogical one in three steps: 1. To motivate the rules of natural deduction through the standard meaning explanations of the connectives and quantifiers that give rise to the introduction rules. 2. To determine the elimination rules by the general inversion principle. 3. To arrive at the rules of sequent calculus by the translation of 1 and normal instances in 2. Somewhat surprisingly, the inversion principle turned out to be more than a pedagogical device, namely a very useful tool in research, as we shall point out below.

10.3 From Semantical Explanations to Rules of Proof

One half of natural deduction, the introduction rules, is a formalization of the BHK provability conditions. Thus, we can say that Gentzen was the one who took the step of extracting a rule system from semantical explanations.[2] These developments led by 1970 or so to the remarkable *computational semantics* of intuitionistic logic, an idea developed further in intuitionistic type theory. Formal proofs are coded as functions and steps of normalization become interpreted as steps of computation of these functions. Strong normalization was also established around 1970 see Prawitz 1971, and becomes interpreted as the termination of the computations, and the uniqueness of normal form as the uniqueness of values of the functions.

Thirty years after Gentzen, and well before the computational semantics was understood in detail, Saul Kripke gave another semantics for intuitionistic logic in terms of *possible worlds*. In classical propositional logic, there is a situation at hand in which to the atomic formulas are assigned truth values that determine the truth values of compound formulas. In Kripke's semantics, these situations are indexed by the worlds, denoted $w, o, r, \ldots$ with $\mathcal{W}$ standing for the collection of all possible worlds, and the notation $w \Vdash P$ standing for the "forcing relation": *atom* P *holds in world* w. This machinery gains strength when the idea of possible worlds is put into use, with the intuition that there is an initial world w_0 in which some thing or other, possibly nothing at all, is known about the atomic facts P, Q, $R, \ldots$, and

[2] If that was his way, which is by no means certain as discussed in von Plato (2012).

that information comes in in the form of added atomic facts, in new worlds $o, r, \ldots$ related to the present one through an *accessibility relation* $w \leqslant o$. The accessibility relation is assumed to have the following properties:

1. There is an *initial world* w_0 such that $w_0 \leqslant w$ for any w in $\mathcal{W}$.
2. The accessibility relation is *reflexive*: $w \leqslant w$ for any w in $\mathcal{W}$.
3. The accessibility relation is *transitive*: If $w \leqslant o$ and $o \leqslant r$, then $w \leqslant r$ for any w, o, r in $\mathcal{W}$.

It is further required that no information be lost, i.e., that the forcing relation be *monotonic*: If $w \Vdash A$ and $w \leqslant o$, then $o \Vdash A$. For compound formulas, forcing is defined inductively, as in the semantical clauses for the connectives:

1. $w \Vdash A \,\&\, B$ whenever $w \Vdash A$ and $w \Vdash B$.
2. $w \Vdash A \vee B$ whenever $w \Vdash A$ or $w \Vdash B$.
3. $w \Vdash A \supset B$ whenever from $w \leqslant o$ and $o \Vdash A$ follows $o \Vdash B$.
4. $w \Vdash \neg A$ whenever from $w \leqslant o$ and $o \Vdash A$ follows $o \Vdash C$ for any C.
5. $w \Vdash C$ for any C if $w \Vdash A$ and $w \Vdash \neg A$ for some A.

This definition will work for intuitionistic logic with a primitive notion of negation. With a defined notion of negation, clause 4 is left out and clause 5 can be put as: no world forces $\bot$. It then happens that proofs of the properties of the forcing relation have to rely on somewhat awkward meta-level reasonings. For example, for $w \Vdash \bot \supset C$, one needs: From $w \leqslant o$ and $o \Vdash \bot$ follows $o \Vdash C$. This is the case because $o \Vdash \bot$ is false.

Under the above clauses 1–5 for compound formulas, the forcing relation for a world w becomes *trivial*, in the sense that w forces all formulas, whenever $w \Vdash A$ and $w \Vdash \neg A$ for some A. It is natural to pose the requirement of *nontriviality: No world must force all formulas. Validity* of a formula A can now be defined as "truth in all possible worlds," or more formally, as $w \Vdash A$ for an arbitrary w.

The correspondence between the inductive clauses of forcing and the provability conditions of natural deduction is straightforward, as a couple of examples show:

For conjunction, one direction of the semantical clause is: If $w \Vdash A$ and $w \Vdash B$, also $w \Vdash A \,\&\, B$. Therefore, if w is arbitrary and the premisses A and B of rule $\&I$ are assumed valid, also its conclusion $A \,\&\, B$ is. In the other direction, the clause is that if $w \Vdash A \,\&\, B$, then $w \Vdash A$ and $w \Vdash B$. Therefore, if the premiss $A \,\&\, B$ of rules $\&E_1$ and $\&E_2$ is valid, also the conclusions A and B are.

For the rule of implication introduction, the definition of validity has to be extended: B is *forced under assumptions* Γ *in world* w whenever from $w \leqslant o$ and $o \Vdash A$ for each A in Γ follows $o \Vdash B$. If w is arbitrary, B is *valid under assumptions* Γ.

The clause for implication is in one direction: If from $w \leqslant o$ and $o \Vdash A$ follows $o \Vdash B$, also $w \Vdash A \supset B$. Therefore, if the premiss B of rule $\supset I$ is valid under the assumption A, i.e., if from $o \Vdash A$ follows $o \Vdash B$, also the conclusion $A \supset B$ of the rule is valid by the clause. In the other direction, assume $w \Vdash A \supset B$ and $w \Vdash A$. By the semantical clause, $o \Vdash B$ whenever $w \leqslant o$ and $o \Vdash A$. In particular, setting w for o, we have that if $w \leqslant w$ and $w \Vdash A$, also $w \Vdash B$. The first condition holds by the reflexivity of the accessibility relation, the second by assumption. Therefore, if the premisses of rule $\supset E$ are valid, also the conclusion is.

The lesson from the above correspondence between syntax and semantics is that one direction of a semantical clause corresponds to an introduction rule, the other direction to an elimination rule.

In perfect analogy to the proof terms of typed lambda-calculus that lead to the computational semantics of intuitionistic logic, we can make the semantics of possible worlds for intuitionistic logic formal, by including these worlds and the forcing relation as parts of a system of rules: Formulas come with *labels* $w, o, r, \ldots$ with the forcing relation written compactly as $w : A$, and the accessibility relation $w \leqslant o$ is a new type of atomic formula. The rules for conjunction and implication are, directly from the semantical clauses:

$$\frac{w : A \quad w : B}{w : A \& B}\,\&I \qquad \frac{w : A \& B}{w : A}\,\&E_1 \qquad \frac{w : A \& B}{w : B}\,\&E_2$$

$$\frac{\begin{array}{c} \overset{1}{w \leqslant o}, \overset{1}{o : A} \\ \vdots \\ o : B \end{array}}{w : A \supset B}\,{\supset}I,1 \qquad \frac{w \leqslant o \quad w : A \supset B \quad o : A}{o : B}\,{\supset}E$$

In rule $\supset I$, the label o has to be arbitrary, i.e., an eigenvariable of the rule.

Accessibility relations are now a part of the formal calculus and their properties have to be represented. To this end, we use the well-developed machinery of *proof analysis*, i.e., of the extension of logical calculi by rules that represent mathematical axioms. The rules can be written in the style of natural deduction as:

$$\frac{w : A \quad w \leqslant o}{o : A}\,Mon \qquad \frac{}{w_0 \leqslant w}\,Init \qquad \frac{}{w \leqslant w}\,Ref \qquad \frac{w \leqslant o \quad o \leqslant r}{w \leqslant r}\,Tr$$

If a semantics is going to be more than just suggestive words, the notion of *proof of validity* has to be considered instead of mere validity. An example from the Kripke semantics for intuitionistic logic shows that proofs of validity can turn out to be essentially the same as formal proofs by syntactic rules:

An example of a semantical proof of validity. $\Vdash A \supset (B \supset A \& B)$. Let w be arbitrary and assume $w \leqslant o$ and $o \Vdash A$. To show $o \Vdash B \supset A \& B$, assume $o \leqslant r$ and $r \Vdash B$. By monotonicity, $r \Vdash A$, so by definition, $r \Vdash A \& B$. Therefore $o \Vdash B \supset A \& B$, and finally $w \Vdash A \supset (B \supset A \& B)$.

Reproduction by the rules of formal semantics.

$$\cfrac{\cfrac{\cfrac{\cfrac{\overset{2}{o : A} \quad \overset{1}{o \leqslant r}}{r : A}\,Mon \qquad \overset{1}{r : B}}{r : A \& B}\,\&I}{o : B \supset A \& B}\,{\supset}I,1}{w : A \supset (B \supset A \& B)}\,{\supset}I,2$$

In the upper instance of rule $\supset I$, the accessibility relation $o \leqslant r$ is closed together with the assumption $r : B$. In the lower instance of rule $\supset I$, the assumption $o : A$ is closed, but the associated accessibility relation $w \leqslant o$ is not used in the derivation. It is closed vacuously.

Translation to a formal derivation in natural deduction. Given a formal proof of validity, it can be translated by an easy algorithm into a formal derivation in natural deduction: First delete all labels and accessibility relations. Now instances of rules *Init, Ref*, and *Tr* have disappeared. Next delete the repetitions that rule *Mon* has left. The result for the above example is:

$$\dfrac{\dfrac{\dfrac{\overset{2}{A} \quad \overset{1}{B}}{A \,\&\, B}\,\&I}{B \supset A \,\&\, B}\,\supset I{,}1}{A \supset (B \supset A \,\&\, B)}\,\supset I{,}2$$

The approach to labelled deduction with the internalization of the Kripke semantics has been developed in the literature in several forms, based on either natural deduction, sequent calculi, or tableau systems. Closest to the approach illustrated here are the works of Simpson (1994) that uses natural deduction and Viganó (2000), based on sequent calculus but with frame rules that correspond to frame properties that do not contain disjunctions in positive parts.

10.4 An Intuitionistic Sequent Calculus with Invertible Rules

Kripke's most fundamental discovery was perhaps the correspondence between conditions on the accessibility relation and axioms of systems of logic. For example, if to the conditions of reflexivity and transitivity of intuitionistic logic the condition of symmetry is added, the possible worlds collapse into one equivalence class and the logic becomes classical. By the correspondence, logical systems between the intuitionistic and classical ones can be captured either by suitable axioms, such as Dummett's axiom $(A \supset B) \vee (B \supset A)$, or by a suitable "frame condition" on the accessibility relation, the linearity condition $w \leqslant o \vee o \leqslant w$ in this case. However, as is seen, the condition employs the same connective $\vee$ as the axiom. A similar problem was met when Gentzen wanted to reason about provability in natural deduction, and his solution was to distinguish between an internal implication $A \supset B$ and an external derivability relation $A \vdash B$ ($A \rightarrow B$ in Gentzen's notation). A similar method is possible here: With frame property *Tr*, a two-premiss "logic-free rule" was used that acts on the atomic premisses $w \leqslant o$ and $o \leqslant r$, to give the atomic conclusion $w \leqslant r$, with no interference with the logical operations of conjunction and implication that would otherwise be used in the expression of the axiom of transitivity. Thus, we have the correspondence between a "logical" and a "logic-free" derivation of $w \leqslant r$

from the assumptions $w \leqslant o$ and $o \leqslant r$, the former with an instance of the transitivity axiom:

$$\dfrac{w \leqslant o \& o \leqslant r \supset w \leqslant r \qquad \dfrac{w \leqslant o \quad o \leqslant r}{w \leqslant o \& o \leqslant r}\,\&I}{w \leqslant r}\,\supset E \qquad\qquad \dfrac{w \leqslant o \quad o \leqslant r}{w \leqslant r}\,Tr$$

More generally, those quantifier-free frame properties that do not contain essential disjunctions, i.e., disjunctions in positive parts, can be converted to additional rules of natural deduction of the type of *Tr*. No mixing of logical properties is produced. The method of conversion of axioms into "logic-free" additional rules had been already developed successively in Negri (1999), Negri and von Plato (1998), and Negri and von Plato (2001, ch. 8), when the first author realized the possibility of converting frame properties of modal logic into rules. This earlier work covered those cases in which frame properties are expressed by universal formulas. The much wider class of *geometric implications*, including typical existence axioms, was covered in Negri (2003).

The limitation on disjunction inherent to additional rules in the style of natural deduction is surpassed if a multisuccedent sequent calculus is used. The logical rules of the labelled sequent calculus are written with shared contexts as in the *G3*-calculi, to support root-first proof search:

$$\dfrac{w:A, w:B, \Gamma \to \Delta}{w:A\&B, \Gamma \to \Delta}\,L\& \qquad \dfrac{\Gamma \to \Delta, w:A \quad \Gamma \to \Delta, w:B}{\Gamma \to \Delta, w:A\&B}\,R\&$$

$$\dfrac{w:A, \Gamma \to \Delta \quad w:B, \Gamma \to \Delta}{w:A \vee B, \Gamma \to \Delta}\,L\vee \qquad \dfrac{\Gamma \to \Delta, w:A, w:B}{\Gamma \to \Delta, w:A \vee B}\,R\vee$$

$$\dfrac{w:A \supset B, \Gamma \to \Delta, w:A \quad w:B, \Gamma \to \Delta}{w:A \supset B, \Gamma \to \Delta}\,L\supset \qquad \dfrac{w \leqslant o, o:A, \Gamma \to \Delta, o:B}{\Gamma \to \Delta, w:A \supset B}\,R\supset$$

$$\dfrac{}{w:\bot, \Gamma \to \Delta}\,L\bot$$

Contrary to unlabelled sequent calculus, rule $R\supset$ has the context Δ also in the premiss. The label o in the rule has to be arbitrary, i.e., an eigenvariable of the rule.

The frame rules of intuitionistic logic in the notation of labelled sequent calculus are:

$$\dfrac{o:A, w:A, w \leqslant o, \Gamma \to \Delta}{w:A, w \leqslant o, \Gamma \to \Delta}\,Mon$$

$$\dfrac{w \leqslant w, \Gamma \to \Delta}{\Gamma \to \Delta}\,Ref \qquad \dfrac{w \leqslant r, w \leqslant o, o \leqslant r, \Gamma \to \Delta}{w \leqslant o, o \leqslant r, \Gamma \to \Delta}\,Tr$$

The calculus has initial sequents of the form $w : P, \Gamma \to \Delta, w : P$ with P an atomic formula.

To obtain a calculus with strong structural properties, the rule of monotonicity is left out in favour of initial sequents with in-built monotonicity *Init*, of the form

$$w \leqslant o, w : P, \Gamma \rightarrow \Delta, o : P$$

The sequent calculus thus obtained and called *G3I* has all structural rules admissible and, moreover, contraction is admissible with the property that a step of contraction preserves the height of derivation. (This is the reason for the repetition of the atoms from the conclusion in the premiss.) As a result, steps of root-first proof search that would produce a duplication are not permitted. By completeness, a formula A is provable in intuitionistic logic if and only if for a label w, the sequent $\rightarrow w : A$ is derivable in the calculus. It follows that the sequent calculus version of the rule *Init* that produces an initial label is not needed as an explicit rule of *G3I*.

Even if *G3I* does not have the restriction of a single-succedent premiss in rule $R\supset$, as if by miracle the calculus does not become classical: An attempt at a root-first derivation of the law of excluded middle gives

$$\dfrac{\dfrac{x \leqslant y, y : P \rightarrow x : P, y : \bot}{\rightarrow x : P, x : \neg P}\ {\scriptstyle R\supset,\, y\, fresh}}{\rightarrow x : P \vee \neg P}\ {\scriptstyle R\vee}$$

The eigenvariable condition is $y \neq x$ by which no initial sequent is reached.

The most remarkable feature of the labelled sequent calculus for intuitionistic logic is the *invertibility of all of its rules*, a property encountered earlier only with unlabelled classical sequent calculi. By this invertibility, the rules preserve countermodels and a terminal node in a failed proof search defines a Kripke countermodel which is automatically a countermodel to the conclusion. In the example above the countermodel is given by the following

$$\begin{array}{c} \bullet\, y \Vdash P \\ \uparrow \\ \bullet\, x \nVdash P \end{array}$$

The parallel proof search/countermodel construction works in full generality for the *G3K*-based modal labelled calculi (Negri 2009).

To obtain a classical calculus, a rule of symmetry is added to the frame rules of *G3I*. Alternatively, no accessibility relation is considered and all the rules, including those for implication, are obtained by labelling in a "flat" way all the rules of *G3c*, as in the propositional part of the calculus *G3K*.

10.5 Labelled Modal Calculi

Turning now to modal logic, the inductive definition of forcing of a modal formula in a possible world w follows from the basic idea of Kripke semantics, which is to define necessity relative to a world w simply as that which happens to hold in all worlds accessible from w, as in:

$w \Vdash \Box A$ *whenever for all* o, from $w \leqslant o$ follows $o \Vdash A$.

The definition gives:

If $o : A$ can be derived for an arbitrary o accessible from w, then $w : \Box A$ can be derived.

Formally, we have the rule in natural deduction style:

$$\dfrac{\begin{array}{c} \overset{1}{w \leqslant o}, \Gamma \\ \vdots \\ o : A \end{array}}{w : \Box A}\ {\scriptstyle \Box I,1}$$

The condition is that o does not occur in Γ. By generalizing the rule to an arbitrary conclusion, that is one in which $o : A$ comes together with an arbitrary succedent Δ, it becomes the sequent calculus rule

$$\dfrac{w \leqslant o, \Gamma \to \Delta, o : A}{\Gamma \to \Delta, w : \Box A}\ {\scriptstyle R\Box}$$

In the rule, the arbitrariness of o becomes the variable condition that o must not occur in Γ, Δ.

The inversion principle is stated in Negri and von Plato (2001, p. 6) in the form *"Whatever follows from the direct grounds for deriving a proposition must follow from that proposition."* Through this principle, one has that consequences of the derivability of $o : A$ from an arbitrary o accessible from w are consequences of $w : \Box A$. Then, similarly to the determination of the "lower level" rule of general implication elimination, we find that whatever follows from $o : A$ must already follow from $w : \Box A$ and $w \leqslant o$, that is, we have the general elimination rule

$$\dfrac{w : \Box A \quad w \leqslant o \quad \begin{array}{c} \overset{1}{o : A} \\ \vdots \\ w : C \end{array}}{w : C}\ {\scriptstyle \Box E,1}$$

If the major premiss is an assumption, the rule can be written in sequent calculus notation, as:

$$\dfrac{\Gamma \to \Delta, w \leqslant o \quad o : A, \Gamma \to \Delta}{w : \Box A, \Gamma \to \Delta}\ {\scriptstyle L\Box'}$$

The rule can be equivalently given as a one-premiss rule in the following form

$$\dfrac{o : A, w : \Box A, w \leqslant o, \Gamma \to \Delta}{w : \Box A, w \leqslant o, \Gamma \to \Delta}\ {\scriptstyle L\Box}$$

The recipe for "meaning in use" is: Meaning-conferring introduction rules are scrutinized under the inversion principle, to obtain general forms of elimination rules. The normal instances of these rules have direct translations to sequent calculus.

The inductive clause for the possibility operator $\Diamond$ is:

$w : \Diamond A$ *whenever for some* o, $w \leqslant o$ *and* $o : A$.

The rules for $\Diamond$ are obtained from the semantic explanation analogously to those of $\Box$. They are:

$$\frac{w \leqslant o, o : A, \Gamma \to \Delta}{w : \Diamond A, \Gamma \to \Delta} L\Diamond \qquad \frac{w \leqslant o, \Gamma \to \Delta, w : \Diamond A, o : A}{w \leqslant o, \Gamma \to \Delta, w : \Diamond A} R\Diamond$$

In rule $L\Diamond$, o is an eigenvariable that corresponds to the existential quantifier in the inductive clause.

In all, the labelled calculi are constructed so that they are equivalent to corresponding axiomatic calculi. More precisely, because the language includes the accessibility and forcing relations, they are conservative extensions of the axiomatic calculi (see Negri 2005 for details).

Properties of the accessibility relation such as reflexivity and transitivity correspond to modal axioms, as in:

	Axiom	Frame property
T	$\Box A \supset A$	$\forall w\ w \leqslant w$ reflexivity
4	$\Box A \supset \Box\Box A$	$\forall wor(w \leqslant o \,\&\, o \leqslant r \supset w \leqslant r)$ transitivity
E	$\Diamond A \supset \Box\Diamond A$	$\forall wor(w \leqslant o \,\&\, w \leqslant r \supset o \leqslant r)$ Euclideanness
B	$A \supset \Box\Diamond A$	$\forall wo(w \leqslant o \supset o \leqslant w)$ symmetry
D	$\Box A \supset \Diamond A$	$\forall w \exists o\ w \leqslant o$ seriality
W	$\Box(\Box A \supset A) \supset \Box A$	no infinite R-chains + transitivity

Let us take as another example the determination of the rules for the "actuality operator" @ from the semantic explanation. The formula $@A$, read *actually A true at world w*, expresses that A is true at the actual world w_a. The forcing notation is:

$w \Vdash @A$ *whenever* $w_a \Vdash A$.

Now we can read out from the semantical explanation, similarly to the modalities of necessity and possibility, the introduction and elimination rules:

$$\frac{w_a : A}{w : @A} @I \qquad \frac{w : @A}{w_a : A} @E$$

The formulation in terms of labelled sequent calculus is:

$$\frac{w_a : A, \Gamma \to \Delta}{w : @A, \Gamma \to \Delta} L@ \qquad \frac{\Gamma \to \Delta, w_a : A}{\Gamma \to \Delta, w : @A} R@$$

An axiomatization of modal systems augmented by the actuality operator has been provided by Hodes (1984), as an extension of first-order S5, and shall not be recalled here. It is straightforward to verify that all the axioms are derivable in the labelled sequent calculus for actuality here obtained as an extension of the basic modal system with reflexivity, transitivity and symmetry of the accessibility relation plus the rules for actuality. For example, axiom $@(A \supset B) \supset (@A \supset @B)$ is derived as follows:

$$\dfrac{\dfrac{\dfrac{\dfrac{\dfrac{w_a : A \to w_a : A \quad w_a : B \to w_a : B}{w_a : A, w_a : A \supset B \to w_a : B}\, L\supset}{w_a : A, w : @(A \supset B) \to w_a : B}\, L@}{w : @A, w : @(A \supset B) \to w : @B}\, R@}{w : @(A \supset B) \to w : @A \supset @B}\, R\supset}{\to w : @(A \supset B) \supset (@A \supset @B)}\, R\supset$$

The labelled approach allows for a fine distinction between various notions of logical consequence that can be adopted: *actualistic* logical consequence is logical consequence relative to the actual world, whereas *universal* (or strong) consequence is relative to an arbitrary world.

The contraction-free labelled sequent calculi were first developed for modal and related logics (Negri 2005), but are not limited to them. They can be applied equally well to create proof systems for pure predicate logic, for example, and for the intermediate logics that were mentioned above. Such logical systems are typically characterized by frame conditions that are added to those of intuitionistic logic, until the conditions of classical logic are reached. This idea is carried through in Dyckhoff and Negri (2012) in which intermediate logical systems are obtained by adding to the labelled calculus for intuitionistic logic rules that correspond to frame conditions. The uniformity provided by the labelled calculi leads to a simple syntactic proof of soundness and faithfulness of the embedding of a wide class of intermediate logics into their modal companions.

10.6 Completeness and Decidability

The connection between derivations in natural deduction and proofs of validity in Kripke semantics is close and suggestive of a completeness theorem. The unification of the semantic and syntactic dimension in labelled sequent calculi leads to such uniform, simple, and direct proofs of completeness for modal logics. Strangely enough, the style of completeness proof that was favored in the literature on modal logic since the late 1960s has been the Henkin-style completeness proof, even if Kripke's 1963 proof of completeness was based on a direct construction of countermodels for failed tableau proofs. Apparently, as documented in Negri (2009), the reasons behind this turn are to be found in negative reviews of Kripke's paper. The review by

Kaplan (1966) contained also a sketch of an alternative, Henkin-style, completeness proof which became the standard until the present days. Labelled sequent calculi, however, allow to recover the original explicit character of Kripke's completeness proof, without the insufficiency in formalization that was lamented by the early reviewers.

The direct proof for labelled sequent calculi is obtained through a Schütte-style argument: all the rules (logical rules and frame rules) of the calculus for a given modal logic are applied, root-first, from a given logical sequent $\Gamma_0 \to \Delta_0$ labelled uniformly with an arbitrary label w. In this way a big tree is built. If all the branches lead to initial sequents or instances of $L\bot$, then the sequent is derivable. Otherwise it may happen that at some stage no rule is applicable and the sequent is neither initial nor an instance of $L\bot$, or that the construction goes on forever. In the two latter cases, a countermodel is built. If the search stops at a sequent $\Gamma \to \Delta$ because no rule is any longer applicable, a countermodel is built by considering all the worlds in Γ, related to each other through the accessibility relations in Γ, and the valuation that forces in w all the atomic formulas for which $w : P$ is in Γ and that does not force in w atomic formulas for which $w : Q$ is in Δ. In the case of an infinite process, by König's lemma, the tree has an infinite branch of sequents $\Gamma_i \to \Delta_i$. Again, the countermodel is built directly by taking as possible worlds all the labels and all the relations in the antecedents Γ_i, with a valuation that forces on the world w the formulas for which $w : P$ is in some of the Γ_i and does not force on w those for which $w : Q$ is in some Δ_i. An inductive argument then shows that the valuation has the property of forcing on w all the formulas A for which $w : A$ is in one of the antecents and no formula B for which $w : B$ is in some of the succedents. A countermodel to $\Gamma_0 \to \Delta_0$ is thus found.

The completeness proof can be turned into a constructive proof of decidability whenever the potentially infinite growth of the search tree can be truncated. The finite countermodel is not extracted from an infinite one, but is built directly from a proof search which has at least a truncated branch. Rather than describing the procedure in general, we illustrate it with an example. First, observe that a check of derivability for a formula A is equivalent to a check of validity. We can thus start with applying root-first the rules of the labelled calculus for intuitionistic logic for the sequent $\to w : \neg\neg A \supset A$, where w is an arbitrary label, as follows (in applications of $L\supset$ the derivable right premiss is omitted):

$$
\dfrac{\dfrac{\dfrac{\dfrac{\dfrac{\dfrac{\begin{array}{c}\vdots\\ w \leqslant o, o \leqslant o, o \leqslant r, r \leqslant l, l : A, r : A, o : \neg\neg A \to o : A, o : \neg A, r : \bot, l : \bot\end{array}}{w \leqslant o, o \leqslant o, o \leqslant r, r : A, o : \neg\neg A \to o : A, o : \neg A, r : \bot, r : \neg A}\,R\supset}{w \leqslant o, o \leqslant o, o \leqslant r, r : A, o : \neg\neg A \to o : A, o : \neg A, r : \bot}\,L\supset}{w \leqslant o, o \leqslant o, o : \neg\neg A \to o : A, o : \neg A}\,R\supset}{w \leqslant o, o \leqslant o, o : \neg\neg A \to o : A}\,L\supset}{w \leqslant o, o : \neg\neg A \to o : A}\,Ref}{\to w : \neg\neg A \supset A}\,R\supset
$$

Clearly, the proof search goes on indefinitely, but there are two ways to see already at this point that it does not lead to a derivation. The first is strictly proof-theoretic and consists in appealing to structural properties of the labelled calculus, namely height-preserving admissibility of substitution for labels (here r/l) and height preserving admissibility of contraction. By these two properties, the above search would yield, together with reflexivity, a shortening of the purported derivation, which contradicts the quest for a minimal one. Alternativelly, and probably more convincingly, we observe that l is a *looping label*, i.e., a label of a formula that already appeared earlier in the search. We obtain a finite countermodel already from this segment of the infinite branch by taking as worlds the labels w, o, r, l with the accessibility relations $w \leqslant o$, $o \leqslant r$, $r \leqslant l$ which are in the search tree, plus the accessibility that witnesses the loop, namely $l \leqslant r$, and their transitive closures plus reflexivities. The valuation is defined by the forcing of A in r and l but not in o. It is clear that $x \nVdash \neg\neg A \supset A$, so a finite countermodel has been found.

Countermodel constructions inspired by the above methodology are used to obtain decision procedures for modal logics with transitive and serial accessibility relations such such as the logic of Priorean linear time (Boretti and Negri 2009) and several classes of intuitionistic multi-modal logics (Garg et al. 2012). The general results guarantee that the frame that arises from the truncated failed proof search gives indeed a countermodel to the conclusion of the failed proof-search, with no need to check that the endformula is not valid in it.

References

Boretti, B., & Negri, S. (2009). Decidability for Priorean linear time using a fixed-point labelled calculus. In M. Giese & A. Waaler (Eds.), *Automated Reasoning with Analytic Tableaux and Related Methods* (pp. 108–122). Berlin: Springer.

Dyckhoff, R. (1988). Implementing a simple proof assistant. In *Workshop on Programming for Logic Teaching: Proceedings 23/1988* (pp 49–59). University of Leeds: Centre for Theoretical Computer Science.

Dyckhoff, R., & Negri, S. (2012). Proof analysis in intermediate logics. *Archive for Mathematical Logic*, *51*(1–2), 71–92.

Garg, D., Genovese, V., & Negri, S. (2012). Countermodels from sequent calculi in multi-modal logics. In *LICS 2012, IEEE computer society* (pp. 315–324).

Gentzen, G. (1934–1935). Untersuchungen Über das logische Schließen I. *Mathematische Zeitschrift*, *39*(2), 176–210.

Gentzen, G. (1936). The consistency of elementary number theory. In M. E. Szabo (Ed.), *The Collected Papers of Gerhard Gentzen* (pp. 132–213). Amsterdam: North-Holland Publishing Company.

Hakli, R., & Negri, S. (2012). Does the deduction theorem fail for modal logic? *Synthese*, *187*(3), 849–867.

Heyting, A. (1930). Die formalen Regeln der intuitionistischen Logik. In *Sitzungsberichte der Preussischen Akademie der Wissenschaften* (pp. 42–56)., Physikalisch- mathematische Klasse.

Heyting, A. (1931). Die intuitionistische Grundlegung der Mathematik. *Erkenntnis*, *2*(1), 106–115.

Hilbert, D., & Bernays, P. (1934). *Grundlagen der Mathematik* (Vol. 1). Berlin: Springer.

Hodes, H. (1984). Axioms for actuality. *Journal of Philosophical Logic*, *13*(1), 27–34.

Joachimski, F., & Matthes, R. (2003). Short proofs of normalization for the simply-typed λ-calculus, permutative conversions and Gödel's T. *Archive for Mathematical Logic*, *42*(1), 59–87.
Kaplan, D. (1966). Review of Kripke. *The Journal of Symbolic Logic*, *31*(1), 120–122.
Kripke, S. A. (1963). Semantical analysis of modal logic I. *Zeitschrift für mathematische Logik und Grundlagen der Mathematik*, *9*(5–6), 67–96.
Lopez-Escobar, E. (1999). Standardizing the N systems of Gentzen. In X. Caicedo & C. Montenegro (Eds.), *Models, Algebras, and Proofs* (pp. 411–434). New York: Dekker.
Martin-Löf, P. (1984). *Intuistionistic Type Theory*. Napoli: Bibliopolis.
Moriconi, E., & Tesconi, L. (2008). On inversion principles. *History and Philosophy of Logic*, *29*(2), 103–113.
Negri, S. (1999). Sequent calculus Proof Theory of intuitionistic apartness and order relations. *Archive for Mathematical Logic*, *38*(8), 521–547.
Negri, S. (2003). Contraction-free sequent calculi for geometric theories, with an application to Barr's theorem. *Archive for Mathematical Logic*, *42*(4), 389–401.
Negri, S. (2005). Proof analysis in modal logic. *Journal of Philosophical Logic*, *34*(5–6), 507–544.
Negri, S. (2009). Kripke completeness revisited. In G. Primiero & S. Rahman (Eds.), *Acts of knowledge—History, Philosophy and Logic* (pp. 247–282). London: College Publications.
Negri, S., & von Plato, J. (1998). Cut elimination in the presence of axioms. *The Bulletin of Symbolic Logic*, *4*(4), 418–435.
Negri, S., & von Plato, J. (2001). *Structural Proof Theory*. Cambridge: Cambridge University Press.
Negri, S., & von Plato, J. (2011). *Proof Analysis: A contribution to Hilbert's Last Problem*. Cambridge: Cambridge University Press.
von Plato, J. (2000). A problem of normal form in natural deduction. *Mathematical Logic Quarterly*, *46*(1), 121–124.
von Plato, J. (2001). Natural deduction with general elimination rules. *Archive for Mathematical Logic*, *40*(7), 541–567.
von Plato, J. (2006). Normal form and existence property for derivations in Heyting arithmetic. *Acta Philosophica Fennica*, *78*, 159–163.
von Plato, J. (2008). Gentzen's proof of normalization for natural deduction. *The Bulletin of Symbolic Logic*, *14*(2), 240–244.
von Plato, J. (2012). Gentzen's proof systems: byproducts in a work of genius. *The Bulletin of Symbolic Logic*, *18*(3), 313–367.
Prawitz, D. (1965). *Natural Deduction: Proof-theoretical Study*. Stockholm: Almqvist & Wicksell.
Prawitz, D. (1971). Ideas and results in Proof Theory. In J. E. Fenstad (Ed.), *Proceedings of the Second Scandinavian Logic Symposium* (pp. 235–307). Amsterdam: North-Holland Publishing Company.
Schroeder-Heister, P. (1984a). Generalized rules for quantifiers and the completeness of the intuitionistic operators & ∨, ⊃, ⅄, ∀, ∃. In M. Richter, et al. (Eds.), *Computation and Proof Theory* (Vol. 1104, pp. 399–426)., Lecture Notes in Mathematics Berlin: Springer.
Schroeder-Heister, P. (1984b). A natural extension of natural deduction. *The Journal of Symbolic Logic*, *49*(4), 1284–1300.
Simpson, A. (1994). *Proof Theory and Semantics for Intuitionistic Modal Logic*. PhD thesis, School of Informatics, University of Edinburgh.
Tennant, N. (1992). *Autologic*. Edinburgh: Edinburgh University Press.
Viganó, L. (2000). *Labelled Non-classical Logics*. Dordrecht: Kluwer Academic Publishers.

Chapter 11
Fusing Quantifiers and Connectives: Is Intuitionistic Logic Different?

Peter Pagin

Abstract A paper by Dag Westerståhl and myself twenty years ago introduced operators that are both connectives and quantifiers. We introduced two binary operators that are classically interdefinable: one that fuses conjunction and existential quantification and one that fuses implication and universal quantification. We called the system PFO. A complete Gentzen-Prawitz style Natural Deduction axiomatization of classical PL was provided. For intuitionistic PL, however, it seemed that existential quantification should be fused with disjunction rather than with conjunction. Whether this was true, and if so why, were questions not answered at the time. Also, it seemed that there is no uniform definition of such a disjunctive-existential operator in classical PFO. This, too, remained a conjecture. In this paper, I return to these previously unresolved questions, and resolve them.

Keywords Anaphora · Axiomatization · Fusing operators · Intuitionistic logic · Natural deduction · Natural language · PFO · Prawitz · Predicate logic · Uniform definition · Unselective binding.

11.1 Background

The motivation for this paper comes from a question that arose in connection with an earlier paper by Dag Westerståhl and myself: Pagin and Westerståhl (1993). By now, the work on that paper dates back 20 years. The motivation for that paper, in turn, was a puzzle in natural language semantics, concerning so-called *donkey sentences*, like

$$\text{If Pedro owns a donkey, he beats it.} \tag{11.1}$$

Equation (11.1) has a straightforward formalization into Predicate Logic (PL):

P. Pagin (✉)
Department of Philosophy, Stockholm University,
Stockholm 10691, Sweden
e-mail: peter.pagin@philosophy.su.se

H. Wansing (ed.), *Dag Prawitz on Proofs and Meaning*, Outstanding Contributions to Logic 7, DOI 10.1007/978-3-319-11041-7_11

$$\forall x((Dx \mathbin{\&} O(P, x)) \to B(P, x)) \tag{11.2}$$

where 'D' abbreviates 'donkey', 'O' 'owns', and 'B' 'beats'. The problem is that we would like the formalization to work compositionally, and it doesn't.[1] Syntactically, (11.1) is an indicative conditional. Compositionally, an indicative conditional 'If A, B' should be translated into PL as '$A^* \to B^*$', where 'A^*' is the translation of 'A' and 'B^*' the translation of 'B'. But the translation into PL of the antecedent of (11.1) is

$$\exists x(O(P, x)) \tag{11.3}$$

and if we follow the simple translation pattern for the conditional we end up with

$$\exists x(O(P, x)) \to B(P, x). \tag{11.4}$$

This is not a sentence, since the variable in the consequent is free. If we enlarge the scope of the quantifier, we get an existential sentence, not a conditional, and this renders the translation flawed.

Among the solutions to this problem, two methods modified PL to ensure a compositional translation. One was DPL (*Dynamic Predicate Logic*), by Groenendijk and Stokhof (1991). DPL employed a non-standard semantics by which the extension of a PL formula is not a set of assignment functions, but a set of *pairs* of assignment functions. This allowed a so-called "dynamic" interpretation of quantifiers and connectives.

The other was PFO (*Predicate Logic with Flexibly Binding Operators*), by Pagin and Westerståhl (1993). Our version of PL had three distinctive features:

1. Logical operators are dyadic: they have two immediate subformulas.
2. Binding is unselective: any variable that occurs in both subformulas is bound.
3. The binding priority is outside-in: the *outermost* operator that can bind a variable does, and cancels binding by embedded operators.

We introduced two operators, '[,]' and '(,)'. The first fuses the conditional and the dyadic universal quantifier. The second fuses conjunction and the dyadic existential quantifier. For example,

$[Px, Qy]$	translates as	$Px \to Qy$,
(Px, Qy)	translates as	$Px \wedge Qy$,
(Px, Qxz)	translates as	$\exists x(Px \wedge Qxz)$,
$[Pxy, Qxy]$	translates as	$\forall x \forall y(Pxy \to Qxy)$.

[1] This problem was introduced into the modern literature by Geach (1962).

In the third line the variable x becomes existentially bound because of occurring in both immediate subformulas. Similarly, in the fourth line, the variables x and y become universally bound. This illustrates the unselective binding.

Outside-in binding is illustrated by the fact that

$$[Px, (Qx, Rxy)] \qquad \text{translates as} \qquad \forall x(Px \to (Qx \wedge Rxy)).$$

The potential binding of x in (Qx, Rxy) is *cancelled* by the circumstance that x is "already" bound by the universal-conditional operator that has larger scope.

It is this outside-in binding direction that allows a compositional translation of the donkey sentence. The antecedent, 'Pedro owns a donkey', is straightforwardly translated as

$$(Dx, O(P, x)) \tag{11.5}$$

The indicative conditional 'if, then' is simply translated by [,]. This gives us as a translation of (11.1):

$$[(Dx, O(P, x)), B(P, x)] \tag{11.6}$$

Equation (11.6), in turn, translates into standard PL notation as (11.2), exactly as desired. The existentially quantified translation of the antecedent of (11.1) reduces to conjunction when embedded in the conditional. And this parallels exactly what takes place in natural English.

It turned out that the semantics for these operators could be given by adding a parameter to the truth definition: a set of already bound variables in the recursive application of the truth predicate. For unembedded formulas, this set is empty, and at every recursive step of applying a truth definition clause, variables bound at that step get added to the set. This will be spelled out in the Sect. 11.2.

It also turned out to be straightforward to provide a Natural Deduction proof theory for these operators. With the addition of the absurdity constant $\bot$, negation $\neg A$ could be defined as $[A, \bot]$. Then, in classical logic, the existential–conjunctive operator could be defined by means of the universal-conditional operator and absurdity. This allows a complete axiomatization of classical PL by means of an introduction and an elimination rule for [,], together with the classical absurdity rules. This will be spelled out in Sect. 11.4.[2]

The puzzle that emerged came to light when we turned to considering an *intuitionistic* version of PFO. To achieve this we needed to add disjunction. Within classical PFO, a disjunction operator '$\{A, B\}$' could be defined as

$$\{A, B\} = [\neg A, B] = [[A, \bot], B].$$

[2] It also turned out to be straightforward to extend the system to treating sub-sentential expression in a flexibly binding version of Montague Grammar. All that was needed was to subject the λ operator to the same binding principles. See Pagin and Westerståhl (1994).

This fuses disjunction with universal quantification. However, this definition is not intuitionistically acceptable. Although it is straightforward to derive the standard introduction rule for disjunction from it, the standard elimination rule cannot be derived without applying double negation elimination. Moreover, accepting this definition together with the standard elimination rule and intuitionistic negation already yields classical logic (cf. Sect. 11.5).

It seemed rather that in the intuitionistic version, disjunction should be fused with *existential* quantification. This was the puzzle. Why would disjunction be fused with universal quantification classically, but with existential quantification intuitionistically? We did not know the answer to this question at the time. One of two purposes of the present paper is to to provide it. The second concerns the question whether an operator that fuses disjunction with existential quantification can be *uniformly* defined in classical PFO. We had conjectured that it couldn't.

In Sect. 11.2, the syntax and the semantics of classical PFO is presented, and in Sect. 11.3, the relation between PFO and classical predicate logic. In Sect. 11.4 I shall set out the Natural Deduction proof theory of PFO, and in Sect. 11.5 the intuitionistic version of PFO. The presentation in these sections draws heavily on the material in Pagin and Westerståhl (1993). In Sect. 11.6 I consider the possibility of an alternative intuitionistic PFO, with a different treatment of disjunction. It will turn out to be possible. In Sect. 11.7, finally, I turn to the question of defining an alternative disjunction in classical PFO. It will turn out that a fusion of disjunction and existential quantification is uniformly definable in classical PFO, but that no such definition is possible with only single occurrences of the immediate subformulas.

11.2 Syntax and Semantics of PFO

11.2.1 Syntax

The vocabulary of PFO consists of *non-logical symbols*, variables, the identity symbol $=$, the absurdity symbol $\bot$, as in standard notation.

Definition 11.1 *Formulas*

(a) Atomic formulas are formulas
(b) If ϕ and ψ are formulas, so are (ϕ, ψ) and $[\phi, \psi]$.

The notion of a *subformula* of a formula is defined in the obvious way. Free and bound variable occurrences are introduced by first giving an inductive definition of what it means for an occurrence of a variable x to be *surface bound* (*depth bound*) in ϕ.

Definition 11.2 *Surface bound and depth bound variable occurrences*

(a) No occurrence of a variable x is *surface bound* or *depth bound* in an atomic formula.

(b) An occurrence of a variable x is *surface bound* in (ϕ, ψ) if x occurs in both ϕ and ψ. Similarly for $[\phi, \psi]$.
(c) An occurrence of x is *depth bound* in (ϕ, ψ) if it is not *surface bound* in (ϕ, ψ), but is *surface bound* in a subformula of (ϕ, ψ). Similarly for $[\phi, \psi]$.

Definition 11.3 *Bound and free variable occurrences* An occurrence of a variable x is *bound* in ϕ if it is either surface or depth bound in ϕ, otherwise it is *free* in ϕ. A formula is a *sentence* if no variable occurs free in it.

Proposition 11.1 (Pagin and Westerståhl (1993)) *If one occurrence of x is free (surface bound, depth bound) in ϕ, then so are all other occurrences of x in ϕ.*

Given a formula ϕ, we let Var_ϕ be the set of variables occurring in ϕ, and $Free_\phi$ ($Sbound_\phi$, $Dbound_\phi$) the set of free (surface bound, depth bound) variables in ϕ. By Proposition 11.1, the latter three sets form a partition of Var_ϕ. Also, let $Bound_\phi = Sbound_\phi \cup Dbound_\phi$.

Definition 11.4 *Quantified variables*
A variable x is *quantified at* a subformula ψ of ϕ, if x is surface bound in ψ but does not occur in ϕ outside of ψ.

Every bound variable of ϕ is quantified at a unique subformula of ϕ (for example, if x is surface bound in ϕ, it is quantified at ϕ in ϕ). But if ϕ in turn is a subformula of θ, x may be quantified at ψ in ϕ although not quantified at ψ in θ. This is the sense in which quantification can be *cancelled* in PFO subformulas.

11.2.2 Semantics

A *model* $\mathcal{M}$ consists, as usual, of a nonempty set M and an interpretation function I assigning suitable denotations to non-logical symbols. Let *Var* be the set of variables. An *M-assignment* is a function f from *Var* to M. We define what it means for f to *satisfy* a formula ϕ in $\mathcal{M}$.

Since quantification in subformulas may be cancelled, the usual ternary satisfaction relation cannot be defined directly by an induction going "from inside and out". However, we can first define inductively a satisfaction relation between *four* things: an assignment f, a formula ϕ, a model $\mathcal{M}$, and a subset X of *Var*; X is to be a set of "marked" variables that cannot be quantified again in subformulas of ϕ.

Given $\mathcal{M}$ with and interpretation function I and an M-assignment f, the *value* $t^{I,f}$ of a term t is defined as usual. If $a_1, \ldots, a_k \in M$ then $f(x_i/a_i)_{1\le i\le k}$ is the assignment which is like f except that a_i is assigned to x_i, $1 \le i \le k$. Also, $\{x_i/a_i\}_{1\le i\le k}$ stands for any M-assignment which assigns a_i to x_i. (The subscript with the condition $1 \le i \le k$ will usually be omitted.)

Definition 11.5 Let $\mathcal{M}$, f and X be as above.

(a) $\mathcal{M}, X \models_f Pt_1, \ldots, t_n \iff \langle t_1^{I,f}, \ldots, t_n^{I,f} \rangle \in P^I$

(b) $\mathcal{M}, X \models_f t_1 = t_2 \iff t_1^{I,f} = t_2^{I,f}$

(c) Not $\mathcal{M}, X \models_f \bot$

Let $(Var_\phi \cap Var_\psi) - X = \{x_1, \ldots, x_k\}$ and let $f' = f(x_i/a_i)$. Then:

(d) $\mathcal{M}, X \models_f (\phi, \psi) \iff$ there are $a_1, \ldots, a_k \in M$ such that

$\mathcal{M}, X \cup \{x_1, \ldots, x_k\} \models_{f'} \phi$ and $\mathcal{M}, X \cup \{x_1, \ldots, x_k\} \models_{f'} \psi$

(e) $\mathcal{M}, X \models_f [\phi, \psi] \iff$ for all $a_1, \ldots, a_k \in M$,

if $\mathcal{M}, X \cup \{x_1, \ldots, x_k\} \models_{f'} \phi$, then $\mathcal{M}, X \cup \{x_1, \ldots, x_k\} \models_{f'} \psi$.

Definition 11.6 $\mathcal{M} \models_f \phi \iff \mathcal{M}, \varnothing \models_f \phi$

If $(Var_\phi \cap Var_\psi) - X$ is empty, i.e., if ϕ and ψ have no variables in common, or if their common variables are all in X, then (ϕ, ψ) behaves as conjunction, and $[\phi, \psi]$ as material implication. One easily establishes

Lemma 11.2 *If f and g are $\mathcal{M}$-assignments which agree on $Free_\phi \cup X$, then $\mathcal{M}, X \models_f \phi$ iff $\mathcal{M}, X \models_g \phi$.*

Corollary 11.3 *If ϕ is a sentence and f, g any two M-assignments, then $M, X \models_f \phi$ iff $M, X \models_g \phi$.*

Thus, the following definition makes sense.

Definition 11.7 If ϕ is a sentence, $\mathcal{M} \models \phi$ iff for some f, $\mathcal{M} \models_f \phi$.

We state another lemma for later use; it follows almost directly from the truth definition:

Lemma 11.4 *If $Var_\phi - X = Var_\phi - Y$, then $\mathcal{M}, X \models_f \phi$ iff $\mathcal{M}, Y \models_f \phi$.*

Negation can be introduced as usual: $\neg\phi \stackrel{\text{def}}{=} [\phi, \bot]$. It follows that

$$\mathcal{M}, X \models_f \neg\phi \iff \text{not } \mathcal{M}, X \models_f \phi$$

The truth definition gives the formula (ϕ, ψ) the same truth conditions as $\neg[\phi, \neg\psi]$ (whether or not ϕ and ψ have common variables), so we may use only [,] , $\bot$ and the comma as logical symbols in PFO.

11.3 PFO and Predicate Logic

Consider, for simplicity, PFO with (,) as a defined operator, and Predicate Logic PL with $=$, $\bot$, $\to$, $\forall$ as primitive logical symbols. For a PL-formula ϕ, FV(ϕ) is the set of free variables in ϕ.

To translate from PFO to PL we simply follow the definition of satisfaction for PFO.

Definition 11.8 For a PFO-formula ϕ and $X \subseteq Var$, define inductively the PL-formula $\phi^{+,X}$ by

(a) $\phi^{+,X} = \phi$, if ϕ is atomic,
(b) $[\phi, \psi]^{+,X} = \forall x_1, \ldots, x_n(\phi^{+,X\cup\{x_1,\ldots,x_n\}} \to \psi^{+,X\cup\{x_1,\ldots,x_n\}})$

where $x_1, \ldots, x_n$ are the elements of $Sbound_{[\phi,\psi]} - X$ (in some fixed order).

Definition 11.9 $\phi^+ = \phi^{+,\varnothing}$

A straightforward induction shows

Proposition 11.5 (Pagin and Westerståhl (1993)) $\mathcal{M}, X \models_f \phi \iff \mathcal{M}, X \models_f^{PL} \phi^{+,X}$

Clearly, the same variables occur in ϕ and $\phi^{+,X}$. Concerning the *free* variables, we have the following

Proposition 11.6 (Pagin and Westerståhl (1993)) *If* $\mathrm{Free}_\phi \subseteq X \subseteq \mathrm{Var}_\phi$, *then* $FV(\phi^{+,X}) = X$. *In particular, if* ϕ *is a sentence, so is* ϕ^+.

For translating in the opposite direction, we define:

Definition 11.10 If ϕ is a PL-formula, define the PFO-formula ϕ^* inductively as follows:

(a) $\phi^* = \phi$, if ϕ is atomic
(b) $(\phi \to \psi)^* = [\phi^*, \psi^*]$
(c) $(\forall x\phi)^* = [x = x , \psi^*]$

Note that $(Px \to Qx)^* = [Px, Qx]$, which is equivalent to $\forall x(Px \to Qx)$, so free variables can become bound in this translation, and the meaning of formulas is not in general preserved. However, if we 'mark' the free variables of ϕ, meaning will be preserved for *strict* formulas:

Definition 11.11 A PL-formula is called *strict*, if (i) no variable occurs both free and bound in it, (ii) all quantifiers use distinct variables, and (iii) there is no vacuous quantification.

Every PL-formula is of course logically equivalent to a strict PL-formula. In PFO, strictness is built into the syntax (Proposition 11.1).

Proposition 11.7 (Pagin and Westerståhl (1993)) *For a strict PL-formula* ϕ, $\mathcal{M} \models^{PL}_{f} \phi$ *iff* $\mathcal{M}, FV(\phi) \models_{f} \phi^*$.

A variable occurs in ϕ iff it occurs in ϕ^*. Also, if ϕ is strict, $Free_{\phi^*} \subseteq \mathrm{FV}(\phi)$. Thus, if ϕ is a sentence, so is ϕ^*, and we get

Corollary 11.8 *If* ϕ *is a strict PL-sentence, then* $\mathcal{M} \models^{PL}_{f} \phi$ *iff* $\mathcal{M} \models_{f} \phi^*$.

Note that $(\forall x Px)^* = (x = x \to Px)^*$, so the function * is not one-one. But this is the only type of exception. In fact, one easily proves that * is a *bijection* from the set of PL-formulas (strict or not) with no subformulas of the form $x = x \to \psi$ to the set of PFO- formulas.

If ϕ is a PFO-sentence, ϕ^+ is strict, so we have

$$\mathcal{M}, X \models_{f} \phi \iff \mathcal{M}, X \models_{f} \phi^{+*}$$

although usually ϕ and ϕ^{+*} are distinct. Likewise, if ϕ is a strict PL-sentence, $\phi \leftrightarrow \phi^{*+}$ is valid.

11.4 Natural Deduction

PFO allows a rather compact formulation of natural deduction, with just one introduction rule and one elimination rule, plus rules for negation and identity. Here deducibility involves *sentences*, so we shall need to assume that there is always a sufficient number of individual constants around to perform instantiations. In the two rules below, $[\phi, \psi]$ is assumed to be a sentence with $Var_{\phi} \cap Var_{\psi} = \{x_1, \ldots, x_n\}$, and $\phi(t_1, \ldots, t_n)$ is the result of simultaneously replacing *all* occurrences (free or bound) of x_i in ϕ by t_i. The rules are presented in the usual informal way. $\phi^{\dagger}$ marks that the assumption ϕ has been discharged.

$$[\,,\,]\text{-INTR} \quad \frac{\begin{array}{c}\phi(t_1, \ldots, t_n)^{\dagger} \\ \vdots \\ \psi(t_1, \ldots, t_n)\end{array}}{[\phi, \psi]}$$

where $\{x_1, \ldots, x_n\} = Var_{\phi} \cap Var_{\psi}$ and $t_1, \ldots, t_n$ do not occur in ϕ, ψ or open assumptions in the derivation of $\psi(t_1, \ldots, t_n)$, except $\phi(t_1, \ldots, t_n)$.

$$[\,,\,]\text{-ELIM} \quad \frac{\phi(t_1, \ldots, t_n) \qquad [\phi, \psi]}{\psi(t_1, \ldots, t_n)}$$

where $\{x_1, \ldots, x_n\} = Var_\phi \cap Var_\psi$ and $t_1, \ldots, t_n$ are closed terms.

We also have the rule for classical negation:

$$\text{NEG}\quad \dfrac{\begin{matrix}\neg\phi^\dagger \\ \vdots \\ \bot\end{matrix}}{\phi}$$

To this we add the two rules for identity:

$$\text{ID-axiom}\quad \dfrac{}{t = t}$$

$$\text{ID-rule}\quad \dfrac{t_1 = t_2 \qquad \phi}{\phi'}$$

where ϕ' results from ϕ by replacing some occurrences of t_1 by t_2.

Let $\Gamma \vdash_f \phi$ mean that there is a derivation of ϕ with open assumptions in Γ. Then we have

Proposition 11.9 (Pagin and Westerståhl (1993)) Completeness theorem*: If* $\Gamma \models \phi$*, then* $\Gamma \vdash \phi$.

It is clear that the Gentzen-Prawitz reduction step (Prawitz 1965) goes through: an [,]-INTR step immediately followed by the corresponding ELIM step reduces to the derivation immediately preceding both steps:

$$\dfrac{\dfrac{\begin{matrix}\phi(t_1, \ldots, t_n) \\ \Pi \\ \psi(t_1, \ldots, t_n)\end{matrix}}{[\phi, \psi]} \qquad \phi(t_1, \ldots, t_n)}{\psi(t_1, \ldots, t_n)} \quad \triangleright \quad \begin{matrix}\phi(t_1, \ldots, t_n) \\ \Pi \\ \psi(t_1, \ldots, t_n)\end{matrix}$$

Rules for the conjunctive-existential operator (,) can be added:

$$(\ ,\ )\text{-INTR}\quad \dfrac{\phi(t_1, \ldots, t_n) \qquad \psi(t_1, \ldots, t_n)}{(\phi, \psi)}$$

where $\{x_1, \ldots, x_n\} = Var_\phi \cap Var_\psi$ and $t_1, \ldots, t_n$ are closed terms.

$$
(\,,\,)\text{-ELIM}\quad \dfrac{(\phi,\psi) \qquad \begin{array}{c}\phi(t_1,\ldots,t_n)^{\dagger} \quad \psi(t_1,\ldots,t_n)^{\dagger}\\ \vdots \\ \theta\end{array}}{\theta}
$$

where $\{x_1,\ldots,x_n\} = Var_{\phi} \cap Var_{\psi}$ and $t_1,\ldots,t_n$ do not occur in ϕ, ψ or θ, nor in open assumptions in the derivation of θ, except possibly in $\phi(t_1,\ldots,t_n)$ and $\psi(t_1,\ldots,t_n)$.

The rules for (,) are derivable from the rules for [,] and negation. Remember that we define (ϕ,ψ) as $\neg[\phi,\neg\psi]$. Here is a derivation of (,)-INTR:

$$
\dfrac{\dfrac{\dfrac{[\phi,\ [\psi,\bot]]^1 \qquad \phi(t_1,\ldots,t_n)}{[\psi(t_1,\ldots,t_n),\ \bot]}\;[\,,\,]\text{-ELIM} \qquad \psi(t_1,\ldots,t_n)}{\bot}\;[\,,\,]\text{-ELIM}}{[[\phi,\ [\psi,\bot]],\ \bot]}\;1,[\,,\,]\text{-INTR}
$$

where the superscript indicates the assumption discharged at the inference step with the corresponding number. The following derivation of (,)-ELIM (slightly amended here) is given in Pagin and Westerståhl (1993):

$$
\dfrac{\dfrac{\dfrac{\dfrac{\begin{array}{c}\phi(t_1,\ldots,t_n)^2 \quad \psi(t_1,\ldots,t_n)^1\\ \vdots\\ \theta\end{array} \qquad [\theta,\bot]^3}{\bot}\;[\,,\,]\text{-ELIM}}{[\psi(t_1,\ldots,t_n),\ \bot]}\;1,[\,,\,]\text{-INTR}}{[\phi,\ [\psi,\ \bot]]}\;2,[\,,\,]\text{-INTR} \qquad [[\phi,\ [\psi,\ \bot]],\ \bot]}{\dfrac{\bot}{\theta}\;3,\text{NEG}}\;[\,,\,]\text{-ELIM}
$$

11.5 Intuitionistic PFO

PFO does not have a primitive disjunction, but the operator { , } defined by

$$
\{\phi,\psi\} \stackrel{\text{def}}{=} [[\phi,\ \bot],\psi]
$$

appears to be what is needed for formalization of disjunction in natural language. Note that $\{\phi,\psi\}$ means that *for all* $x_1,\ldots,x_n$, ϕ *or* ψ, where $x_1,\ldots,x_n$ are the variables common to ϕ and ψ.

This definition is intuitionistically unacceptable.[3] To see this, recall that the standard elimination rule for disjunction must come out valid:

$$\dfrac{\phi \vee \psi \qquad \begin{matrix} \phi^{\dagger} \\ \vdots \\ \theta \end{matrix} \qquad \begin{matrix} \psi^{\dagger} \\ \vdots \\ \theta \end{matrix}}{\theta}$$

Hence, by the proposed definition, we would have as a valid argument

$$\dfrac{[[\phi,\ \bot], \phi] \qquad \begin{matrix} \phi^{\dagger} \\ \vdots \\ \phi \end{matrix} \qquad \begin{matrix} \phi^{\dagger} \\ \vdots \\ \phi \end{matrix}}{\phi} \tag{11.7}$$

But we also have

$$\dfrac{\dfrac{\dfrac{[[\phi,\ \bot],\ \bot] \qquad [\phi,\ \bot]^{\dagger}}{\bot}}{\phi}\ \text{NEG}_I}{[[\phi,\ \bot], \phi]}\ [\ ,\ ]\text{-INTR} \tag{11.8}$$

Here in the penultimate step, we rely on the intuitionistic absurdity rule NEG_I, i.e., *ex falso quodlibet*. Combining (11.8) and (11.7), in that order, we derive the rule of double negation elimination, and hence the classical absurdity rule, relying only on the intuitionistic absurdity rule, the elimination rule for disjunction, and the proposed definition of disjunction.

Therefore, to get an intuitionistic version of PFO, it appears that we need to fuse *existential* quantification and disjunction into one operator. Thus, we introduce the operator

$$\langle \phi, \psi \rangle$$

meaning that *for some* $x_1, \ldots, x_n$, ϕ *or* ψ (i.e., with a corresponding clause added to the PFO truth definition), and with the following introduction and elimination rules:

[3] The following observations are added in the present paper.

$$\langle\,,\rangle\text{-INTR}\quad \frac{\phi(t_1,\ldots,t_n)}{\langle\phi,\psi\rangle} \qquad \frac{\psi(t_1,\ldots,t_n)}{\langle\phi,\psi\rangle}$$

where $\{x_1,\ldots,x_n\} = Var_\phi \cap Var_\psi$ and $t_1,\ldots,t_n$ are closed terms.

$$\langle\,,\rangle\text{-ELIM}\quad \frac{\langle\phi,\psi\rangle \qquad \begin{matrix}\phi(t_1,\ldots,t_n)^\dagger \\ \vdots \\ \theta\end{matrix} \qquad \begin{matrix}\psi(t_1,\ldots,t_n)^\dagger \\ \vdots \\ \theta\end{matrix}}{\theta}$$

where $x_1,\ldots,x_n = Var_\phi \cap Var_\psi$ and $t_1,\ldots,t_n$ do not occur in ϕ, ψ or θ, nor in open assumptions in the derivation of θ, except possibly in $\phi(t_1,\ldots,t_n)$ and $\psi(t_1,\ldots,t_n)$.

Intuitionistic PFO, PFO_I, has a natural deduction system consisting of the introduction and elimination rules for [,], (,), ⟨ , ⟩ , the identity rules, and the intuitionistic absurdity rule,

$$\mathrm{NEG}_I \quad \frac{\bot}{\phi}$$

instead of the classical one. That this is really a system for intuitionistic first-order logic can be shown proof-theoretically. Let $\mathrel{|\!\!\!\frac{PI}{\quad}}$ be the derivability relation of PFO_I, and $\mathrel{|\!\!\!\frac{SI}{\quad}}$ the derivability relation for standard intuitionistic predicate logic. Extend the translations $^+$ and * of Sect. 11.3 so that, where $\{x_1,\ldots,x_n\} = (Var_\phi \cap Var_\psi) - X$,

$$\langle\phi,\psi\rangle^{+,X} = \exists x_1,\ldots,\exists x_n(\phi^{+,X\cup\{x_1,\ldots,x_n\}} \vee \psi^{+,X\cup\{x_1,\ldots,x_n\}})$$

and

$$(\phi \vee \psi)^* = \langle\phi^*,\psi^*\rangle.$$

Let $\phi \approx \psi$, where ϕ, ψ are PL_I-formulas, mean that ϕ is strict and results from ψ by bound variable changes and elimination of vacuous quantification, and let $\Gamma \approx \Delta$ mean that $\approx$ is a 1-1 relation between the sets Γ and Δ. Then it can be shown that for each derivation in the one system there is a corresponding derivation in the other system, in the following sense:

Proposition 11.10 (Pagin and Westerståhl (1993))

(a) If $\Gamma \mathrel{|\!\!\!\frac{PI}{\quad}} \phi$, *then* $\Gamma \mathrel{|\!\!\!\frac{SI}{\quad}} \phi^+$
(b) $\Gamma \mathrel{|\!\!\!\frac{SI}{\quad}} \phi$ *iff there are* $\Gamma_0 \approx \Gamma$ *and* $\phi_0 \approx \phi$ *such that* $\Gamma_0 \mathrel{|\!\!\!\frac{SI}{\quad}} \phi_0$
(c) If Γ, ϕ *are strict and* $\Gamma \mathrel{|\!\!\!\frac{SI}{\quad}} \phi$, *then* $\Gamma^* \mathrel{|\!\!\!\frac{PI}{\quad}} \phi^*$
(d) If $\Gamma^{+*} \mathrel{|\!\!\!\frac{PI}{\quad}} \phi^{+*}$, *then* $\Gamma \mathrel{|\!\!\!\frac{PI}{\quad}} \phi$

11.6 An Alternative Intuitionistic PFO?

It is now time to note an error of reasoning in Sect. 11.5. We noted that the proper disjunction operator for intuitionistic logic can not be defined as $[[\phi, \bot], \psi]$. From that we (tentatively) inferred that in the intuitionistic version of PFO, disjunction must be be fused with existential quantification. But that does not follow. It has not been ruled out that we *can* fuse disjunction and universal quantification, only that it must be given its own rules. Whether this can be done is what we shall investigate in this section. To this end we shall define a new operator $\langle\!\langle\, , \rangle\!\rangle$ with its own introduction and elimination rules. The operator $\langle\!\langle\, , \rangle\!\rangle$ is given by the following rules:

$$\langle\!\langle\, , \rangle\!\rangle\text{-INTR} \qquad \frac{\begin{array}{c}\vdots\\ \phi(t_1, \ldots, t_n)\end{array}}{\langle\!\langle \phi, \psi \rangle\!\rangle} \qquad \frac{\begin{array}{c}\vdots\\ \psi(t_1, \ldots, t_n)\end{array}}{\langle\!\langle \phi, \psi \rangle\!\rangle}$$

where $\{x_1, \ldots, x_n\} = Var_\phi \cap Var_\psi$ and $t_1, \ldots, t_n$ are closed terms and $t_1, \ldots, t_n$ do not occur in ϕ or in open assumptions in the derivation of $\phi(t_1, \ldots, t_n)$. Similarly for ψ in the second rule.

$$\langle\!\langle\, , \rangle\!\rangle\text{-ELIM} \qquad \frac{\langle\!\langle \phi, \psi \rangle\!\rangle \qquad \begin{array}{c}\phi(t_1, \ldots, t_n)^\dagger\\ \vdots\\ \theta\end{array} \qquad \begin{array}{c}\psi(t_1, \ldots, t_n)^\dagger\\ \vdots\\ \theta\end{array}}{\theta}$$

In this case, there are no restrictions on the choice of the terms $t_1, \ldots, t_n$, or their occurrence in the derivation, which reflects the *universal instantiation* aspect of the rule, and distinguishes $\langle\!\langle\, , \rangle\!\rangle$-ELIM from $\langle\, , \rangle$-ELIM.

It is clear that $\langle\!\langle\, , \rangle\!\rangle$ satisfies the Gentzen-Prawitz reduction requirement:

$$\frac{\dfrac{\begin{array}{c}\Pi_1\\ \phi(t_1, \ldots, t_n)\end{array}}{\langle\!\langle \phi, \psi \rangle\!\rangle} \qquad \begin{array}{c}\phi(t_1, \ldots, t_n)^\dagger\\ \Pi_2\\ \theta\end{array} \qquad \begin{array}{c}\psi(t_1, \ldots, t_n)^\dagger\\ \Pi_3\\ \theta\end{array}}{\theta} \quad \rhd \quad \begin{array}{c}\Pi_1\\ \phi(t_1, \ldots, t_n)\\ \Pi_2\\ \theta\end{array}$$

We shall use PI' to refer to the system of deduction containing the rules for $[\, ,]$, $(\, ,)$, and $\langle\!\langle\, , \rangle\!\rangle$, the identity rules, and the intuitionistic absurdity rule. We need then to ascertain that PI' really is Intuitionistic Predicate Logic. For that purpose we modify the translation functions $^+$ and * from Sects. 11.3 and 11.5 as follows:

Definition 11.12 For a PI′-formula ϕ and $X \subseteq Var$, define inductively the SI-formula $\phi^{+,X}$ by

(a) $\phi^{+,X} = \phi$, if ϕ is atomic,
(b) $[\phi, \psi]^{+,X} = \forall x_1, \ldots, x_n(\phi^{+,X\cup\{x_1,\ldots,x_n\}} \rightarrow \psi^{+,X\cup\{x_1,\ldots,x_n\}})$
(c) $(\phi, \psi)^{+,X} = \exists x_1, \ldots, \exists x_n(\phi^{+,X\cup\{x_1,\ldots,x_n\}} \wedge \psi^{+,X\cup\{x_1,\ldots,x_n\}})$
(d) $\langle\!\langle \phi, \psi \rangle\!\rangle^{+,X} = \forall x_1, \ldots, x_n(\phi^{+,X\cup\{x_1,\ldots,x_n\}} \vee \psi^{+,X\cup\{x_1,\ldots,x_n\}})$

where $x_1, \ldots, x_n$ are the elements of $Sbound_{[\phi,\psi]} - X$ (in some fixed order).

Definition 11.13 If ϕ is a SI-formula, define the PI′-formula ϕ^* inductively by:

(a) $\phi^* = \phi$, if ϕ is atomic
(b) $(\phi \rightarrow \psi)^* = [\phi^*, \psi^*]$
(c) $(\forall x\phi)^* = [x = x , \psi^*]$
(d) $(\exists x\phi)^* = (x = x, \phi^*)$
(e) $(\phi \vee \psi)^* = \langle\!\langle \phi^*, \psi^* \rangle\!\rangle$

Given these definitions, we can state the equivalence of PI′ and SI as the following theorem:

Theorem 11.11 *(a) If* $\Gamma \overset{PI'}{\vdash} \phi$, *then* $\Gamma \overset{SI}{\vdash} \phi^+$
(b) If Γ, ϕ *are strict and* $\Gamma \overset{SI}{\vdash} \phi$, *then* $\Gamma^* \overset{PI}{\vdash} \phi^*$
(c) If $\Gamma^{+*} \vdash_{PI'} \phi^{+*}$, *then* $\Gamma \vdash_{PI'} \phi$.

Proof The strategy is as in the proof of Proposition 11.10, given in outline in Pagin and Westerståhl (1993).

For (a) the proof proceeds by induction over the length of derivations. Here we shall carry out the induction steps for the $\langle\!\langle \, , \rangle\!\rangle$ rules. So assume first that we have a derivation in PI′ with an application of $\langle\!\langle \, , \rangle\!\rangle$ – INTR:

$$\begin{array}{c} \Gamma \\ \vdots \\ \dfrac{\phi(t_1, \ldots, t_n)}{\langle\!\langle \phi, \psi \rangle\!\rangle} \end{array}$$

where $t_1, \ldots, t_n$ do not occur in any formula in Γ. By this assumption, we also have $\Gamma \overset{PI'}{\vdash} \phi(t_1, \ldots, t_n)$, and therefore, by the induction hypothesis,

$$\Gamma^+ \overset{SI}{\vdash} \phi(t_1, \ldots, t_n)^+.$$

From $\phi(t_1, \ldots, t_n)^+$ way may infer in SI

$$\phi(t_1, \ldots, t_n)^+ \vee \psi(t_1, \ldots, t_n)^+$$

by disjunction introduction. Since $t_1, \ldots, t_n$ do not occur in Γ, they do not occur in Γ^+ either. So the conditions for universal introduction is met for all these terms. As noted in the proof of 11.10, we need the general fact that translation is preserved if we also remove marked variables when those variables are replaced by closed terms: for any FI′ formula $\xi = \xi(x_1, \ldots, x_n y_1, \ldots, y_m)$

$$\xi^{+,\{x_1,\ldots,x_n,y_1,\ldots,y_m\}}[t_1, \ldots, t_n/x_1, \ldots, x_n] = \xi(t_1, \ldots, t_n y_1, \ldots, y_m)^{+,\{y_1,\ldots,y_m\}} \tag{\#}$$

Then, in n steps of $\forall$-introduction we will have

$$\begin{aligned} &\forall x_1, \ldots, \forall x_n(\phi(t_1, \ldots, t_n)^+[x_1, \ldots, x_n/t_1, \ldots, t_n] \vee \\ &\psi^+(t_1, \ldots, t_n)[x_1, \ldots, x_n/t_1, \ldots, t_n]) = \\ &\forall x_1, \ldots, \forall x_n(\phi^{+,\{x_1,\ldots,x_n\}} \vee \psi^{+,\{x_1,\ldots,x_n\}}) = \langle\langle \phi, \psi \rangle\rangle^+ \end{aligned} \tag{$\star$}$$

using (#). Hence, $\Gamma^+ \vdash^{SI} \langle\langle \phi, \psi \rangle\rangle^+$.

Assume that the last step is an application of $\langle\langle\ ,\ \rangle\rangle$-ELIM:

$$\dfrac{\begin{matrix}\Gamma_1 \\ \vdots \\ \langle\langle \phi, \psi \rangle\rangle\end{matrix} \qquad \begin{matrix}\Gamma_2 \;\; \phi(t_1, \ldots, t_n)^\dagger \\ \vdots \\ \theta\end{matrix} \qquad \begin{matrix}\Gamma_3 \;\; \psi(t_1, \ldots, t_n)^\dagger \\ \vdots \\ \theta\end{matrix}}{\theta}$$

where $x_1, \ldots, x_n = Var_\phi \cap Var_\psi$. By the induction hypothesis, we have that

$\Gamma_1^{+,\varnothing} \vdash^{SI} \langle\langle \phi, \psi \rangle\rangle^{+,\varnothing}$,
$\Gamma_2^{+,\varnothing} \cup \{\phi(t_1, \ldots, t_n)^{+,\varnothing}\} \vdash^{SI} \theta^{+,\varnothing}$, and
$\Gamma_3^{+,\varnothing} \cup \{\psi(t_1, \ldots, t_n)^{+,\varnothing}\} \vdash^{SI} \theta^{+,\varnothing}$.

Since

$$\langle\langle \phi, \psi \rangle\rangle^{+,\varnothing} = \forall x_1, \ldots, \forall x_n(\phi^{+,\{x_1,\ldots,x_n\}} \vee \psi^{+,\{x_1,\ldots,x_n\}}),$$

we can perform universal instantiation on $x_1, \ldots, x_n$, and in n steps arrive at

$\Gamma_1^+ \vdash^{SI} \phi(t_1, \ldots, t_n)^+ \vee \psi(t_1, \ldots, t_n)^+$.

again applying (#) to take "unmark" a variable at each step. In virtue of this, the following derivation, ending with $\vee$-elimination, is valid in SI:

$$\dfrac{\begin{matrix}\Gamma_1^+ \\ \vdots \\ \phi(t_1, \ldots, t_n)^+ \vee \psi(t_1, \ldots, t_n)^+\end{matrix} \qquad \begin{matrix}\Gamma_2^+ \;\; \phi(t_1, \ldots, t_n)^{+\dagger} \\ \vdots \\ \theta^+\end{matrix} \qquad \begin{matrix}\Gamma_3^+ \;\; \psi(t_1, \ldots, t_n)^{+\dagger} \\ \vdots \\ \theta^+\end{matrix}}{\theta^+}$$

Therefore, $\Gamma_1^+ \cup \Gamma_2^+ \cup \Gamma_3^+ \mid\!\!\overset{SI}{\longrightarrow} \theta^+$. This completes the induction.

For (b), we again perform induction over the length of derivations. Assume that $\Gamma \overset{SI}{\vdash} \phi$. We may assume that ϕ and all formulas in Γ are strict (cf. Proposition 11.10b). If there is a derivation of ϕ from Γ in SI, then there is a *normal* derivation of ϕ from Γ in SI (in the sense of Prawitz, 1965, ch. IV). By the Subformula Property of normal derivations, each formula in this derivation is a subformula of either ϕ or some formula in Γ. Since ϕ and Γ are strict, it follows that each formula in the derivation is strict. We show by induction over the length of the normal derivation that if $\Gamma \overset{SI}{\vdash} \phi$, then $\Gamma^* \overset{PI'}{\vdash} \phi^*$.

Here, we perform the $\vee$-elimination step. It is the mirror image of the $\langle\!\langle\, , \rangle\!\rangle$-ELIM step in (a). We have in SI

$$\dfrac{\begin{array}{c}\Gamma_1\\ \vdots \\ \phi \vee \psi\end{array} \qquad \begin{array}{c}\Gamma_2 \;\; \phi^\dagger\\ \vdots \\ \theta\end{array} \qquad \begin{array}{c}\Gamma_3 \;\; \psi^\dagger\\ \vdots \\ \theta\end{array}}{\theta}$$

Using the induction hypothesis, in analogy to the proof of the other direction, and the fact that $(\phi \vee \psi)^* = \langle\!\langle \phi^*, \psi^* \rangle\!\rangle$, we arrive at the following derivation in PI$'$.

$$\dfrac{\begin{array}{c}\Gamma_1^*\\ \vdots \\ \langle\!\langle \phi^*, \psi^* \rangle\!\rangle\end{array} \qquad \begin{array}{c}\Gamma_2^* \;\; \phi^{*\dagger}\\ \vdots \\ \theta^*\end{array} \qquad \begin{array}{c}\Gamma_3^* \;\; \psi^{*\dagger}\\ \vdots \\ \theta^*\end{array}}{\theta^*}$$

By the strictness assumption, ϕ^* and ψ^* have no variables in common, and so no variables become surface bound in $\langle\!\langle \phi^*, \psi^* \rangle\!\rangle$. Hence, $\Gamma_1^* \cup \Gamma_2^* \cup \Gamma_3^* \overset{PI'}{\vdash} \theta^*$. This completes the induction.

(c) One proves by induction over the complexity of ϕ that $\phi^{+*} \overset{PI'}{\vdash} \phi$ and that $\phi \overset{PI'}{\vdash} \phi^{+*}$.

This shows that we may well fuse conjunction and universal quantification in an alternative intuitionistic PFO. So in this, respect, intuitionistic logic is not different from classical logic.

11.7 Defining Disjunction in Classical PFO

We saw in the beginning of Sect. 11.5 that the disjunction operator { , } which is definable in classical PFO, is not intuitionistically acceptable. For disjunction, whether fused with existential or with universal quantification, primitive rules are needed. But the operator $\langle\ ,\ \rangle$ of Sect. 11.5, which fuses disjunction and existential quantification, can be added to classical logic as well. If $\langle\ ,\ \rangle$ is classically but not intuitionistically definable, we have discovered a new difference between the two logics.

As observed in Pagin and Westerståhl (1993), for every formula $\langle\phi, \psi\rangle$, where ϕ and ψ are PFO formulas, and where $\{x_1, \ldots, x_n\} = \mathit{Var}_\phi \cap \mathit{Var}_\psi$, there is an equivalent PFO formula:

$$(x_1 = x_1, (\ldots, (x_n = x_n, [[\phi, \bot], \psi]))). \tag{11.9}$$

Each $x_i = x_i$ conjunct has the effect of both introducing existential quantification, jointly with $[[\phi, \bot], \psi]$, and *cancelling* the universal quantification in $[[\phi, \bot], \psi]$ itself. Since there is one such subformula for each variable in $\mathit{Var}_\phi \cap \mathit{Var}_\psi$, the subformula $[[\phi, \bot], \psi]$ reduces to a conditional, and since the antecedent is negated, a conditional that is equivalent to $\phi \vee \psi$. That this works is immediate from the inspection of an example.

This method does not, however, provide a *uniform* definition of $\langle\cdot, \cdot\rangle$ in PFO. If in (11.9) we replace the surface bound variables in the subformula $[[\phi, \bot], \psi]$ (only) by $y_1, \ldots, y_n$, the result is no longer equivalent to $\langle\phi, \psi\rangle$.

Definition 11.14 An interpreted operator $\dot{[}\ ,\ \dot{]}$ is *uniformly definable* in PFO iff there is a PFO schema $C(A, B)$, where C is a PFO context and A, B are schematic letters, such that

(a) For any PFO formulas ϕ and ψ, $C(\phi, \psi)$ is a PFO formula, and
(b) For any model $\mathcal{M}$, assignment f and variable set X, and for any PFO formulas ϕ, ψ

$$\mathcal{M}, X \models_f C(\phi, \psi) \quad \text{iff} \quad \mathcal{M}, X \models_f \dot{[}\phi, \psi)\dot{]}$$

It was conjectured in Pagin and Westerståhl (1993, note 16,) that $\langle\cdot, \cdot\rangle$ is not uniformly definable in PFO, in this sense. It turns out, however, that this conjecture was false, and that $\langle\cdot, \cdot\rangle$ can be defined as follows:

$$\langle\phi, \psi\rangle \overset{\text{def}}{=} [\ [[[\phi, \bot], \psi]\ ,\ [[[\phi, \bot], \psi], \bot]],\ \bot\ ]$$

To verify the definition, we translate the right-hand, abbreviated ξ, side into PL. Let $\mathit{Var}_\phi \cap \mathit{Var}_\psi = X = \{x_1, \ldots, x_n\}$. We have

$$\begin{aligned}
\xi^{+} &= [\,[[[\phi,\bot],\psi]\,,\,[[[\phi,\bot],\psi],\bot]],\ \bot\,]^{+} \\
&= \neg([\,[[\phi,\bot],\psi]\,,\,[[[\phi,\bot],\psi],\bot]\,])^{+} \\
&= \neg\forall x_1,\ldots,\forall x_n([[\phi,\bot],\psi]^{+,X} \to [\,[[\phi,\bot],\psi],\ \bot\,]^{+,X}) \\
&= \neg\forall x_1,\ldots,\forall x_n((\neg\phi^{+,X}) \to \psi^{+,X}) \to \neg(\neg\phi^{+,X} \to \psi^{+,X}))) \\
&\equiv \exists x_1,\ldots,\exists x_n((\phi^{+,X} \vee \psi^{+,X}) \wedge \neg\neg(\phi^{+,X} \vee \psi^{+,X})) \\
&\equiv \exists x_1,\ldots,\exists x_n(\phi^{+,X} \vee \psi^{+,X})
\end{aligned}$$

after PL transformations in the last two steps.

There is, however, a weaker version of the conjecture which is true: there is no uniform definition of $\langle\phi,\psi\rangle$ in PFO having a *single* occurrence each of ϕ and ψ. We shall now provide a proof of this. The theorem builds on a lemma concerning occurrences of formulas and variables and a lemma concerning an hereditary property of occurrences.

Definition 11.15 A formula ϕ has *positive* occurrence in a context $C(A)$ iff for any model $\mathcal{M}$, assignment f and variable set $X \supseteq Var_{\phi}$, $\mathcal{M}, X \models_{f} C(\phi)$ follows either from $\mathcal{M}, X \models_{f} \phi$ or from $\mathcal{M}, X \models_{f} \neg\phi$.

A formula ϕ has *negative* occurrence in a context $C(A)$ iff for any model $\mathcal{M}$, assignment f and variable set $X \supseteq Var_{\phi}$, $\mathcal{M}, X \models_{f} \neg C(\phi)$ follows either from $\mathcal{M}, X \models_{f} \phi$ or from $\mathcal{M}, X \models_{f} \neg\phi$.

An occurrence that is neither positive nor negative is said to be *neutral.*

By the corresponding definition for PL, disjuncts have positive occurrences in disjunctions, and so have negated disjuncts, such as $\neg\phi$ in $(\neg\phi) \vee \psi$. Conjuncts have negative occurrences in conjunctions, and so have negated conjuncts.

Definition 11.16 An occurrence of a variable x in a PFO formula ϕ is *universal* iff the prenex normal form of ϕ^{+} is $\ldots\forall x\ldots(\phi^{+,Var_{\phi}})$. An occurrence of a variable x in a PFO formula ϕ is *existential* iff the prenex normal form of ϕ^{+} is $\ldots\exists x\ldots(\phi^{+,Var_{\phi}})$.

Thus, x in $[Fx, Gx]$ has universal occurrence, since the prenex normal form of $[Fx, Gx]^{+}$ is $\forall x(Fx \to Gx)$. Similarly, x in $[[Fx, Gx], \bot]$ has existential occurrence, since the prenex normal form of $[[Fx, Gx], \bot]^{+}$ is $\exists x(\neg(Fx \to Gx))$. For brevity I shall say "is positive", "is universal", etc.

Lemma 11.12 *Let ξ be a PFO formula, $[\phi,\psi]$ a subformula of ξ, and x a variable that is* quantified *in $[\phi,\psi]$ in ξ. Then if ϕ and ψ are* positive *in ξ, x is* universal *in ξ, and if ϕ and ψ are* negative *in ξ, then x is* existential *in ξ.*

Proof We prove the lemma by induction over formula complexity, where the complexity of a formula ψ is that number of occurrences of [,] in ψ. Atomic formulas thus have complexity 0. The proposition is trivially true for atomic formulas, since no variables are bound in (unembedded) atomic formulas. So assume that the proposition is true of formulas of complexity up to k.

Then let ξ be a PFO formula of complexity $k+1$, $[\phi, \psi]$ a subformula of ξ and x a variable that is quantified in $[\phi, \psi]$. We have three cases: i) $[\phi, \psi] = \xi$, or $\xi = [\xi_1, \xi_2]$, and (ii) $[\phi, \psi]$ is a subformula of ξ_1, or (iii) $[\phi, \psi]$ is a subformula of ξ_2.

In case (i) it is immediate from the semantics that both ϕ and ψ are positive in ξ, and that x is universal in ξ.

In case (ii), neither ϕ nor ψ can be negative in ξ if the truth value of ξ_2 depends on the choice of model. In order to falsify ξ, ξ_1 must be made true and ξ_2 false. From the assumption that x is quantified in $[\phi, \psi]$, which is a subformula of ξ_1, it follows that x occurs both in ϕ and in ψ, and that neither ϕ nor ψ occurs in ξ_2, since if either of them occurred there, x would be quantified (and hence surface bound) in ξ itself. Since neither ϕ nor ψ occurs in ξ_2, it is impossible for a truth value of either to falsity ξ.

But if x_2 is $\bot$, or some other logically false formula not containing x, ϕ and ψ may be negative in ξ. If so, they are positive in ξ_1. By the induction hypothesis, x is then universal in ξ_1. Where $Y = \{x_1, \ldots, x_n\}$, we will then have

$$\xi^+ = \forall x_1, \ldots, \forall x_n(\ldots, \forall x, \ldots, (\xi_1^{+,X\cup Y}) \rightarrow \xi_2^{+,Y})$$

where X is $Sbound_{\xi_1}$ (we have taken account of variables Sbound in ξ_1 but not of those surface bound in ξ_2). Since the prenex normal form of $\forall x(Ax) \rightarrow p$ is $\exists x(Ax \rightarrow p)$, we get as prenex normal form

$$\xi^+ = \forall x_1, \ldots, \forall x_n, Q_{n+1}x_{n+1}, \ldots, \exists x, \ldots, Q_{n+m}x_{n+m}(\xi_1^{+,X\cup Y} \rightarrow \xi_2^{+,Y})$$

where the $\{x_{n+1}, \ldots, x_{n+j}, x, x_{n+j+1}, \ldots, x_{n+m}\} = X$. The Q_i are appropriate first-order quantifiers. Since x is existentially quantified in the prenex normal form of ξ^+, it is existential in ξ.

It may be that neither ϕ nor ψ is positive in ξ (if $\xi_1 = [\phi, \psi]$, then ϕ must be true and ψ false in order to render ξ_1 false and thereby ξ true). It may be that one or both are positive in ξ.

Assume that both ϕ and ψ are positive in ξ. Then both ϕ and ψ must be negative in ξ_1, since only the falsity of ξ_1 guarantees the truth of ξ. By the induction hypothesis, x is existential in ξ_1. That is,

$$\xi_1^+ = \ldots \exists x \ldots (\xi^{+,X})$$

where X is $Sbound_{\xi_1}$. Then

$$\xi^+ = \forall x_1, \ldots, \forall x_n(\ldots \exists x \ldots (\xi_1^{+,X\cup Y} \rightarrow \xi_2^{+,Y})),$$

where $Y = \{x_1, \ldots, x_n\}$ is $Sbound_{\xi}$. Since x does not occur in ξ_2, the prenex normal form of ξ^+ will be of the format

$$\forall x_1, \ldots, \forall x_n, Q_{n+1}x_{n+1}, \ldots, \forall x, \ldots Q_{n+m}x_{n+m}(\xi_1^{+,X\cup Y} \rightarrow \xi_2^{+,Y})$$

where the $\{x_{n+1}, \ldots, x_{n+j}, x, x_{n+j+1}, \ldots, x_{n+m}\} = X$. The Q_i are appropriate first-order quantifiers. Since x is universally quantified in the prenex normal form of ξ^+, it is universal in ξ.

That completes case (ii). Case (iii) is analogous.

We will also need an observation about positive occurrences. Consider the example

$$[[\phi, \psi], \psi] \tag{11.10}$$

Since $\phi \models [[\phi, \psi], \psi]$, ϕ is positive in (11.10). The position itself of ϕ in (11.10) is not enough to guarantee the truth of (11.10), but together with the distribution of other subformulas it is. If ψ is true, (11.10) is true, and so ψ is positive in (11.10). If ψ is false, $[\phi, \psi]$ is false (since ϕ is assumed true), and hence again (11.10) is true. Since (11.10) is true in case the antecedent is false, $[\phi, \psi]$ is positive in (11.10) as well.

This holds generally: if a subformula ξ of ψ has a subformula ϕ that is positive in ψ, then ξ is positive in ψ as well. This is the topic of the following observation.

Lemma 11.13 *Let $C(A)$ be a contingent PFO sentence schema. That is, there is a sentence ϕ such that neither $\models C(\phi/A)$, nor $\models [C(\phi/A), \bot]$. Let A have a single occurrence in $C(A)$, and let $C'(A)$ be a sub-schema of C. If it holds for any PFO formula ϕ that ϕ is positive in $C(\phi/A)$, then it also holds for any ϕ that $C'(\phi/A)$ is positive in $C(\phi/A)$.*

Proof It is enough to prove the claim for immediate sub-contexts, where

$$C'(A) = [A, \psi], \quad \text{or} \quad C'(A) = [\psi, A].$$

The Lemma then follows by induction over the depth of embedding.

We consider the case where $C'(A) = [A, \psi]$. The other case is symmetric. Assume that the Lemma is false. Then there is a formula ϕ that is positive in $C(\phi/A)$ (from now on abbreviated to C), while $[\phi, \psi]$ is not. We may choose ϕ so that it is logically independent of ψ and of any formula ξ outside $[\phi, \psi]$ in C.

Let $V(\phi)$ be the true-making value of ϕ. It is truth if $\phi \models C$ and falsity if $\neg\phi \models C$.

By assumption there is a model $\mathcal{M}$ where C is false. Hence, in $\mathcal{M}$, no subformula ξ outside $[\phi, \psi]$ that is positive in C has its true-making value in $\mathcal{M}$. Call models with this property λ models. Again, by the assumption of logical independence of ϕ from other subformulas of C, there is another λ model $\mathcal{M}'$ which is like $\mathcal{M}$ except that ϕ has $V(\phi)$ in $\mathcal{M}'$. Hence, also $\mathcal{M}' \models C$.

Now, if $V(\phi)$ is falsity, then $\mathcal{M}' \models [\phi, \psi]$. Since $\mathcal{M}'$ is a λ model, the truth of C in $\mathcal{M}'$ is not determined by subformulas ξ of C outside $[\phi, \psi]$, and so the truth of C in $\mathcal{M}$ must depend on the truth of $[\phi, \psi]$ in $\mathcal{M}'$, and likewise in all λ models where ϕ is false. Hence, $[\phi, \psi]$ is positive in C.

If $V(\phi)$ is truth, then either ψ is true in $\mathcal{M}'$, so that $\mathcal{M}' \models [\phi, \psi]$, or else ψ is false in $\mathcal{M}'$, so that $\mathcal{M}' \models [[\phi, \psi], \bot]$. If it always the one or always the other, in every λ model where ϕ is true, then again $[\phi, \psi]$ is positive in C.

Could it instead be that ψ is true is true in some λ models where ϕ is true and false in others? In that case, $[\phi, \psi]$ is not positive in C, but depends on the truth value of some other subformula C_2? Since ϕ is positive in C it must then be the case that there is a subformula

$$B = [C_1(\phi, \psi), C_2] \quad \text{or} \quad D = [C_2, C_1(\phi, \psi)]$$

which is positive in C, whose value does not depend on the value of ψ, and such that $C_1(\phi, \psi)$ is *not* positive in C. But this cannot be. For, in the case of B, if the true-making value of B is truth, then C_2 is positive in C, contradicting the assumption that $\mathcal{M}'$ is a λ model. And if the true-making value of B is falsity, then $C_1(\phi, \psi)$ is positive in C, contradicting the assumption about C_1. The D case is analogous.

Theorem 11.14 *$\langle \phi, \psi \rangle$ is not uniformly definable in classical PFO with a single occurrence of ϕ and ψ.*

Proof Suppose $C(A, B)$ is a PFO schema such that for any PFO formulas ϕ, ψ, $C(\phi, \psi)$ has one occurrence each of ϕ and ψ, and is equivalent to $\langle \phi, \psi \rangle$. Then there must be a subformula

$$\xi = [C_1(\phi/\psi), C_2(\psi/\phi)]$$

where the variables $x_1, \ldots, x_n$ shared between ϕ and ψ get *quantified*. (The slash notation here means that ϕ xor ψ occurs.) For assume that x_i occurs outside ξ in $C(\phi, \psi)$. Substitute y_i for x_i in ϕ and ψ, giving us ϕ' and ψ', respectively. The interpretation of $\langle \phi', \psi' \rangle$, as unembedded, is the same as that of $\langle \phi, \psi \rangle$ (a change of bound variables), but the interpretation of $C(\phi', \psi')$, as unembedded, is not the same as that of $C(\phi, \psi)$ (since only some of the bound variables were replaced). Hence, we do not have a uniform definition.

Let $x \in Var_\phi \cap Var_\psi$. For the definition to be adequate, x must by be existential in $C(\phi, \psi)$. Again, for the definition to be adequate, ϕ and ψ must be positive in ξ. Assume that ϕ occurs in C_1 and ψ in C_2 (the other case is symmetric).

From the assumption of adequacy, by Lemma 11.13, both $C_1(\phi)$ and $C_2(\psi)$ are positive in $C(\phi, \psi)$. By Lemma 11.12, x is then *universal* in $C(\phi, \psi)$. But since adequacy requires x to be *existential* in $C(\phi, \psi)$, the definition cannot be adequate.

The basic reason why the one-occurrence kind of uniform definition does not always work is precisely the fusion of connectives and quantifiers that defines PFO: negate an operator and you switch *both* the connective and the quantifier at the same time. To compensate for this, multiple occurrences are needed, under different negation embeddings, so that the negation effect on the quantifiers and on the connectives can be separated.

Still, since $\langle\ ,\ \rangle$ is uniformly definable in classical PFO, this is a respect in which classical and intuitionistic logic do differ.

References

Geach, P. T. (1962). *Reference and generality*. Ithaca: Cornell University Press.

Groenendijk, J., & Stokhof, M. (1991). Dynamic predicate logic. *Linguistics and Philosophy*, *14*, 39–100.

Pagin, P., & Westerståhl, D. (1993). Predicate logic with flexibly binding operators and natural language semantics. *Journal of Logic, Language and Information*, *2*, 89–128.

Pagin, P & Westerståhl, D. (1994). Flexible variable-binding and Montague grammar. In P. Dekker & M. Stokhof (Eds.), *Proceedings of the ninth Amsterdam Colloquium*(pp. 519–525)

Prawitz, D. (1965). *Natural deduction. A proof-theoretic study*. Stockholm: Almqvist and Wiksell International. Republished by Dover Publications, Mineola, NY, 2006.

Chapter 12
On Constructive Fragments of Classical Logic

Luiz Carlos Pereira and Edward Hermann Haeusler

Abstract In the late twenties and early thirties of the last century several results were obtained concerning relations between classical logic (CL) and intuitionistic logic. Glivenko, Kolmogorov, Gödel, Gentzen and Kuroda, this last appeared in 1950, provided well-known interpretations of classical logic into intuitionistic logic, in this way transferring constructive aspects to the fragments on which these interpretations are based. The aim of the present paper is to investigate the constructive behavior of other fragments of CL and of fragments of classical S4. We shall be mainly concerned with the fragments $\{\neg, \wedge, \bot, \forall\}$, $\{\neg, \wedge, \bot, \exists\}$, $\{\rightarrow\}$, $\{\neg, \wedge, \bot, \diamond\}$, and $\{\neg, \wedge, \bot, \Box\}$. Our general approach will be exclusively proof-theoretical.

Keywords Classical logic · Constructive interpretations · Proof theory · Intuitionistic logic

12.1 Introduction

In the late twenties and early thirties of last century several results were obtained concerning some relations between classical logic (CL) and intuitionistic logic (IL), and between classical arithmetic (PA) and intutionistic arithmetic (HA). In 1927 Glivenko proved two important results relating classical propositional logic (CPL) to intuitionistic propositional logic (IPL). Glivenko's first result shows that A is a theorem of CPL iff $\neg\neg A$ is a theorem of IPL. His second result establishes that we

Luiz Carlos Pereira would like to thank CNPq and FAPERJ/PRONEX for their financial support. Edward Hermann Haeusler would like to thank CNPq Universal 483460/2011-7. Both authors would like to thank to profs. Paulo Augusto Veloso and Bruno Lopes Vieira.

L.C. Pereira (✉)
Departamento de Filosofia, PUC-Rio, Rio de Janeiro, Brazil
e-mail: luiz@inf.puc-rio.br

E.H. Haeusler
Departamento de Informática, PUC-Rio, Rio de Janeiro, Brazil
e-mail: hermann@inf.puc-rio.br

H. Wansing (ed.), *Dag Prawitz on Proofs and Meaning*, Outstanding Contributions to Logic 7, DOI 10.1007/978-3-319-11041-7_12

cannot distinguish CPL from IPL with respect to theorems of the form $\neg A$. In 1925 Kolmogorov proved that CPL could be translated into IPL (see Kolmogorov (1925)). In 1933 Gödel defined an interpretation of PA into HA (see Gödel (1933)), and in the same year Gentzen defined a new interpretation of PA into HA (see Gentzen (1974)). These interpretations/translations/embeddings were defined as functions from the language of PA into some fragment of the language of the HA that preserve some important properties, like theoremhood. In fact, Gentzen's and Gödel's results encapsulate a stronger result. Let us call a formula A stable iff $\vdash_{IL} A \leftrightarrow \neg\neg A$, and let us call a theory T atomically stable iff every atomic formula in T is stable.

Theorem: Let T be any classical first order theory formulated in the fragment $\{\neg, \forall, \wedge\}$. If T is atomically stable then every theorem of T is also an intuitionistic theorem.

The interpretations defined by Gödel and Gentzen differ only with respect to the fragment of the language of HA: Gödel interprets the implication sign "$\rightarrow$" in terms of "$\neg$" and "$\wedge$", while Gentzen keeps it in the image language. This small syntactical difference has an important consequence, since the interpretation of implication in terms of negation and conjunction allows Gödel to obtain the following nice result as a preparatory step for the definition of his interpretation function (see Gödel (1933)):

Theorem: Let A be a formula in the fragment $\{\neg, \wedge\}$. Then $\vdash_{CL} A \Leftrightarrow \vdash_{IL} A$.

Proof Every theorem A in the fragment $\{\neg, \wedge\}$ has a cannonical form: $\exists B_1, \ldots, B_k$ such that $A \leftrightarrow \neg B_1 \wedge \ldots \wedge \neg B_k$. The result then follows directly from Glivenko's theorem for classical propositional logic.

The immediate effect of this result is that the fragment $\{\neg, \wedge\}$ is insufficient to distinguish the class of classical propositional theorems from the class of intuitionistic propositional theorems. A nice way to put Gödel's result is: we can do classical propositional logic without classical logic!

The aim of the present paper is to investigate the constructive behavior of other fragments of CL and of fragments of classical S4. We shall be mainly concerned with the fragments $\{\neg, \wedge, \bot, \forall\}$, $\{\neg, \wedge, \bot, \exists\}$, $\{\rightarrow\}$, $\{\neg, \wedge, \bot, \diamond\}$, and $\{\neg, \wedge, \bot, \Box\}$. Our general approach will be exclusively proof-theoretical. For a very preliminary portuguese version of some of the results of this article see Pereira (2008).

12.2 Some Proof-Theoretical Results

Before we start discussing some fragments of classical logic, let us recapitulate some basic proof-theoretical results.

In 1965 Prawitz proved the normalization theorem for classical first order logic (see Prawitz (1965)). Prawitz normalization strategy can be roughly described as follows:

1. Restrict the language of classical first order logic to the fragment $\{\neg, \wedge, \rightarrow, \bot, \forall\}$. Of course nothing is lost with this restriction.

2. Reduce all applications of the classical absurd rule to atomic applications, i.e., to applications with atomic conclusions.
3. Apply your favorite normalization strategy for intuitionistic first order logic.

The moral behind Prawitz' normalization strategy is: if we need classical reasoning at all, we need it just at the atomic level (very small sins!).

Another normalization strategy (not so well-known as Prawitz') is due to Jonathan Seldin (see Seldin (1986, 1989)). This strategy consists of the following steps:

1. Restrict the language of classical first order logic to the fragment $\{\neg, \wedge, \vee, \rightarrow, \bot, \exists\}$. Again, nothing is lost with this restriction.
2. Show that every derivation Π of $\Gamma \vdash A$ in the fragment can be transformed into a derivation Π' of $\Gamma \vdash A$ such that Π' contains at most one application of the classical absurd rule, and in case this application does occur, it is the last rule applied in Π'.
3. Apply your favorite normalization strategy for intuitionistic first order logic.

The moral behind Seldin's normalization strategy is: if we need classical reasoning at all, we need it just once (just one last sin—maybe a big one!).

Glivenko's theorems and Gödel's theorem for the fragment $\{\neg, \wedge\}$ of CPL are trivial consequences of Seldin's normalization strategy. In fact, this strategy allows a nice extension of Glivenko's theorems to first order logic.

Theorem 12.1 *(Glivenko) Let $\neg A$ be a classical theorem in the fragment $\{\neg, \wedge, \vee, \rightarrow, \bot, \exists\}$. It follows that $\neg A$ is an intuitionistic theorem.*

12.3 The Fragment $\{\neg, \wedge, \bot, \forall\}$

If we add the universal quantifier to the fragment $\{\neg, \wedge\}$, then we can obviously distinguish classical logic from intuitionistic logic: the formula

$$\neg(\forall x \neg(\neg P(x) \wedge \neg Q(x)) \wedge \forall x \neg Q(x) \wedge \forall x P(x))$$

which is a form of the disjunctive syllogism in the fragment $\{\neg, \wedge, \bot, \forall\}$, is a classical theorem, but not an intuitionistic one. Although the fragment $\{\neg, \wedge, \forall\}$ is not "intuitionistic", there are several nice constructive results that can be obtained for it. Our first result shows that negation is constructively "involutive" with respect to theorems, that is, if $\neg\neg A$ is an intuitionistic theorem in the fragment, then A is also an intuitionistic theorem.

Theorem 12.2 *(Involution) If $\neg\neg A$ is an intuitionistic theorem in the fragment $\{\neg, \wedge, \bot, \forall\}$, then A is also a theorem in the same fragment.*

Proof By induction on A. Basis: Trivial. Induction step:

1. A is $\neg B$. Directly from the fact that $\vdash_{IL} \neg\neg\neg B \rightarrow \neg B$.

2. A is $(B \wedge C)$. The result follows directly from the fact that

$$\vdash_{IL} (\neg\neg(B \wedge C) \rightarrow (\neg\neg B \wedge \neg\neg C))$$

and the induction hypothesis.
3. A is $\forall x B(x)$. The result follows directly from the fact that

$$\vdash_{IL} (\neg\neg\forall x B(x) \rightarrow \forall x \neg\neg B(x))$$

and the induction hypothesis.

Theorem 12.3 *If* $\forall x \neg\neg A(x)$ *is an intuitionistic theorem in the fragment* $\{\neg, \wedge, \bot, \forall\}$, *then* $\neg\neg\forall x A(x)$ *is also a theorem in the same fragment.*

Proof $\vdash_{IL} \forall x \neg\neg A(x)$ implies $\vdash_{IL} \neg\neg A(x)$. By Theorem 12.2 we have $\vdash_{IL} A(x)$ and by $\forall$-introduction $\vdash_{IL} \forall x A(x)$, finally $\vdash_{IL} \neg\neg\forall x A(x)$.

A restricted form of Glivenko's second theorem can be obtained for the fragment $\{\neg, \wedge, \bot, \forall\}$.

Theorem 12.4 *Let* $A(x)$ *be a formula in the fragment* $\{\neg, \wedge\}$. *If* $\vdash_{CL} \neg\forall x A(x)$, *then* $\vdash_{IL} \neg\forall x A(x)$.

Proof By Seldin's normalization strategy we known that from any proof Π of $\exists x \neg A(x)$ in classical logic, there is a proof $\exists x \neg A(x)$ with at most one application of $\bot$-classical rule, the last one. So, if $\exists x \neg A(x)$ is a classical theorem, then there is an intuitionistic derivation of $\bot$ from $\neg\exists x \neg A(x)$. From $\neg\forall x A(x) \vdash_{CL} \exists x \neg A(x)$ and the fact that $\vdash_{CL} \neg\forall x A(x)$, there is an intuitionistic derivation of $\bot$ from $\neg\exists x \neg A(x)$. Since there is an intuitionistic derivation of $\neg\exists x \neg A(x)$ from $\forall x A(x)$, there is a an intuitionistic derivation of $\bot$ from $\forall x A(x)$, and hence there is an intuitionistic proof of $\neg\forall x A(x)$.

12.4 The Fragment $\{\neg, \wedge, \bot, \exists\}$

As in the case of the fragment $\{\neg, \wedge, \bot, \forall\}$, we can easily show that the fragment is not intuitionistic, a nice example being $\exists x \neg(\neg P(x) \wedge \exists x P(x))$. However, the fragment $\{\neg, \wedge, \bot, \exists\}$ is interesting as the following results show.

Theorem 12.5 *Let* $A(x)$ *be quantifier free. Then, if* $\vdash_{CL} \exists x A(x)$, *then* $\vdash_{IL} \exists x A(x)$.

Proof By induction on A. Basis: A is atomic. Trivial.
Induction hypothesis:

1. A is $(B \wedge C)$. Directly from the induction hypothesis and

$$\vdash_{IL} \exists x(B(x) \wedge C(x)) \leftrightarrow (\exists x B(x) \wedge \exists x C(x))$$

2. A is $\neg B(x)$. We know that $\vdash_{CL} \exists x \neg B(x)$ implies $\vdash_{CL} \neg \forall x B(x)$, and hence by Theorem 12.4, $\vdash_{IL} \neg \forall x B(x)$. We have that $\forall x B(x) \vdash_{Int} \bot$. From this fact and the form of a normal derivation of $\bot$ from $\forall x B(x)$, we know that there are $a_1, \ldots, a_n$, such that, $B(a_1), \ldots, B(a_n) \vdash_{IL} \bot$, and hence, $\exists x B(x) \vdash_{IL} \bot$, by a series of $\exists - Elim$ applications. This entails that $\vdash_{CL} \forall x \neg B(x)$, since $\neg \exists x B(x)$ is classically equivalent to $\forall x \neg B(x)$. From the fact that $\forall x \neg B(x) \vdash_{CL} \neg B(x)$, we have that $\vdash_{CL} \neg B(x)$, and hence $\vdash_{IL} \neg B(x)$ from which we can finally have $\vdash_{IL} \exists x \neg B(x)$.

Theorem 12.6 *Let A be a sentence in the fragment* $\{\neg, \wedge, \bot, \exists\}$ *such that no quantifier occurs in the scope of any quantifier. Then, if* $\vdash_{CL} A$, *then* $\vdash_{IL} A$.

Proof By induction on A. Basis: A is atomic. Trivial.
Induction step:

1. A is $\neg B$ for some B. The result follows directly from Glivenko's theorem for the fragment $\{\neg, \wedge, \bot, \exists\}$.
2. A is $(B \wedge C)$, for some B and C. The result follows directly from the induction hypothesis.
3. A is $\exists x B(x)$, the result is obtained directly from Theorem 12.5.

The moral behind this result is: no classical logic without the iteration of quantifiers!

12.5 The Fragment {→}

The fragment $\{\rightarrow\}$ is very interesting because, as it is well-known, it is sufficient to distinguish classical propositional logic from its intuitionistic counterpart. The obvious example of a theorem classically proved, but not intuitionistically proved is Peirce's formula $(((A \rightarrow B) \rightarrow A) \rightarrow A)$. It is obvious that the system consisting only of introduction and elimination rules for $\rightarrow$ is not classically complete (not even intuitionistically complete). A complete natural deduction system for classical implicational logic can be obtained through the addition of a rule corresponding to Peirce's formula to the usual rules for implication.

Let us call the rule below P-rule (Peirce's rule)

$$\begin{array}{c} [(A \rightarrow B)] \\ | \\ \dfrac{A}{A} \end{array}$$

The natural deduction system obtained through the addition of the P-rule to the rules for implication satisfies Prawitz' normalization strategy, which means that every application of the P-rule can be reduced to an atomic application, i.e., an application with atomic conclusion (see Zimmermann (2002)). It also satisfies a form of Seldin's strategy (see Pereira et al. (2010)). From Seldin's strategy we can proof the following version of Glivenko's theorem for the implicational fragment of classical logic. We use $\vdash_{NI-Imp} A$ to denote that A is a theorem of the implicational fragment of intuitionistic logic.

Theorem 12.7 *(Glivenko) Let A be a formula in the implicational fragment and let* $\{p_1, \ldots, p_n\}$ *be the set of atomic formulae occurring in A. Then,*$\vdash_{CL} A$ *if and only if*$\vdash_{NI-Imp} (A \to p_1) \to ((A \to p_2) \ldots ((A \to p_n) \to A) \ldots)$.

Glivenko for the implicational fragment gives us a translation based on implication, which is quite natural, given that the actual point of disagreement depends on implication.

12.6 The Modal Case

Given that the main tools we use depend on proof-theoretical results, let us start with some proof-theory for modal logic. Can we obtain for classical S4 the same proof-theoretical results already obtained for classical propositional logic? Do we have Prawitz' normalization strategy? Do we have Seldin's normalization strategy? Consider the fragment $C'_{S4} = \{\neg, \wedge, \to, \bot, \Box\}$ (as usual, nothing is lost!). Can we show that every application of $\bot$-classical in this fragment can be reduced to atomic applications, i.e., applications with atomic conclusions? Unfortunately the answer is negative. The application of $\bot$-classical in the derivation below cannot be reduced to a simpler application.

$$\dfrac{\dfrac{\dfrac{[\neg\Box A] \qquad \neg\Box A \to p}{p} \qquad \neg p}{\bot}}{\Box A}$$

In fact, Medeiros has showed that the derivation

$$\dfrac{\dfrac{\dfrac{\dfrac{\dfrac{\dfrac{[\neg\Box A] \qquad \neg\Box A \to p}{p} \qquad \neg p}{\bot}}{\Box A}}{A}}{C \to A}}{\Box(C \to A)}$$

is a counter-example to normalization for S4 formalized as in Prawitz (1965) (see Medeiros (2006)).

Do we have Seldin's normalization strategy for classical S4? Consider the fragment $C^*_m = \{\neg, \wedge, \vee, \rightarrow, \bot, \diamond\}$ (again, nothing is lost!). As in the case of the existential quantifier we can show that:

1. every derivation Π of $\Gamma \vdash A$ in C^* can be transformed into a derivation Π' of $\Gamma \vdash A$ such that Π' has at most one application of $\bot$-classical that can only occur as the last ruled in Π';
2. And then we can use our favorite intuitionistic strategy.

The moral again is: If you need to use classical reasoning at all in the fragment $\{\neg, \wedge, \vee, \rightarrow, \bot, \diamond\}$, you need it just once! — The last rule we use!

Can we extend the results obtained for first order logic to classical S4? We have just seen that Prawitz' normalization procedure does not work for the fragment $\{\neg, \wedge, \vee, \rightarrow, \bot, \Box\}$. Seldin's normalization procedure does not work for the $\Box$. We can define a new reduction in order to try to obtain normalization for the fragment $\{\neg, \wedge, \vee, \rightarrow, \bot, \Box\}$, but the new reduction does not fit Prawitz' normalization strategy!

We could also try to use a promotion-like rule for $\Box$ in the definition of a new formalization for classical S4:

$$\dfrac{\Box B_1 \ldots \Box B_k \qquad \begin{matrix}[\Box B_1] \ldots [\Box B_k] \\ | \\ C\end{matrix}}{[C]}$$

But Medeiro's counter example applies to the new system too: we can't reduce all applications of $\bot$-classical to atomic applications in the new system with the given restrictions. Our proof of Theorem 1 (involution) strictly depends on the result that $(\neg\neg\Box A \rightarrow \Box\neg\neg A)$ (reduction of the complexity of applications of the classical absurd rule).

12.7 Simpson's Natural Deduction for Intuitionistic S4

A possible solution to these difficulties takes us to annotated/labelled modal systems as used by Alex Simpson in his Ph.D. Thesis. The main idea is to bring the modal accessibility relation into the syntax (possible world semantics is explicitly reflected by the syntax!). The introduction rule for the $\Box$ can be formulated as:

$$\dfrac{\begin{matrix}[xRy] \\ | \\ y : A\end{matrix}}{x : \Box A}$$

(*) y does not occur on any other open assumption on which $y : A$ depends.

Elimination rule for $\Box$:

$$\frac{x : \Box A \qquad xRy}{y : A}$$

All other rules in Simpson's Natural Deduction (see Simpson (1993)) have to deal with the accessibility relation, as it can be seen in Fig. 12.1. Rule $(\diamond E)$ has the following restriction: y must be different from both x and z and must not occur in any open assumption upon which $z : \psi$ depends on than the distinguished occurrences of $y : \phi$ and xRy.

Adding the $\bot$-classical rule to Simpson's (labelled) Natural Deduction we obtain classical modal system. We can then show that:

1. Every application of $\bot$-classical in the fragment $\{\neg, \wedge, \rightarrow, \bot, \Box\}$ can be reduced to an atomic application.

$$\frac{x : \bot}{y : \phi}(\bot E)$$

$$\frac{x : \phi \quad x : \psi}{x : \phi \wedge \psi}(\wedge I) \qquad \frac{x : \phi \wedge \psi}{x : \phi}(\wedge E1) \qquad \frac{x : \phi \wedge \psi}{x : \psi}(\wedge E2)$$

$$\frac{x : \phi}{x : \phi \vee \psi}(\vee I1) \qquad \frac{x : \psi}{x : \phi \vee \psi}(\vee I2) \qquad \frac{x : \phi \vee \psi \quad \begin{array}{c}[x : \phi] \\ \vdots \\ y : \theta\end{array} \quad \begin{array}{c}[x : \psi] \\ \vdots \\ y : \theta\end{array}}{y : \theta}(\vee E)$$

$$\frac{\begin{array}{c}[x : \phi] \\ \vdots \\ x : \psi\end{array}}{x : \phi \rightarrow \psi}(\rightarrow I) \qquad \frac{x : \phi \rightarrow \psi \quad x : \phi}{x : \psi}(\rightarrow E)$$

$$\frac{y : \phi \quad xRy}{x : \diamond\phi}(\diamond I) \qquad \frac{x : \diamond\phi \quad \begin{array}{c}[y : \phi]\ [xRy] \\ \vdots \\ z : \psi\end{array}}{z : \psi}(\diamond E)$$

Fig. 12.1 Simpson's natural deduction system

2. Every derivation Π of $\Gamma \vdash A$ in $C^\star$, in the fragment $\{\neg, \wedge, \rightarrow, \bot, \diamond\}$, can be transformed into a derivation Π' of $\Gamma \vdash A$ such that Π' has at most one application of $\bot$-classical that can only occur as the last ruled in Π'.

$$\begin{array}{c}[x:\neg\Box A]\\ \Pi\\ \dfrac{x:\bot}{x:\Box A}\end{array}$$

reduces to

$$\begin{array}{c}\dfrac{\dfrac{\dfrac{[x:\Box A]^k \qquad [xRy]^n}{y:A} \qquad [y:\neg A]^m}{y:\bot}}{\dfrac{x:\bot}{[x:\neg\Box A]}\,k}\\ \Pi\\ \dfrac{\dfrac{\dfrac{x:\bot}{y:\bot}}{y:\neg\neg A}\,m}{x:\Box\neg\neg A}\,n\end{array}$$

where y is new.
We can show that

$$\dfrac{\dfrac{\dfrac{\dfrac{\dfrac{\dfrac{\dfrac{\dfrac{[x:\Box A]^1 \qquad [xRy]^n}{y:A} \qquad [y:\neg A]^2}{y:\bot}}{x:\bot}}{x:\neg\Box A}\,1 \qquad x:\neg\neg\Box A}{x:\bot}}{y:\bot}}{y:\neg\neg A}\,2}{\dfrac{x:\Box\neg\neg A}{x:\neg\neg\Box A \rightarrow \Box\neg\neg A}}\,3$$

The reduction below shows one case of the extension of Seldin's strategy to the fragment $\{\neg, \wedge, \rightarrow, \bot, \diamond\}$.

$$\dfrac{\begin{array}{c}[y:\neg A]^2\\ \Pi\\ \dfrac{y:\bot}{y:A}\,2\end{array} \qquad xRy}{x:\diamond A}$$

reduces to:

$$
\begin{array}{c}
\dfrac{\dfrac{\dfrac{\dfrac{\dfrac{[y:A]^1 \qquad xRy}{x:\Diamond A} \qquad [x:\neg\Diamond A]^2}{x:\bot}}{y:\bot}}{y:\neg A}\;1 \\
\Pi \\
\dfrac{\dfrac{y:\bot}{x:\bot}}{x:\Diamond A}\;2
\end{array}
$$

12.8 The Fragment $\{\neg, \wedge, \bot, \Box\}$

As in the case of first order, this fragment is sufficient to distinguish classical S4 from (any reasonable version of) intuitionistic S4,[1] and the formula $\neg(\Box\neg(\neg P \wedge \neg Q) \wedge \Box\neg Q \wedge \neg\Box P)$ is a nice example of a classical theorem that is not an intuitionistic one. Again, as in the case of first order logic, we can obtain the following results:

Theorem 12.8 *Let A be a formula in the fragment $\{\neg, \wedge, \bot, \Box\}$. Then, $\vdash_{IS4} x : \neg\neg A$ iff $\vdash_{IS4} x : A$.*

Proof Trivial induction on the complexity of A.

Theorem 12.9 *Let A be a formula in the fragment $\{\neg, \wedge, \bot, \Box\}$. Then, $\vdash_{IS4} x : \Box\neg\neg A$ iff $\vdash_{IS4} x : \neg\neg\Box A$*

Proof Directly from the involution of negation in this fragment.

12.9 The Fragment $\{\neg, \wedge, \bot, \Diamond\}$

This fragment is also sufficient to distinguish classical S4 from intuitionistic logic S4 (IS4), and the formula $\Diamond\neg(\neg A \wedge \Diamond A)$ is a nice example of a classical theorem that is not an intuitionistic one. As in the case of $\Box$, we can obtain some constructive results:

Theorem 12.10 *Let A be a formula in $\{\neg, \wedge, \bot\}$. Then, $\vdash_{S4} \Diamond A$ iff $\vdash_{IS4} x : \Diamond A$, for arbitrary x.*

[1] It is true that, different from the classical case, we don't have a "canonical" intuitionistic S4: there are several ways to combine the modal accessibility relation with the partial order required by the semantics for the propositional operators. Without any loss of generality, we are assuming here that the intuitionistic version of S4 we are considering is Simpson's.

Proof $\vdash_{S4} \diamond A \Rightarrow \vdash_{S4} \neg\Box\neg A \Rightarrow \vdash_{IS4} x : \neg\Box\neg A$. From the last fact we know that there is a derivation in $IS4$ of $x : \bot$ from $x : \Box\neg A$. By the (normal) form of this last derivation there is a subderivation $y_1 : \neg A, xRy_1, \ldots, y_n : \neg A, xRy_n \vdash_{IS4} z : \bot$, and hence by a series of $\diamond$-Elim we have $x : \diamond\neg A \vdash_{IS4} z : \bot$ and by an $\bot$-Elim we obtain that $x : \diamond\neg A \vdash_{IS4} z : \bot$. Thus $\vdash_{IS4} x : \neg\diamond\neg A$ that implies $\vdash_{S4} \neg\diamond\neg A \Rightarrow \vdash_{S4} \Box A \Rightarrow \vdash_{S4} A \vdash_{IS4} x : A \vdash_{IS4} x : \diamond A$.

Theorem 12.11 *Let A be a formula in* $\{\neg, \wedge, \bot, \diamond\}$ *such that no modality occurs in the scope of another modality. Then* $\vdash_{S4} A$ *iff* $\vdash_{IS4} x : A$*, for arbitrary* x.

Proof By induction on the complexity of A. Basis: Atomic. Trivial Inductive step:

- A is $\neg B$ for some B. The result is consequence of Glyvenko second theorem for $\{\neg, \wedge, \bot, \diamond\}$.
- A is $(B \wedge C)$. Directly from the inductive hypothesis.
- A is $\diamond B$. Directly from Theorem 12.10.

Theorem 12.11 has a nice conceptual meaning: there is no classical logic in the fragment $\{\neg, \wedge, \bot, \diamond\}$ without iteration of modalities!

12.10 Conclusion and Some Suggestions for Future Work

We have shown that several fragments of CL and classical S4 have interesting constructive properties: in some fragments negation is constructively involutive, in other fragments double negation can be internalized and externalized. These results are directly related to different normalization strategies: Glivenko's results are trivial consequences of Seldin's strategy, and double-negation translation is a natural consequence of Prawitz' strategy. There are some interesting directions worth investigating:

1. Although the relation between CL and IL is quite well known, the relation between CL and some intermediary logics is completely unknown. In fact, several intermediary logics are still waiting for proof-theoretical-friendly formalizations. This direction is specially interesting in the case of Arithmetic: we know that Constant-domain Arithmetic collapses into Peano's Arithmetic, but there are other interesting "intermediary" Arithmetics whose relation to PA are waiting to be studied.
2. Our proof-theoretical analysis of the implicational fragment is based on the P-rule, but there are other ways to give complete formalizations for Classical Implicational Logic (see Gordeev (1987)). Would it be possible to extract from the proof-theory of these formalizations interesting translations from Classical Implicational Logic into Intuitionistic/Minimal Implicational Logic?
3. An interesting result satisfied by the fragment $\{\neg, \wedge, \bot, \exists\}$ establishes that classical logic in this fragment requires "iteration" of quantifiers and, for this reason, in a certain sense Aristotelian logic is constructive. It would be interesting to explore

this result in order to give a syntactical proof (via cut-elimination/normalization) for the decidability of the monadic calculus

4. We have seen how to use annotated/labelled systems in order to solve some difficulties related to the rule of □-Introduction, but it would certainly be interesting to find new restrictions for the application of a Prawitz' style □-introduction.

References

Gentzen, G. (1974). Über das Verhältnis zwischen intuitionistischer und klassischer Arithmetik. *Archiv für Mathematische Logik, 16*, 119–132.

Gödel, K. (1933). Zur intuitionistischen Arithmetik und Zahlentheorie. *Ergebnisse eines mathematischen Kolloquiums*, *4*, 34–38.

Gordeev, L. (1987). On cut elimination in the presence of Peirce rule. *Archive for Mathematical Logic*, *26*(1), 147–164.

Kolmogorov, A. (1925). Sur le principe de tertium non datur. *Met Sbornik*, *32*, 646–667. (translation in van Heijenoort).

Medeiros, M. (2006). A new S4 classical modal logic in natural deduction. *Journal of Symbolic Logic*, *71*(3), 799–809.

Pereira, L., Haeusler, E., Costa, V., & Sanz, W. (2010). A new normalization strategy for the implicational fragment of classical propositional logic. *Studia Logica*, *96*(1), 95–108.

Pereira, L. C., Haeusler, E. H., & Madeiros, M. (2008). Alguns resultados sobre fragmentos com negação da lógica clássica. *O que nos faz pensar*, *23*, 105–111.

Prawitz, D. (1965). *Natural Deduction: A Proof-Theoretical Study*. Stockholm: Almqvist & Wicksell.

Seldin, J. (1986). On the proof-theory of the intermediate logic MH. *Journal of Symbolic Logic*, *51*(3), 626–647.

Seldin, J. P. (1989). Normalization and excluded middle I. *Studia Logica*, *48*(2), 193–217.

Simpson A (1993) The proof theory and semantics of intuitionistic modal logic. Ph.D. thesis, Cambridge, revised September 1994

Zimmermann, E. (2002). Peirce's rule in natural deduction. *Theoretical Computer Science*, *275*(1-2), 561–574.

Chapter 13
General-Elimination Harmony and Higher-Level Rules

Stephen Read

Abstract Michael Dummett introduced the notion of harmony in response to Arthur Prior's tonkish attack on the idea of proof-theoretic justification of logical laws (or analytic validity). But Dummett vacillated between different conceptions of harmony, in an attempt to use the idea to underpin his anti-realism. Dag Prawitz had already articulated an idea of Gerhard Gentzen's into a procedure whereby elimination-rules are in some sense functions of the corresponding introduction-rules. The resulting conception of general-elimination harmony ensures that the rules are transparent in the meaning they confer, in that the elimination-rules match the meaning the introduction-rules confer. The general-elimination rules which result may be of higher level, in that the assumptions discharged by the rule may be of (the existence of) derivations rather than just of formulae. In many cases, such higher-level rules may be "flattened" to rules discharging only formulae. However, such flattening is often only possible in the richer context of so-called "classical" or realist negation, or in a multiple-conclusion environment. In a constructivist context, the flattened rules are harmonious but not stable.

Keywords Harmony · 'Tonk' · Generalized-elimination rules · Higher-level rules · Flattening · Reduction · Local completeness · Classical logic · Intuitionistic logic · Negation · Gentzen · Dummett · Prawitz

---

Invited paper for *Dag Prawitz on Proofs and Meaning*, edited by Heinrich Wansing, in the *Studia Logica* series *Trends in Logic*. I owe a substantial intellectual debt to Dag Prawitz, to whose writings on proof theory I was first introduced by John Mayberry at Bristol, and which have inspired and guided me repeatedly throughout my career. This work is supported by Research Grant AH/F018398/1 (Foundations of Logical Consequence) from the Arts and Humanities Research Council, UK.

S. Read (✉)
Department of Philosophy, University of St Andrews, Scotland, Edgecliffe,
The Scores, St Andrews, Fife KY16 9AL, UK
e-mail: slr@st-andrews.ac.uk

H. Wansing (ed.), *Dag Prawitz on Proofs and Meaning*, Outstanding Contributions to Logic 7, DOI 10.1007/978-3-319-11041-7_13

13.1 Analytic Validity

In a famous article published in December, Arthur Prior (1960) attacked the idea that

> there are inferences whose validity arises solely from the meanings of certain expressions occurring in them.

He argued that validity must instead be based on truth-preservation, not on meaning. To illustrate what he considered to be the problem with what he dubbed analytic validity, Prior introduced a new connective 'tonk' with the rules:

$$\frac{\alpha}{\alpha \text{ tonk } \beta}\ \text{tonk-I} \qquad\qquad \frac{\alpha \text{ tonk } \beta}{\beta}\ \text{tonk-E}$$

By chaining together an application of tonk-I with one of tonk-E, we can apparently derive any proposition (β) from any other (α). This is clearly absurd and disastrous. How can one possibly *define* such an inference into existence?

We may agree with Prior that 'tonk' had not been given any recognisable meaning by these rules. Rather, whatever meaning tonk-introduction had conferred on the neologism 'tonk' was then contradicted by Prior's tonk-elimination rule. But we might respond to Prior by claiming that if rules were set down for a term which did properly confer meaning on it, then certain inferences would be "analytic" in virtue of that meaning. What constraints must rules satisfy in order to confer a coherent meaning on the terms involved?

Dummett (1973) introduced the term 'harmony' for this constraint: in order for the rules to confer meaning on a term, two aspects of its use must be in harmony. Those two aspects are the grounds for an assertion as opposed to the consequences we are entitled to draw from such an assertion. Those whom Prior was criticising, Dummett claimed, had committed the "error" of failing to appreciate

> the interplay between the different aspects of 'use', and the requirement of harmony between them (p. 396 Dummett 1973).
>
> If the linguistic system as a whole is to be coherent, there must be a harmony between these two aspects (p. 221 Dummett 1991).

In appealing to this connection between the grounds, or introduction-rules, and the consequences, or elimination-rules, Dummett was following the lead of Dag Prawitz, who in turn was following out an idea of Gerhard Gentzen's, in a famous and much-quoted passage where he says that

> the E-inferences are, through certain conditions, *unique* consequences of the respective I-inferences (p. 81 Gentzen 1969).

In a series of articles on the "foundations of a general proof theory" published in the early 1970s, (Prawitz 1973, 1974, 1975) set out to find a characterization of validity of argument independent of model theory, as typified by Tarski's truth-preservationist account of logical consequence. Following Gentzen's idea in the passage cited above, Prawitz accounts an argument or derivation valid by virtue of the meaning or definition of the logical constants encapsulated in the I-rules. Suppose we take the I-rules as given. Then any argument (or in the general case, argument-schema) is valid if there is a "justifying operation" ultimately articulating the argument into the application of I-rules to atomic sentences[1]:

> The main idea is this: while the introduction inferences represent the form of proofs of compound formulas by the very meaning of the logical constants ... and hence preserve validity, other inferences have to be justified by the evidence of operations of a certain kind
>
> (p. 234 Prawitz 1973).

What Prawitz does, in fact, is frame his E-rules in such a way that such a justification is possible. Given a set of I-rules for a connective (in general, there may be several, as in the familiar case of '$\vee$'), the E-rules (again, there may be several, as in the case of '$\wedge$') which are justified by the meaning so conferred are those which will permit an operation of Prawitz' kind. The principle underlying this procedure is called by Prawitz (1965), following Paul Lorenzen, the "inversion principle". Prawitz refers to (Lorenzen 1955), and more particularly to Hermes (1959, p. 65), for the full statement of the principle. Prawitz (1965, p. 33) writes[2]:

> Let α be an application of an elimination rule that has B as consequence. Then, deductions that satisfy the sufficient condition [...] for deriving the major premiss of α, when combined with deductions of the minor premisses of α (if any), already 'contain' a deduction of B; the deduction of B is thus obtainable directly from the given deductions without the addition of α.

Each E-rule is harmoniously justified by satisfying the constraint that whenever its premises are provable (by application of one of the I-rules), the conclusion is derivable (by use of the assertion-conditions framed in the I-rule), that is, it is admissible (*zulässig*: Lorenzen (1955, p. 30); Hermes (1959, p. 63). Francez and Dyckhoff (2012) introduced the term "General-Elimination Harmony" for the form which this procedure accords to the E-rules.[3]

13.2 General-Elimination Harmony

Suppose there are m I-rules for a connective 'δ', each with n_i premises, $0 \leq i \leq m$[4]:

$$\frac{\pi_{i1} \quad \dots \quad \pi_{in_i}}{\delta\vec{\alpha}}\ \delta\text{-I}_i$$

Here $\delta\vec{\alpha}$ is a formula with main connective 'δ'. Each π_{ij}, $0 \leq j \leq n_i$, may be a wff (as in $\wedge$I), or a derivation of a wff from certain assumptions which are discharged

[1] See also Prawitz (2006) and Read (2014).

[2] Cf. Schroeder-Heister (2006), Schroeder-Heister (2007).

[3] For an extended discussion of GE-harmony, see Read (2010).

[4] m may be zero, as Prawitz (1973, p. 243) notes is the case for the absurdity constant, $\perp$, which has no grounds for its assertion.

by the rule (as in $\rightarrow$I). In accordance with the inversion principle, this set of I-rules justifies $\prod_{i=1}^{m} n_i$ E-rules, each of the form:[5]

$$\dfrac{\delta\vec{\alpha} \qquad \begin{matrix}[\pi_{1j_1}]\\ \vdots \\ \gamma\end{matrix} \qquad \cdots \qquad \begin{matrix}[\pi_{mj_m}]\\ \vdots \\ \gamma\end{matrix}}{\gamma}\ \delta\text{-E}$$

Each minor premise derives γ from one of the grounds, π_{ij_i}, in the ith rule for asserting $\delta\vec{\alpha}$.

The justification is this: the GE-procedure ensures that one can infer γ from $\delta\vec{\alpha}$ whenever one can infer γ from one of the grounds for assertion of $\delta\vec{\alpha}$. Consequently, the actual assertion of $\delta\vec{\alpha}$ is an unnecessary detour:

$$\dfrac{\dfrac{\begin{matrix}\vdots\\ \pi_{i1}\end{matrix} \quad \cdots \quad \begin{matrix}\vdots\\ \pi_{in_i}\end{matrix}}{\delta\vec{\alpha}}\ \delta\text{-I} \qquad \begin{matrix}[\pi_{1j_1}]\\ \vdots \\ \gamma\end{matrix} \qquad \cdots \qquad \begin{matrix}[\pi_{mj_m}]\\ \vdots \\ \gamma\end{matrix}}{\gamma}\ \delta\text{-E}$$

converts to

$$\begin{matrix}\vdots\\ \pi_{ij_i}\\ \vdots\\ \gamma\end{matrix}$$

Having one minor premise in each E-rule drawn from among the premises for each I-rule ensures that, whichever I-rule justified assertion of $\delta\vec{\alpha}$ (here it was the ith), one of its premises can be paired with one of the minor premises to remove the unnecessary application of δ-I immediately followed by δ-E.

The idea behind harmony is that the elimination-rule should allow one to infer all and only what is justified by the meaning conferred by the introduction-rule. The above procedure certainly shows that the E-rule permits one to infer no more than is so justified. But it does not show that the rule permits inference of everything that is justified by the meaning so conferred. The idea that the E-rule should not only not be too weak but also not too strong was called by Dummett (1991 Chap. 13), "stability". For example, consider the Curry-Fitch rules for $\Diamond$ (possibility)[6]:

[5] If $m = 0$, the empty product predicts one E-rule, to infer an arbitrary conclusion from $\bot$. If $n_i = 0$ for some i, the product is 0. E.g., if we introduce $\top$ by an I-rule with no premises (even if we give alternative, more restrictive, grounds for its assertion), $\top$ is a tautology, and nothing can be inferred from it which is not already provable.

[6] See Curry (1950 Chap. V), Fitch (1952 Chap. 3), Prawitz (1965 Chap. VI).

$$\frac{\alpha}{\Diamond\alpha}\ \Diamond\text{-I} \qquad \text{and} \qquad \frac{\Diamond\alpha \quad \begin{matrix}[\alpha]\\ \vdots \\ \gamma\end{matrix}}{\gamma}\ \Diamond\text{-E}$$

provided that in the case of $\Diamond$-E, every assumption on which the minor premise γ depends, apart from α (the so-called parametric formulae), is modal (that is, has the form $\Box\beta$) and γ is co-modal (that is, has the form $\Diamond\beta$). These rules are not stable: the (unrestricted) rule $\Diamond$-I does not justify the restriction put on $\Diamond$-E. $\Diamond$-I appears to say that $\Diamond\alpha$ just means α—that is, $\Diamond\alpha$ is assertible just when α is. But the model theory shows that the rules do define possibility. Quite how they interact to do so is far from obvious.[7]

Dummett and Prawitz (and others) make a yet stronger claim: that an inference is not justified if the rules are not harmonious.[8] For example, Dummett (1991, p. 299) claims that classical logic, with classical negation, is incoherent since it cannot be given harmonious rules. In my view, this asks too much of harmony and the constraints on the rules it invokes. An example from Dummett (1991, p. 291) is the minimal theory of negation. Dummett's introduction-rule for negation:

$$\frac{\begin{matrix}[\alpha]\\ \vdots \\ \neg\alpha\end{matrix}}{\neg\alpha}\ \neg\text{-I}$$

does not justify the intuitionistically valid elimination-rule:

$$\frac{\neg\alpha \quad \alpha}{\beta}\ \text{EFQ}$$

Nonetheless, the intuitionistic E-rule is still valid, and between them the two rules give the intuitionistic theory of negation. EFQ simply is not justified by the I-rule which Dummett proposes. To discern the meaning which these rules define it is not enough just to look at the I-rule. One must look at the elimination-rule too, just as one had to do with the inharmonious rules for $\Diamond$. What harmony and stability can do for us is ensure that the I- and E-rules confer the same meaning, and so ensure that the meaning is transparent in the grounds for assertion, that is, the I-rule. We can see this by considering some particular cases.

[7] To obtain harmonious rules, the right response is, of course, not to strengthen $\Diamond$-E to match $\Diamond$-I, but to find some way to weaken $\Diamond$I. For one possible solution, see Read (2008).

[8] See, e.g., Dummett (1991, pp. 286–287), qualified only by the remark: "when [the] rules are held completely to determine the meanings of the logical constants." Cf. Prawitz (1985, p. 138), Schroeder-Heister (2006, p. 532), Tennant (2013 Sect. 11).

Here is an example, perhaps over-familiar, but suitably revealing. Given as I-rule:

$$\frac{\alpha \qquad \beta}{\alpha \wedge \beta}\ \wedge\mathrm{I}$$

the inversion principle yields two generalized $\wedge$-E rules, assuming $\wedge$-I to exhaust the grounds for asserting $\alpha \wedge \beta$ (so $m = 1$ and $n_1 = 2$):

$$\frac{\alpha \wedge \beta \qquad \begin{array}{c}[\alpha] \\ \vdots \\ \gamma\end{array}}{\gamma}\ \wedge\text{-E}_1 \qquad \text{and} \qquad \frac{\alpha \wedge \beta \qquad \begin{array}{c}[\beta] \\ \vdots \\ \gamma\end{array}}{\gamma}\ \wedge\text{-E}_2$$

The generalized $\wedge$-E rules yield the more familiar $\wedge$-E rules of Simp(lification) immediately as instances, by letting γ be α and β respectively:

$$\frac{\alpha \wedge \beta \qquad \begin{array}{c}[\alpha] \\ \vdots \\ \alpha\end{array}}{\alpha}\ \wedge\text{-E}_1 \qquad \text{and} \qquad \frac{\alpha \wedge \beta \qquad \begin{array}{c}[\beta] \\ \vdots \\ \beta\end{array}}{\beta}\ \wedge\text{-E}_2$$

which reduce to

$$\frac{\alpha \wedge \beta}{\alpha}\ \mathrm{Simp}_1 \qquad \text{and} \qquad \frac{\alpha \wedge \beta}{\beta}\ \mathrm{Simp}_2$$

given that we can always derive γ from γ, for all γ. Conversely, $\wedge$-E$_1$ follows from Simp$_1$, that is, that if there is a derivation of γ from α, then γ follows from $\alpha \wedge \beta$:

$$\begin{array}{c}\dfrac{\alpha \wedge \beta}{\alpha} \\ \vdots \\ \gamma\end{array}$$

and the same for $\wedge$-E$_2$.

Dyckhoff and Francez, Schroeder-Heister and others, have a single form of the generalized rule[9]:

$$\frac{\alpha \wedge \beta \qquad \begin{array}{c}\underbrace{[\alpha] \quad [\beta]} \\ \vdots \\ \gamma\end{array}}{\gamma}\ \wedge\text{-GE}$$

[9] See, e.g., Francez and Dyckhoff (2012, p. 615), Schroeder-Heister (1984, p. 1294), Negri and Plato (2001, p. 7).

To see that $\wedge$-GE is equivalent to the conjunction of $\wedge$-E$_1$ and $\wedge$-E$_2$, let us replace the two-dimensional representation of the derivation of γ from α and β by the linear form $\alpha, \beta \Rightarrow \gamma$. Then we can derive each of $\wedge$-E$_1$ and $\wedge$-E$_2$ from $\wedge$-GE:

$$\dfrac{\alpha \wedge \beta \quad \dfrac{\alpha \Rightarrow \gamma}{\alpha, \beta \Rightarrow \gamma}\text{ K (Weakening)}}{\gamma}\ \wedge\text{-GE}$$

and the same for β. Conversely,

$$\dfrac{\alpha \wedge \beta \quad \dfrac{\alpha \wedge \beta \quad \alpha, \beta \Rightarrow \gamma}{\beta \Rightarrow \gamma}\ \wedge\text{-E}_1}{\gamma}\ \wedge\text{-E}_2$$

What this shows is $\alpha \wedge \beta, \alpha \wedge \beta \Rightarrow \gamma$, and $\wedge$-GE follows by Contraction (W).

Thus we have two competing forms of $\wedge$-E, though they are equivalent, given Contraction and Weakening. But in the absence of W and K, which is the right form? Recall the additive and multiplicative left-rules for $\wedge$ and $\otimes$ in linear logic[10]:

$$\dfrac{\alpha, \Gamma \Rightarrow \Theta}{\alpha \wedge \beta, \Gamma \Rightarrow \Theta}\ \wedge\Rightarrow \qquad \dfrac{\beta, \Gamma \Rightarrow \Theta}{\alpha \wedge \beta, \Gamma \Rightarrow \Theta}\ \wedge\Rightarrow \qquad \dfrac{\alpha, \beta, \Gamma \Rightarrow \Theta}{\alpha \otimes \beta, \Gamma \Rightarrow \Theta}\ \otimes\Rightarrow$$

Clearly, $\wedge$-GE gives the multiplicative E-rule for $\otimes$, whereas $\wedge$-E$_1$ and $\wedge$-E$_2$ give the correct E-rules for additive $\wedge$. In the presence of W and K, the additive/multiplicative distinction is erased, but to give the rules in their proper form, we need to give separate E-rules for $\wedge$, each corresponding to one of the premises in $\wedge$-I:

$$\dfrac{\alpha \wedge \beta \quad \alpha \Rightarrow \gamma}{\gamma}\ \wedge\text{-E}_1 \qquad \text{and} \qquad \dfrac{\alpha \wedge \beta \quad \beta \Rightarrow \gamma}{\gamma}\ \wedge\text{-E}_2$$

confirming the correctness of the GE-procedure.

$\wedge$-I, $\wedge$-E$_1$ and $\wedge$-E$_2$ are harmonious. But are they stable? That is, do $\wedge$-E$_1$ and $\wedge$-E$_2$ allow one to derive all the consequences that the meaning encapsulated in $\wedge$-I justifies? Davies and Pfenning (2001, p. 560) call harmony, "local soundness", that the E-rules allow one to derive no more than the I-rule justifies; that they allow one to derive no less, they dub "local completeness". They write:

> Local completeness ensures that we can recover all information present in a connective: there is some way to apply the elimination rules so we can reconstitute a proof of the original proposition using its introduction rules.

But as Dummett (1991, pp. 288–289) showed, the restricted $\vee$-rules of quantum logic satisfy that condition:

$$\dfrac{\alpha \vee \beta \quad \dfrac{[\alpha]^1}{\alpha \vee \beta}\ \vee\text{I}_1 \quad \dfrac{[\beta]^2}{\alpha \vee \beta}\ \vee\text{I}_2}{\alpha \vee \beta}\ \vee\text{E}_\text{Q}(1,2)$$

[10] See, e.g., Girard et al. (1989, p. 152).

where no undischarged assumptions are allowed in the minor premises of $\vee E_Q$. Yet the quantum $\vee$-rules are clearly incomplete, in that they do not allow us to prove the distribution of '$\wedge$' over '$\vee$'.

The test is too weak—we need to recover not just the original proposition, but its grounds. But how can we recover the original grounds, *viz* α or β, from $\alpha \vee \beta$? In multiple-conclusion reasoning, it is straightforward. We need only derive the sequent $\alpha \vee \beta \Rightarrow \alpha, \beta$. In single-conclusion systems, we can at best show that $\neg\alpha, \alpha \vee \beta \vdash \beta$ and $\neg\beta, \alpha \vee \beta \vdash \alpha$, that is, that if one of the grounds for asserting $\alpha \vee \beta$ fails, then, given that $\alpha \vee \beta$ holds, the other ground must hold. In general, we can show that we can derive π_{kl} from $\delta\vec{\alpha}$ for every $k \leq m$ and $l \leq n_k$, given the falsity of π_{ij} for each $i \neq k$ and some $j \leq n_i$. The general-elimination rules are not only in harmony with the I-rules, they are moreover, stable.

13.3 Flattening the Rules

As a second example, consider the I-rule for implication:

$$\dfrac{\begin{array}{c}[\alpha]\\ \vdots\\ \beta\end{array}}{\alpha \to \beta}\to\text{-I} \qquad \text{that is,} \qquad \dfrac{\alpha \Rightarrow \beta}{\alpha \to \beta}\to\text{-I}$$

inferring (an assertion of the form) $\alpha \to \beta$ from (a derivation of) β, permitting the discharge of (zero or more occurrences of) α. Whatever form $\to$-E has, there must be the appropriate justificatory operation of which Prawitz spoke. That is, we should be able to infer from an assertion of $\alpha \to \beta$ no more (and no less) than we could infer from whatever warranted assertion of $\alpha \to \beta$. We can represent this as follows[11]:

$$\dfrac{\alpha \to \beta \qquad \begin{array}{c}\left[\begin{array}{c}[\alpha]\\ \vdots\\ \beta\end{array}\right]\\ \vdots\\ \gamma\end{array}}{\gamma}\to\text{-E} \qquad \text{that is,} \qquad \dfrac{\alpha \to \beta \qquad \begin{array}{c}[\alpha \Rightarrow \beta]\\ \vdots\\ \gamma\end{array}}{\gamma}\to\text{-E}$$

Thus, if we can infer γ from assuming the existence of a derivation of β from α, we can infer γ from $\alpha \to \beta$.[12]

[11] The form of representation here is inspired by Gentzen's notation in the draft of his dissertation, (Gentzen 1932).

[12] Note that in $\alpha \Rightarrow \beta$, the assumption α is closed, either by a rule discharging the assumption (e.g., $\to$-I) or by a derivation of α (e.g., in the proof of the minor premise of $\to$-E).

What does

$$\begin{array}{c}[\alpha \Rightarrow \beta]\\ \vdots\\ \gamma\end{array} \qquad \text{that is,} \qquad \begin{array}{c}\left[\begin{array}{c}[\alpha]\\ \vdots\\ \beta\end{array}\right]\\ \vdots\\ \gamma\end{array}$$

mean? It says that we have a derivation of γ on the assumption that we have a derivation of β from α. Hence, if we have a derivation of α, we may assume we are able to derive β, from which we derive γ. That is,

$$\dfrac{\alpha \to \beta \qquad \begin{array}{c}\left[\begin{array}{c}[\alpha]\\ \vdots\\ \beta\end{array}\right]\\ \mathcal{D}\\ \gamma\end{array}}{\gamma}\,\to\text{-E}$$

means that we can connect a derivation of α with a derivation of γ from β as follows:

$$\dfrac{\alpha \to \beta \qquad \begin{array}{c}\mathcal{D}'\\ \left[\begin{array}{c}\alpha\\ \vdots\\ \beta\end{array}\right]\\ \mathcal{D}''\\ \gamma\end{array}}{\gamma}\,\to\text{-E}$$

$\to$-E is an example of what Schroeder-Heister (1984) called "higher-level" rules, where what is assumed (and discharged) is a rule (here the inference of β from α) rather than just a formula. We can "flatten" the rule by separating the derivation $\mathcal{D}'$ of α from the derivation $\mathcal{D}''$ of γ from β[13]:

$$\dfrac{\alpha \to \beta \qquad \begin{array}{c}\mathcal{D}'\\ \alpha\end{array} \qquad \begin{array}{c}[\beta]\\ \mathcal{D}''\\ \gamma\end{array}}{\gamma}\,\to\text{-E}'$$

There are now no "higher-level" assumptions, just minor premises α and γ, where in $\to$-E$'$ any assumption of the form β used to derive γ may be discharged.

Another way to think of this move appeals to the sequent calculus formulation, as before. The minor premise of $\to$-E reads: $(\alpha \Rightarrow \beta) \Rightarrow \gamma$. Using Gentzen's $\Rightarrow$-left rule, we have

$$\dfrac{\alpha \to \beta \qquad \dfrac{\Rightarrow \alpha \qquad \beta \Rightarrow \gamma}{(\alpha \Rightarrow \beta) \Rightarrow \gamma}\,\Rightarrow\text{-left}}{\gamma}\,\to\text{-E}$$

[13] Dyckhoff (1988) was possibly the first to propose this formulation, which we can also find in, e.g., von Plato (2001, p. 545). Dyckhoff rejected it for reasons summarized in Dyckhoff (2013).

Thus our generalized E-rule for $\rightarrow$ reads:

$$\dfrac{\alpha \rightarrow \beta \qquad \alpha \qquad \begin{array}{c}[\beta]\\ \vdots \\ \gamma\end{array}}{\gamma}\ \rightarrow\text{-E}'$$

Letting $\gamma = \beta$ and again assuming we can derive β from itself, we obtain the familiar rule of Modus Ponendo Ponens (MPP):

$$\dfrac{\alpha \rightarrow \beta \qquad \alpha}{\beta}\ \text{MPP}$$

Conversely, given the premises of $\rightarrow$-E$'$, we can derive γ using MPP:

$$\begin{array}{c}\dfrac{\alpha \rightarrow \beta \quad \alpha}{\beta}\ \text{MPP}\\ \vdots \\ \gamma\end{array}$$

Similar considerations arise in the obvious introduction rule for equivalence $\leftrightarrow$, which requires both that β be derivable from α and vice versa:

$$\dfrac{\begin{array}{c}[\alpha]\\ \vdots \\ \beta\end{array} \qquad \begin{array}{c}[\beta]\\ \vdots \\ \alpha\end{array}}{\alpha \leftrightarrow \beta}\ \leftrightarrow\text{-I}$$

Here $m = 1$ and $n_1 = 2$, so there are two E-rules each with one minor premise:

$$\dfrac{\alpha \leftrightarrow \beta \qquad \begin{array}{c}[\alpha \Rightarrow \beta]\\ \vdots \\ \gamma\end{array}}{\gamma}\ \leftrightarrow\text{-E}_1 \qquad \dfrac{\alpha \leftrightarrow \beta \qquad \begin{array}{c}[\alpha \Rightarrow \beta]\\ \vdots \\ \gamma\end{array}}{\gamma}\ \leftrightarrow\text{-E}_2$$

Each simplifies by flattening of the rules and moves similar to those with the generalized rule for $\rightarrow$-E, to obtain:

$$\dfrac{\alpha \leftrightarrow \beta \quad \alpha}{\beta}\ \leftrightarrow\text{-E}_1' \qquad \dfrac{\alpha \leftrightarrow \beta \quad \beta}{\alpha}\ \leftrightarrow\text{-E}_2'$$

Suppose we now introduce a novel connective which disjoins the grounds for asserting $\alpha \leftrightarrow \beta$ instead of conjoining them:

$$\dfrac{\begin{array}{c}[\alpha]\\ \vdots \\ \beta\end{array}}{\alpha \odot \beta}\ \odot\text{-I}_1 \qquad \dfrac{\begin{array}{c}[\beta]\\ \vdots \\ \alpha\end{array}}{\alpha \odot \beta}\ \odot\text{-I}_2$$

Here we have two I-rules each with one premise ($m = 2$ and $n_1 = n_2 = 1$), so there will be one E-rule with two minor premises:

$$\dfrac{\alpha \odot \beta \qquad \begin{matrix}[\alpha \Rightarrow \beta] \\ \vdots \\ \gamma\end{matrix} \qquad \begin{matrix}[\beta \Rightarrow \alpha] \\ \vdots \\ \gamma\end{matrix}}{\gamma}\ \odot\text{-E}$$

Flattening of the rules yields:

$$\dfrac{\alpha \odot \beta \quad \alpha \quad \begin{matrix}[\beta] \\ \vdots \\ \gamma\end{matrix} \quad \beta \quad \begin{matrix}[\alpha] \\ \vdots \\ \gamma\end{matrix}}{\gamma}\ \odot\text{-E}'$$

and the major premise seems redundant. We can prove γ directly from the minor premises (indeed, twice over).

This may seem puzzling, but in fact, reflection shows that it is to be expected, at least from a classical perspective. The two cases of $\odot$-I show that $\alpha \odot \beta$ means that either β is derivable from α (possibly given other assumptions) or α is derivable from β. That is a classical tautology: $(\alpha \to \beta) \vee (\beta \to \alpha)$. Take the following intuitionistic negation rules:

$$\dfrac{\begin{matrix}[\alpha] \\ \vdots \\ \beta\end{matrix} \qquad \begin{matrix}[\alpha] \\ \vdots \\ \neg\beta\end{matrix}}{\neg\alpha}\ \mathcal{R} \qquad\qquad \dfrac{\neg\alpha \qquad \alpha}{\beta}\ \mathcal{V}$$

(These are Gentzen's negation rules $\mathcal{R}$ and $\mathcal{V}$, which we will look at in the next section.) With them, we can prove $\neg\neg(\alpha \odot \beta)$:

$$\dfrac{\dfrac{\dfrac{\dfrac{}{\neg(\alpha \odot \beta)}\ (1) \qquad \dfrac{\dfrac{}{\alpha}\ (2)}{\alpha \odot \beta}\ \odot\text{-I}_2(3)}{\beta}\ \mathcal{V}}{\alpha \odot \beta}\ \odot\text{-I}_1(2) \qquad \dfrac{}{\neg(\alpha \odot \beta)}\ (1)}{\neg\neg(\alpha \odot \beta)}\ \mathcal{R}(1)$$

This will yield a classical proof of $\alpha \odot \beta$ by extending the above proof by an application of DN (double-negation elimination), or replacing the application of $\mathcal{R}$ by classical *reductio*:

$$\dfrac{\begin{matrix}[\neg\alpha] \\ \vdots \\ \beta\end{matrix} \qquad \begin{matrix}[\neg\alpha] \\ \vdots \\ \neg\beta\end{matrix}}{\alpha}\ \mathcal{CR}$$

The example of $\odot$ illustrates a general problem affecting the flattening procedure. In the case of $\rightarrow$ and $\leftrightarrow$, the flattened rule is easily shown to be as strong as the higher-level rule. But in the case of $\odot$ and other connectives, this is not true. The flattened rules, though harmonious, are not in general stable. Schroeder-Heister (2014) raises the issue, identifying two kinds of problem case. Take a case of δ-E:

$$\dfrac{\delta\vec{\alpha} \quad \begin{matrix}[\pi_{1j_1}] \\ \mathcal{D}_1 \\ \gamma\end{matrix} \quad \cdots \quad \begin{matrix}[\beta_1 \Rightarrow \beta_2] \\ \mathcal{D}_i \\ \gamma\end{matrix} \quad \cdots \quad \begin{matrix}[\pi_{mj_m}] \\ \mathcal{D}_m \\ \gamma\end{matrix}}{\gamma}\ \delta\text{-E}$$

where discharged assumption π_{ij_i} is of higher level, assuming a derivation of β_2 from β_1. First, in any application, the assumption may have been discharged vacuously; that is, there may be a proof of γ not depending on the assumption of a derivation of β_2 from β_1 at all. Secondly, in the derivation of γ from the higher-level assumption of a derivation of β_2 from β_1, there may be some assumption made in the derivation of β_1 which is only discharged subsequent to the use of the higher-level assumption:

$$\begin{matrix}\mathcal{D}'_i \\ \beta_1 \\ \vdots \\ \beta_2 \\ \mathcal{D}''_i \\ \gamma\end{matrix}$$

That is, β_1 may depend on some assumption ϵ on which β_2, but not γ, also depends. So ϵ is discharged in the course of $\mathcal{D}''_i$. In a derivation using the flattened rule, however, ϵ is left undischarged in $\mathcal{D}'_i$ and is not available for use in $\mathcal{D}''_i$, which consequently is no longer a derivation.

We can see the problem plainly if we apply Davies and Pfenning's test for local completeness to $\alpha \odot \beta$. With the higher-level E-rule, a detour is easily introduced:

$$\dfrac{\alpha \odot \beta \quad \dfrac{[\alpha \Rightarrow \beta]}{\alpha \odot \beta}\ \odot\text{-I}_1 \quad \dfrac{[\beta \Rightarrow \alpha]}{\alpha \odot \beta}\ \odot\text{-I}_2}{\alpha \odot \beta}\ \odot\text{-E}$$

But with the flattened E-rule, the detour cannot be achieved:

$$\dfrac{\alpha \odot \beta \quad \alpha \quad \dfrac{[\beta]}{\alpha \odot \beta}\ \odot\text{-I}_1 \quad \beta \quad \dfrac{[\alpha]}{\alpha \odot \beta}\ \odot\text{-I}_2}{\alpha \odot \beta}\ \odot\text{-E}'$$

In the sub-derivation of $\alpha \odot \beta$ from β, α has to be discharged vacuously, as does β in the sub-derivation of $\alpha \odot \beta$ from α; and the re-introduction of $\alpha \odot \beta$ requires sub-proofs of α and β, which will generally not be possible.

It therefore seems that, in general, the higher-level rule is stronger than its flattened version. At least, this is clearly so in intuitionistic logic. Take the rules for the novel connective c_2 introduced in Schroeder-Heister (2014):

$$\frac{\begin{array}{c}[\alpha_1]\\ \vdots\\ \alpha_2\end{array}}{c_2(\alpha_1,\alpha_2,\alpha_3,\alpha_4)}\;c_2\text{-I}_1 \qquad\qquad \frac{\begin{array}{c}[\alpha_3]\\ \vdots\\ \alpha_4\end{array}}{c_2(\alpha_1,\alpha_2,\alpha_3,\alpha_4)}\;c_2\text{-I}_2$$

Considerations of GE-harmony yield a single higher-level E-rule with two minor premises:

$$\frac{c_2(\alpha_1,\alpha_2,\alpha_3,\alpha_4) \qquad \begin{array}{c}[\alpha_1\Rightarrow\alpha_2]\\ \vdots\\ \gamma\end{array} \qquad \begin{array}{c}[\alpha_3\Rightarrow\alpha_4]\\ \vdots\\ \gamma\end{array}}{\gamma}\;c_2\text{-E}$$

and the corresponding flattened rule:

$$\frac{c_2(\alpha_1,\alpha_2,\alpha_3,\alpha_4) \qquad \alpha_1 \quad \begin{array}{c}[\alpha_2]\\ \vdots\\ \gamma\end{array} \qquad \alpha_3 \quad \begin{array}{c}[\alpha_4]\\ \vdots\\ \gamma\end{array}}{\gamma}\;c_2\text{-E}'$$

Let $c_2(\vec{\alpha})$ abbreviate $c_2(\alpha_1,\alpha_2,\alpha_3,\alpha_4)$, and $\vee(\vec{\alpha})$ abbreviate $(\alpha_1\to\alpha_2)\vee(\alpha_3\to\alpha_4)$. With the higher-level rule, c_2-E, we can show by intuitionistically acceptable means that $c_2(\vec{\alpha}) \dashv\vdash (\alpha_1\to\alpha_2)\vee(\alpha_3\to\alpha_4)$:

$$\dfrac{c_2(\vec{\alpha}) \qquad \dfrac{\dfrac{\dfrac{\overline{\alpha_1}^{\,(1)} \qquad \overline{\alpha_1\Rightarrow\alpha_2}^{\,(2)}}{\alpha_2}}{\alpha_1\to\alpha_2}\;\to\text{-I}_1(1)}{\vee(\vec{\alpha})} \qquad \dfrac{\dfrac{\dfrac{\overline{\alpha_3}^{\,(3)} \qquad \overline{\alpha_3\Rightarrow\alpha_4}^{\,(4)}}{\alpha_4}}{\alpha_3\to\alpha_4}\;\to\text{-I}_2(3)}{\vee(\vec{\alpha})}}{(\alpha_1\to\alpha_2)\vee(\alpha_3\to\alpha_4)}\;c_2\text{-E}(2,4)$$

Conversely,

$$\dfrac{(\alpha_1\to\alpha_2)\vee(\alpha_3\to\alpha_4) \qquad \dfrac{\dfrac{\overline{\alpha_1}^{\,(1)} \qquad \overline{\alpha_1\to\alpha_2}^{\,(2)}}{\alpha_2}}{c_2(\vec{\alpha})}\;c_2\text{-I}(1) \qquad \dfrac{\dfrac{\overline{\alpha_3}^{\,(3)} \qquad \overline{\alpha_3\to\alpha_4}^{\,(4)}}{\alpha_4}}{c_2(\vec{\alpha})}\;c_2\text{-I}(3)}{c_2(\vec{\alpha})}\;\vee\text{-E}(2,4)$$

With the flattened rules, however, it is not possible to derive $(\alpha_1\to\alpha_2)\vee(\alpha_3\to\alpha_4)$ from $c_2(\vec{\alpha})$ using intuitionistically valid rules. But with classical *reductio*, $\mathcal{CR}$, it is possible:

$$\dfrac{\dfrac{c_2(\vec{\alpha})\quad \dfrac{\dfrac{\dfrac{\dfrac{\dfrac{\overline{\neg\alpha_1}^{(2)}\quad \overline{\alpha_1}^{(3)}}{\alpha_2}\,\mathcal{V}}{\alpha_1\to\alpha_2}\,{\to}\text{-I}(3)}{\vee(\vec{\alpha})}\quad \overline{\neg\vee(\vec{\alpha})}^{(1)}}{\alpha_1}\,\mathcal{CR}(2)\quad \dfrac{\dfrac{\overline{\alpha_2}^{(4)}}{\alpha_1\to\alpha_2}}{\vee(\vec{\alpha})}\quad \dfrac{\dfrac{\dfrac{\dfrac{\overline{\neg\alpha_3}^{(5)}\quad \overline{\alpha_3}^{(6)}}{\alpha_4}\,\mathcal{V}}{\alpha_3\to\alpha_4}\,{\to}\text{-I}(6)}{\vee(\vec{\alpha})}\quad \overline{\neg\vee(\vec{\alpha})}^{(1)}}{\alpha_3}\,\mathcal{CR}(5)\quad \dfrac{\dfrac{\overline{\alpha_4}^{(7)}}{\alpha_3\to\alpha_4}}{\vee(\vec{\alpha})}}{\vee(\vec{\alpha})}\,c_2\text{-E}'(4,7)\quad \overline{\neg\vee(\vec{\alpha})}^{(1)}}{(\alpha_1\to\alpha_2)\vee(\alpha_3\to\alpha_4)}\,\mathcal{CR}(1)$$

There is more than the *consequentia mirabilis* role of $\mathcal{CR}$ in play here (that is, to infer α from a demonstration that $\neg\alpha$ leads to contradiction). There is also much use of K and W in the multiple and vacuous discharge of assumptions in $\mathcal{CR}$ and $\to$-I.

$\odot$ is a special case of c_2, with $\alpha_1 = \alpha_4$ and $\alpha_2 = \alpha_3$, so we have a proof using classical *reductio* with the flattened E-rule that $(\alpha \to \beta) \vee (\beta \to \alpha) \dashv\vdash \alpha \odot \beta$. The reason $\alpha \odot \beta$ is classically derivable, and that $c_2(\vec{\alpha}) \dashv\vdash (\alpha_1 \to \alpha_2) \vee (\alpha_3 \to \alpha_4)$ is that the classical negation rules yield the full classical theory of implication, as in the multiple-conclusion sequent calculus, and the classical left-implication sequent calculus rule $\to$-left is invertible, that is, if $\Gamma, \alpha \to \beta \Rightarrow \Delta$ is derivable, so are $\Gamma \Rightarrow \alpha, \Delta$ and $\Gamma, \beta \Rightarrow \Delta$.[14] The classical negation rules of natural deduction and sequent calculus allow single-conclusion systems to mimic multiple-conclusion (at least to this extent)[15] by "parking" the negations of the parametric succedent formulae as antecedent formulae (i.e., assumptions). Consider the following multiple-conclusion sequent calculus proof that $c_2(\vec{\alpha}) \vdash \vee(\vec{\alpha})$:

$$\dfrac{\dfrac{\dfrac{\dfrac{\alpha_1\Rightarrow\alpha_1}{\alpha_1\Rightarrow\alpha_1,\alpha_2}}{\Rightarrow\alpha_1,\alpha_1\to\alpha_2}}{\Rightarrow\alpha_1,\vee(\vec{\alpha})}\quad \dfrac{\dfrac{\dfrac{\alpha_2\Rightarrow\alpha_2}{\alpha_2,\alpha\Rightarrow\alpha_2}}{\alpha_2\Rightarrow\alpha\to\alpha_2}}{\alpha_2\Rightarrow\vee(\vec{\alpha})}\quad \dfrac{\dfrac{\dfrac{\alpha_3\Rightarrow\alpha_3}{\alpha_3\Rightarrow\alpha_3,\alpha_4}}{\Rightarrow\alpha_3,\alpha_3\to\alpha_4}}{\Rightarrow\alpha_3,\vee(\vec{\alpha})}\quad \dfrac{\dfrac{\dfrac{\alpha_4\Rightarrow\alpha_4}{\alpha_4,\alpha_4\Rightarrow\alpha_4}}{\alpha_4\Rightarrow\alpha_4\to\alpha_4}}{\alpha_4\Rightarrow\vee(\vec{\alpha})}}{c_2(\vec{\alpha})\Rightarrow\vee(\vec{\alpha})}\,c_2\text{-left}$$

The rule c_2-left used here reads:

$$\dfrac{\Gamma\Rightarrow\alpha_1,\Delta\quad \Gamma,\alpha_2\Rightarrow\Delta\quad \Gamma\Rightarrow\alpha_3,\Delta\quad \Gamma,\alpha_4\Rightarrow\Delta}{\Gamma,c_2\Rightarrow\Delta}\,c_2\text{-left}$$

The rules for negation in Gentzen's LK, his classical sequent calculus, are:

$$\dfrac{\Gamma\Rightarrow\Delta,\alpha}{\neg\alpha,\Gamma\Rightarrow\Delta}\,\neg\text{-left}\qquad\qquad \dfrac{\alpha,\Gamma\Rightarrow\Delta}{\Gamma\Rightarrow\Delta,\neg\alpha}\,\neg\text{-right}$$

With these rules, within the multiple-conclusion system LK we can establish the following derived rule:

$$\dfrac{\neg\alpha,\Gamma\Rightarrow\Delta}{\Gamma\Rightarrow\Delta,\alpha}\,\mathcal{C}\qquad \text{with proof:}\qquad \dfrac{\dfrac{\neg\alpha,\Gamma\Rightarrow\Delta}{\Gamma\Rightarrow\Delta,\neg\neg\alpha}\quad \dfrac{\dfrac{\alpha\Rightarrow\alpha}{\Rightarrow\alpha,\neg\alpha}}{\neg\neg\alpha\Rightarrow\alpha}}{\Gamma\Rightarrow\Delta,\alpha}\,\text{Cut}$$

[14] See, e.g., Proposition 3.5.4 (vi) in Troelstra and Schwichtenberg (2000, p. 79).

[15] But see Murzi and Hjortland (2009).

If we now move to a single-conclusion sequent calculus, we can use $\mathcal{C}$ (with Δ empty)[16] to derive two further classical single-conclusion negation rules, $\mathcal{CT}$ (i.e., contraposition) and $\mathcal{CM}$ (i.e., *consequentia mirabilis*):

$$\dfrac{\Gamma,\neg\alpha \Rightarrow \beta}{\Gamma,\neg\beta \Rightarrow \alpha}\,\mathcal{CT} \qquad \text{with proof:} \qquad \dfrac{\dfrac{\Gamma,\neg\alpha \Rightarrow \beta}{\Gamma,\neg\beta,\neg\alpha \Rightarrow}\,\neg\text{-left}}{\Gamma,\neg\beta \Rightarrow \alpha}\,\mathcal{C}$$

$$\dfrac{\Gamma,\neg\alpha \Rightarrow \alpha}{\Gamma \Rightarrow \alpha}\,\mathcal{CM} \qquad \text{with proof:} \qquad \dfrac{\dfrac{\dfrac{\Gamma,\neg\alpha \Rightarrow \alpha}{\Gamma,\neg\alpha,\neg\alpha \Rightarrow}\,\neg\text{-left}}{\Gamma,\neg\alpha \Rightarrow}\,\mathrm{W}}{\Gamma \Rightarrow \alpha}\,\mathcal{C}$$

Then we can establish $c_2(\vec{\alpha}) \vdash \vee(\vec{\alpha})$ in single-conclusion sequent calculus using $\mathcal{CT}$ and $\mathcal{CM}$:

$$\dfrac{\dfrac{\dfrac{\dfrac{\dfrac{\dfrac{\dfrac{\alpha_1 \Rightarrow \alpha_1}{\neg\alpha_2,\alpha_1 \Rightarrow \alpha_1}}{\neg\alpha_1,\alpha_1 \Rightarrow \alpha_2}\,\mathcal{CT}}{\neg\alpha_1 \Rightarrow \alpha_1 \to \alpha_2}}{\neg\alpha_1 \Rightarrow \bigvee(\vec{\alpha})}}{\neg\bigvee(\vec{\alpha}) \Rightarrow \alpha_1}\,\mathcal{CT} \quad \dfrac{\dfrac{\dfrac{\alpha_2 \Rightarrow \alpha_2}{\alpha_2,\alpha_1 \Rightarrow \alpha_2}}{\alpha_2 \Rightarrow \alpha_1 \to \alpha_2}}{\alpha_2 \Rightarrow \bigvee(\vec{\alpha})} \quad \dfrac{\dfrac{\dfrac{\dfrac{\dfrac{\alpha_3 \Rightarrow \alpha_3}{\neg\alpha_4,\alpha_3 \Rightarrow \alpha_3}}{\neg\alpha_3,\alpha_3 \Rightarrow \alpha_4}\,\mathcal{CT}}{\neg\alpha_3 \Rightarrow \alpha_3 \to \alpha_4}}{\neg\alpha_3 \Rightarrow \bigvee(\vec{\alpha})}}{\neg\bigvee(\vec{\alpha}) \Rightarrow \alpha_3}\,\mathcal{CT} \quad \dfrac{\dfrac{\dfrac{\alpha_4 \Rightarrow \alpha_4}{\alpha_4,\alpha_3 \Rightarrow \alpha_4}}{\alpha_4 \Rightarrow \alpha_3 \to \alpha_4}}{\alpha_4 \Rightarrow \bigvee(\vec{\alpha})}}{\neg\bigvee(\vec{\alpha}), c_2(\vec{\alpha}) \Rightarrow \bigvee(\vec{\alpha})}\,c_2\text{-left}}{c_2(\vec{\alpha}) \Rightarrow (\alpha_1 \to \alpha_2) \vee (\alpha_3 \to \alpha_4)}\,\mathcal{CM}$$

This proof exhibits essentially the same proof architecture as the earlier proof of the same result in classical natural deduction.

13.4 Negation and Inconsistency

Often, $\neg\alpha$ is treated by definition as $\alpha \to \bot$, where $\bot$ is governed solely by an elimination-rule, from $\bot$ infer anything:

$$\dfrac{\bot}{\alpha}\,\bot\mathrm{E}$$

Gentzen (1932) treated '$\neg$' as primitive. As introduction-rule, he took *reductio ad absurdum*, as noted in Sect. 13.3:

$$\dfrac{\begin{matrix}[\alpha] \\ \vdots \\ \beta\end{matrix} \qquad \begin{matrix}[\alpha] \\ \vdots \\ \neg\beta\end{matrix}}{\neg\alpha}\,\mathcal{R}$$

[16] If we require the succedent to be non-empty, we can capture "empty" succedent with an instance of $\bot$.

What elimination-rule does this justify? We can infer from $\neg\alpha$ whatever (all and only that which) we can infer from its grounds. There is one I-rule with two premises ($m = 1, n_1 = 2$), so there will be two E-rules, one for each premise of the I-rule:

$$\dfrac{\neg\alpha \qquad \begin{array}{c}\left[\begin{array}{c}[\alpha]\\ \vdots\\ \beta\end{array}\right]\\ \vdots\\ \gamma\end{array}}{\gamma}\,\neg\mathrm{E}_1 \qquad \text{and} \qquad \dfrac{\neg\alpha \qquad \begin{array}{c}\left[\begin{array}{c}[\alpha]\\ \vdots\\ \neg\beta\end{array}\right]\\ \vdots\\ \gamma\end{array}}{\gamma}\,\neg\mathrm{E}_2$$

Flattening the rules as before, where we infer γ from assuming the existence of derivations, respectively, of β and of $\neg\beta$ from α, we obtain:

$$\dfrac{\neg\alpha \qquad \alpha \qquad \begin{array}{c}[\beta]\\ \vdots\\ \gamma\end{array}}{\gamma} \qquad \text{and so} \qquad \begin{array}{c}\dfrac{\neg\alpha \qquad \alpha}{\beta}\\ \vdots\\ \gamma\end{array}$$

and

$$\dfrac{\neg\alpha \qquad \alpha \qquad \begin{array}{c}[\neg\beta]\\ \vdots\\ \gamma\end{array}}{\gamma} \qquad \text{and so} \qquad \begin{array}{c}\dfrac{\neg\alpha \qquad \alpha}{\neg\beta}\\ \vdots\\ \gamma\end{array}$$

The second of these is simply a special case of the first, and so we have justified Gentzen's form of *Ex Falso Quodlibet* as the matching E-rule for '$\neg$': [17]

---

[17] Recall from Sect. 13.2 Dummett's minimal I-rule for '$\neg$':

$$\dfrac{\alpha \Rightarrow \neg\alpha}{\neg\alpha}\,\neg\text{-I}$$

GE-considerations justify as E-rule:

$$\dfrac{\neg\alpha \qquad \begin{array}{c}[\alpha \Rightarrow \neg\alpha]\\ \vdots\\ \gamma\end{array}}{\gamma}$$

which flattens to

$$\dfrac{\neg\alpha \qquad \alpha}{\neg\alpha}$$

But this adds nothing to what was already derivable using $\neg$-I.:

$$\frac{\neg\alpha \qquad \alpha}{\beta}\,\mathcal{V}$$

The account of negation given by $\mathcal{R}$ and $\mathcal{V}$ is intuitionistic. But similar arguments extend this account to classical negation, by setting it in a multiple-conclusion framework.[18] $\mathcal{R}$ generalizes to a multiple-conclusion rule as:

$$\frac{\begin{matrix}[\alpha]\\ \vdots \\ \beta,\Delta\end{matrix} \qquad \begin{matrix}[\alpha]\\ \vdots \\ \neg\beta,\Delta\end{matrix}}{\neg\alpha,\Delta}\,\mathcal{R}_m$$

from which the inversion principle yields the pair of higher-order E-rules:

$$\frac{\neg\alpha,\Gamma \qquad \begin{matrix}[\alpha \Rightarrow \beta]\\ \vdots \\ \Delta\end{matrix}}{\Gamma,\Delta} \qquad \text{and} \qquad \frac{\neg\alpha,\Gamma \qquad \begin{matrix}[\alpha \Rightarrow \neg\beta]\\ \vdots \\ \Delta\end{matrix}}{\Gamma,\Delta}$$

which flatten and simplify as before to

$$\frac{\neg\alpha,\Gamma \qquad \alpha,\Delta}{\Gamma,\Delta}\,\mathcal{V}_m$$

From $\mathcal{R}_m$ and $\mathcal{V}_m$ we can derive double-negation elimination, and so justify $\mathcal{CR}$, as derived rules:

$$\dfrac{\neg\neg\alpha \qquad \dfrac{\dfrac{\dfrac{}{\alpha}\,(1)}{\alpha,\alpha}\,\mathrm{K} \qquad \dfrac{\dfrac{}{\alpha}\,(1)}{\neg\alpha,\alpha}\,\mathrm{K}}{\neg\alpha,\alpha}\,\mathcal{R}_m(1)}{\alpha}\,\mathcal{V}_m$$

Finally, consider the one-place connective, $\bullet$, whose single introduction-rule has one hypothetical premise:

$$\frac{\begin{matrix}[\bullet\alpha]\\ \vdots \\ \alpha\end{matrix}}{\bullet\,\alpha}\,\bullet\text{-I}$$

GE-harmony yields as E-rule in the usual way:

$$\frac{\bullet\alpha \qquad \begin{matrix}[\bullet\alpha \Rightarrow \alpha]\\ \vdots \\ \gamma\end{matrix}}{\gamma}\,\bullet\text{-E} \qquad \text{which flattens to} \qquad \frac{\bullet\alpha \qquad \bullet\alpha}{\alpha}\,\bullet\text{-E}$$

[18] See Read (2000, pp. 149–150).

where $\bullet\alpha$ is both major and minor premise. $\bullet$ satisfies the inversion principle, whereby

$$\dfrac{\dfrac{\begin{matrix}[\bullet\alpha]\\ \mathcal{D}\\ \alpha\end{matrix}}{\bullet\alpha}\,\bullet\text{-I} \qquad \begin{matrix}\mathcal{D}'\\ \bullet\alpha\end{matrix}}{\alpha}\,\bullet\text{-E} \qquad \text{converts to} \qquad \begin{matrix}\mathcal{D}'\\ \bullet\alpha\\ \mathcal{D}\\ \alpha\end{matrix}$$

$\bullet\alpha$ is a formal Curry paradox, for $\bullet$ introduces inconsistency, in fact, triviality, since we can prove α, for any α:

$$\dfrac{\dfrac{\dfrac{\overline{\bullet\alpha}^{\,1} \quad \overline{\bullet\alpha}^{\,1}}{\alpha}\,\bullet\text{-E}}{\bullet\alpha}\,\bullet\text{-I}\,[1] \qquad \dfrac{\dfrac{\overline{\bullet\alpha}^{\,2} \quad \overline{\bullet\alpha}^{\,2}}{\alpha}\,\bullet\text{-E}}{\bullet\alpha}\,\bullet\text{-I}\,[2]}{\alpha}\,\bullet\text{-E}$$

(Note, however, the use of Contraction in each application of $\bullet$-I.) The proof fails to normalize, since clearly, if we try to remove the maximum formula $\bullet\alpha$ in the left-hand premise of the final use of $\bullet$-E, we obtain just the same proof again. How can we prevent this? Should it be prevented?

One proposal is Dummett's complexity constraint:

> The minimal demand we should make on an introduction rule intended to be self-justifying is that its form be such as to guarantee that, in any application of it, the conclusion will be of higher complexity than any of the premisses and than any discharged hypothesis. We may call this the 'complexity condition'.

Although this rules out $\bullet$, and classical *reductio*, it also rules out apparently innocuous rules such as Gentzen's $\mathcal{R}$ above, and even Dummett's own $\neg$-I rule for minimal negation. The moral to draw is that GE-harmony is not designed to rule out anything, but to ensure that the E-rules add no more (and no less) to whatever meaning is given by the assertion-conditions encapsulated in the I-rule(s). The I-rule for $\bullet$ already shows its inconsistency, which in turn justifies $\bullet$-E. Harmony does not import inconsistency, but serves to make it transparent.

13.5 Conclusion

Michael Dummett introduced the notion of harmony in response to Arthur Prior's tonkish attack on the idea of proof-theoretic justification of logical laws (or analytic validity). Dummett developed the notion of harmony in different ways, in an attempt to use the idea to underpin his anti-realism. One of these ways (so-called "intrinsic harmony") drew on work by Dag Prawitz, in which he articulated an idea of Gerhard Gentzen's into a procedure based on Lorenzen's inversion principle whereby elimination-rules are in the appropriate sense functions of the corresponding introduction-rules. Roy Dyckhoff and Nissim Francez coined the term "general-elimination harmony" for the relationship created by this procedure.

GE-harmony ensures that meaning is given solely, and hence transparently, by the assertion-conditions encapsulated in the I-rule(s), in such a way that the E-rule(s) add no more and no less to whatever meaning is given by the assertion-conditions encapsulated in the I-rule(s). The E-rules which result may be of higher level, permitting the discharge of rules as well as formulae. Such rules can be flattened to rules in which only formulae are discharged, but only in the context of classical *reductio*, or a multiple-conclusion format, can we be sure that the flattened rules will be equivalent to the higher-level rules. In general, the flattened rules are weaker than the I-rule warrants, and so are not in Dummett's term, stable.

References

Curry, H. (1950). *A theory of formal deducibility*. Notre Dame: Indiana: University of Notre Dame Press.

Davies, R., & Pfenning, F. (2001). A modal analysis of staged computation. *Journal of the ACM*, *48*, 555–604.

Dummett, M. (1973). *Frege: philosophy of language*. London: Duckworth.

Dummett, M. (1991). *Logical basis of metaphysics*. London: Duckworth.

Dyckhoff, R. (1988). Implementing a simple proof assistant. In J. Derrick & H. Lewis (Eds.), *Proceedings of the workshop on programming for logic teaching, leeds 1987, Centre for theoretical computer science and departments of pure mathematics and philosophy* (pp. 49–59). Leeds: University of Leeds.

Dyckhoff, R. (2013). General[-ised] elimination rules, http://rd.host.cs.st-andrews.ac.uk/talks/2013/Tuebingen-PTS2-talk.pdf.

Fitch, F. (1952). *Symbolic logic: an introduction*. New York: The Ronald Press Co.

Francez, N., & Dyckhoff, R. (2012). A note on harmony. *Journal of Philosophical Logic*, *41*, 613–28.

Gentzen G (1932) Untersuchungen über das logische Schliessen. Manuscript 974:271 in Bernays Archive, Eidgenössische Technische Hochschule Zürich.

Gentzen, G. (1969). Investigations concerning logical deduction. In M. Szabo (Ed.), *The Collected Papers of Gerhard Gentzen* (pp. 68–131). Amsterdam: North-Holland.

Girard, J. Y., Taylor, P., & Lafont, Y. (1989). *Proofs and types*. Cambridge: Cambridge University Press.

Hermes, H. (1959). Zum Inversionsprinzip der operativen Logik. In A. Heyting (Ed.), *Constructivity in Mathematics* (pp. 62–68). Amsterdam: North-Holland.

Lorenzen, P. (1955). *Einführung in die operative Logik und Mathematik*. Berlin: Springer.

Murzi, J., & Hjortland, O. (2009). Inferentialism and the categoricity problem: reply to Raatikainen. *Analysis*, *69*, 480–88.

Negri, S., & von Plato, J. (2001). *Structural proof theory*. Cambridge: Cambridge UP.

Prawitz, D. (1965). *Natural Deduction*. Stockholm: Almqvist and Wiksell.

Prawitz, D. (1973). Towards the foundation of a general proof theory. In P. Suppes, L. Henkin, A. Joja, & G. Moisil (Eds.), *Logic, methodology and philosophy of science IV: proceedings of the 1971 international congress* (pp. 225–50). Amsterdam: North-Holland.

Prawitz, D. (1974). On the idea of a general proof theory. *Synthese*, *27*, 63–77.

Prawitz, D. (1975). Ideas and results in proof theory. In J. Fenstad (Ed.), *Proceedings of the second scandinavian logic symposium* (pp. 235–50). Amsterdam: North-Holland.

Prawitz, D. (1985). Remarks on some approaches to the concept of logical consequence. *Synthese*, *62*, 153–171.

Prawitz, D. (2006). Meaning approached via proofs. *Synthese*, *148*, 507–524.

Prior, A. (1960). The runabout inference ticket. *Analysis*, *21*, 38–39.

Read, S. (2000). Harmony and autonomy in classical logic. *Journal of Philosophical Logic*, *29*, 123–154.

Read, S. (2008). Harmony and modality. In C. Dégremont, L. Kieff, & H. Rückert (Eds.), *Dialogues, logics and other strange things: essays in honour of Shahid Rahman* (pp. 285–303). London: College Publications.

Read, S. (2010). General-elimination harmony and the meaning of the logical constants. *Journal of Philosophical Logic*, *39*, 557–76.

Read, S. (2014). Proof-theoretic validity. In O. Hjortland (Ed.), *Caret C*. Foundations of Logical Consequence: Oxford University Press.

Schroeder-Heister, P. (1984). A natural extension of natural deduction. *Journal of Symbolic Logic*, *49*, 1284–1300.

Schroeder-Heister, P. (2006). Validity concepts in proof-theoretic semantics. *Synthese*, *148*, 525–571.

Schroeder-Heister, P. (2007). Generalized definitional reflection and the inversion principle. *Logica Universalis*, *1*, 355–76.

Schroeder-Heister, P. (2014). Generalized elimination inference, higher-level rules, and the implications-as-rules interpretation of the sequent calculus. In E. Haeusler, L. C. Pereira, & V. de Palva (Eds.), *Advances in natural deduction* (pp. 1–29). Berlin: Springer.

Tennant, N. (n.d.) Inferentialism, logicism, harmony, and a counterpoint. To appear in ed. Alex Miller, *Essays for Crispin Wright: Logic, Language and Mathematics* (in preparation for Oxford University Press: Volume 2 of a two-volume Festschrift for Crispin Wright, co-edited with Annalisa Coliva), http://people.cohums.ohio-state.edu/tennant9/crispin_rev.pdf.

Troelstra, A., & Schwichtenberg, H. (2000). *Basic proof theory* (2nd ed.). Cambridge: Cambridge University Press.

von Plato, J. (2001). Natural deduction with general elimination rules. *Archive for Mathematical Logic*, *40*, 521–47.

Chapter 14
Hypothesis-Discharging Rules in Atomic Bases

Tor Sandqvist

Abstract This paper investigates the idea, familiar *inter alia* from Prawitz, that an inference is to be deemed *valid*, relative to a basis of inference rules for atomic sentences, just in case every extension of that basis supporting the premisses of the inference also supports its conclusion. Specifically, we try to carry out this idea in a setting where atomic bases are allowed to contain rules that license the discharging of hypotheses. The results are mixed. While the ensuing concept of validity appears to be an extensionally adequate one, from a conceptual point of view the theory is deemed unsatisfactory in that certain inferences come out valid, as it were, for the wrong reason. So, for instance, an atomic rule contained in a basis will qualify as valid relative to that basis, but not simply in virtue of the fact that it occurs there; its validity also depends on the assumption that all bases conform to a particular format, thus lending a peculiarly holistic character to the theory.

Keywords Assumption · Atomic inference rule · Discharge · Extension · Hypothesis · Inference · Validity

14.1 Introduction

In standard classical truth-conditional semantics, the meanings of sentential logical operators are specified by describing how the truth conditions of logically compound sentences are determined by those of their immediate subsentences, and ultimately by the truth conditions of logically atomic sentences. For the purposes of logical investigation, the truth conditions of logical atoms are ordinarily considered as primitively given, and of little interest in themselves; what matters are such properties of sentences as remain constant over all atomic valuations—in the first place, of course, constant truth, i.e. logical validity.

T. Sandqvist (✉)
Department of Philosophy and History of Technology,
School of Architecture and the Built Environment,
Royal Institute of Technology (KTH), 100 44 Stockholm, Sweden
e-mail: tor.sandqvist@abe.kth.se

H. Wansing (ed.), *Dag Prawitz on Proofs and Meaning*, Outstanding Contributions to Logic 7, DOI 10.1007/978-3-319-11041-7_14

Our aim in this article will be to do something similar within the framework of proof-theoretical semantics. In general terms, the project may be described as follows. The central semantic notion, corresponding to the concept of *truth* in the classical framework, is the notion of *correctness of an inference*. A system of inference rules dealing with logically atomic sentences is considered as given, and on the basis of these atomic rules a notion of correctness of inference among sentences in general is defined. Just as in the truth-conditional case, the expansion of the semantic notion from the atomic fragment to the language as a whole is required to be *conservative*; that is, just as a classical truth evaluation of the full language on the basis of a valuation of atomic sentences is expected to leave truth values of atoms as they are, so, in our present context, the assessment of an inference among atoms as correct or incorrect should not change when inferences between sentences in general are evaluated on the basis of those between atoms.

The ambition will be not merely to produce an inference relation with a satisfactory extension, but to construe this relation as a natural outgrowth of its atomic foundation, determined by the meanings of logically compound sentences, as specified by a simple semantic clause for each logical operator—again in analogy with classical truth-functional semantics. As we shall find, the difficulty of obtaining such a construction varies greatly depending on what the general form of the underlying atomic inference rules is taken to be. If atomic rules are assumed always to take the form of *production rules*, proceeding from premisses to conclusion without altering the set of hypothetical assumptions under which reasoning is taking place, then, as will be seen in Sect. 14.2, the task is a straightforward one. But once this restriction is relaxed and atomic rules are allowed to discharge hypotheses (in the manner of, e.g., the standard rule of $\supset$ introduction), problems arise. Three different attempts are made in Sects. 14.3, 14.4, and 14.5, none of them more than partially successful. The question of what to make of this unsatisfactory state of affairs is briefly discussed in Sect. 14.6, but no firm conclusion will be reached; if the paper has any merit, it is that of directing attention to a puzzle, rather than offering a solution.

The question we are ultimately addressing is this: What is it for an inference among sentences in a language possessing logical vocabulary to be *valid*, relative to a set of rules governing atomic inferences? The investigation is akin to, and inspired by, Dag Prawitz's quest, in Prawitz (1971, 1973), and elsewhere, for a philosophically illuminating general notion of *validity of an argument*, but differs from it in a couple of respects. In the first place, unlike Prawitz, whose investigations have typically centered on various transformation operations on argument structures, in the name of technical simplicity I shall for the most part be abstracting from the internal workings of arguments, focussing chiefly on the relationship between their premisses and conclusions. Secondly, and less obviously, there is probably a difference in our respective construals of the notion of an *atomic basis*. Whereas Prawitz is usually inclined to think of the rules in an atomic basis primarily as *meaning-defining introduction rules* for atomic sentences (analogous to those for logical compounds) in terms of which other inference rules are to be justified—including, perhaps, other rules dealing with atoms—my conception of an atomic basis is rather that of a *totality* of atomic rules accepted, for whatever reasons, by an agent or linguistic collective at a particular time.

In the present study, our attention will be confined to a single logical operator: the conditional connective $\supset$.

14.2 Type-1 Bases

In this section we shall be working under the assumption that rules governing the use of logically atomic sentences never allow the discharging of temporary hypotheses. As described in the Introduction, this will enable us in a straightforward way to extend a given set of atomic rules to a consequence relation pertaining to a language of arbitrarily deeply nested conditionals. In order to provide a foundation for further development in subsequent sections, this comparatively simple matter will here be given a somewhat elaborate exposition.

Imagine, then, a linguistic practice involving a language devoid of any logical vocabulary, but subject to various rules, each of the form:

(1) *If you are in a position to assert all members of P, then you may assert q,*

where P is a finite set of sentences and q a sentence; when $P = \emptyset$, the rule is simply a license to assert q. A rule of this kind will be referred to as a *rule of type 1*. For technical purposes, it will be convenient to identify it with the ordered pair (P, q), henceforth to be referred to as "$P \Rightarrow q$". We will adopt the usual convention of suppressing set-theoretical notation, writing "P, Q" for "$P \cup Q$", "P, r" for "$P \cup \{r\}$", "$\Rightarrow q$" for "$\emptyset \Rightarrow q$", etc.

By a *basis of type 1* we shall understand any set of type-1 rules; i.e. any relation between finite sets of sentences and sentences in our logic-free language. For a given type-1 basis $\mathbf{B}$, we specify inductively a set $A_{\mathbf{B}}$ to be intuitively thought of as the set of sentences assertable on the strength of $\mathbf{B}$:

(2) If $P \Rightarrow q \in \mathbf{B}$ and $P \subseteq A_{\mathbf{B}}$ then $q \in A_{\mathbf{B}}$.
(3) Only as required by (2) is any object a member of $A_{\mathbf{B}}$.

For instance, if $\mathbf{B}$ contains the rules $\Rightarrow a$, $a \Rightarrow b$, and $a, b \Rightarrow c$, then (2) implies that $a \in A_{\mathbf{B}}$ since $\Rightarrow a \in \mathbf{B}$, whence $b \in A_{\mathbf{B}}$ since $a \Rightarrow b \in \mathbf{B}$, meaning that $c \in A_{\mathbf{B}}$ since $a, b \Rightarrow c \in \mathbf{B}$.

If we agree to call a set X *closed under* $P \Rightarrow q$ if it satisfies the condition that $q \in X$ if $P \subseteq X$, and *closed under* $\mathbf{B}$ if it is closed under every rule in $\mathbf{B}$, then $A_{\mathbf{B}}$ may be equivalently characterized as the smallest set closed under $\mathbf{B}$.

Now a set's being closed under $\mathbf{B}$ may well guarantee its being closed under various sequents not themselves in $\mathbf{B}$. For instance, if $\mathbf{B}$ contains $a \Rightarrow b$ and $b \Rightarrow c$, any set X closed under $\mathbf{B}$ will necessarily be closed under $a \Rightarrow c$ as well. This sort of implicit endorsement of a sequent $P \Rightarrow q$ on the part of a basis $\mathbf{B}$ can be enunciated either, as in the preceding sentence and in (4) below, by means of quantification over *sets of sentences*, or, as in (5), by quantifying over *bases*. That is to say, the following conditions are equivalent:

(4) Every set closed under $\mathbf{B}$ is closed under $P \Rightarrow q$.
(5) For every basis $\mathbf{C} \supseteq \mathbf{B}$, $A_{\mathbf{C}}$ is closed under $P \Rightarrow q$.

That (4) implies (5) is evident from the fact that $A_{\mathbf{C}}$ will be a set closed under $\mathbf{B}$. For the converse implication, assume (5), and consider the basis $\mathbf{B}' = \mathbf{B} \cup \{\Rightarrow p \mid p \in P\}$. By hypothesis $A_{\mathbf{B}'}$ is closed under $P \Rightarrow q$, and clearly $P \subseteq A_{\mathbf{B}'}$; hence $q \in A_{\mathbf{B}'}$. To see that (4) holds good, consider an arbitrary set X closed under $\mathbf{B}$ and assume that $P \subseteq X$. Then X is closed under $\mathbf{B}'$, and so must be a superset of $A_{\mathbf{B}'}$, whence $q \in X$, as required.

When this condition expressed in different ways by (4) and (5) obtains, we shall write "$P \Vdash_{\mathbf{B}} q$". In the special case where P is empty, obviously $\Vdash_{\mathbf{B}} q$ iff $q \in A_{\mathbf{B}}$.

One intuitively appealing way of conceiving of atomic bases is as representations of possible states of information, subject to expansion by new rules as observations are made and insights gained. Under such a conception, formulation (5) makes plain why a speaker of our imagined language who considers himself subject to all rules in a basis $\mathbf{B}$ might take an interest in the question whether $P \Vdash_{\mathbf{B}} q$ for certain nonempty P. Even if you are not today in a position to assert all members of P, in planning for future action it may well be useful to be able to recognize in advance that, should your body of knowledge ever expand so as to bring you into such a position, at that point q is sure to be assertable as well.

What can be usefully *recognized* can be meaningfully *said*. Accordingly, we imagine, the speakers of our primitive language now introduce a logical operator, the *conditional*, subject to the rule: $\Vdash_{\mathbf{B}} p \supset q$ iff $p \Vdash_{\mathbf{B}} q$. That is to say, $p \supset q$ is assertable on the strength of a basis just in case that basis supports the inference from p to q in the sense variously expressed by (4) and (5).

Since by definition a basis only deals with logically atomic sentences, condition (4) only makes sense as applied to atoms. A criterion along the lines of (5), by contrast, can be applied to sentences of arbitrary complexity—that is, to arbitrarily deeply nested conditionals. In this way we arrive at what may be described as an inductively characterizable assertion-conditional semantics for such sentences; where (upper-case) italic letters stand for (finite sets of) atomic sentences, and (upper-case) Greek letters for (finite sets of) sentences in general:

(6) If $P \Rightarrow q \in \mathbf{B}$, and $\Vdash_{\mathbf{B}} p$ for every p in P, then $\Vdash_{\mathbf{B}} q$.
(7) If Φ is nonempty, and $\Vdash_{\mathbf{C}} \psi$ for every $\mathbf{C} \supseteq \mathbf{B}$ such that $\Vdash_{\mathbf{C}} \varphi$ for every φ in Φ, then $\Phi \Vdash_{\mathbf{B}} \psi$.
(8) If $\varphi \Vdash_{\mathbf{B}} \psi$ then $\Vdash_{\mathbf{B}} \varphi \supset \psi$.
(9) Only as required by (6)–(8) does the relation $\Vdash$ obtain between any objects.

(Note how, in (7), $\Vdash_{\mathbf{C}}$ plays a role analogous to that of $A_{\mathbf{C}}$ in (5).)

Define the *degree* of an atomic sentence as 1, the *degree* of a conditional $\varphi \supset \psi$ as 1 plus the sum of the degrees of φ and ψ, and the *degree* of a set-sentence pair (Φ, ψ) as the sum of the degrees of ψ and the members of Φ. Any doubts as to the formal legitimacy of our definition of $\Vdash$ should be dispelled by the observation that, firstly, clause (6) taken by itself makes for an unproblematic inductive definition of $\Vdash$ as applied to individual sentences of degree 1, and secondly, in either one of clauses (7) and (8) the degree of the set-sentence pair generated exceeds that of any pair figuring in the conditions of application, meaning that the extension of $\Vdash$ among pairs of any degree $n > 1$ is fully determined by its extension among pairs of degrees less than n. Indeed, an equivalent definition, cast in recursive rather than inductive form, can be given as follows:

(10) $\Vdash_{\mathbf{B}} q$ iff every set closed under **B** contains q.
(11) Where Φ is nonempty, $\Phi \Vdash_{\mathbf{B}} \psi$ iff $\Vdash_{\mathbf{C}} \psi$ for every $\mathbf{C} \supseteq \mathbf{B}$ such that $\Vdash_{\mathbf{C}} \varphi$ for every φ in Φ.
(12) $\Vdash_{\mathbf{B}} \varphi \supset \psi$ iff $\varphi \Vdash_{\mathbf{B}} \psi$.

Whichever formulation one prefers, the relation $\Vdash$ thus defined is quite well-behaved. To begin with, it is trivially *monotonic* in the sense that $\Vdash_{\mathbf{B}} \subseteq \Vdash_{\mathbf{C}}$ whenever $\mathbf{B} \subseteq \mathbf{C}$. Put differently,

(13) if $\Phi \Vdash_{\mathbf{B}} \psi$ and $\mathbf{B} \subseteq \mathbf{C}$ then $\Phi \Vdash_{\mathbf{C}} \psi$.

By (11) and (13), it holds of any Φ, empty or not, that

(14) $\Phi \Vdash_{\mathbf{B}} \psi$ iff $\Vdash_{\mathbf{C}} \psi$ for every $\mathbf{C} \supseteq \mathbf{B}$ such that $\Vdash_{\mathbf{C}} \varphi$ for every φ in Φ.

Moreover, for any **B**, $\Vdash_{\mathbf{B}}$ is easily seen to be closed under the standard rules of Conditional Proof and Modus Ponens, alias $\supset$-introduction and $\supset$-elimination.

(15) If $\Theta, \varphi \Vdash_{\mathbf{B}} \psi$ then $\Theta \Vdash_{\mathbf{B}} \varphi \supset \psi$.
(16) If $\Theta \Vdash_{\mathbf{B}} \varphi \supset \psi$ and $\Theta \Vdash_{\mathbf{B}} \varphi$ then $\Theta \Vdash_{\mathbf{B}} \psi$.

Proof of (15). Assume that $\Theta, \varphi \Vdash_{\mathbf{B}} \psi$. In order to show that $\Theta \Vdash_{\mathbf{B}} \varphi \supset \psi$, consider any $\mathbf{C} \supseteq \mathbf{B}$ such that $\Vdash_{\mathbf{C}} \theta$ for every θ in Θ; by (14) it will suffice to show that $\Vdash_{\mathbf{C}} \varphi \supset \psi$, i.e. that $\varphi \Vdash_{\mathbf{C}} \psi$. So we consider an arbitrary $\mathbf{D} \supseteq \mathbf{C}$ such that $\Vdash_{\mathbf{D}} \varphi$. By monotonicity of $\Vdash$, $\Vdash_{\mathbf{D}} \theta$ for every θ in Θ. Since by hypothesis $\Theta, \varphi \Vdash_{\mathbf{B}} \psi$ and $\mathbf{B} \subseteq \mathbf{D}$, it follows that $\Vdash_{\mathbf{D}} \psi$, as required.

Proof of (16). Again we use (14). If $\Theta \Vdash_{\mathbf{B}} \varphi \supset \psi$ and $\Theta \Vdash_{\mathbf{B}} \varphi$, then, for every $\mathbf{C} \supseteq \mathbf{B}$ such that $\Vdash_{\mathbf{C}} \theta$ for every θ in Θ, we have $\Vdash_{\mathbf{C}} \varphi$ and $\Vdash_{\mathbf{C}} \varphi \supset \psi$, whence $\varphi \Vdash_{\mathbf{C}} \psi$, whence $\Vdash_{\mathbf{C}} \psi$, as required.

A further property of $\Vdash$ worth noting is this: For any basis **B**, closure under any rule $P \Rightarrow q$ in **B** is guaranteed, not only of $A_{\mathbf{B}}$, the set of atomic sentences holding *categorically* under **B**, but also, for any finite set Θ, of the set of sentences holding under **B** *conditionally on* Θ. That is to say,

(17) If $P \Rightarrow q \in \mathbf{B}$, and $\Theta \Vdash_{\mathbf{B}} p$ for every p in P, then $\Theta \Vdash_{\mathbf{B}} q$.

In other words, inference according to a rule of **B** remains justified under any set of hypothetical assumptions. The proof is trivial: for any $\mathbf{C} \supseteq \mathbf{B}$, if $\Vdash_{\mathbf{C}} \theta$ for every θ in Θ, then $\Vdash_{\mathbf{C}} p$ for every p in P, whence $\Vdash_{\mathbf{C}} q$ as required since $P \Rightarrow q \in \mathbf{C}$.

A final fact, which has already been mentioned but is worth pointing out explicitly as a point in favour of $\Vdash$, is the equivalence of conditions (5) and (4), which (since always $p \in A_{\mathbf{B}}$ iff $\Vdash_{\mathbf{B}} p$) may be restated as follows:

(18) $P \Vdash_{\mathbf{B}} q$ iff every set closed under **B** is closed under $P \Rightarrow q$;

or, equivalently, $P \Vdash_{\mathbf{B}} q$ iff q belongs to every set that includes P and is closed under **B**. $\Vdash_{\mathbf{B}}$, in other words, in its atomic fragment coincides with what one would naturally describe as derivability by means of the rules in **B**; $\Vdash$, that is, possesses the virtue of *conservativeness* discussed in the Introduction.

Thus far, our investigation has yielded no surprises: conservativeness, monotonicity, closure under $\supset$-introduction and $\supset$-elimination, and preservation of atomic rules under hypothetical assumptions are precisely what we should expect of a semantic entailment relation. The semantic theory itself, too, follows traditional lines. It is, in effect, a stripped-down version of Prawitz's (1971, Sect. A.1) notion of a (weakly) valid argument, restricted to the conditional connective and divested of all features relating to argument structures and their transformations. So far, so good.

14.3 A Failed Attempt at Generalization

Comparing clauses (6) and (8) from our inductive definition of $\Vdash$ one observes an important structural difference. According to that definition, the only way in which the assertion of an atomic sentence q can be justified on the strength of a basis is in virtue of some set of sentences P all of whose members are themselves *categorically assertable*. By contrast, a conditional $\varphi \supset \psi$ counts as assertable whenever its succedent ψ is *hypothetically inferable* from its antecedent φ. In the context of clause (8), our treatment of atomic sentences begins to look somewhat restrictive. Perhaps in the name of structural uniformity we ought to try to generalize our framework so as to allow for the possibility that an atomic sentence too may count as assertable in virtue of certain purely hypothetical inferences.

Informally, the general format of an atomic rule of this nature may be indicated as follows.

(19) *If you are in a position to infer* p_1 *from hypothetical premisses* R_1*, and* p_2 *from hypothetical premisses* R_2*, and ... and* p_n *from hypothetical premisses* R_n*, then you may assert* q,

where n is allowed to be 0 (in which case the rule is simply a license to assert q) and any of the R_i is allowed to be empty (in which case the condition in question is just that the speaker be in a position to assert p_i; if all of the R_i are empty we are, in effect, left with an instance of (1)). Formally, we shall let (19) be represented by the ordered pair $(\{R_1 \Rightarrow p_1, R_2 \Rightarrow p_2, \ldots, R_n \Rightarrow p_n\}, q)$, or, as we shall prefer to write it, $\mathbf{P} \Rightarrow q$, where $\mathbf{P} = \{R_1 \Rightarrow p_1, \ldots, R_n \Rightarrow p_n\}$. An ordered pair of this sort, whose first component is a finite set of type-1 rules and whose second component is an atomic sentence, will be referred to as a *rule of type 2*. A *basis of type 2* is a set of type-2 rules; such sets will be referred to as "$\mathcal{B}$", "$\mathcal{C}$", etc.[1]

What might a generalization to type-2 bases of the inductive definition consisting of clauses (6) through (9) look like? A *prima facie* reasonable idea is to keep (7) and (8) as they are and modify (6) so as to deal with type-2 rules in the most straightforward way possible:

(20) If $\mathbf{P} \Rightarrow q \in \mathcal{B}$, and $R \Vdash_{\mathcal{B}} p$ for every $R \Rightarrow p$ in $\mathbf{P}$, then $\Vdash_{\mathcal{B}} q$.
(21) If Φ is nonempty, and $\Vdash_{\mathcal{C}} \psi$ for every $\mathcal{C} \supseteq \mathcal{B}$ such that $\Vdash_{\mathcal{C}} \varphi$ for every φ in Φ, then $\Phi \Vdash_{\mathcal{B}} \psi$.
(22) If $\varphi \Vdash_{\mathcal{B}} \psi$ then $\Vdash_{\mathcal{B}} \varphi \supset \psi$.
(23) Only as required by (20)–(22) does the relation $\Vdash$ obtain between any objects.

But as natural as this may seem, we now run into a difficulty. In contrast to the case of type-1 bases, it is by no means obvious that the above four clauses amount to a legitimate inductive definition. Put differently, it is not clear that there exists a *smallest* relation $\Vdash$ satisfying (20)–(22). The argument put forward on pages 316 and 5 for the legitimacy of the earlier definition cannot be extended to this latter set of clauses, since (20), the clause for individual atomic sentences, now potentially makes use of cases of $\Vdash$ generated by (21).

The nature of the difficulty is more clearly brought out if, disregarding conditionals for the time being, we restrict our attention to atomic sentences. Then (22) is omitted and (21) becomes:

(21′) If P is nonempty, and $\Vdash_{\mathcal{C}} q$ for every $\mathcal{C} \supseteq \mathcal{B}$ such that $\Vdash_{\mathcal{C}} p$ for every p in P, then $P \Vdash_{\mathcal{B}} q$.

The concern, again, is whether there even exists such a thing as the smallest relation $\Vdash$ satisfying (20) and (21′). We will not attempt in the present paper to settle this issue. We can, however, state with certainty that *if* there exists a smallest relation satisfying (20) and (21′), then that relation will not be montonic. For, as is proved in the Appendix, *there exists a relation satisfying (20) and (21′) which is not a superrelation of any relation satisfying (20), (21′), and the monotonicity condition (24) below.*

(24) If $P \Vdash_{\mathcal{B}} q$ and $\mathcal{B} \subseteq \mathcal{C}$ then $P \Vdash_{\mathcal{C}} q$.

This finding rules out any chance of there being a relation that is at once monotonic and the smallest relation satisfying (20) and (21′). With respect to our proposal to adopt clauses (20) through (23) as a philosophically satisfactory generalization of (6) through (9), this must be considered a fatal objection. For (24) is surely a *sine qua non* if $\Vdash$ is supposed to capture the intuitive notion of an inference's being justified solely on the strength of the rules contained in a basis. We are going to have to try another approach.

[1] Our use of '$\Rightarrow$' is essentially adopted from Schroeder-Heister (1984). Our concept of a type-1 rule may be described in the terminology of that work as a *rule of level 1 or 2* (to wit, one dealing with atoms); similarly, a type-2 rule is a(n atomic) *rule of level 1, 2, or 3*.

14.4 A Two-Tiered Approach

In Sect. 14.2, taking $\Vdash$ to be inductively generated by clauses (6) through (8), we observed that (15)–(17) would then fall out as fairly immediate consequences. The latter rules differ from the definitional clauses in that each one of them features an arbitrarily large "passive" context set Θ of formulas. Now there is nothing to prevent us from using a statement involving such context formulas as a generative clause in an inductive definition. The following clauses define what we shall refer to as the relation of *derivability* among atomic sentences in a type-2 basis $\mathcal{B}$, symbolically $\vdash_{\mathcal{B}}$:

(25) $T, p \vdash_{\mathcal{B}} p$.
(26) If $\mathbf{P} \Rightarrow q \in \mathcal{B}$, and $T, R \vdash_{\mathcal{B}} p$ for every $R \Rightarrow p$ in $\mathbf{P}$, then $T \vdash_{\mathcal{B}} q$.
(27) Only as required by (25) and (26) does the relation $\vdash$ obtain between any objects.

It is a simple exercise to show that

(28) if $T \vdash_{\mathcal{B}} p$ and $T \subseteq U$ then $U \vdash_{\mathcal{B}} p$.

As the reader may convince herself ;lwith the aid of (28), $\vdash_{\mathcal{B}}$ may be equivalently characterized as the relation that obtains between A and b just in case it is possible, in the atomic natural-deduction system consisting of all and only rules of the form

$$\frac{\begin{matrix}[R_1] \\ p_1\end{matrix} \quad \dots \quad \begin{matrix}[R_n] \\ p_n\end{matrix}}{q}$$

with $((R_1 \Rightarrow p_1), \dots, (R_n \Rightarrow p_n)) \Rightarrow q \in \mathcal{B}$, to construct a derivation in which all undischarged premisses are members of A and the conclusion is b. In adopting clause (26) we have in effect decided to read (19) as implicitly providing for an arbitrary set T of additional hypotheses: *If [given some background set T of hypotheses] you are in a position* [...] *then you may assert q [conditionally on those hypotheses].*

$\vdash_{\mathcal{B}}$, then, is the atomic inference relation generated by $\mathcal{B}$, and our job, as specified in Sect. 14.1, is to extend it in a conservative manner to an inference relation pertaining to the entire language. In the present and following sections we are going to investigate two candidate methods of achieving this effect. The resulting relations (to be labelled $\rhd_{\mathcal{B}}$ and $\blacktriangleright_{\mathcal{B}}$, respectively) will both be found to be well-behaved from an extensional point of view. (As it happens, they turn out to be extensionally equivalent.) But conceptually, it will be argued, each leaves something to be desired.

In preparation for these investigations, we record a few properties of the atomic derivability relation $\vdash$. Firstly, if $\vdash_{\mathcal{B}} u$ for every $u \in U$, then, by a straightforward inductive argument using (28), $T - U \vdash_{\mathcal{B}} q$ whenever $T \vdash_{\mathcal{B}} q$; hence, putting P for T and P for U,

(29) if $P \vdash_{\mathcal{B}} q$, and $\vdash_{\mathcal{B}} p$ for every p in P, then $\vdash_{\mathcal{B}} q$.

By an equally simple inductive argument, $T, U \vdash_{\mathcal{B}} q$ whenever $T \vdash_{\mathcal{B} \cup \{\Rightarrow u | u \in U\}} q$. Hence

(30) if $\vdash_{\mathcal{C}} q$ for every $\mathcal{C} \supseteq \mathcal{B}$ such that $\vdash_{\mathcal{C}} p$ for every p in P, then $P \vdash_{\mathcal{B}} q$,

since $\mathcal{B} \cup \{\Rightarrow u \mid p \in P\}$ is such a $\mathcal{C}$.

By (29), (30), and the obvious fact that $\vdash_{\mathcal{B}} \subseteq \vdash_{\mathcal{C}}$ whenever $\mathcal{B} \subseteq \mathcal{C}$, it follows that

(31) $P \vdash_{\mathcal{B}} q$ iff $\vdash_{\mathcal{C}} q$ for every $\mathcal{C} \supseteq \mathcal{B}$ such that $\vdash_{\mathcal{C}} p$ for every p in P,

a fact which conversely implies each of (29) and (30).

On top of our atomic-derivability relation $\vdash$, a relation $\rhd$ of valid inference relative to a basis, pertaining to the whole language, may now be recursively defined as follows.[2]

(32) $\rhd_{\mathcal{B}} q$ iff $\vdash_{\mathcal{B}} q$.
(33) Where Φ is nonempty: $\Phi \rhd_{\mathcal{B}} \psi$ iff $\rhd_{\mathcal{C}} \psi$ for every $\mathcal{C} \supseteq \mathcal{B}$ such that $\rhd_{\mathcal{C}} \varphi$ for every φ in Φ.
(34) $\rhd_{\mathcal{B}} \varphi \supset \psi$ iff $\varphi \rhd_{\mathcal{B}} \psi$.

In Sect. 14.2 the relation $\Vdash$ was found to have a number of desirable properties: it is monotonic, we saw, in the sense of (13); it is conservative in the sense given by (18); and for any basis $\mathcal{B}$, the consequence relation $\Vdash_{\mathcal{B}}$ supports reasoning in accordance with the standard introduction and elimination rules for $\supset$ ((15) and (16), respectively) as well as any of the atomic rules to be found in $\mathcal{B}$ (as generalized in (17)). We shall now verify that $\rhd$, too, possesses all of these virtues.

Clearly $\rhd$ is monotonic: For single atomic formulas, by (32) the monotonicity of $\rhd$ follows from that of $\vdash$; in cases where (33) applies, monotonicity is immediate since a larger $\mathcal{B}$ means quantification over a smaller set of extensions; and for conditionals, (34) refers us to (33).

Next, $\rhd_{\mathcal{B}}$ amounts to a conservative extension of the derivability relation $\vdash_{\mathcal{B}}$ to the whole language: for any finite set P of atomic sentences, and any atomic sentence q,

(35) $P \rhd_{\mathcal{B}} q$ iff $P \vdash_{\mathcal{B}} q$.

This holds by (32) if P is empty, by (33) and (31) otherwise.

$\rhd_{\mathcal{B}}$ is also easily seen to be closed under Conditional Proof and Modus Ponens:

(36) If $\Theta, \varphi \rhd_{\mathcal{B}} \psi$ then $\Theta \rhd_{\mathcal{B}} \varphi \supset \psi$.
(37) If $\Theta \rhd_{\mathcal{B}} \varphi \supset \psi$ and $\Theta \rhd_{\mathcal{B}} \varphi$ then $\Theta \rhd_{\mathcal{B}} \psi$.

The proofs are entirely parallel to those of (15) and (16).

Lastly, in a less straightforward way, $\rhd_{\mathcal{B}}$ can be seen to be closed, in the context of any set of assumptions Θ, under the atomic rules in $\mathcal{B}$:

(38) If $\mathbf{P} \Rightarrow q \in \mathcal{B}$, and $\Theta, R \rhd_{\mathcal{B}} p$ for every $R \Rightarrow p$ in $\mathbf{P}$, then $\Theta \rhd_{\mathcal{B}} q$.

[2] This is essentially the notion of validity studied in Sect. 4 of de Campos Sanz and Piecha (2014).

Assume the hypotheses. In order to show that $\Theta \rhd_{\mathcal{B}} q$, suppose in addition that $\mathcal{C} \supseteq \mathcal{B}$ and $\rhd_{\mathcal{C}} \theta$ for every θ in Θ; whether Θ contains any elements or not, it will be sufficient for our purposes to establish that $\rhd_{\mathcal{C}} q$. For any $\mathcal{D} \supseteq \mathcal{C}$ such that $\rhd_{\mathcal{D}} r$ for every r in R, $\rhd_{\mathcal{D}} p$ since $\Theta, R \rhd_{\mathcal{B}} p$ and by monotonicity $\rhd_{\mathcal{D}} \theta$ for every θ in Θ. By definition of $\rhd$, it follows that $\vdash_{\mathcal{D}} p$ for every $\mathcal{D} \supseteq \mathcal{C}$ such that $\vdash_{\mathcal{D}} r$ for every r in R. Hence, by (30) (putting $\mathcal{C}$, $\mathcal{D}$, R, r, p for $\mathcal{B}, \mathcal{C}, P, p, q$, respectively), $R \vdash_{\mathcal{C}} p$. Since this holds for every $R \Rightarrow p$ in **P**, we may infer that $\vdash_{\mathcal{C}} q$, i.e. $\rhd_{\mathcal{C}} q$, as required.

Extensionally, then, everything seems in order. But comparing the above proof of (38) with the arguments given for (36) and (37), we notice something of a discrepancy. While (36) and (37) may be described as fairly immediate consequences of the definition of $\rhd$, in proving (38) we had to make use of the special property of $\vdash$ we are referring to as (30). Specifically, in proving (38) for a basis $\mathcal{B}$, we had to appeal to the fact that (30) holds not only for $\mathcal{B}$ itself, but for an arbitrary extension $\mathcal{C}$ of $\mathcal{B}$; a fact that depends on the assumption that the relation of atomic derivability in such an extension will always take the form indicated by our definition of $\vdash$. In other words, $\mathcal{B}$'s satisfying (38) depends on certain restrictions on the ways in which $\mathcal{B}$ might be allowed to expand; an idea that tallies badly with an intuitive conception of $\rhd_{\mathcal{B}}$ as a relation of inferability obtaining solely in virtue of the rules in $\mathcal{B}$.

For a concrete illustration of how (38) might be disrupted if our concept $\vdash$ of derivability were to be liberalized so as to allow for non-standard kinds of derivation rule, consider three atoms a, b and c, and let $\mathcal{B}$ be the basis containing the single rule $(a \Rightarrow b) \Rightarrow c$. $\mathcal{B}$ itself is an ordinary type-2 basis, and according to our official definition of $\vdash$, $\vdash_{\mathcal{B}}$ may be described as the relation $\vdash_*$ inductively generated by the clauses:

(39) For any T and any p: $T, p \vdash_* p$.
(40) For any T: if $T, a \vdash_* b$ then $T \vdash_* c$.

Like every basis, by (33) and (34) $\mathcal{B}$ supports inference from $a \supset b$ and a to b; that is to say, $a \supset b, a \rhd_{\mathcal{B}} b$. By (38), therefore, we expect it to hold true that $a \supset b \rhd_{\mathcal{B}} c$; and given our official definition of $\vdash$, this will indeed be the case. But now consider a basis $\mathfrak{C}$ comprising, in addition to (39) and (40), the non-standard clause:

(41) If $\vdash_* a$ then $\vdash_* b$.

This clause is non-standard since, owing to its omission of the context set T, it cannot be cast in the form of a type-2 rule. Now the presence of (41) ensures that every extension $\mathfrak{D}$ of $\mathfrak{C}$ such that $\rhd_{\mathfrak{D}} a$, i.e. such that $\vdash_{\mathfrak{D}} a$, will also have the property that $\vdash_{\mathfrak{D}} b$, i.e. $\rhd_{\mathfrak{D}} b$; thus $a \rhd_{\mathfrak{C}} b$, i.e. $\rhd_{\mathfrak{C}} a \supset b$. Yet it is not the case that $\vdash_{\mathfrak{C}} c$, since inspection of (39) through (41) reveals that any P and q such that $P \vdash_{\mathfrak{C}} q$ must satisfy the condition: $P \neq \emptyset$ and if $q = b$ or $q = c$ then $b \in P$ or $c \in P$. (Any set-sentence pair introduced in accordance with (39) satisfies this condition, and the property is preserved by (40) and vacuously preserved by (41).) Since, then, $\rhd_{\mathfrak{C}} a \supset b$ but $\not\rhd_{\mathfrak{C}} c$, we may conclude that $a \supset b \not\rhd_{\mathcal{B}} c$ despite the fact that $(a \Rightarrow b) \Rightarrow c \in \mathcal{B}$ and $a \supset b, a \rhd_{\mathcal{B}} b$—and so, as advertised, we have a counterexample to (38).

Is this a problem? What we have found—we repeat—is not that (38) fails on our officially suggested definitions of validity ($\rhd$) and atomic derivability ($\vdash$); only that it *would* fail if our notion of atomic derivability were loosened so as to make room for certain kinds of rules that we are not actually permitting. Nevertheless, there is something unsatisfactory about the situation. Unlike the theory of Sect. 14.2, the present account cannot claim to capture the idea that rules of inference contained in a basis are valid relative to that basis solely in virtue of belonging to it; the fact that certain other, formally possible rules are absent from all allowable extensions plays an essential role as well.

14.5 Alternative Two-Tiered Approach

Moved (I believe) by concerns similar to ours, de Campos Sanz et al. (2013) suggest an alternative way of extending an atomic derivability relation to a language possessing a conditional connective. Subjected to a minor simplification and notationally adapted to the present context, the definition runs as follows. (As before, latin letters signify logical atomicity.)

(42) $P \blacktriangleright_{\mathcal{B}} q$ iff $P \vdash_{\mathcal{B}} q$.
(43) Where Φ is a nonempty set of compound sentences: Φ, $P \blacktriangleright_{\mathcal{B}} \psi$ iff$P \blacktriangleright_{\mathcal{C}} \psi$ for every $\mathcal{C} \supseteq \mathcal{B}$ such that $\blacktriangleright_{\mathcal{C}} \varphi$ for every φ in Φ.
(44) $P \blacktriangleright_{\mathcal{B}} \varphi \supset \psi$ iff $P, \varphi \blacktriangleright_{\mathcal{B}} \psi$.

Let us take a closer look at this relation and see whether it avoids the shortcomings observed in $\rhd$.

To begin with, it is not hard to establish that, from an extensional point of view, $\blacktriangleright$ is equivalent to $\rhd$:

(45) $\Phi \blacktriangleright_{\mathcal{B}} \psi$ iff $\Phi \rhd_{\mathcal{B}} \psi$.

To prove this, begin by showing—a straightforward matter—that

(46) $\Theta, \Phi \rhd_{\mathcal{B}} \psi$ iff $\Phi \rhd_{\mathcal{C}} \psi$ for every $\mathcal{C} \supseteq \mathcal{B}$ such that $\rhd_{\mathcal{C}} \theta$ for every θ in Θ.

From (46), equally straightforwardly, we infer the converse of (36), so that

(47) $\Theta \rhd_{\mathcal{B}} \varphi \supset \psi$ iff $\Theta, \varphi \rhd_{\mathcal{B}} \psi$.

(45) may now be established by induction on the degree of (Φ, ψ). Assume that it holds for all bases and all set-formula pairs of degrees less than that of (Φ, ψ), and consider three cases according to the definition of $\blacktriangleright$ in (42) through (44). Where 'IH' abbreviates 'by induction hypothesis':

$$P \blacktriangleright_{\mathcal{B}} q \overset{(42)}{\Leftrightarrow} P \vdash_{\mathcal{B}} q$$
$$\overset{(35)}{\Leftrightarrow} P \rhd_{\mathcal{B}} q;$$
$$\Phi, P \blacktriangleright_{\mathcal{B}} \psi \overset{(43)}{\Leftrightarrow} P \blacktriangleright_{\mathcal{C}} \psi \text{ for every} \mathcal{C} \supseteq \mathcal{B} \text{ such that } \blacktriangleright_{\mathcal{B}} \varphi \text{ for every} \varphi \text{ in } \Phi$$
$$\overset{(\text{IH})}{\Leftrightarrow} P \rhd_{\mathcal{C}} \psi \text{ for every } \mathcal{C} \supseteq \mathcal{B} \text{ such that } \rhd_{\mathcal{B}} \varphi \text{ for every} \varphi \text{ in } \Phi$$
$$\overset{(46)}{\Leftrightarrow} \Phi, P \rhd_{\mathcal{B}} \psi;$$
$$P \blacktriangleright_{\mathcal{B}} \varphi \supset \psi \overset{(44)}{\Leftrightarrow} P, \varphi \blacktriangleright_{\mathcal{B}} \psi$$
$$\overset{(\text{IH})}{\Leftrightarrow} P, \varphi \rhd_{\mathcal{B}} \psi$$
$$\overset{(47)}{\Leftrightarrow} P \rhd_{\mathcal{B}} \varphi \supset \psi.$$

The extensional equivalence of $\rhd$ and $\blacktriangleright$ does not necessarily mean that the latter shares the conceptual deficiencies of the former. The source of our dissatisfaction with $\rhd$ was not that it gave the wrong answer to the question of what entails what, but that it seemed to misallocate the *source* of that entailment, construing it as dependent on factors of which, intuitively, it ought to be independent. $\rhd_{\mathcal{B}}$, we saw, is closed under Modus Ponens, Conditional Proof, and the rules in $\mathcal{B}$. Extensionally, this is as it should be; but closure under the rules in $\mathcal{B}$ was seen to depend in a problematic way on the properties of $\vdash$. We shall now show that $\blacktriangleright$ avoids this problem, but at the price of instead rendering closure under Modus Ponens problematic in a similar way.

Closure of $\blacktriangleright_{\mathcal{B}}$ under the rules in $\mathcal{B}$ is established as follows.

(48) If $\mathbf{P} \Rightarrow q \in \mathcal{B}$, and $\Theta, R \blacktriangleright_{\mathcal{B}} p$ for every $R \Rightarrow p$ in $\mathbf{P}$, then $\Theta \blacktriangleright_{\mathcal{B}} q$.

Set $\Theta = S \cup \Sigma$, where the members of S are all atomic, and the members of Σ all composite. Assume the hypotheses; in order to see that $\Sigma, S \blacktriangleright_{\mathcal{B}} q$, consider any $\mathcal{C} \supseteq \mathcal{B}$ such that $\blacktriangleright_{\mathcal{C}} \sigma$ for every σ in Σ. For such a $\mathcal{C}$ it must hold that $S, R \blacktriangleright_{\mathcal{C}} p$, i.e. $S, R \vdash_{\mathcal{C}} p$, for every $R \Rightarrow p$ in $\mathbf{P}$; hence $S \vdash_{\mathcal{C}} q$, i.e. $S \blacktriangleright_{\mathcal{C}} q$, as required.

As for Conditional Proof:

(49) If $\Theta, \varphi \blacktriangleright_{\mathcal{B}} \psi$ then $\Theta \blacktriangleright_{\mathcal{B}} \varphi \supset \psi$.

Where Σ and S are as before, assume again that $\mathcal{C} \supseteq \mathcal{B}$ and $\blacktriangleright_{\mathcal{C}} \sigma$ for every σ in Σ.

Case 1: φ is atomic. Then we immediately have $S, \varphi \blacktriangleright_{\mathcal{C}} \psi$, whence $S \blacktriangleright_{\mathcal{C}} \varphi \supset \psi$, as required.

Case 2: φ is composite. Let $\mathcal{D} \supseteq \mathcal{C}$ and $\blacktriangleright_{\mathcal{D}} \varphi$; then $S \blacktriangleright \psi$; hence $S, \varphi \blacktriangleright_{\mathcal{C}} \psi$, whence again $S \blacktriangleright_{\mathcal{C}} \varphi \supset \psi$.

When it comes to Modus Ponens, however, no similarly straightforward line of reasoning suggests itself. Of course, closure of $\blacktriangleright_{\mathcal{B}}$ under Modus Ponens follows immediately from (37) and (45); it certainly holds in general that

(50) if $\Theta \blacktriangleright_{\mathcal{B}} \varphi \supset \psi$ and $\Theta \blacktriangleright_{\mathcal{B}} \varphi$ then $\Theta \blacktriangleright_{\mathcal{B}} \psi$.

But in citing (45) in this way we are relying on (31), on which our proof of (45) depends. To see how (31) comes to be invoked in a concrete case, consider why it is that $a \supset b, b \supset c, a \blacktriangleright_{\mathcal{B}} c$ for any $\mathcal{B}$, a, b, and c. Consider any $\mathcal{C} \supseteq B$ such that $\blacktriangleright_{\mathcal{C}} a \supset b$ and $\blacktriangleright_{\mathcal{C}} b \supset c$, i.e. such that $a \blacktriangleright_{\mathcal{C}} b$ and $b \blacktriangleright_{\mathcal{C}} c$, i.e. such that $a \vdash_{\mathcal{C}} b$ and $b \vdash_{\mathcal{C}} c$. By (31) thrice applied it follows that $a \vdash_{\mathcal{C}} c$, i.e. $a \blacktriangleright_{\mathcal{C}} c$, as required.

Just as in the case of $\triangleright$, things change if non-standard rules of derivation are allowed into the game. Set $\mathcal{B} = \emptyset$, so that $\vdash_{\mathcal{B}}$ is the relation $\vdash_*$ generated by the single clause

(51) $T, p \vdash_* p$.

Let $\mathfrak{C}$ comprise, in addition to (51), the non-standard clauses

(52) $a \vdash_* b$,
(53) $b \vdash_* c$.

Then $\blacktriangleright_{\mathfrak{C}} a \supset b$ and $\blacktriangleright_{\mathfrak{C}} b \supset c$. Yet obviously clauses (51) through (53) alone will not bring it about that $a \vdash_* c$; hence $a \not\blacktriangleright_{\mathfrak{C}} c$, and so $a \supset b, b \supset c, a \not\blacktriangleright_{\mathcal{B}} c$.

So once again, while the actual well-behavedness of our proposed validity relation is not in question, that good behaviour is seen to depend not only on the inference rules to be found *within* atomic bases, but also on an advance assurance that certain kinds of formally possible rules will always be kept *out* of them.

14.6 Conclusion

We have tried, without success, to find a notion of inferability relative to an atomic basis which would

(i) generate a conservative extension of the atomic inference relation holding in the basis in question;
(ii) accommodate type-2 atomic rules;
(iii) be closed under standard introduction and elimination rules for the conditional connective, as well as under the atomic rules in each respective basis; and
(iv) represent closure under these rules as resulting entirely from the *presence* of rules in that basis, and not as depending on general *restrictions* on the forms that rules in extensions of the basis might take.

Of course, our failure in this paper to come up with a theory meeting all of these requirements does not amount to a conclusive proof that no such theory is possible. Nevertheless, let us consider the *desiderata* in turn, to see what giving either one of them up might amount to.

Firstly, we could abandon (i). As suggested in the Introduction, the importance of achieving conservativeness may depend on one's philosophical conception of atomic basis; if the rules in a basis are thought of as meaning-defining introduction rules for atomic sentences, it is not unreasonable to think that some atomic rules not present in the basis—such as, for instance, the corresponding elimination rules—might nevertheless qualify as valid. Yet in proof-theoretically informed discussions concerning the justification of deductive reasoning, it is not uncommon to regard conservativeness with respect to a previously given terminology as a non-negotiable condition on the introduction of new logical vocabulary—conservativeness, to wit, in regard not only of what is categorically provable, but of what is inferable from what. (See, for instance, (Dummett 1978, p. 302) or (Tennant 1997, p. 318)) It would be unsatisfactory to have to conclude that a notion of logically valid inference has no place in such contexts.

Alternatively, we might give up on (ii) and stick to the relation $\Vdash$ from Sect. 14.2 (or, what amounts in effect to the same thing, with Prawitz's (1971) notion of (weak) validity). As demonstrated in Sandqvist (2009), this semantic theory renders not only intuitionistic logic valid, but classical logic as well; whether this is a good or a bad thing I leave it to the reader to decide. But in any event, such a concept of atomic basis seems somewhat restrictive. As we asked in Sect. 14.3: If conditionals can be accepted on the strength of hypothetical reasoning, might not the same hold true of some atomic sentences?

Forgoing (iii) seems too radical a measure to merit serious consideration.

How about (iv)? If, as seems likely, the relation $\rhd$ alias $\blacktriangleright$ should turn out to be extensionally satisfactory, maybe this is good enough? Our chief objection to $\rhd_{\mathcal{B}}$ was the fact that its closure under the rules in $\mathcal{B}$, in the sense given by (38), depended upon certain properties of $\vdash$; similarly, $\blacktriangleright_{\mathcal{B}}$ was rejected because its closure under Modus Ponens was seen to depend on such properties. Against requirement (iv), it might not unreasonably be argued that (30), and even more so (29), are fairly innocuous considered as properties of a relation $\vdash$ of derivability; and, as we have seen, as long as these conditions are fulfilled, requirement (iii) will in fact be met. They are, however, substantive conditions; it is possible to concoct systems of atomic deduction that violate them. The idea that (as in the case of $\blacktriangleright$) such a fundamental rule as Modus Ponens should depend for its validity on certain assumptions regarding the permissible forms of future modes of reasoning seems unattractive; it lends the resulting meaning theory a measure of holism which, *ceteris paribus*, one would rather do without.

Appendix

The purpose of this Appendix is to substantiate the claim, made in Sect. 14.3, that, whether or not there exists such a thing as *the smallest relation satisfying Clauses (20) and (21′)*, any such smallest relation fails to satisfy (24).

Consider the following conditions (corresponding to (20), (21′) and (24), respectively) on an arbitrary three-place relation $\mathfrak{X}$ between sets of atomic sentences, type-2 bases, and individual atomic sentences.

(R) If $\mathbf{P} \Rightarrow q \in \mathcal{B}$, and $R\ \mathfrak{X}_{\mathcal{B}}\ p$ for every $R \Rightarrow p$ in $\mathbf{P}$, then $\mathfrak{X}_{\mathcal{B}}\ q$.
(E) If $\mathfrak{X}_{\mathcal{C}}\ q$ for every $\mathcal{C} \supseteq \mathcal{B}$ such that $\mathfrak{X}_{\mathcal{C}}\ p$ for every p in P, then $P\ \mathfrak{X}_{\mathcal{B}}\ q$.
(M) If $P\ \mathfrak{X}_{\mathcal{B}}\ q$ and $\mathcal{C} \supseteq \mathcal{B}$ then $P\ \mathfrak{X}_{\mathcal{C}}\ q$.

We begin by stating that $\vdash$ *is the smallest relation* $\mathfrak{X}$ *satisfying* (R), (E), *and* (M) (where $\vdash$ is as defined in Sect. 14.4). That $\vdash$ does indeed satisfy the three conditions has been established in Sect. 14.4. And that $\vdash$ is included in any $\mathfrak{X}$ satisfying the conditions follows by induction from the fact that any such $\mathfrak{X}$ will also satisfy clauses (25) and (26), with $\mathfrak{X}$ in place of $\vdash$. The details are left to the reader.

Next, we show that *there exists a relation* $\mathfrak{S}$ *which satisfies* (R) *and* (E) *yet does not include* $\vdash$.

Such a relation $\mathfrak{S}$ can be constructed as follows. Pick two distinct atoms a and b, and define, for any $\mathcal{B}$:

$$\begin{array}{lrcl} & \mathfrak{S}_{\mathcal{B}}\, b & \Leftrightarrow & \text{if} \vdash_{\mathcal{B}} a \text{ then } \vdash_{\mathcal{B}} b. \\ \text{Where } p \neq b, & \mathfrak{S}_{\mathcal{B}}\, p & \Leftrightarrow & a \vdash_{\mathcal{B}} p. \\ \text{Where } P \text{ is nonempty,} & P\, \mathfrak{S}_{\mathcal{B}}\, q & \Leftrightarrow & \mathfrak{S}_{\mathcal{C}}\, q \text{ for every } \mathcal{C} \supseteq \mathcal{B} \text{ such that} \\ & & & \mathfrak{S}_{\mathcal{C}}\, p \text{ for every } p \text{ in } P. \end{array}$$

(The "if—then" in the first clause is to be understood materially, so that, whenever $\nvdash_{\mathcal{B}} a$, vacuously $\mathfrak{S}_{\mathcal{B}}\, b$.)

This $\mathfrak{S}$ satisfies (E) by definition. To see that it satisfies (R), note first that for any $\mathcal{B}$ and any p,

(i) if $a \vdash_{\mathcal{B}} b$ then $\mathfrak{S}_{\mathcal{B}}\, p$, and
(ii) if $\mathfrak{S}_{\mathcal{B}}\, p$ and $\vdash_{\mathcal{B}} a$ then $\vdash_{\mathcal{B}} p$.

(i) holds by definition where $p \neq b$, and, where $p = b$, in virtue of the fact that $\vdash_{\mathcal{B}} b$ if $a \vdash_{\mathcal{B}} b$ and $\vdash_{\mathcal{B}} a$. (ii), similarly, holds by definition where $p = b$, and where $p \neq b$ in virtue of the fact that $\vdash_{\mathcal{B}} p$ if $a \vdash_{\mathcal{B}} p$ and $\vdash_{\mathcal{B}} a$.

Now suppose that $\mathbf{P} \Rightarrow q \in \mathcal{B}$, and moreover that $R\, \mathfrak{S}_{\mathcal{B}}\, p$ for every $R \Rightarrow p$ in $\mathbf{P}$. For any such $R \Rightarrow p$, consider the basis

$$\mathcal{B}_{R,a} = \mathcal{B} \cup \{\Rightarrow r \mid r \in R\} \cup \{\Rightarrow a\}.$$

For any r in R it holds that $\vdash_{\mathcal{B}_{R,a}} r$, whence $a \vdash_{\mathcal{B}_{R,a}} r$, whence $\mathfrak{S}_{\mathcal{B}_{R,a}}\, r$ by (i). Since by hypothesis $R\, \mathfrak{S}_{\mathcal{B}}\, p$ it follows that $\mathfrak{S}_{\mathcal{B}_{R,a}}\, p$; and since moreover $\vdash_{\mathcal{B}_{R,a}} a$, by (ii) we may infer that $\vdash_{\mathcal{B}_{R,a}} p$, meaning that $a, R \vdash_{\mathcal{B}} p$.

Since this is true of every $R \Rightarrow p$ in $\mathbf{P}$, and by hypothesis $\mathbf{P} \Rightarrow q \in \mathcal{B}$, it follows that $a \vdash_{\mathcal{B}} q$, whence $\mathfrak{S}_{\mathcal{B}}\, q$ by (i); this completes the verification of (R).

It remains to show that $\vdash$ is not included in $\mathfrak{S}$. Pick two formulas c and d distinct from a and b, and consider the basis

$$\mathcal{C} = \{(\Rightarrow a, \ \Rightarrow b) \Rightarrow c, (a \Rightarrow c) \Rightarrow d\}.$$

By induction according to the definition of $\vdash$, it is easily verified that, in general, $P \vdash_{\mathcal{C}} q$ only if q is a truth-functional consequence of $P \cup \{a \supset (b \supset c), (a \supset c) \supset d\}$. Therefore $\nvdash_{\mathcal{C}} a$ and $a \nvdash_{\mathcal{C}} d$, whence $\mathfrak{S}_{\mathcal{C}}\, b$ but $\not\mathfrak{S}_{\mathcal{C}}\, d$, whence $b\, \not\mathfrak{S}_{\mathcal{C}}\, d$. Yet clearly $b \vdash_{\mathcal{C}} d$ since $b, a \vdash_{\mathcal{C}} c$; hence $\vdash$ is not a subrelation of $\mathfrak{S}$.

Thus, since $\vdash$ is the smallest $\mathfrak{X}$ satisfying (R), (E), and (M), it follows that, as claimed on page 319, $\mathfrak{S}$ is not a superrelation of any relation satisfying these conditions.

References

de Campos Sanz, W., Piecha, T. & Schroeder-Heister, P. (2013). Constructive semantics, admissibility of rules and the validity of Peirce's law. *Logic Journal of the IGPL*, 22, 297–308.

de Campos Sanz, W., & Piecha, T. (2014). A critical remark on the BHK interpretation of implication. In P.E. Bour, G. Heinzmann, W. Hodges & P. Schroeder-Heister (Eds.), *14th CLMPS 2011 Proceedings, forthcoming in Philosophia Scientiae*, 18(3).

Dummett, M. (1978). The justification of deduction. In *Truth and other enigmas* (pp. 290–318). Cambridge: Harvard University Press.

Prawitz, D. (1971). Ideas and results in proof theory. In J.E. Fenstad (Ed.), *Proceedings of the second Scandinavian logic symposium*, (pp. 235–307). Amsterdam: North-Holland Publishing Company.

Prawitz, D. (1973). Towards a foundation of a general proof theory. In P. Suppes, L. Henkin, A. Joja & G. C. Moisil (Eds.), *Logic, Methodology and Philosophy of Science IV*, (pp. 225–250). Amsterdam: North-Holland Publishing Company.

Sandqvist, T. (2009). Classical logic without bivalence. *Analysis*, *69*, 211–218.

Schroeder-Heister, P. (1984). A natural extension of natural deduction. *Journal of Symbolic Logic*, *49*(4), 1284–1300.

Tennant, N. (1997). *The taming of the true*. Oxford: Clarendon Press.

Chapter 15
Harmony in Proof-Theoretic Semantics: A Reductive Analysis

Peter Schroeder-Heister

Abstract We distinguish between the *foundational analysis* of logical constants, which treats all connectives in a single general framework, and the *reductive analysis*, which studies general connectives in terms of the standard operators. With every list of introduction or elimination rules proposed for an n-ary connective c, we associate a certain formula of second-order intuitionistic propositional logic. The formula corresponding to given introduction rules expresses the *introduction meaning*, the formula corresponding to given elimination rules the *elimination meaning* of c. We say that introduction and elimination rules for c are in *harmony* with each other when introduction meaning and elimination meaning match. Introduction or elimination rules are called *flat*, if they can discharge only formulas, but not rules as assumptions. We can show that not every connective with flat introduction rules has harmonious flat elimination rules, and conversely, that not every connective with flat elimination rules has harmonious flat introduction rules. If a harmonious characterisation of a connective is given, it can be explicitly defined in terms of the standard operators for implication, conjunction, disjunction, falsum and (propositional) universal quantification, namely by its introduction meaning or (equivalently) by its elimination meaning. It is argued that the reductive analysis of logical constants implicitly underlies Prawitz's (1979) proposal for a general schema for introduction and elimination rules.

Dag Prawitz's (1979) article was one of the main starting points of my Dr. phil. thesis, whose external examiner Dag became in 1981. His work on proofs and meaning has been a great source of inspiration to me ever since. I am extremely happy to be able to contribute the present paper to this volume dedicated to his work. It may be read as an extended commentary which further pursues the approach Dag initiated with his article.—This work was carried out within the French-German ANR-DFG project "Hypothetical Reasoning: Its Proof-Theoretic Analysis" (HYPOTHESES) (DFG Schr 275/16-2). I should like to thank Luca Tranchini for very helpful discussions on the topic of this paper, and Thomas Piecha and Dag Prawitz for many helpful remarks. I am also grateful to an anonymous reviewer of the *Review of Symbolic Logic* for comments and suggestions on the paper by Olkhovikov and the author (2014a) that have been useful for the revision of the current paper.

P. Schroeder-Heister (✉)
Wilhelm-Schickard-Institut für Informatik, Universität Tübingen,
Sand 13, 72076 Tübingen, Germany
e-mail: psh@uni-tuebingen.de

H. Wansing (ed.), *Dag Prawitz on Proofs and Meaning*, Outstanding Contributions to Logic 7, DOI 10.1007/978-3-319-11041-7_15

Keywords Proof-theoretic semantics · Proof-theoretic harmony · Logical connectives · Generalised rules · Functional completeness · Conservativeness · Uniqueness

15.1 Introduction: Reductive Proof-Theoretic Semantics

The proof-theoretic semantics of logical constants is predominantly concerned with the meaning of the standard connectives, which in intuitionistic propositional logic are implication ($\rightarrow$), conjunction ($\wedge$), disjunction ($\vee$) and absurdity ($\bot$) (see Schroeder-Heister 2012a). Even if we confine ourselves to intuitionistic logic, and here to the propositional case, this is a severe limitation. It is natural to ask how one should deal with arbitrary n-ary propositional connectives. For example, in a natural deduction framework, one should discuss what introduction and elimination rules for such connectives look like, and what it means that these rules are in harmony with each other, a requirement standardly made in proof-theoretic semantics. These questions will be the subject of this paper. The seminal paper on this topic within the framework of natural deduction is Prawitz (1979).[1]

Unlike Prawitz, we shall not try to formulate a general schema for elimination rules given certain introduction rules. We do not attempt the reverse procedure either—starting from eliminations and trying to formulate a general schema for introductions. We shall rather propose general schemas both for introduction and for elimination rules, and then formulate a criterion that tells when such rules are in harmony with each other. This criterion will not be based on the syntactic form of these rules but on their content. Certain elimination rules will be harmonious with given introduction rules not because they have a specific form which is developed from that of the introduction rules, or vice versa. We shall instead associate with each set of introduction rules for a connective c the introduction meaning c^I of c according to these rules, which describes the content of the introduction rules. Likewise, with each set of elimination rules for c we shall associate the elimination meaning c^E of c according to these rules, which describes the content of the elimination rules. We call the introduction and elimination rules for c harmonious when c^I and c^E are equivalent.

This presupposes that we have a language at our disposal in which we can express introduction and elimination meanings c^I and c^E, and a deductive system in which we can establish their equivalence. As such a language and system we use intuitionistic propositional logic **PL**, sometimes including universal propositional quantifiers, i.e., second-order intuitionistic propositional logic **PL2**. The introduction and elimination meanings c^I and c^E are formulas of **PL** (and **PL2**, respectively). This means that we take the logic of the standard intuitionistic operators for granted and explain the meaning of arbitrary n-ary connectives with respect to them. Therefore we call this approach a *reductive* analysis of logical constants.

[1] For the framework of the sequent calculus, the seminal paper is von Kutschera (1968), most results of which can be carried over to the natural-deduction framework.

Being reductive, our approach differs from *foundational* approaches, according to which the introduction and elimination rules for arbitrary n-ary connectives as well as those for the standard operators are accommodated in a single basic framework. Such foundational approaches, which have been proposed, for example, by von Kutschera (1968) and Schroeder-Heister (1984), carry a certain conceptual and notational overhead with them, which is not needed for all purposes. Many results can be established within the reductive framework which nonetheless pertain to the foundational frameworks. The results on flattening mentioned below are of this kind. A foundational approach corresponding to the reductive one of this paper is carried out in Schroeder-Heister (2014a).

Based on a schema S that describes the general form of introductions, and another schema S' that describes the general form of eliminations, each for an n-ary constant c, and using the introduction and elimination meanings c^I and c^E of c with respect to these schemas, we can ask questions such as the following:

1. Given certain introduction rules for c satisfying the schema S, and certain elimination rules for c satisfying the schema S', are these introduction and elimination rules in harmony with each other? If they are not in harmony with each other, when do they guarantee at least conservativeness? When do they guarantee the uniqueness of c?
2. Given certain introduction rules for c satisfying the schema S, are there elimination rules for c satisfying the schema S', such that these introduction and elimination rules are in harmony with each other?
3. Conversely, given certain elimination rules for c satisfying the schema S', are there introduction rules for c satisfying the schema S, such that these intruduction and elimination rules are in harmony with each other?

Answers to these questions are facilitated by the great technical advantages of our reductive approach. As we can express the strengh of introduction and elimination rules directly in terms of propositional formulas without referring to the rules themselves, we gain access to the apparatus of standard (second-order) propositional logic with all its well-established methods. We can use such methods to prove results about the possible forms of formulas intuitionistically equivalent to c^I or to c^E. This enables us in particular to establish negative results about the shape of harmonious introduction and elimination rules, i.e., results telling us that rules of certain restricted forms are *not* appropriate as introduction or elimination rules. This is important for the discussion of so-called "general" elimination rules in the sense of Dyckhoff, Tennant, Lopez-Escobar and von Plato (see Schroeder-Heister 2014b). More generally, these results constrain the structure of introduction and elimination rules in the foundational framework with rules of higher levels (Schroeder-Heister 1984).[2] This is due to the fact that entities in the foundational framework such as higher-level rules have an implicational counterpart in standard propositional logic, so that negative results about the latter carry over to the former. Our central result here are the non-flattening theorems, according to which there are not always elimination rules in

[2] And, correspondingly, those of von Kutschera's (1968) framework.

harmony with given introduction rules, or introduction rules in harmony with given elimination rules, which are *flat* in the sense that only formulas (and not 'higher' entities such as rules) can be discharged as assumptions.[3]

Traditionally, in investigations of n-ary connectives, proof-theoretic semantics has been concerned with the question: Given introduction rules of a certain form, how do we have to frame elimination rules such that they are in harmony with the introductions? This question is certainly related to our questions above (in particular to the second one), but covers only part of the problem. The formulation of harmonious elimination rules which are obtained as unique syntactic functions of introduction rules[4] gives us a certain principle of harmony. However, it cannot tell us anything about the relation between introductions and eliminations which are not of the form considered. A connective such as Prior's (1960) *tonk* would simply be ill-defined, as its given elimination rules are stronger than the harmonious elimination rule based on its given introduction rule. In our reductive framework, which is completely symmetric with respect to introduction and elimination rules (as it starts with independent general schemas for both of them), *tonk* has a well defined introduction meaning $tonk^I$ as well as a well-defined elimination meaning $tonk^E$ whose mutual relationship we can investigate (see below Sect. 15.5 and Table 15.3).

In Sects. 15.2–15.5 we give independent schemas for introduction and elimination rules for n-ary propositional operators c and define and characterize introduction and elimination meanings c^I and c^E with respect to them. We define harmony in terms of c^I and c^E and relate it to Belnap's (1962) criteria of conservativeness and uniqueness. We then report some positive and negative results concerning the form of harmonious elimination and introduction rules starting from introduction or elimination rules, respectively. The negative results mainly concern the fact of whether rules can be flattened in a certain way.

When introduction and elimination meanings c^I and c^E of c coincide, then c can be defined by either c^I or c^E. This relates our result to the investigation of the functional completeness of logical connectives. Here, in the intuitionistic framework, "functional completeness" means the expressive completeness in analogy to truth-functional completeness in classical logic.[5] We shall deal with this topic in Sect. 15.6. In Sect. 15.7 we relate our approach to Prawitz's (1979) paper. We argue that his approach is best regarded as a reductive rather than a foundational approach, even if it was not intended as such.

The final Sect. 15.8 sketches some problems a more foundational approach faces in comparison to the reductive approach advocated here. It is claimed that modus

---

[3] The term 'flattening' has been coined by Read (see Read 2014, this volume).

[4] Already Gentzen spoke of eliminations as "functions" of introductions in the frequently quoted passage of Gentzen (1934/35, p. 189) that Prawitz (1965) first drew attention to.

[5] It is convenient to use the term "functional completeness" to distinguish this matter from semantic completeness, which is an entirely different issue. The term "functional" is definitely not perfect, but evokes the right associations. One might think of rules as transforming proofs into proofs, and therefore of "proof functions", in the intuitionistic case.

ponens and the two projections as elimination rules for implication and conjunction, respectively, are fundamental and cannot be superseded by more general rules.

15.2 Introduction and Elimination Rules

We consider introduction and elimination rules for n-ary connectives c in a natural deduction framework. As the general form of an introduction rule for c we propose the following:

$$(c\mathrm{I})\quad \frac{\begin{matrix}[\Gamma_1] \\ s_1\end{matrix} \quad \cdots \quad \begin{matrix}[\Gamma_\ell] \\ s_\ell\end{matrix}}{c(p_1,\ldots,p_n)}\quad , \tag{15.1}$$

where $s_1,\ldots,s_\ell$ are propositional variables and the Γ_i are (possibly empty) lists of propositional variables, which can be discharged at the application of $(c\mathrm{I})$. As a limiting case we allow for $\ell = 0$ (which covers the case of the truth constant $\top$). All propositional variables occurring in the rule must be among $p_1,\ldots,p_n$. Schema (15.1) corresponds to the schema proposed in Prawitz (1979).

Evidently, the introduction rules for the standard intuitionistic connectives $\wedge, \vee, \rightarrow$

$$\frac{p_1 \quad p_2}{p_1 \wedge p_2} \qquad \frac{p_1}{p_1 \vee p_2} \qquad \frac{p_2}{p_1 \vee p_2} \qquad \frac{\begin{matrix}[p_1] \\ p_2\end{matrix}}{p_1 \rightarrow p_2}$$

fall under this schema, with ℓ being 2 in the case of conjunction and 1 in the case of disjunction and implication, and with the Γ_i being empty in the case of conjunction and disjunction, and Γ_1 consisting just of p_1 in the case of implication.

Our schema for introduction rules is quite restricted. We do not, for example, allow for any connective occurring above the inference line. This means that we cannot, for example, characterize negation by an introduction rule referring to absurdity in its premiss

$$(\neg\,\mathrm{I})\quad \frac{\begin{matrix}[p_1] \\ \bot\end{matrix}}{\neg p_1}\ .$$

However, for the point we want to make our schema is sufficient. It is easy to extend it to the case where connectives are introduced in a certain order, where an 'earlier' connective can be used to define a 'later' one.[6] The fact that we do not consider 'extra

[6] Prawitz (1979) works in a more general context, allowing for dependencies between connectives. More involved is the problem of self-referential operators, the premisses of whose introduction rules may contain the operator being introduced. We do not discuss this problem here (see Schroeder-Heister 2012b). Tranchini (2014) has pointed out that our reductive approach is not capable of dealing with this sort of phenomena, and that a notion of 'rule equivalence' is needed, in contradistinction to our approach which according to Tranchini is based on 'formula equivalence'. A definition of harmony which is nearer to the level of rules and which would correspond to this notion of rule

variables' beyond $p_1, \ldots, p_n$ in the premisses of (cI) is another restriction, which is not relevant for the points we want to make here. We should like to remark, however, that introduction rules with extra variables in premisses are a neglected topic in proof-theoretic semantics. They represent an interesting and significant extension of the means of expression available, which corresponds to introducing existentially understood variables into the meaning of connectives.

There may be more than one introduction rule of the form (15.1) for c (as it is the case with disjunction). We assume that they are given as a finite list, where, as a limiting case, the empty list of introduction rules is permitted. This covers the absurdity constant $\bot$, which has no introduction rule. A connective which plays a prominent role in our investigations is the ternary operator $\star$ with the following two introduction rules

$$(\star\,\mathrm{I})\quad \frac{\begin{matrix}[p_1]\\ p_2\end{matrix}}{\star(p_1, p_2, p_3)} \qquad \frac{p_3}{\star(p_1, p_2, p_3)}\ . \tag{15.2}$$

As our general schema for an elimination rule for an n-ary connective c we propose the following:

$$(c\mathrm{E})\quad \frac{c(p_1, \ldots, p_n) \qquad \begin{matrix}[\Gamma_1]\\ s_1\end{matrix} \quad \ldots \quad \begin{matrix}[\Gamma_\ell]\\ s_\ell\end{matrix}}{q}\ , \tag{15.3}$$

where $s_1, \ldots, s_\ell, q$ are propositional variables and the Γ_i are (possibly empty) lists of propositional variables. $c(p_1, \ldots, p_n)$ is called the *major premiss* of (cE), the remaining premisses are called the *minor premisses* of (cE). We allow for $\ell = 0$, in which case minor premisses are lacking. We do not impose any restriction on the propositional variables occurring in (cE). They may (and will normally) comprise $p_1, \ldots, p_n$, but any number of propositional variables beyond $p_1, \ldots, p_n$ may be present. This generalizes the fact that in elimination rules such as $\vee$-elimination

$$(\vee\mathrm{E})\quad \frac{p_1 \vee p_2 \qquad \begin{matrix}[p_1]\\ r\end{matrix} \qquad \begin{matrix}[p_2]\\ r\end{matrix}}{r} \tag{15.4}$$

the additional propositional variable r is used as minor premiss and conclusion. Our motivation for proposing (15.3) as elimination schema is that we should be able to choose anything whatsoever as possible consequence of $c(p_1, \ldots, p_n)$, which means that the minor premisses and the conclusion should not be constrained in any way. The lack of this restriction makes our schema more general than elimination schemas derived from given introduction rules such as those in Prawitz (1979) and

equivalence, is proposed in Olkhovikov and Schroeder-Heister (2014b). In the German translation of Prawitz (1979), Prawitz himself considers (or at least mentions the possibility of) self-referential connectives, for example a connective defined in terms of its own negation.

Schroeder-Heister (1984) where $s_1, \ldots, s_\ell, q$ are identical, i.e., represented by a single variable r.[7]

It is absolutely crucial to realize that we are formulating (15.3) as an independent schema in its own right, i.e., without any reference to potential introduction rules for c. (15.3) is our general schema for an arbitrary elimination rule, not a general schema for elimination rules given certain introduction rules. All proposals for general elimination schemas that can be found in the literature on proof-theoretic semantics consider a schema that is generated from introduction rules given beforehand, thus (explicitly or implicitly) following Gentzen's (1934/35) idea that elimination rules are "functions" of introduction rules.

Evidently, the elimination rules for the standard intuitionistic connectives $\wedge$, $\vee$, $\rightarrow$, $\bot$ are of the form (15.3): The rule of $\vee$-elimination has just been stated. In the two $\wedge$-elimination rules

$$(\wedge\mathrm{E})\ \frac{p_1 \wedge p_2}{p_1} \qquad \frac{p_1 \wedge p_2}{p_2} \tag{15.5}$$

ℓ is zero and q is p_1 or p_2, respectively. In modus ponens

$$(\rightarrow\mathrm{E})\ \frac{p_1 \rightarrow p_2 \qquad p_1}{p_2} \tag{15.6}$$

we have that ℓ is 1, s_1 is p_1 and q is p_2, with Γ_1 being empty. In the case of absurdity, the rule of *ex falso quodlibet*

$$(\bot\mathrm{E})\ \frac{\bot}{q} \tag{15.7}$$

leaves q unchanged with ℓ being zero. Note that a general schema for elimination rules in which $s_1, \ldots, s_\ell, q$ are identical, cannot accommodate these rules.

There may be more than one elimination rule for c, as is the case with conjunction. We suppose that elimination rules for a connective c are given as a finite list, where, as a limiting case, we allow for the empty list of elimination rules. This covers the verity constant $\top$, which has no elimination rule.

Our elimination schema (cE) is restricted in a way similar to the introduction schema (cI): No operators are permitted to occur in it except the c in the major premiss. This restriction could be released, but in its given form the schema is sufficient for the points we want to make. Note, however, that in the elimination case, there is no restriction corresponding to the 'no-extra-variable' constraint. If we disallowed extra variables beyond $p_1, \ldots, p_n$ in (cE), we would not be able to formulate, e.g., the elimination rule for disjunction. This means that for the topic discussed in this paper it is crucial to consider variables in elimination rules which are understood universally. When translating elimination rules into standard logic, this will lead us to use not just intuitionistic propositional logic **PL**, but intuitionistic propositional logic with universal propositional quantification **PL2**.

---

[7] However, it is less general than these other schemas in that here Γ_i may only contain propositional variables.

A connective which will play a prominent role in the following, is the ternary operator $\circ$, which has a single elimination rule:

$$(\circ\,\mathrm{E})\quad \frac{\circ(p_1, p_2, p_3) \qquad \begin{array}{c}[p_1]\\ p_2\end{array}}{p_3}\,. \tag{15.8}$$

15.3 Introduction Meaning and Elimination Meaning

In what follows, we take intuitionistic propositional logic based on the standard connectives $\wedge, \vee, \rightarrow, \bot$ for granted. We call this system *standard propositional logic* **PL**. We use formulas of this logic to express the intended meaning of n-ary connectives c, for which introduction or elimination rules are given. This means that we do not deal with the justification of the introduction and elimination rules for these standard connectives, nor with inversion and harmony principles and the like for them. Our enterprise is of a *reductive* kind, reducing problems associated with arbitrary n-ary connectives c to problems that only have to do with the standard connectives. If introduction and/or elimination rules for c are specified, by **PL+cI** we denote the system **PL** extended with the introduction rules for c, by **PL+cE** the system **PL** extended with the elimination rules for c and by **PL+cIE** the system **PL** extended with both the introduction and elimination rules for c. When we consider derivability in any of these systems, it will always be clear from the context, from which language the formulas involved are drawn, e.g., whether they contain c or not.

Suppose an introduction rule (cI) for an n-ary connective c is given according to (15.1). Then the intended meaning c^I of c according to this introduction rule, in short: the *introduction meaning* of c, can be expressed by translating the premisses of (cI) into a standard propositional formula. Let $\bigwedge \Gamma_i$ denote the conjunction of all elements of Γ_i. We define c^I to be the formula

$$(\bigwedge \Gamma_1 \rightarrow s_1) \wedge \;\ldots\; \wedge (\bigwedge \Gamma_\ell \rightarrow s_\ell) \tag{15.9}$$

Then the rule $\dfrac{c^I}{c(p_1, \ldots, p_n)}$ is derivable in **PL** extended with the rule (cI), and (cI) is derivable in **PL** extended with the rule $\dfrac{c^I}{c(p_1, \ldots, p_n)}$.[8] If we have k introduction rules $(c\mathrm{I})_1, \ldots, (c\mathrm{I})_k$ for c, then the introduction meaning c^I of c is defined to be

[8] We call a rule R *derivable* in a formal system K, if applications of R can be eliminated from all derivations in K, i.e., if $\Gamma \vdash_{K+R} \varphi$ implies, $\Gamma \vdash_K \varphi$ for any formula φ and any set of assumptions Γ. This corresponds to the usual definition of derivability of rules when R does not discharge assumptions, but includes the case of assumption-discharging rules. Note that we request the eliminability of R under arbitrary assumptions Γ. (Otherwise we would be defining the notion of admissibility of a rule.)

$$c_1^I \vee \ldots \vee c_k^I ,$$

where for each introduction rule $(c\mathrm{I})_i$, the formula c_i^I is specified as in (15.9). If $k = 0$, then c^I is absurdity $\bot$, and there is no introduction rule. We note as a fact:

Fact I1: *The rule* $\dfrac{c^I}{c(p_1, \ldots, p_n)}$ *is derivable in* **PL+cI**, *i.e. in* **PL** *extended with* $(c\mathrm{I})_1, \ldots, (c\mathrm{I})_k$, *and each introduction rule* $(c\mathrm{I})_i$ *is derivable in* **PL** *extended with the rule* $\dfrac{c^I}{c(p_1, \ldots, p_n)}$.

Using this fact we can conclude that in the context of introduction rules, c can be replaced with c^I. More precisely, let Γ' and φ' result from a set of formulas Γ and a formula φ by simultaneously replacing every occurrence of c with c^I. This is done by replacing every subformula of Γ and φ of the form $c(\varphi_1, \ldots, \varphi_n)$ with $c^I[p_1, \ldots, p_n/\varphi_1, \ldots, \varphi_n]$, where $[p_1, \ldots, p_n/\varphi_1, \ldots, \varphi_n]$ denotes the simultaneous substitution of $\varphi_1, \ldots, \varphi_n$ for $p_1, \ldots, p_n$. Then we can show that

$$\Gamma \vdash_{\mathbf{PL+cI}} \varphi \text{ implies } \Gamma' \vdash_{\mathbf{PL}} \varphi' .$$

In other words: In contexts, where only introductions for c are available, c and c^I behave identically. We note this as a fact:

Fact I2: *If* Γ' *and* φ' *result from* Γ *and* φ *by replacing* c *with* c^I, *then* $\Gamma \vdash_{\mathbf{PL+cI}} \varphi$ *implies* $\Gamma' \vdash_{\mathbf{PL}} \varphi'$.

Another way of exhibiting the equivalence between c^I and c with respect to introduction rules is by saying that for any set Γ of formulas not containing c,

$$\Gamma \vdash_{\mathbf{PL+cI}} c(p_1, \ldots, p_n) \text{ iff } \Gamma \vdash_{\mathbf{PL}} c^I .$$

This equivalence, which immediately follows from the two previous facts, expresses that the introduction meaning c^I of c is the weakest formula in the language without c which by using the introduction rules for c allows one to infer $c(p_1, \ldots, p_n)$. We might also say that in assertion position, i.e., on the right side of the turnstile, c^I is equally strong as c.[9] We note this as a fact:

Fact I3: *If* Γ *does not contain c, then:* $\Gamma \vdash_{\mathbf{PL+cI}} c(p_1, \ldots, p_n)$ iff $\Gamma \vdash_{\mathbf{PL}} c^I$.

In the elimination case the situation is slightly more complicated. In an elimination rule for c variables beyond $p_1, \ldots, p_n$ can be present. In the following, these variables are also called *extra variables*. A typical example of an extra variable is the variable

[9] Popper (1947, see Schroeder-Heister 2005) was the first to characterize logical constants in terms of maximality and minimality conditions. Tennant (1978, p. 74) used them as the basic ingredient of a principle of harmony.

r in the following formulation of the standard elimination rule for disjunction (15.4). It is obvious that the extra variables in elimination rules have a universal meaning. Correspondingly, we extend our means of expression by considering not just standard intuitionistic propositional logic, but this logic together with universal propositional quantification. We call this system **PL2**.[10] Correspondingly, we use the abbreviations **PL2+cI**, **PL2+cE** and **PL2+cIE**, if in addition introduction, elimination, or both introduction and elimination rules for c are available in the system.

Suppose an elimination rule (cE) is given for an n-ary connective c according to (15.3). Then the intended meaning c^E of c according to this elimination rule, in short: the *elimination meaning* of c, is obtained as follows. We remove the major premiss $c(p_1, \ldots, p_n)$ from (cE) and translate the 'rest' of the rule, which tells what can be inferred from $c(p_1, \ldots, p_n)$ according to (cE), into a formula of **PL2**. Thus we define c^E to be the formula

$$\overline{\forall}(((\bigwedge \Gamma_1 \to s_1) \wedge \ldots \wedge (\bigwedge \Gamma_\ell \to s_\ell)) \to q)\,. \tag{15.10}$$

Here $\overline{\forall}$ universally quantifies all extra variables in c^E.[11] Then the rule $\dfrac{c(p_1, \ldots, p_n)}{c^E}$ is derivable in **PL2** extended with the rule (cE), and (cE) is derivable in **PL2** extended with the rule $\dfrac{c(p_1, \ldots, p_n)}{c^E}$. If we have k elimination rules $(c\text{E})_1, \ldots, (c\text{E})_k$ for c, then the elimination meaning c^E of c is defined to be

$$c_1^E \wedge \ldots \wedge c_k^E\,,$$

where for each elimination rule $(c\text{E})_i$, the formula c_i^E is specified as in (15.10). If $k = 0$, then c^E is verity $\top$. We note as a fact:

Fact E1: *The rule* $\dfrac{c(p_1, \ldots, p_n)}{c^E}$ *is derivable in* **PL2+cE**, *i.e. in* **PL2** *extended with* $(c\text{E})_1, \ldots, (c\text{E})_k$, *and each elimination rule* $(c\text{E})_i$ *is derivable in* **PL2** *extended with the rule* $\dfrac{c(p_1, \ldots, p_n)}{c^E}$.[12]

Using this fact we can conclude that in the context of elimination rules, c can be replaced with c^E. More precisely, let Γ' and φ' result from Γ and φ by simultaneously replacing every occurrence of c with c^E. This is done by replacing every subformula

---

[10] To express the meanings of elimination rules, we can restrict ourselves to the case of prenex formulas, i.e., formulas quantified only from outside. More involved forms of quantification might be considered, but are not needed here. We also do not use the fact that by using propositional quantification and implication, all intuitionistic connectives become definable (see Prawitz 1965).

[11] More precisely, $\overline{\forall}$ stands for $\forall r_1 \ldots \forall r_j$, where $\{r_1, \ldots, r_j\}$ is the set of those variables occurring in c^E, which are different from any variable in $\{p_1, \ldots, p_n\}$.

[12] Note that for this statement propositional quantification is not really needed, as we are treating all rules as schemas, which means that universal quantification could remain implicit just by the usage of free propositional variables.

of Γ and φ of the form $c(\varphi_1, \ldots, \varphi_n)$ with $c^E[p_1, \ldots, p_n/\varphi_1, \ldots, \varphi_n]$. Then we can show that

$$\Gamma \vdash_{\mathbf{PL2+cE}} \varphi \text{ implies } \Gamma' \vdash_{\mathbf{PL2}} \varphi' .$$

In other words: In contexts, where only eliminations for c are available, c and c^E behave identically. We note as a fact:

Fact E2: *If* Γ' *and* φ' *result from* Γ *and* φ *by replacing* c *with* c^E*, then:* $\Gamma \vdash_{\mathbf{PL2+cE}} \varphi$ implies $\Gamma' \vdash_{\mathbf{PL2}} \varphi'$.

Another way of exhibiting the equivalence between c^E and c with respect to elimination rules is by saying that for any set $\Gamma \cup \{\varphi\}$ of formulas not containing c,

$$c(p_1, \ldots, p_n), \Gamma \vdash_{\mathbf{PL2+cE}} \varphi \text{ iff } c^E, \Gamma \vdash_{\mathbf{PL2}} \varphi .$$

This equivalence, which immediately follows from the two previous facts, expresses that the elimination meaning c^E of c is the strongest formula in the language without c, which, by using the elimination rules for c, can be inferred from $c(p_1, \ldots, p_n)$. We might also say that in assumption position, i.e., on the left side of the turnstile, c^E is equally strong as c. We note this as a fact:

Fact E3: *If* Γ *and* φ *do not contain c, then:*
$c(p_1, \ldots, p_n), \Gamma \vdash_{\mathbf{PL2+cE}} \varphi$ iff $c^E, \Gamma \vdash_{\mathbf{PL2}} \varphi$.

15.4 Harmony, Conservativeness, Uniqueness

Given introduction rules for c, we have defined the introduction meaning of c to be the formula c^I, which is a formula in standard intuitionistic propositional logic **PL**. Given elimination rules for c, we have defined the elimination meaning of c to by the formula c^E, which is a prenex formula in **PL2**, i.e., in standard intuitionistic propositional logic with propositional quantification. If both introduction and elimination rules are provided for c, we say that they are in *harmony* with each other, if introduction meaning and elimination meaning of c match, i.e., if in **PL2** we can show:

$$c^I \dashv\vdash c^E . \tag{15.11}$$

Splitting up harmony into its two directions, we can say the following. Suppose the introduction meaning of c entails its elimination meaning:

$$c^I \vdash c^E . \tag{15.12}$$

Then we have conservativeness of the introduction and elimination rules for c over **PL**. We state this as a result:

Conservativeness Lemma *Suppose that in* **PL2** *it holds that $c^I \vdash c^E$. Suppose that in* **PL+cIE** *we have that $\Delta \vdash \varphi$ for a set of formulas Δ and a formula φ which do not contain c. Then $\Delta \vdash \varphi$ holds already in* **PL**.

Proof sketch We use a normalisation argument. In the introduction context we can, according to **Fact I2**, replace c with c^I, and in the elimination context we can, according to **Fact E2**, replace c with c^E. If c occurs both in introduction and elimination context, i.e., as a maximal formula, we replace it with c^I followed by a derivation in **PL2** of c^E from c^I. Thus $\Delta \vdash \varphi$ holds is **PL2**. Using the normalisability of **PL2** and the conservativeness of **PL2** over **PL**, we obtain our result. □

That this proof needs to rely on the heavy machinery of normalisation of **PL2** (i.e., Girard's system **F**; see, e.g., Girard et al. 1989, and Prawitz 1971) is due to the fact that we describe introduction and elimination meanings in abstract terms, and here the latter by means of a second-order formula. If we deal with a concrete system with harmonious rules, for example rules following a general schema for eliminations as in Prawitz (1979) or Schroeder-Heister (1984), then normalisation and conservativeness can be proven directly in the system under consideration.

Conversely, suppose the elimination meaning of c entails its introduction meaning:

$$c^E \vdash c^I \,. \tag{15.13}$$

Then c is uniquely characterised in the following sense.

Uniqueness Lemma *If we extend* **PL2** *with the introduction rules for c and with the elimination rules for a duplicate c' of c (in the joint language containing both c and c'), we can show*

$$c'(p_1, \ldots, p_n) \vdash c(p_1, \ldots, p_n).$$

If introduction and elimination rules for both c and c' are available, this gives us the equivalence

$$c'(p_1, \ldots, p_n) \dashv\vdash c(p_1, \ldots, p_n) \,.$$

Proof This follows immediately from **Fact I1** and **Fact E1** above, which give us the derivability of the rules $\dfrac{c'(p_1, \ldots, p_n)}{c^E}$ and $\dfrac{c^I}{c(p_1, \ldots, p_n)}$. □

Conservativeness and uniqueness are the two conditions Belnap (1962) considered to be crucial for the inferential definition of a connective (see also Došen and Schroeder-Heister 1985). Similar conditions appear under different names in proof-theoretic semantics, for example validity and stability (Dummett 1991) or (local) soundness and (local) completeness (Francez and Dyckhoff 2012). There are considerable differences between these and related notions, in particular as to whether they are understood locally (refer to applications of rules) or globally (refer to the behaviour of the logical system as a whole). In this paper, we understand the two conditions in an abstract way, by relating the logically coded introduction and elimination meanings of a connective, rather than using its introduction and elimination rules themselves (in contradistinction, for example, to the foundational analyses of Francez and Dyckhoff 2012, and Schroeder-Heister 2014a).

As we have defined harmony and, correspondingly, conservativeness and uniqueness in terms of derivability in **PL2**, one might ask[13] whether any of these relations is decidable. As **PL2** is undecidable, the immediate answer is negative. However, the conservativeness direction (15.12) is in fact decidable, as c^E is a prenex formula whose quantifiers can be represented by free variables, so that conservativeness becomes a derivability problem in the decidable system **PL**. The uniqueness direction (15.13) may contain quantifiers on the hypothesis side, which cannot be eliminated. There, as we would guess, the undecidability of **PL2** comes into effect.[14]

15.5 The Existence of Harmonious Rules

If the introduction and elimination rules for an n-ary connective c are in harmony with each other, we also say that the elimination rules are *harmonious* for the introduction rules and the introduction rules are *harmonious* for the elimination rules. For many sets of introduction rules there are harmonious elimination rules and vice versa. Table 15.1 gives the introduction and elimination rules of the standard connectives $\wedge$, $\vee$, $\rightarrow$, $\bot$ and $\top$ together with their respective introduction and elimination meanings. These rules are all in harmony with each other. For $\wedge$, $\rightarrow$ and $\top$ this is trivial, as the respective introduction and elimination meanings are identical. In the case of disjunction and absurdity, we can easily show in **PL2** that

$$p_1 \vee p_2 \dashv\vdash \forall r(((p_1 \rightarrow r) \wedge (p_2 \rightarrow r)) \rightarrow r)$$

and

$$\bot \dashv\vdash \forall r\, r \ .$$

In Table 15.2 we consider forms of conjunction and implication with alternative elimination rules. Here we have harmony as well. For example, in oder to prove it for the introduction and elimination rules for $\times$ we can easily show in **PL2** that

$$p_1 \times p_2 \dashv\vdash \forall r_1 r_2 r(((p_1 \rightarrow r_1) \wedge (p_2 \rightarrow r_2) \wedge ((r_1 \wedge r_2) \rightarrow r)) \rightarrow r) \ .$$

Table 15.3 presents some further connectives. The connectives c_1 and c_2 have the same introduction rules but different elimination rules. We nevertheless have harmony in both cases, since in **PL2** we can show that

---

[13] Heinrich Wansing posed this question.

[14] More precisely, we do not have an argument at hand showing that the derivability of a quantifier-free formula from a prenex formula in **PL2** is decidable.

Table 15.1 Introduction and elimination rules together with introduction and elimination meanings for the standard connectives

	Introduction rules	Elimination rules
	Introduction meaning	Elimination meaning
a	$\dfrac{p_1 \qquad p_2}{p_1 \wedge p_2}$	$\dfrac{p_1 \wedge p_2}{p_1} \qquad \dfrac{p_1 \wedge p_2}{p_2}$
	$p_1 \wedge p_2$	$p_1 \wedge p_2$
b	$\dfrac{p_1}{p_1 \vee p_2} \qquad \dfrac{p_2}{p_1 \vee p_2}$	$\dfrac{p_1 \vee p_2 \qquad \begin{matrix}[p_1]\\ r\end{matrix} \qquad \begin{matrix}[p_2]\\ r\end{matrix}}{r}$
	$p_1 \vee p_2$	$\forall r(((p_1 \rightarrow r) \wedge (p_2 \rightarrow r)) \rightarrow r)$
c	$\dfrac{\begin{matrix}[p_1]\\ p_2\end{matrix}}{p_1 \rightarrow p_2}$	$\dfrac{p_1 \rightarrow p_2 \qquad p_1}{p_2}$
	$p_1 \rightarrow p_2$	$p_1 \rightarrow p_2$
d	No I rule	$\dfrac{\bot}{r}$
	$\bot$	$\forall r \; r$
e	$\dfrac{}{\top}$	No E rule
	$\top$	$\top$

In all cases introduction and elimination rules are in harmony with each other

$$(p_1 \wedge p_2) \vee p_3 \dashv\vdash \forall r((((p_1 \wedge p_2) \rightarrow r) \wedge (p_3 \rightarrow r)) \rightarrow r)$$

as well as

$$(p_1 \wedge p_2) \vee p_3 \dashv\vdash \forall r(((p_1 \rightarrow r) \wedge (p_3 \rightarrow r)) \rightarrow r) \wedge \forall r(((p_2 \rightarrow r) \wedge (p_3 \rightarrow r)) \rightarrow r) \tag{15.14}$$

Note that we take the intuitionistic logic of the standard connectives for granted, which means in particular that we assume that the standard structural rules are at our disposal. Otherwise we would not, for example, be able to show (15.14), for which we essentially need distribution or $\vee$ over $\wedge$:

$$(p_1 \wedge p_2) \vee p_3 \dashv\vdash (p_1 \vee p_3) \wedge (p_2 \vee p_3)$$

which is not available as a general law when thinning or contraction are restricted.[15]

[15] It depends on whether these connectives are read additively or multiplicatively. This point is in particular relevant, if (like, e.g., Read 2010, 2014, this volume) one prefers more than one elimination rule in cases such as $\bullet$ (see Table 15.2).

Table 15.2 Introduction and elimination rules together with introduction and elimination meanings for various connectives related to the standard ones

	Introduction rules	Elimination rules
	Introduction meaning	Elimination meaning
a	$\dfrac{p_1 \quad p_2}{p_1 \& p_2}$	$\dfrac{p_1 \& p_2 \quad \begin{matrix}[p_1, p_2]\\ r\end{matrix}}{r}$
	$p_1 \wedge p_2$	$\forall r(((p_1 \wedge p_2) \to r) \to r)$
b	$\dfrac{p_1 \quad p_2}{p_1 \bullet p_2}$	$\dfrac{p_1 \bullet p_2 \quad \begin{matrix}[p_1]\\ r\end{matrix}}{r} \qquad \dfrac{p_1 \bullet p_2 \quad \begin{matrix}[p_2]\\ r\end{matrix}}{r}$
	$p_1 \wedge p_2$	$\forall r((p_1 \to r) \to r) \wedge \forall r((p_2 \to r) \to r)$
c	$\dfrac{p_1 \quad p_2}{p_1 \times p_2}$	$\dfrac{p_1 \times p_2 \quad \begin{matrix}[p_1]\\ r_1\end{matrix} \quad \begin{matrix}[p_2]\\ r_2\end{matrix} \quad \begin{matrix}[r_1, r_2]\\ r\end{matrix}}{r}$
	$p_1 \wedge p_2$	$\forall r_1 r_2 r(((p_1 \to r_1) \wedge (p_2 \to r_2) \wedge ((r_1 \wedge r_2) \to r)) \to r)$
d	$\dfrac{\begin{matrix}[p_1]\\ p_2\end{matrix}}{p_1 \supset p_2}$	$\dfrac{p_1 \supset p_2 \quad p_1 \quad \begin{matrix}[p_2]\\ r\end{matrix}}{r}$
	$p_1 \to p_2$	$\forall r((p_1 \wedge (p_2 \to r)) \to r)$

In all cases introduction and elimination rules are in harmony with each other

Obviously, the rules given for the connectives $\&\supset$ and *tonk* are not harmonious, since neither

$$p_1 \wedge (p_1 \to p_2) \dashv\vdash p_1 \to p_2$$

nor

$$p_1 \dashv\vdash p_2$$

holds in **PL2**.

The n-ary connectives *ci* and *ce* represent connectives of a general form. In the case of *ci* we have harmony, provided the introduction rules do not discharge any assumption, but are just productions. Likewise, the rules for *ce* are harmonious, provided the elimination rules do not discharge any assumption. This means that, if the introduction rules of *ci* are of the form stated in Table 15.3, then there is always a harmonious elimination rule. If the elimination rules for *ce* are of the form stated, then there is always a harmonious introduction rule.

Table 15.3 Introduction and elimination rules together with introduction and elimination meanings for further connectives

	Introduction rules / Introduction meaning	Elimination rules / Elimination meaning
a	$\dfrac{p_1 \qquad p_2}{c_1(p_1, p_2, p_3)} \qquad \dfrac{p_3}{c_1(p_1, p_2, p_3)}$	$\dfrac{c_1(p_1, p_2, p_3) \qquad \begin{matrix}[p_1, p_2]\\ r\end{matrix} \qquad \begin{matrix}[p_3]\\ r\end{matrix}}{r}$
	$(p_1 \wedge p_2) \vee p_3$	$\forall r((((p_1 \wedge p_2) \rightarrow r) \wedge (p_3 \rightarrow r)) \rightarrow r)$
b	$\dfrac{p_1 \qquad p_2}{c_2(p_1, p_2, p_3)} \qquad \dfrac{p_3}{c_2(p_1, p_2, p_3)}$	$\dfrac{c_2(p_1, p_2, p_3) \qquad \begin{matrix}[p_1]\\ r\end{matrix} \qquad \begin{matrix}[p_3]\\ r\end{matrix}}{r}$ $\dfrac{c_2(p_1, p_2, p_3) \qquad \begin{matrix}[p_2]\\ r\end{matrix} \qquad \begin{matrix}[p_3]\\ r\end{matrix}}{r}$
	$(p_1 \wedge p_2) \vee p_3$	$\forall r(((p_1 \rightarrow r) \wedge (p_3 \rightarrow r)) \rightarrow r)$ $\wedge \forall r(((p_2 \rightarrow r) \wedge (p_3 \rightarrow r)) \rightarrow r)$
c	$\dfrac{p_1 \qquad \begin{matrix}[p_1]\\ p_2\end{matrix}}{p_1 \mathbin{\&\!\supset} p_2}$	$\dfrac{p_1 \mathbin{\&\!\supset} p_2 \qquad p_1}{p_2}$
	$p_1 \wedge (p_1 \rightarrow p_2)$	$p_1 \rightarrow p_2$
d	$\dfrac{p_1}{p_1 \; tonk \; p_2}$	$\dfrac{p_1 \; tonk \; p_2}{p_2}$
	p_1	p_2
e	$\dfrac{\Delta_1}{ci(p_1, \ldots, p_n)} \quad \cdots \quad \dfrac{\Delta_m}{ci(p_1, \ldots, p_n)}$ Δ_i of the form $q_{i1} \ldots q_{i\ell_i}$ with $\{q_{i1}, \ldots, q_{i\ell_i}\} \subseteq \{p_1, \ldots, p_n\}$	$\dfrac{ci(p_1, \ldots, p_n) \qquad \begin{matrix}[\Delta_1]\\ r\end{matrix} \quad \ldots \quad \begin{matrix}[\Delta_m]\\ r\end{matrix}}{r}$
	$\bigwedge \Delta_1 \vee \ldots \vee \bigwedge \Delta_m$	$\forall r((\bigwedge \Delta_1 \rightarrow r) \wedge \ldots \wedge (\bigwedge \Delta_m \rightarrow r) \rightarrow r)$
f	$\dfrac{\begin{matrix}[\Delta_1]\\ q_1\end{matrix} \qquad \cdots \qquad \begin{matrix}[\Delta_m]\\ q_m\end{matrix}}{ce(p_1, \ldots, p_n)}$	$\dfrac{ce(p_1, \ldots, p_n) \qquad \Delta_1}{q_1} \quad \cdots \quad \dfrac{ce(p_1, \ldots, p_n) \qquad \Delta_1}{q_m}$ Δ_i of the form $q_{i1} \ldots q_{i\ell_i}$ with $\{q_{i1}, \ldots, q_{i\ell_i}\} \cup \{q_1, \ldots, q_m\} \subseteq \{p_1, \ldots, p_n\}$
	$(\bigwedge \Gamma_1 \rightarrow q_1) \wedge \ldots \wedge (\bigwedge \Gamma_m \rightarrow q_m)$	$(\bigwedge \Gamma_1 \rightarrow q_1) \wedge \ldots \wedge (\bigwedge \Gamma_m \rightarrow q_m)$

However, not for every given set of introduction rules there are harmonious elimination rules, and not for every given set of elimination rules there are harmonious introduction rules. An example of a connective with given introduction rules, for which there are no harmonious elimination rules is the connective $\star$, whose introduction rules ($\star$ I) are given in (15.2). Its introduction meaning $\star^I$ is $(p_1 \rightarrow p_2) \vee p_3$. If there were harmonious elimination rules for $\star$, then, according to the definitions in Sect. 15.3, its elimination meaning $\star^E$ would have to be of the form $\star_1^E \wedge \ldots \wedge \star_k^E$, where each $\star_i^E$ is of the form $\overline{\forall}(((\bigwedge \Gamma_1 \rightarrow s_1) \wedge \ldots \wedge (\bigwedge \Gamma_\ell \rightarrow s_\ell)) \rightarrow q)$. However, in Olkhovikov and Schroeder-Heister (2014a) it could be demonstrated that no formula of this form is equivalent to $\star^I$ in **PL2**.

If we allow for connectives already defined to occur in introductions and eliminations, then there are harmonious elimination rules for $\star$. The trivial $\star$ elimination rule would be the single rule

$$\frac{\star(p_1, p_2, p_3)}{(p_1 \rightarrow p_2) \vee p_3}$$

which assumes that implication and disjunction are already being given. An alternative elimination rule only assumes that implication is available:

$$(\star\,\mathrm{E})\ \frac{\star(p_1, p_2, p_3) \qquad \begin{matrix}[p_1 \rightarrow p_2]\\ r\end{matrix} \qquad \begin{matrix}[p_3]\\ r\end{matrix}}{r}\ .$$

The elimination meaning $\star^E$ of $\star$ according to this elimination rule is

$$\forall r((((p_1 \rightarrow p_2) \rightarrow r) \wedge (p_3 \rightarrow r)) \rightarrow r)$$

which can easily be shown to be equivalent in **PL2** to $\star^I$:

$$(p_1 \rightarrow p_2) \vee p_3 \dashv\vdash \forall r((((p_1 \rightarrow p_2) \rightarrow r) \wedge (p_3 \rightarrow r)) \rightarrow r)\ .$$

Instead of assuming implication to be a connective already defined, we could extend the apparatus of natural deduction by using rules of higher levels, i.e., rules that can discharge not only formulas but also rules which are used as assumptions, as described in Schroeder-Heister (2014a). In such a framework the elimination rule for $\star$ would take the form

$$(\star\,\mathrm{E})\ \frac{\star(p_1, p_2, p_3) \qquad \begin{matrix}[p_1 \Rightarrow p_2]\\ r\end{matrix} \qquad \begin{matrix}[p_3]\\ r\end{matrix}}{r}\ .$$

Here $p_1 \Rightarrow p_2$ represents the *rule* which allows one to pass over from p_1 to p_2. It is assumed as an assumption in the subderivation of the left minor premiss and is discharged at the application of ($\star$ E). The result by Olkhovikov and Schroeder-Heister (2014a) can then be read as showing that $\star$ does not have flat elimination rules, where, following a terminology proposed by Read (2014, this volume), an elimination rule is called *flat*, if it does not allow one to discharge rules, but only formulas. Flat rules are rules of the kind considered in standard (non-extended) natural deduction.[16]

Non-flattening theorem for elimination rules *The connective ⋆ does not have flat elimination rules.*

If we allow for rules of higher levels, then every set of introduction rules for an n-ary connective c has harmonious elimination rules. In fact, only one single elimination rule is needed, which we call the *generalised* or *canonical* elimination rule. It is constructed as follows: We associate with every introduction rule for c of the form

$$(c\mathrm{I})\quad \frac{\begin{matrix}[\Gamma_1] & & [\Gamma_\ell]\\ s_1 & \dots & s_\ell\end{matrix}}{c(p_1,\dots,p_n)}$$

a list Δ of rules

$$(\Gamma_1 \Rightarrow s_1),\dots,(\Gamma_\ell \Rightarrow s_\ell)$$

representing the premisses of this introduction rule (note that the double arrow is used to linearly denote rules rather than implications). If there are m introduction rules for c, we obtain m such lists $\Delta_1,\dots,\Delta_m$. Then the canonical elimination rule for c has the form

$$(c\mathrm{E})_{GEN}\quad \frac{c(p_1,\dots,p_n)\qquad \begin{matrix}[\Delta_1] & & [\Delta_m]\\ r & \dots & r\end{matrix}}{r}\ . \tag{15.15}$$

This schema is devised such as to guarantee that introduction and elimination meaning of c match. If Δ_i is $(\Gamma_1 \Rightarrow s_1),\dots,(\Gamma_\ell \Rightarrow s_\ell)$, let Δ_i^{PROP} be its propositional translation $(\bigwedge\Gamma_1 \to s_1),\dots,(\bigwedge\Gamma_\ell \to s_\ell)$. Then the elimination meaning of c is defined as

$$\forall r(((\bigwedge\Delta_1^{PROP} \to r)\wedge\dots\wedge(\bigwedge\Delta_m^{PROP} \to r))\to r)\ .$$

This corresponds to the definition of elimination meaning in Sect. 15.3 with the only difference that we cannot just form the conjunction of the elements of the Δ_i, as they are not necessarily formulas, but have to propositionally translate these elements into conjunction-implication-formulas, if they are rules. Then we can easily prove in **PL2** that introduction meaning (which is defined as before in Sect. 15.3) and elimination meaning of c match:

$$c_1^I \vee \dots \vee c_m^I \dashv\vdash \forall r(((\bigwedge\Delta_1^{PROP} \to r)\wedge\dots\wedge(\bigwedge\Delta_m^{PROP} \to r))\to r)$$

[16] The general elimination schema Francez and Dyckhoff (2012) propose is flat and therefore not harmonious in the sense of the definition of harmony proposed in Sect. 15.4. However, their point on local soundness and completeness is independent of this schema and applies to harmonious rules in our sense.

We simply have to use that c_i^I is identical to $\bigwedge \Delta_i^{PROP}$.

In fact, if we allow for implication-conjunction formulas to occur as assumptions in elimination rules, we can obtain the same result without having to rely on rules as assumptions. Instead of the general elimination schema (15.15) we could use the following schema, which results from (15.15) by replacing lists of assumptions rules Δ_i with lists of their propositional translations Δ_i^{PROP}:

$$(cE)_P \quad \frac{c(p_1, \ldots, p_n) \qquad \begin{matrix}[\Delta_1^{PROP}] \\ r\end{matrix} \quad \ldots \quad \begin{matrix}[\Delta_m^{PROP}] \\ r\end{matrix}}{r} \quad . \tag{15.16}$$

We call it the *Prawitz schema* for generalised elimination rules, as it corresponds to the schema proposed in Prawitz (1979). For further discussion of this issue see Sect. 15.7.

Our result even pertains to the case in which the introduction rules are not flat, i.e., may be of higher levels. In that case, the elements of Δ_i may be rules which discharge assumptions. For example, consider the following quaternary operator c with the following two introduction rules:

$$(c\,\mathrm{I}) \quad \frac{\begin{matrix}[p_1 \Rightarrow p_2] \\ p_3\end{matrix}}{c(p_1, p_2, p_3, p_4)} \qquad \frac{p_4}{c(p_1, p_2, p_3, p_4)} \quad .$$

According to the general schema $(c\mathrm{E})_{GEN}$ the corresponding canonical (and thus harmonious) elimination rule is

$$(c\,\mathrm{E}) \quad \frac{c(p_1, p_2, p_3, p_4) \qquad \begin{matrix}[(p_1 \Rightarrow p_2) \Rightarrow p_3] \\ r\end{matrix} \qquad \begin{matrix}[p_4] \\ r\end{matrix}}{r} \quad .$$

In general it holds that, if we pass from given introduction rules to the corresponding canonical elimination rule, the level always goes up by one step, as the premisses of the introduction rules then occur as dischargeable assumptions of minor premisses in the canonical elimination rule. This cannot be avoided, i.e., we can always construct a connective whose introduction rules are of level n, without there being harmonious elimination rules of level n or below. This generalised non-flattening result is proved in Olkhovikov and Schroeder-Heister (2014b, Theorem 1)]:

Generalised non-flattening theorem for elimination rules *Suppose the schema for introduction rules is limited to rules of maximal level n. Then we can always find a connective satisfying such a schema, whose elimination schema cannot be of level n or below, i.e. must be at least of level* $n+1$. *In fact, we can choose the* $(n+1)$*-place connective with the introduction meaning*

$$((\ldots(p_1 \to p_2)\ldots \to p_{n-1}) \to p_n) \vee p_{n+1},$$

which is a generalisation of ⋆.

If we start with elimination rules, we have an analogous situation: Not all connectives which are specified by given elimination rules have harmonious introduction rules. Consider negation $\neg$ with the elimination rule:

$$(\neg E)\ \frac{\neg p_1 \qquad p_1}{r}\ .$$

According to our definition, its elimination meaning $\neg^E$ is $\forall r(p_1 \to r)$. However, there is no set of introduction rules for $\neg$ such that the introduction meaning $\neg^I$ is equivalent to $\neg^E$. This is simply a consequence of the (almost) trivial fact that negation $\neg$ cannot be defined in terms of implication and conjunction.

However, if we allow for connectives already defined to occur in introduction rules, we can easily give an appropriate introduction rule for $\neg$:

$$(\neg \mathrm{I})\ \frac{\begin{matrix}[p_1]\\ \bot\end{matrix}}{\neg p_1}\ .$$

The introduction meaning $\neg^I$ of $\neg$ is now $p_1 \to \bot$, which is interderivable in **PL2** with its elimination meaning $\neg^E$:

$$p_1 \to \bot \dashv\vdash \forall r(p_1 \to r)$$

by using the absurdity rule (*ex falso quodlibet*), which is the elimination rule for $\bot$. Another example is the ternary connective $\circ$ with the elimination rule ($\circ$ E) given in (15.8). Its elimination meaning $\circ^E$ is $(p_1 \to p_2) \to p_3$. If there were harmonious introduction rules for $\circ$, its introduction meaning $\circ^I$ could be described by a disjunction of formulas $\circ^I_1 \vee \ldots \vee \circ^I_k$, where each formula $\circ^I_i$ would be of the form $(\bigwedge \Gamma_1 \to s_1) \wedge \ldots \wedge (\bigwedge \Gamma_\ell \to s_\ell)$. It can be shown, however, that $\circ^E$ is never equivalent in **PL2** to a disjunction of formulas of this form. The proof of this fact can be found in Olkhovikov and Schroeder-Heister (2014a).

If we allow for connectives, which are already defined, to occur in introduction rules, we could equip $\circ$ with the trivial introduction rule

$$\frac{(p_1 \to p_2) \to p_3}{\circ(p_1, p_2, p_3)}$$

or alternatively with

$$(\circ\ \mathrm{I})\ \frac{\begin{matrix}[p_1 \to p_2]\\ p_3\end{matrix}}{\circ(p_1, p_2, p_3)}\ .$$

The introduction meaning $\circ^I$ according to this introduction rule is $(p_1 \to p_2) \to p_3$, which is identical to its elimination meaning. If we use higher-level rules, we can write the introduction rule for c as

$$(\circ\,\mathrm{I})\ \frac{\begin{array}{c}[p_1 \Rightarrow p_2]\\ p_3\end{array}}{\circ(p_1, p_2, p_3)}$$

The result by Olkhovikov and Schroeder-Heister (2014a) then says that $\circ$ does not have flat introduction rules.

Non-flattening theorem for introduction rules *The connective $\circ$ does not have flat introduction rules.*

However, even if we allow for rules of higher levels, not every set of elimination rules for an n-ary connective c has corresponding harmonious introduction rules. This is due to the fact that in (cE) propositional variables beyond $p_1, \ldots, p_n$ may occur, which, as schematic variables, have a universal meaning and correspondingly enter the elimination meaning c^E as universally bound. If we want to turn the content of such an elimination inference into the premiss of an introduction rule, we have to devise a binding mechanism at the structural level. We need not only rules as assumptions, but also bound variables in the premisses of rules. For that to achieve we define the general schema of an introduction rule for c to be of the form

$$(c\mathrm{I})\ \frac{\begin{pmatrix}[\Gamma_1]\\ s_1\end{pmatrix}_{\overline{r_1}} \quad \ldots \quad \begin{pmatrix}[\Gamma_\ell]\\ s_\ell\end{pmatrix}_{\overline{r_\ell}}}{c(p_1, \ldots, p_n)}\ .$$

Here the $\overline{r_i}$ are lists of propositional variables different from $p_1, \ldots, p_n$, which cannot be substituted (as can $p_1, \ldots, p_n$), but which in the subproofs of s_i from Γ_i are treated like constants ('parameters' or 'free variables' in a different terminology).

Assuming this extension of natural deduction with quantified higher-level rules (described in detail in Schroeder-Heister 2014a), we can construct introduction rules, which are in harmony with given elimination rules for c of the form (cE) as given in (15.3). In fact, only one single introduction rule, is needed, which we call the *generalised* or *canonical* introduction rule. It is constructed as follows: We associate with every elimination rule of the form (15.3)

$$(c\mathrm{E})\ \frac{c(p_1, \ldots, p_n) \quad \begin{array}{c}[\Gamma_1]\\ s_1\end{array} \quad \ldots \quad \begin{array}{c}[\Gamma_\ell]\\ s_\ell\end{array}}{q}\ ,$$

a list Δ of rules

$$(\Gamma_1 \Rightarrow s_1), \ldots, (\Gamma_\ell \Rightarrow s_\ell)$$

representing the premisses of this elimination rule. From this we construct the pattern

$$\left(\begin{array}{c}[\Delta]\\ q\end{array}\right)_{\{s_1,\ldots,s_\ell,q\}}$$

representing what can be inferred from $c(p_1, \ldots, p_n)$ using this elimination rule. Suppose c has m elimination rules and we have associated m patterns

$$\left(\begin{array}{c}[\Delta_1]\\ q_1\end{array}\right)_{Var_1} \quad \ldots \quad \left(\begin{array}{c}[\Delta_m]\\ q_m\end{array}\right)_{Var_m}$$

with them, respectively. Here Var_i are the sets of variables occurring in the respective patterns beyond $p_1, \ldots, p_n$. Then the canonical introduction rule for c has the following form:

$$(c\mathrm{I})_{GEN}\ \frac{\left(\begin{array}{c}[\Delta_1]\\ q_1\end{array}\right)_{Var_1} \quad \ldots \quad \left(\begin{array}{c}[\Delta_m]\\ q_m\end{array}\right)_{Var_m}}{c(p_1, \ldots, p_n)}\ .$$

It is constructed in such a way that introduction meaning and elimination meaning of c match. If Δ_i is $(\Gamma_1 \Rightarrow s_1), \ \ldots \ , (\Gamma_\ell \Rightarrow s_\ell)$, let Δ_i^{PROP} be its propositional translation $(\bigwedge \Gamma_1 \rightarrow s_1), \ \ldots \ , (\bigwedge \Gamma_\ell \rightarrow s_\ell)$. Then the introduction meaning of c is defined as

$$\overline{\forall}(\bigwedge \Delta_1^{PROP} \rightarrow q_1) \wedge \ldots \wedge \overline{\forall}(\bigwedge \Delta_m^{PROP} \rightarrow q_m)$$

(note that $\overline{\forall}$ binds all variables beyond $p_1, \ldots, p_n$). This corresponds to the definition of introduction meaning in Sect. 15.3 with the only difference that we cannot just take the conjunction of the elements of the Γ_i, as they are not necessarily formulas, but have to propositionally translate these elements into conjunction-implication-formulas, if they are rules. Then elimination meaning (which is defined as before in Sect. 15.3) and introduction meaning of c match, i.e., the following holds in **PL2**:

$$\overline{\forall}(\bigwedge \Delta_1^{PROP} \rightarrow q_1) \wedge \ldots \wedge \overline{\forall}(\bigwedge \Delta_m^{PROP} \rightarrow q_m) \ \dashv\vdash \ c_1^E \wedge \ldots \wedge c_k^E\ .$$

This is actually trivial since c_i^E is identical to $\overline{\forall}(\bigwedge \Delta_I^{PROP} \rightarrow q_m)$.

This result pertains to the case in which the elimination rules are not flat, i.e. may be of higher levels. In that case, the elements of Δ_i may be rules which discharge assumptions. Note however that when passing from eliminations to introductions, not only the level of the rule goes up by one step, but we have to use structural quantification in the premiss of the introduction rule, too, if the elimination rules contain extra variables. Going up one step cannot be avoided, i.e., we can always construct a connective whose elimination rules reach level n, without there being harmonious introduction rules of level n or below (Olkhovikov and Schroeder-Heister 2014b, Theorem 2):

Generalised non-flattening theorem for introduction rules *Suppose the schema for elimination rules is limited to rules of maximal level n. Then we can always find a connective satisfying such a schema, whose introduction schema cannot be of level n or below, i.e. must be at least of level* $n+1$. *In fact, we can choose the* $(n+1)$*-place connective with the elimination meaning* $(\ldots(p_1 \to p_2)\ldots \to p_n) \to p_{n+1}$, *which is a generalisation of* $\circ$.

The fact that in the canonical introduction rule we add some sort of structural quantification leads to a further generalisation. Once in the canonical introduction rule we allow for structural quantification in the premisses, there is no reason in principle why we should not specify introduction rules for a connective by using this sort of quantification in their premisses. In the corresponding harmonious canonical elimination rule this would lead to structural quantification in the assumptions of minor premisses. But if we allow for that, there is no reason why we should not iterate this process and use any sort of embedded (i.e. nested) universal quantification in the specification of connectives. In the end this means that, at the structural level, we would use means of expression which correspond to those available in **PL2** at the logical level.[17]

Concerning the negative results presented, the reader should keep in mind that we are working in an intuitionistic framework throughout. If we used classical second-order propositional logic $\mathbf{PL2}_c$ instead of the intuitionistic system **PL2**, we could always find harmonious rules, as emphasized by Read (2014, this volume). For example, $\star$ could be given the (flat) harmonious elimination rule

$$\frac{\star(p_1, p_2, p_3) \quad p_1 \quad \begin{array}{c}[p_2]\\ r\end{array} \quad \begin{array}{c}[p_3]\\ r\end{array}}{r}\ ,$$

since we can show in $\mathbf{PL2}_c$ that

$$(p_1 \to p_2) \lor p_3 \dashv\vdash \forall r((p_1 \land (p_2 \to r) \land (p_3 \to r)) \to r)\ .$$

15.6 Functional Completeness

From *Fact I1* (Sect. 15.3) we know that

[17] This is discussed in detail in Schroeder-Heister (2014a).—The fact that, as shown by Pitts (1992), **PL2** can be translated into **PL**, cannot be used here, as this translation uses all connectives of **PL** including disjunction, which does not have a structural analogue in our framework. However, this translation might become useful in the context of functional completeness. See Sect. 15.6 and footnote 19.—We have not discussed the issue of structural existential quantification, as this is not relevant for our central topic. In the framework discussed here we need to add universal quantification if from given elimination rules we want to construct harmonious introduction rules. If we allowed for extra variables and thus for implicit existential quantification in the premisses of introduction rules, we would need to add universal quantification when passing to harmonious elimination rules.

$$c^I \vdash c(p_1, \ldots, p_n) \tag{15.17}$$

holds in **PL+cI**, and from *Fact E1* we know that

$$c(p_1, \ldots, p_n) \vdash c^E \tag{15.18}$$

holds in **PL2+cE**. Now suppose that the introduction and elimination rules for c are in harmony with each other, i.e.,

$$c^I \dashv\vdash c^E \tag{15.19}$$

holds in **PL2**. This implies that both

$$c(p_1, \ldots, p_n) \dashv\vdash c^I \tag{15.20}$$

and

$$c(p_1, \ldots, p_n) \dashv\vdash c^E \tag{15.21}$$

hold in **PL2+cIE**, which means that we can regard (15.20) and (15.21) as two explicit definitions of c in **PL2**. Therefore c can be expressed by means of the connectives of **PL2**, which are $\wedge$, $\vee$, $\rightarrow$, $\bot$ and $\forall$. As $\wedge$, $\vee$ and $\bot$ are definable in **PL2** in terms of $\rightarrow$ and $\forall$, we obtain as a result that c can be expressed by using $\rightarrow$ and $\forall$ in the system that comprises both second-order propositional logic and the introduction and elimination rules for c.

For example, our connective $\star$ can be defined either by $(p_1 \rightarrow p_2) \vee p_3$ (its introduction meaning) or by $\forall r((((p_1 \rightarrow p_2) \rightarrow r) \wedge (p_3 \rightarrow r)) \rightarrow r)$ (its elimination meaning). From the latter formula we can eliminate conjunction by rewriting it as $\forall r(((p_1 \rightarrow p_2) \rightarrow r) \rightarrow ((p_3 \rightarrow r) \rightarrow r))$, obtaining a definition of $\star$ in terms of $\forall$ and $\rightarrow$.[18]

This is a reductive version of functional completeness in the sense that constants of standard second-order intuitionistic logic suffice to express all connectives definable by harmonious introduction and elimination rules. It is *reductive* as the standard constants (here $\rightarrow$ and $\forall$) are taken for granted and are conceptually not on the same level as the connective c.

Whereas (15.21) gives rise to a definition of c in the language of **PL2**, which can use propositional quantification in the definiens, the right hand side of (15.20) is a formula of **PL**, which can only contain $\wedge$, $\vee$, $\rightarrow$ and $\bot$ as connectives (see our definition of c^E and c^I in Sect. 15.3). As the derivability relation in (15.20) is that of **PL2+cIE**, (15.20) only yields a definition of c in **PL2**, even if no quantifier occurs in the definiens. However, the following observation shows that only derivability in **PL+cIE** is needed to establish (15.20), so that (15.20) actually is a definition of c in

[18] If we use the standard translation of $s_1 \wedge s_2$ into second-order logic, which is $\forall q((s_1 \rightarrow (s_2 \rightarrow q)) \rightarrow q)$, we would obtain the more complicated formula $\forall r(\forall q(((p_1 \rightarrow p_2) \rightarrow r) \rightarrow ((p_3 \rightarrow r) \rightarrow q) \rightarrow q) \rightarrow r)$. The formula $\forall r(((p_1 \rightarrow p_2) \rightarrow r) \rightarrow ((p_3 \rightarrow r) \rightarrow r))$ is actually the standard second-order translation of $(p_1 \rightarrow p_2) \vee p_3$ which uses the translation of $s_1 \vee s_2$ as $\forall r((s_1 \rightarrow r) \rightarrow ((s_2 \rightarrow r) \rightarrow r))$.

PL in terms of the connectives $\wedge$, $\vee$, $\rightarrow$ and $\bot$. In view of (15.17) we have to show that we do not need second-order quantification to establish $c(p_1, \ldots, p_n) \vdash c^I$. As c^E is a prenex formula of the form $\forall\!\!\forall\varphi$, where $\forall\!\!\forall$ represents a quantifier prefix and φ the quantifier-free kernel, we can, from the derivability of $c^E \vdash c^I$ in **PL2** [see (15.19)] and normalisation for **PL2**, conclude the derivability of $\varphi_1, \ldots, \varphi_n \vdash c^I$ in **PL**, where $\varphi_1, \ldots, \varphi_n$ are certain quantifier-free instances of φ. Furthermore, from $c(p_1, \ldots, p_n)$ we can derive each φ_i in **PL+cE**, which yields the required derivation of c^I from $c(p_1, \ldots, p_n)$ in **PL+cE**.

Therefore we have obtained the following result:

Functional completeness *Any connective c with harmonious introduction and elimination rules can be defined in* **PL** *by its introduction meaning* c^I *and also by its elimination meaning* c^E.

Here the same remark we made after the Conservativeness Lemma in Sect. 15.4 applies: That this proof needs to rely on the heavy machinery of normalisation of **PL2** is due to our description of introduction and elimination meanings in abstract terms by means of second-order formulas. In a concrete system with harmonious rules it would be replaced with a direct syntactic proof using the rules available (see Prawitz 1979; Schroeder-Heister 1984).

Pitts (1992) defines a translation * from **PL2** into **PL**, such that $\Gamma \vdash_{\mathbf{PL2}} \varphi$ entails $\Gamma^* \vdash_{\mathbf{PL}} \varphi^*$, where for every quantifier-free φ, φ^* is identical to φ. Thus, from (15.19), we can conclude that in **PL** the following holds:

$$c^I \dashv\vdash (c^E)^* .$$

This gives us another definition of c, namely as $(c^E)^*$. It might be interesting to check what $(c^E)^*$ looks like for various c^I.[19]

15.7 Prawitz's Account of Functional Completeness

Our reductive approach offers a plausible way of understanding Prawitz's (1979) proof of functional completeness of the standard intuitionistic constants $\wedge$, $\vee$, $\rightarrow$ and $\bot$. Prawitz starts from (15.1) as the general schema for introduction rules of an n-ary connective c. He then associates a corresponding elimination rule for c (there is only a single one) as follows. From an introduction rule of the form (cI) a list Δ^{PROP} of conjunction-implication formulas

$$\bigwedge \Gamma_1 \rightarrow s_1, \;\ldots\;, \bigwedge \Gamma_\ell \rightarrow s_\ell$$

[19] The reference to Pitts (1992) was brought to my attention by an anonymous reviewer of Olkhovikov and Schroeder-Heister (2014a).

is constructed, where $\bigwedge \Gamma_i$ denotes the conjunction of all elements of Γ_i, where $\bigwedge \Gamma_i \rightarrow s_i$ is identified with s_i, if Γ_i is empty. This list represents propositionally the premisses of (cI). If we have m introduction rules of the form (cI), we obtain m such lists $\Delta_1^{PROP}, \ldots, \Delta_m^{PROP}$. Then the elimination rule (cE) has the following form:

$$(c\mathrm{E})_P \ \frac{c(p_1, \ldots, p_n) \quad \begin{matrix}[\Delta_1^{PROP}] \\ r\end{matrix} \quad \ldots \quad \begin{matrix}[\Delta_m^{PROP}] \\ r\end{matrix}}{r} \ ,$$

which is exactly the schema (15.16). This schema is modelled by Prawitz after the pattern of the standard $\vee$ elimination rule. It expresses that everything that can be derived from the premisses of each introduction rule for c can be derived from its conclusion. To put it differently: $c(p_1, \ldots, p_n)$ is the strongest proposition that can be derived from the premisses of each introduction rule for c. In the case of absurdity $\bot$, which has no introduction rule, we obtain the *ex falso quodlibet* as the limiting case of (cE) ($m = 0$, i.e. no minor premisses).

Unfortunately, $(c\mathrm{E})_P$ already uses the connectives $\wedge$ and $\rightarrow$, which means that it cannot be used as a schema covering them. In fact, conjunction is not used in the elimination rule for conjunction, which according to $(c\mathrm{E})_P$ takes the form

$$(\wedge \mathrm{E}_{GEN}) \ \frac{p \wedge q \quad \begin{matrix}[p, q] \\ r\end{matrix}}{r} \ ,$$

but only in more complicated elimination rules. Thus one may view conjunction as defined by this general rule and later refer to it as an already defined connective. However, in the case of implication, Prawitz's elimination rule takes the form

$$\frac{p_1 \rightarrow p_2 \quad \begin{matrix}[p_1 \rightarrow p_2] \\ r\end{matrix}}{r} \ ,$$

which is trivial and therefore useless. From a foundational point of view, Prawitz's meaning-theoretical considerations as well as his proof that every connective can be defined in terms of the four standard connectives $\wedge$, $\vee$, $\rightarrow$ and $\bot$ misses out on implication.

However, if we adopt a reductive view, as we are doing in this paper, we can leave Prawitz's schema as it stands. We take the meaning of the standard connectives (in particular implication) for granted. Prawitz's schema then shows how the meaning of all connectives *except the standard ones* is reduced to the meaning of the standard ones. His completeness proof establishes that every connective which is characterised in a certain way is definable in terms of the standard connectives.

Therefore, from a reductive point of view, Prawitz's approach makes perfect sense. It is less general than the one advanced here in that he is proposing an introduction schema and generating a general elimination schema from it, rather than starting from independent introduction and elimination schemas and investigating their strength.

In defining harmonious elimination rules for given introduction rules, Prawitz does not need to use second-order quantification. In our terminology, Prawitz's work is a reductive approach focussing on the definability of connectives by their introduction meaning.

15.8 Outlook: The Foundational Approach

As mentioned in Sect. 15.2, the standard connectives $\wedge$, $\vee$, $\rightarrow$ and $\bot$ with their standard inference rules fall under the general schemas (15.1) and (15.3) for introduction and elimination inferences for an arbitrary connective c. In this sense they play no special role. However, they are needed and therefore taken for granted when formulating the introduction and elimination meaning of c and the corresponding notion of harmony. In order to establish harmony we need the logic of the standard operators. This is why our approach is reductive and not foundational.

This does not mean that our notion of harmony is not applicable to the standard operators. In fact, in Table 15.1 introduction and elimination meanings were associated with the standard connectives. For example, the formula $p_1 \vee p_2$ is the introduction meaning of $\vee$, the formula $\forall r(((p_1 \rightarrow r) \wedge (p_2 \rightarrow r)) \rightarrow r)$ its elimination meaning, and their equivalence

$$p_1 \vee p_2 \dashv\vdash \forall r(((p_1 \rightarrow r) \wedge (p_2 \rightarrow r)) \rightarrow r) \tag{15.22}$$

establishes that the standard introduction and elimination rules for disjunction as given in Table 15.1 are in harmony with each other. However, to show (15.22) we use the standard introduction and elimination rules for disjunction (plus those for $\wedge$, $\rightarrow$ and $\forall$), supposing that they are appropriate and therefore harmonious in some basic ('primordial') sense. If we chose different rules for the standard connectives which were not 'harmonious' in this basic sense, then (15.22) would perhaps no longer hold. This would not only affect disjunction but any other claim of reductive harmony.

At first sight one might think that this problem affects only disjunction and absurdity, as, in order to establish harmony for them, a real proof in **PL2** as a background logic must be given, using at least one logical rule for these connectives. In the case of conjunction and implication, introduction and elimination meanings are literally identical (see Table 15.1). To prove harmony we just need to rely on the identities

$$p_1 \wedge p_2 \dashv\vdash p_1 \wedge p_2 \qquad\qquad p_1 \rightarrow p_2 \dashv\vdash p_1 \rightarrow p_2$$

rather than on any logical rule of **PL2**. However, this impression is misleading. Let us consider implication. When defining the introduction meaning of a connective according to a given introduction rule, we translated the fact that a premiss depends on an assumption by an implication between the assumption and the premiss. That is, if an introduction rule for c is of the form

$$\frac{\ldots \quad \begin{matrix}[p_1]\\ p_2\end{matrix} \quad \ldots}{c(\ldots)}$$

the relation between assumption p_1 and premiss p_2 is interpreted as $p_1 \to p_2$. When defining the elimination meaning of a connective according to a given elimination rule, we translated the relation between minor premisses and conclusion by an implication as well. That is, in an elimination rule of the form

$$\frac{c(\ldots) \qquad \ldots \; p_1 \; \ldots}{p_2} \tag{15.23}$$

the relationship between p_1 and p_2 was translated by the same implication $p_1 \to p_2$. This means that the dependence on an assumption and the relation between premiss and conclusion of a rule is given the same meaning. This is exactly what standard implication says:

$$\frac{\begin{matrix}[p_1]\\ p_2\end{matrix}}{p_1 \to p_2} \qquad \frac{p_1 \to p_2 \qquad p_1}{p_2}\ . \tag{15.24}$$

According to its introduction rule it expresses the dependence on an assumption, and according to its elimination rule (modus ponens) it expresses the relation between (minor) premiss and conclusion. In this way some fundamental harmony between implication introduction and modus ponens is built into the translation of rules for n-ary connectives to generate their introduction and elimination meanings. Something similar holds for conjunction, where we interpret the fact that p_1 and p_2 occur as two premisses in an introduction rule in the same way as the fact that p_1 and p_2 are the conclusions of two elimination rules, namely by conjunction $\wedge$:

$$\frac{\ldots \; p_1 \; \ldots \; p_2 \; \ldots}{c(\ldots)} \qquad \frac{c(\ldots) \qquad \ldots}{p_1} \qquad \frac{c(\ldots) \qquad \ldots}{p_2}\ . \tag{15.25}$$

This is exactly what standard conjunction says:

$$\frac{p_1 \qquad p_2}{p_1 \wedge p_2} \qquad \frac{p_1 \wedge p_2}{p_1} \qquad \frac{p_1 \wedge p_2}{p_2}\ . \tag{15.26}$$

According to its introduction rule it expresses the association of two premisses, and according to its elimination rules it expresses the association of the conclusions of the two rules.

This shows that in our reductive approach we are implicitly relying on some fundamental harmony inherent in the rules of the standard connectives. How far it is possible to give a foundational analysis of this harmony is another matter. Any tool introduced to analyse and describe this harmony will possibly have to rely on some 'deeper' sort of harmony governing its own reasoning principles. This is a fundamental problem for approaches such as Lorenzen's (1955), von Kutschera's (1968) and our own (Schroeder-Heister 1984) that all deal with structural analogues of implication (especially higher-level rules) in order to deal with logical connectives.[20]

We should, however, already mention a point of specific interest: The standard connectives for implication and conjunction involved in describing the introduction and elimination meanings of logical constants are those with the two projections and modus ponens, respectively, as elimination rules. As explained above, the interpretation of the relation between p_1 and p_2 in (15.23) by means of implication corresponds to modus ponens in (15.24), and the interpretation of the association of the two elimination rules in (15.25) corresponds to the two projections in (15.26). This does not speak against generalised forms of implication or conjunction (in Table 15.2 denoted by & and ⊃), but shows that *modus-ponens-based implication* and *projection-based conjunction* are not only connectives in their own right, but basic connectives that cannot be superseded by others.[21]

References

Belnap, N. D. (1962). Tonk, plonk and plink. *Analysis, 22*, 130–134.

Došen, K., & Schroeder-Heister, P. (1985). Conservativeness and uniqueness. *Theoria, 51*, 159–173.

Dummett, M. (1991). *The logical basis of metaphysics*. London: Duckworth.

Dyckhoff, R. (2009). Generalised elimination rules and harmony. (Manuscript, University of St. Andrews, http://rd.host.cs.st-andrews.ac.uk/talks/2009/GE.pdf)

Dyckhoff, R. (2015). Some remarks on proof-theoretic semantics. In T. Piecha & P. Schroeder-Heister (Eds.), *Advances in proof-theoretic semantics*. Berlin: Springer.

Francez, N., & Dyckhoff, R. (2012). A note on harmony. *Journal of Philosophical Logic, 41*, 613–628.

Gentzen, G. (1934/35). Untersuchungen über das logische Schließen. *Mathematische Zeitschrift, 39*, 176–210, 405–431 [English translation. In: The Collected Papers of Gerhard Gentzen (ed. M. E. Szabo), Amsterdam: North Holland (1969), pp. 68–131].

Girard, J.-Y., Lafont, Y., & Taylor, P. (1989). *Proofs and types*. Cambridge: Cambridge University Press.

Lorenzen, P. (1955). *Einführung in die operative Logik und Mathematik*. Berlin: Springer. 2nd edition 1969.

Olkhovikov, G. K., & Schroeder-Heister, P. (2014a). On flattening elimination rules. *Review of Symbolic Logic, 7*, 60–72.

Olkhovikov, G. K., & Schroeder-Heister, P. (2014b). Proof-theoretic harmony and the levels of rules: General non-flattening results. In E. Moriconi & L. Tesconi (Eds.), *Second Pisa colloquium in logic, language and epistemology* (pp. 245–287). Pisa: Edizioni ETS.

Pitts, A. M. (1992). On an interpretation of second order quantification in first order intuitionistic propositional logic. *Journal of Symbolic Logic, 57*, 33–52.

Popper, K. R. (1947). New foundations for logic. *Mind, 56*, 193–235. (Corrections *Mind, 57*, 1948, 69–70).

Prawitz, D. (1965). *Natural deduction: A proof-theoretical study*. Stockholm: Almqvist & Wiksell. (Reprinted Mineola NY: Dover Publ., 2006).

[20] The problems and merits of a foundational analysis of logical constants in terms of quantified higher-level rules are discussed in (Schroeder-Heister 2014a).

[21] This is a point also reached by Dyckhoff (2009, 2015), on the basis of related considerations.

Prawitz, D. (1971). Ideas and results in proof theory. In J. E. Fenstad (Ed.), *Proceedings of the Second Scandinavian Logic Symposium (Oslo 1970)* (pp. 235–308). Amsterdam: North-Holland.
Prawitz, D. (1979). Proofs and the meaning and completeness of the logical constants. In J. Hintikka, I. Niiniluoto, & E. Saarinen (Eds.), *Essays on Mathematical and Philosophical Logic:* Proceedings of the Fourth Scandinavian Logic Symposium and the First Soviet-Finnish Logic Conference, Jyväskylä, Finland, June 29-July 6, 1976, Dordrecht: Kluwer (pp. 25–40) [revised German translation 'Beweise und die Bedeutung und Vollständigkeit der logischen Konstanten', Conceptus, 16, 1982, 31–44].
Prior, A. N. (1960). The runabout inference-ticket. *Analysis*, *21*, 38–39.
Read, S. (2010). General-elimination harmony and the meaning of the logical constants. *Journal of Philosophical Logic*, *39*, 557–576.
Read, S. (2014). General-elimination harmony and higher-level rules. In H. Wansing (Ed.), *Dag Prawitz on proofs and meaning (= this volume)*. Heidelberg: Springer.
Schroeder-Heister, P. (1984). A natural extension of natural deduction. *Journal of Symbolic Logic*, *49*, 1284–1300.
Schroeder-Heister, P. (2005). Popper's structuralist theory of logic. In I. Jarvie, K. Milford, & D. Miller (Eds.), *Karl Popper: A centenary assessment* (Vol. III, pp. 17–36). Aldershot: Ashgate.
Schroeder-Heister, P. (2012a). Proof-theoretic semantics. In E. Zalta (Ed.), *Stanford Encyclopedia of Philosophy*. Stanford: http://plato.stanford.edu
Schroeder-Heister, P. (2012b). Proof-theoretic semantics, self-contradiction, and the format of deductive reasoning. *Topoi*, *31*, 77–85.
Schroeder-Heister, P. (2014a). The calculus of higher-level rules, propositional quantifiers, and the foundational approach to proof-theoretic harmony. *Studia Logica, 102* (Special issue, ed. Andrzej Indrzejczak, commemorating the 80th anniversary of Gentzen's and Jaśkowski's groundbreaking works on assumption based calculi).
Schroeder-Heister, P. (2014b). Generalized elimination inferences, higher-level rules, and the implications-as-rules interpretation of the sequent calculus. In L. C. Pereira, E. H. Haeusler, & V. de Paiva (Eds.), *Advances in natural deduction: A celebration of Dag Prawitz's work* (pp. 1–29). Heidelberg: Springer.
Tennant, N. (1978). *Natural logic*. Edinburgh: Edinburgh University Press.
Tranchini, L. (2014). Harmony and rule equivalence. In E. Moriconi & L. Tesconi (Eds.), *Second Pisa colloquium in logic, language and epistemology* (pp. 288–299). Pisa: Edizioni ETS.
von Kutschera, F. (1968). Die Vollständigkeit des Operatorensystems $\{\neg, \wedge, \vee, \supset\}$ für die intuitionistische Aussagenlogik im Rahmen der Gentzensemantik. *Archiv für mathematische Logik und Grundlagenforschung*, *11*, 3–16.

Chapter 16
First-Order Logic Without Bound Variables: Compositional Semantics

William W. Tait

Abstract A strict version of compositional semantics would have all composite meaningful expressions be of the form XY, where X and Y are meaningful and the concatenation expresses application of a function (X) to an argument (Y). In proof theory, compositionality is violated because of bound variables both in formulas (quantification) and in deductions (introduction rules). Two applications of typed combinator theory are used to introduce a proof theory for first-order predicate logic without identity in which there are no bound variables.

Keywords Semantics · Compositionality · Schönfinkel · Combinator · Curry · Type · Explicit definition · Reduction · Normal form

An attractive format for semantics is that in which composite expressions are built up from atomic ones by means of the operation of concatenation and the concatenation XY expresses the application of a function denoted by X to an argument denoted by Y. The use of relative pronouns presents an obstacle to this form of compositional semantics, since the reference of a relative pronoun in one component may occur in another. In the standard notation of first-order predicate logic this occurs in the form of variable-binding operations of quantification: in the sentence $\forall x \varphi(x)$, the reference of x in $\forall x$ is in $\varphi(x)$ and neither component has an independent meaning. Frege, in the interests of compositional semantics, was led by this to declare that the open formula $\varphi(x)$ is semantically significant: it simply denotes an 'incomplete object'. We won't discuss here the many reasons for rejecting this very ugly idea, but reject it we will. So the demands of compositional semantics require that we formalize first-order predicate logic without using bound variables.

Of course the use of bound variables is very natural and the means that we use to eliminate them can result in quite complex expressions. Our purpose, therefore,

---

This paper was written in honor of Dag Prawitz for the occasion, now alas long past, of his 70th birthday.

W.W. Tait (✉)
Department of Philosophy and CHSS, University of Chicago,
1115 E. 58th St., Chicago, IL 60637, USA
e-mail: williamtait@mac.com

H. Wansing (ed.), *Dag Prawitz on Proofs and Meaning*, Outstanding Contributions to Logic 7, DOI 10.1007/978-3-319-11041-7_16

is not the practical one of finding the most readable notation: it is the theoretical one of obtaining a compositional semantics. On the other hand, we shouldn't be too humble: although the notation $\varphi(v)$ with free variable v and instances $\varphi(t)$ where t is a term is quite intuitive, the substitutions involved in actual cases, in substituting t for the possibly multiple occurrences of v in $\varphi(v)$, can create long expressions. The elimination procedure will consist in showing that $\varphi(v)$ can be expressed by $\varphi' v$, expressing application of the function φ' to the argument v, where φ' itself is built up in accordance with our function-argument paradigm from the variables, other than v, and constants in $\varphi(v)$. Thus for example, $\forall x\varphi(x)$ is now expressed by $\forall\varphi'$, which again will be seen to express application of a function to an argument. Once φ' is constructed, the substitutions $\varphi' s$, $\varphi' t$, etc., for v in $\varphi' v$ are more easily processed than the corresponding substitutions $\varphi(s)$, $\varphi(t)$, etc. The equivalence of $\varphi' v$ and $\varphi(v)$ consists of a sequence of reduction steps reducing the former to the latter, where each reduction step consists in replacing an expression XY for the application of a function (X) to an argument (Y) by the expression for its value.

In Part I we are going to consider only sentences expressed in an arbitrary first-order language $\mathcal{F}$ without identity and with universal quantification $\forall$ as the only variable-binding operation. It suffices to take as the remaining logical constants just $\rightarrow$ for implication and $\perp$ for the absurd proposition. (Negation $\neg\varphi$ is expressed by $\varphi \rightarrow\perp$.) This of course suffices only so long as we are considering just the classical conception of logic. On the constructive conception, we would need to add conjunction, disjunction, and the further variable-binding operation $\exists$ of existential quantification. The inclusion of these constants in the case of classical logic would not complicate the treatment of bound variables in formulas, only lengthen it; and so we won't bother.

It is not only the semantics of formulas that loses its modularity on account of bound variables. This also happens with deductions: for example, the deduction of $\forall x\varphi(x)$ from a deduction $p(v)$ of $\varphi(v)$ binds the variable in $p(v)$. For any individual term t, $p(t)$ denotes the corresponding deduction of $\varphi(t)$. The other case involves $\rightarrow$-Introduction, where a deduction of $\varphi \rightarrow \psi$ arises from a deduction of ψ from the assumption of φ. So long as we think of deductions as purely syntactical objects, the 'Deduction Theorem' shows how to eliminate $\rightarrow$-introduction. But, if we take the view that deductions denote certain objects, namely proofs, then the assumption of φ should be understood as a variable $v = v_\varphi$ ranging over proofs of φ and the deduction of ψ then again has the form $p(v)$, and v again gets bound when we 'discharge' the assumption to obtain $\varphi \rightarrow \psi$. For any deduction q of φ, $p(q)$ then denotes the result of replacing the assumptions v of φ by its proof q. Again, in both the case of $\forall$-introduction and $\rightarrow$-introduction, we will show how to obtain $p(v)$ from an expression $p' v$, where p' is built up by means of application of function to argument from the variables other than v and constants in $p(v)$. The reduction of $p' v$ to $p(v)$ again is a matter of replacing expressions XY of the application of a function to an argument by the corresponding expression for the values. The proof of the Deduction Theorem turns out to be essentially the construction of p' from $p(v)$. We will discuss the elimination of variable-binding in deductions in Part II of the paper.

I tackled the same problem of eliminating bound variables for the Curry-Howard type theory in Tait (1998). The attempt to relativize that paper to predicate logic doesn't quite work: but something like it does. The problem arises in connection with eliminating bound variables from deductions. The earlier method of eliminating bound variables, applied to deductions in the framework of first-order logic, leads to deductions outside that framework. In order to avoid this, we need to introduce further operations on first-order deductions than those provided by the earlier paper but which, from the higher point of view of Curry-Howard type theory, are not really new.

As we will indicate, Quine addressed the problem of eliminating bound variables both from first-order formulas (Quine 1960) and from deductions (Quine 1966a), but the two are not integrated: in his formalism for first-order deductions, the formulas contain bound variables. Moreover, his treatment of formulas does not provide a compositional semantics for them and his treatment of deductions provides no semantics at all. My aim is a semantics of formulas and deductions in which neither contain bound variables and in which the semantics is compositional in the sense indicated above.

16.1 Preamble

The central idea of this paper was first presented in a lecture in Göttingen in 1920 by Moses Schönfinkel, a Russian member of Hilbert's group in foundations of mathematics from 1914–1924. In 1924 Heinrich Behman, another member of the group, published the lecture under the title "Über die Bausteine der mathematischen Logik", with some added paragraphs of his own.[1] The idea in question was of course the *theory of combinators* or, as we should now say, of *untyped* combinators. We start with an alphabet of atomic symbols, some constants and some variables. These are to include the constants K and S, called *combinators*.[2] The set $\mathcal{W}$ of formulas is the least set containing all the atomic symbols and such that, if X and Y are in $\mathcal{W}$, then so is (XY). Parentheses are necessary here, so we don't yet have our ideal, mentioned in the introduction, of composition simply by concatenation. For $n > 1$ we will write

$$X_1X_2\ldots X_n := (\ldots(X_1X_2)\ldots X_n)$$

[1] There is another paper of Schönfinkel, on special cases of the decision problem. This was prepared and published in 1929 by Paul Bernays. Not much seems to be known about Schönfinkel. Here is what Wikipedia has to say about his life after Göttingen:

> After he left Göttingen, Schönfinkel returned to Moscow. By 1927 he was reported to be mentally ill and in a sanatorium. His later life was spent in poverty, and he died in Moscow some time in 1942. His papers were burned by his neighbors for heating.

[2] Schönfinkel used C instead of K.

(association to the left). We call formulas of the form *KXY* and *SXYZ* *convertible* and call X and $(XZ)(YZ)$ their *values*, respectively, and write

$$KXY\ CONV\ X \qquad SXYZ\ CONV\ (XZ)(YZ).$$

The relation

$$X \succ' Y$$

between formulas is defined to mean that Y is obtained by replacing one occurrence of a convertible part of X by its value. We say that X *reduces to* Y, written

$$X \succeq Y$$

if there is a chain

$$X = Z_0 \succ' \ldots \succ' Z_n = Y$$

with $n \geq 0$ (so that X reduces to itself). Let

$$X \equiv Y$$

mean that X and Y reduce to a common formula. Call a formula *normal* if it contains no convertible parts. If X reduces to Y and Y is normal, we say that Y is a *normal form* of X. Here is a theorem:

Church-Rosser Theorem [Uniqueness of Normal Form]. If X reduces to both Y and Z, then Y and Z reduce to a common formula U. So every formula has at most one normal form (Church and Rosser 1936).

The second part is of course immediate from the first: If X has the two normal forms Y and Z, then they must reduce to a common formula U. Given that Y and Z are normal, we must then have $Y = U = Z$. In consequence of the Church-Rosser Theorem, $\equiv$ is an equivalence relation among formulas.

Actually Church and Rosser proved this for a related formalism, Church's *calculus of lambda conversion* or simply the *lambda calculus* (Church 1941). I'm not sure who first proved it for the theory of untyped combinators; but the usual proof given now, both for combinators and the lambda calculus, is mine (first presented in a seminar on the lambda calculus at Stanford in Spring of 1965). Unlike the original Church-Rosser proof, the argument is quite simple: the idea is to define a weak notion of reduction (1-reduction, below) which implies $\succeq$ and for which the theorem is trivial but such that every reduction is obtained by a sequence of weak reductions. (Once considered in this way, the requisite notion of weak reduction in various extensions of the theory of combinators or the lambda calculus is usually obvious.)

We define the notion that X 1-*reduces* to Y, written (temporarily) $X \longrightarrow Y$, by induction on the number of occurrences of symbols in X:

- $KXY \longrightarrow X$.

- $SXYZ \longrightarrow (XZ)(YZ)$.
- If $X_i = Y_i$ or $X_i \longrightarrow Y_i$ for $i = 1, \ldots, n$, then $X_1 \ldots X_n \longrightarrow Y_1 \ldots Y_n$.

By induction on the number of occurrences of symbols in X, one easily proves:

Lemma *Let X 1-reduce to Y and to Z. Then there is a U such that Y and Z 1-reduce to U:*

$$\begin{array}{ccc} X & \xrightarrow{1-RED} & Y \\ {\scriptstyle 1-RED}\downarrow & & \downarrow{\scriptstyle 1-RED} \\ Z & \xrightarrow[1-RED]{} & U \end{array}$$

□

Since $X \succ Y$ implies X 1-reduces to Y, it is immediate that X reduces to Y if and only if there is a chain

$$X \longrightarrow \cdots \longrightarrow Y.$$

The proof of Church-Rosser then is:

$$\begin{array}{ccccccc} X = X_{11} & \longrightarrow & X_{12} & \longrightarrow \cdots \longrightarrow & & & X_{1m} = Y \\ \downarrow & & \downarrow & & \downarrow & & \downarrow \\ X_{21} & \longrightarrow & X_{22} & \longrightarrow \cdots \longrightarrow & & & X_{2m} \\ \downarrow & & \downarrow & & \downarrow & & \downarrow \\ \vdots & \longrightarrow & \vdots & \longrightarrow & \vdots & \longrightarrow & \vdots \\ \downarrow & & \downarrow & & \downarrow & & \downarrow \\ Z = X_{n1} & \longrightarrow & X_{n2} & \longrightarrow \ldots \longrightarrow & & & X_{nm} = U \end{array}$$

□

So we may think of a formula X as defining a partial function $\bar{X}$ on the set of normal terms in $\mathcal{W}$: for each normal $Y \in \mathcal{W}$, $\bar{X}Y$ is defined and $= Z$ if and only if Z is the (unique) normal form of the formula XY. For example, it was on the basis of this idea that Church represented all the partial recursive functions in his calculus of lambda conversion. On the other hand, just because of this representation, we know *a priori* that not every formula has a normal form. Indeed it is easy to construct an example: let

$$I := SKK$$

Then

$$IX \; CONV \; KX(KX) \succ' \succ' X$$

so that I is the 'identity function'. Now let $Y = SII$. Then

$$YY = SIIY \succ (IY)(IY) \succ \cdots \succ YY \succ \cdots.$$

represents the only reduction chains for YY, and so it has no normal form.

Two distinct normal formulas X and Y may define the same function $\bar{X} = \bar{Y}$. For example, for any normal formula X, $S(KX)I$ is normal and $\overline{S(KX)I} = \bar{X}$. We will see that this particular example is significant and has led to the extension of the class of convertible formulas to include those of the form $S(KX)I$ and taking its value to be X. So, in the definition of 1-reduction, we should add the clause

$$S(KX)I \longrightarrow X.$$

There will be no difficulty extending the proof of the Lemma and so of the Church-Rosser Theorem to admit these conversions, called *η-conversions*. (Schönfinkel did not consider η-conversion.)

The power of the theory of combinators—and why it is essentially equivalent to the calculus of lambda conversion and why it is of interest to us—lies in the following:

Explicit Definition Theorem I (Schönfinkel 1924). Let $X(v)$ be a formula, where v is a variable. There is a formula X' of such that

$$X'v \succeq X(v).$$

Moreover, the atomic symbols in X' exclude v and are either combinators or are in $X(v)$.

Of course it then follows by substitution that, for any formula θ, $\varphi(\theta) \equiv \varphi'\theta$. The proof is by induction on the complexity of $X(v)$:

Case 1 $X(v) = v$. Set $X' = I$.

$$X'v = Iv \succeq v.$$

Case 2 $X(v) = X$ is a formula not containing v. Set $X' = KX$.

$$X'v = KXv \succeq X.$$

Case 3 $X(v) = Y(v)Z(v)$ contains v. Set $X' = SY'Z'$. Then

$$X'v = SY'Z'v \succeq (Y'v)(Z'v) \succeq Y(v)Z(v) = X(v).$$

□

Notice that η-conversion is not used in this construction. Given $X(v)$, we will denote the corresponding X' given by the Explicit Definition Theorem by

$$\Lambda y.X(y)$$

Here y is a *bound variable*; we also use x, z, all with or without subscripts, as bound variables. They are distinct from the free variables that occur in $\mathcal{W}$ and occur only bound by Λ or, later, by quantifiers. They occur only in the context of abbreviations for formulas that contain no bound variables at all: after all, that is our aim, to eliminate bound variables.

$\Lambda yX(y)$ satisfies the principle of *lambda-conversion*

$$(\Lambda yX(y))Y \equiv X(Y)$$

for any formula Y. So far, what we have said holds true whether or not we include η- conversions. Note that $S(KX)I = \Lambda yXy$ and so η-conversion yields the equation

$$\Lambda yXy \equiv X.$$

Remark 1 I don't use the more usual lower case λ here because that is used in Church's calculus of lambda conversion. There is an obvious translation of each system, combinator theory or lambda calculus, into the other; but in either direction the formula of combinator theory may be normal while the corresponding formula of the lambda calculus is not. Thus, if we replace each λ in a formula of the form $\lambda x.t(x)$ of the lambda calculus by Λ, the result is a normal formula in combinator theory, whether or not the original formula is normal. On the other hand, a formula of the theory of combinators can be translated into the lambda calculus by replacing K by $\lambda x\lambda y.x$ and S by $\lambda x\lambda y\lambda z.xz(yz)$. But again, the normal formulas Kv and Suv, for example, translate into non-normal formulas of the lambda calculus.

An important difference between λ and Λ arises in connection with η-conversion. Suppose $\bar{X} = \bar{Y}$, i.e. $Xv \equiv Yv$ for a free variable v not in either X or Y. Then $\lambda zXz \equiv \lambda zYz$ and so by η-conversion, $X \equiv Y$. Thus in the context of the lambda-calculus, distinct normal terms define distinct normal functions. But in the theory of combinators $Xv \equiv Yv$ does not imply $\Lambda zXz \equiv \Lambda zYz$; and so η-conversion in that context is not so natural. □

The Explicit Definition Theorem accomplishes part of what we want: formulas $X(v)$ are obtained from the corresponding formulas $X'v$ by successively replacing certain parts (UV) by their 'values'. But if we were to just apply this to eliminating bound variables from first=order formulas and proofs, it would be a purely formal transformation, with no semantical content. However, by introducing type structure into the theory of combinators, the combinators become semantically meaningful: XY (and now parentheses are unnecessary) denotes the application of a function to one of its arguments. We will in fact need two such type structures: one in Part

I, to eliminate bound variables from formulas, and the other in Part II, to eliminate bound variables in deductions. The types in Part I are built up from atoms by passing from types A and B to the type $A \Rightarrow B$ of functions from objects of type A to objects of type B. The types of the combinators are just the axioms for the theory of implication, when the atoms are regarded as atomic sentences and $\Rightarrow$ is understood as implication. This type structure for combinatorial logic (and the lambda calculus) seems to have been first discussed in Curry and Feys (1958, Chap. 9) and is generally associated with Curry. The type structure in Part II derives from Bill Howard's extension of Curry's idea of propositions as types of objects to the case of propositions expressed in first-order predicate logic. (Howard distributed some notes on this idea in the late 1960s and finally published them in Howard 1980.)

16.2 Part I

1. First, some conventions and notation: we take 1 and 0 to be the truth-values *TRUE* and *FALSE*, respectively, so that

$$2 = \{0, 1\}$$

is the set of truth-values. Let A and B be sets. As we just specified,

$$A \Rightarrow B$$

denotes the set of all functions from A to B. We define

$$A_1 \Rightarrow \cdots \Rightarrow A_n \Rightarrow B$$

for $n > 1$ by induction to be $A_1 \Rightarrow (A_2 \Rightarrow \cdots \Rightarrow A_n \Rightarrow B)$ (association to the right). When $A_1 = \cdots = A_n = A$, we denote this by

$$A \Rightarrow_n B.$$

For the case $n = 0$, we simply define $A \Rightarrow_0 B$ to be B.

The *types* are defined by

- $\mathcal{D}$ and 2 are types.
- If A and B are types, so is $A \Rightarrow B$.

$\mathcal{D}$ is just a formal symbol and types are just syntactical objects. But we are speaking about semantics and so we are assuming that a specific model M of $\mathcal{F}$ is given. So we may think of $\mathcal{D}$ as denoting its domain D_M. In this way every type A becomes a set A_M. An n-ary function constant f of $\mathcal{F}$ denotes in M an object f_M in $\mathcal{D}_M \Rightarrow_n \mathcal{D}_M$. An n-ary relation constant R of $\mathcal{F}$ denotes in M an object $R_M \in \mathcal{D}_M \Rightarrow_n 2$. Notice that an individual constant denotes an element of the domain $\mathcal{D}_M$ of the model and a propositional constant denotes a truth-value.

2. We already are drawing on another idea from Schönfinkel (1924). It is standard to interpret an n-ary function or relation constant for $n > 0$ as a function in

$$\mathcal{D}^n \Rightarrow E$$

where E is either 2 or $\mathcal{D}$ and $\mathcal{D}^n$ is the set of ordered n-tuples of elements of $\mathcal{D}$. Schönfinkle was first to note explicitly that functions of several variables (i.e. $n > 1$) can be reduced to functions of a single variable via the one-to-one correspondence

$$f \mapsto f'$$

between the $f \in D_1 \times \cdots \times D_n \Rightarrow E$ and the $f' \in D_1 \Rightarrow \cdots \Rightarrow D_n \Rightarrow E$ given by

$$f(x_1, \ldots, x_n) = (\ldots (f' x_1) \ldots x_n) = f' x_1 \ldots x_n.$$

By utilizing this correspondence in the interpretation of the function and relation constants of $\mathcal{F}$ (where E is $\mathcal{D}$ and 2, respectively), we avoid having to introduce Cartesian products $E \times F$ into the type set **M**. On the standard interpretation of the function and relation constants, our compositional semantics would be complicated by the fact that we would need two forms of composition: besides the operation XY expressing the application of a function to an argument, we would also need the operation (X, Y) expressing the operation of taking pairs.

3. We introduce a new formalism

$$\mathcal{G}$$

consisting of a set of expressions, called the *formulas* of $\mathcal{G}$, and an assignment to each formula of $\mathcal{G}$ a type.

- The n-ary function constants of $\mathcal{F}$ are atomic formulas in $\mathcal{G}$ of type $\mathcal{D} \Rightarrow_n \mathcal{D}$. Each denotes an object in its type.
- The n-ary relation constants of $\mathcal{F}$ are atomic formulas in $\mathcal{G}$ of type $\mathcal{D} \Rightarrow_n 2$. Each denotes an object in its type.
- Other constants will be specified presently as formulas of $\mathcal{G}$, each with an assigned type and each denoting an object in its type.
- There is an infinite supply of free variables of type $\mathcal{D}$.
- All other formulas of $\mathcal{G}$ are composite. The rule of formation of composite formulas and the determination of their types is simply the rule of *evaluation*—application of a function to an argument: If the formula φ is of type $A \Rightarrow B$ and the formula ψ is of type A, then $\varphi\psi$ is a formula of type B. All formulas are obtained from the constants and free variables by means of this construction.

Notice that, unlike in the theory of untyped combinators, we do not need parentheses around $\varphi\psi$:

Unique Readability Lemma. Every formula φ of $\mathcal{G}$ is a unique concatenation

$$\varphi = \varphi_0\varphi_1 \ldots \varphi_n$$

where $n \geq 0$, φ_0 is a constant and each φ_i is a formula. □

We won't bother to prove this result. The reason why we do not need parentheses is of course the type structure. But unique readability does not imply *easy* readability and so we will sometimes use parentheses to indicate how the formula should be read, even though it is the only way in which it can be read.

The terms and atomic formulas of $\mathcal{F}$ are formulas of $\mathcal{G}$ of types $\mathcal{D}$ and 2, respectively: let f and R be n-ary function and relation constants, respectively, and let $\theta_1, \ldots, \theta_n$ be formulas of type $\mathcal{D}$, then

$$f\theta_1 \ldots \theta_n$$

is of type $\mathcal{D}$ and

$$R\theta_1 \ldots \theta_n$$

is of type 2. The remaining constants include the logical constants:

Absurdity

$$\bot$$

is a constant formula of $\mathcal{G}$ of type 2 and, specifically, is false (i.e. denotes 0).

Implication

$$\rightarrow$$

is a constant formula of $\mathcal{G}$ of type $2 \Rightarrow (2 \Rightarrow 2)$. Specifically, if φ and ψ denote truth-values, then $(\varphi \rightarrow \psi) = \rightarrow \varphi\psi$ is true (i.e. denotes 1) unless φ is true and ψ is false.

The *universal quantifier*

$$\forall$$

is of type

$$(\mathcal{D} \Rightarrow 2) \Rightarrow 2.$$

Specifically, if φ is a closed formula of $\mathcal{G}$ of type $\mathcal{D} \Rightarrow 2$, then $\forall\varphi$ is true just in case $\varphi\theta$ takes the value 1 for all closed formulas θ of $\mathcal{G}$ of type $\mathcal{D}$.

And now we pay the price for compositionality: for the remaining constants of $\mathcal{G}$, we must introduce the *typed combinators*:

- For all types A, the constant $K_{\mathcal{D},A}$ of type

$$A \Rightarrow (\mathcal{D} \Rightarrow A)$$

is a formula of $\mathcal{G}$, defined for $a : A, t : \mathcal{D}$ by

$$K_{\mathcal{D},A}\varphi t \ CONV \ \varphi.$$

- For all types A, B, the constant $S_{\mathcal{D},A,B}$ of type

$$[\mathcal{D} \Rightarrow (A \Rightarrow B)] \Rightarrow [(\mathcal{D} \Rightarrow A) \Rightarrow (\mathcal{D} \Rightarrow B)]$$

is a formula of $\mathcal{G}$ defined for $\varphi : (\mathcal{D} \Rightarrow (A \Rightarrow B), \psi : (\mathcal{D} \Rightarrow A), t : \mathcal{D}$ by

$$S_{\mathcal{D},A,B}\varphi\psi t \ CONV \ (\varphi t)(\psi t).$$

Finally, it is in connection with $\mathcal{G}$ that we need certain instances of η-conversion, but first we need a definition of the identity function on type $\mathcal{D}$: we simply give type structure to the 'identity function' I above, making it the identity function $I_{\mathcal{D}}$ on $\mathcal{D}$. Let $B = [\mathcal{D} \Rightarrow \mathcal{D}]$. Then $S_{\mathcal{D},B,\mathcal{D}}$ has type

$$[\mathcal{D} \Rightarrow (B \Rightarrow \mathcal{D})] \Rightarrow [(\mathcal{D} \Rightarrow B) \Rightarrow (\mathcal{D} \Rightarrow \mathcal{D})]$$

$K_{\mathcal{D},B}$ has type $\mathcal{D} \Rightarrow (B \Rightarrow \mathcal{D})$ and $K_{\mathcal{D},\mathcal{D}}$ has type $\mathcal{D} \Rightarrow B$. So

$$I_{\mathcal{D}} := S_{\mathcal{D},B,\mathcal{D}}K_{\mathcal{D},B}K_{\mathcal{D},\mathcal{D}}$$

has type $\mathcal{D} \Rightarrow \mathcal{D}$ and for v of type $\mathcal{D}$

$$I_{\mathcal{D}}v \succeq K_{\mathcal{D},B}v(K_{\mathcal{D},\mathcal{D}}v) \succeq v.$$

The case of η-conversion we shall need is this: Let φ be a formula of type $\mathcal{D} \Rightarrow B$. Then

$$S_{\mathcal{D},\mathcal{D},B}(K_{\mathcal{D},\mathcal{D}\to B}\varphi)I_{\mathcal{D}} \ CONV \ \varphi.$$

Thus, the typed combinators are actually defined as objects of a certain type by equations that, in untyped combinator theory, are just formal rules of conversion. The realization (in Curry-Feys) that combinator theory could be typed rested on the observation that the formulas of the two untyped conversion relations X *CONV* Y can be 'stratified'—i.e. types can be assigned to X and Y so that the right-hand side of the conversion for KXY respects the type assignment and types can be assigned to X, Y and Z so that the right hand side of the conversion of $SXYZ$ respects the type assignment. This is trivial for K. The stratification of $XY(ZY)$ is almost as easy. (But then, remember the story of Christopher Columbus and the egg.)

Of course nothing in the above discussion depends upon the particular role of $\mathcal{D}$ in the system $\mathcal{G}$. Along with $K_{\mathcal{D},A}$ and $S_{\mathcal{D},A,B}$, we could introduce $K_{C,A}$ and $S_{C,A,B}$ with the corresponding conversion rules for an arbitrary type C and then define the identity function I_C on C. We don't need to do this, however, because the only variables in formulas of $\mathcal{G}$ are of type $\mathcal{D}$. (See the Explicit Definition Theorem II below.)

Now we have specified all of the constants of $\mathcal{G}$, together with their denotations, and so have completed the definition of the set of formulas of $\mathcal{G}$ with their types.

4. The notions of a reduction, normal term, normal form and equivalence $\equiv$ between formulas of $\mathcal{G}$ are defined exactly as in the case of the untyped combinators: just ignore the types. There are two well-known properties of reduction. The first of them we have already proved:

Church-Rosser Theorem [Uniqueness of Normal Form] If φ reduces to ψ and to χ, then ψ and χ reduce to a common formula θ. So, in particular, if ψ and χ are normal, then $\psi = \chi$.

The proof given in the Preamble applies since it has nothing to do with type structure. □

Recall that $\varphi \succ' \psi$ means that ψ is obtained by converting a single part of φ. The other property is the

Well-foundedness Theorem Every reduction sequence

$$\varphi_0 \succ' \varphi_1 \succ' \cdots$$

is finite.

This result, which does depend on type structure, implies that every formula has a (necessarily unique) normal form: simply iterate the process of taking arbitrary simple reductions. By well-foundedness, the process will come to and end with a normal formula. In the case of untyped combinators, we had the counterexample YY, where $Y = SII$. The Well-foundedness Theorem is sometimes called the *Strong Normalization Theorem* because it states not only that every formula has a normal form, but that, no matter what choice one makes in taking successive simple reductions, one will arrive at the normal form.

A simple method of proof, found in a somewhat richer setting in Tait (1963, Appendix B) and (1967), is as follows: define the notion of a *computable formula of type A* by induction on the complexity of A.

- If A is $\mathcal{D}$ or 2, then a formula φ of type A is computable if and only if it is well-founded (i.e. every reduction sequence starting with φ is finite).
- A formula φ of type $A \Rightarrow B$ is computable if and only if $\varphi\psi$ is a computable formula of type B for every computable formula ψ of type A.

Now prove simultaneously by induction on the complexity of A that every computable term of type A is well-founded and that every variable of type A is computable, and then prove by induction on the complexity of the formula φ that, if it is of type A, then it is a computable formula of type A. □

It follows from Church-Rosser and Well-foundedness that the relation $\equiv$ is a decidable equivalence relation on the set of all formulas of $\mathcal{G}$.

5. The proof of the Explicit Definition Theorem is essentially the same as in the untyped case:

Explicit Definition Theorem II Let $\varphi(v)$ be a formula of $\mathcal{G}$ of type B, where v is a free variable of type $\mathcal{D}$. There is a normal formula φ' of $\mathcal{G}$ of type $C \Rightarrow B$ such that

$$\varphi(v) \equiv \varphi' v.$$

φ' does not contain v. It contains only variable and constants that are in φ and combinators. □

Again, this theorem does not depend on the special role of $\mathcal{D}$. By introducing the combinators $K_{C,A}$ and $S_{C,A,B}$ for arbitrary type C, the theorem would hold with $\mathcal{D}$ replaced throughout by C.

Given $\varphi(v)$, we will again denote the corresponding φ' given by the Explicit Definition Theorem II by

$$\Lambda x.\varphi(x).$$

Let $\varphi(v_1, \ldots, v_n)$ be a formula of type B where $v_1, \ldots, v_n$ are distinct free variables of type $\mathcal{D}$. Then by n iterated applications of the Explicit Definition Theorem

$$\Lambda x_1 \ldots \Lambda x_n.\varphi(x_1, \ldots, x_n)$$

is of type

$$\mathcal{D} \Rightarrow_n B$$

and

$$(\Lambda x_1 \ldots \Lambda x_n.\varphi(x_1, \ldots, x_n))t_1 \ldots t_n \equiv \varphi(t_1, \ldots, t_n)$$

for terms t_i of $\mathcal{F}$.

In particular let $\varphi(v)$ be a normal formula of $\mathcal{G}$ of type 2, where v is a free variable of type $\mathcal{D}$. Then $\Lambda x.\varphi(x)$ is of type $\mathcal{D} \Rightarrow 2$. So we may introduce the bound variable notation for the universal quantifier as an abbreviation:

$$\forall x.\varphi(x) := \forall \Lambda x.\varphi(x).$$

$\forall x.\varphi(x)$ will be true just in case $(\Lambda x \in \mathcal{D}.\varphi(x))d \equiv \varphi(d)$ is true for each $d \in D$, just in case it should be.

6. With each formula φ of $\mathcal{F}$, we associate a normal formula φ^* of $\mathcal{G}$ as follows:

- If φ is atomic, then $\varphi^* = \varphi$.
- $(\varphi \rightarrow \psi)^* = \varphi^* \rightarrow \psi^*$
- $(\forall x \varphi(x))^* = \forall \Lambda x[\varphi(x)^*]$ where $\varphi(x)^*$ is the result of substituting x for v in $\varphi(v)^*$.

We will identify the formula φ of $\mathcal{F}$ with the corresponding φ^* in $\mathcal{G}$.

Interpretation Theorem [Interpretation of $\mathcal{F}$ in $\mathcal{G}$].

a. The terms of $\mathcal{F}$ are precisely the normal formulas of $\mathcal{G}$ of type $\mathcal{D}$.
b. The formulas of $\mathcal{F}$ are normal formulas of $\mathcal{G}$ of type 2 and every normal formula of $\mathcal{G}$ of type 2 is equivalent to a formula of $\mathcal{F}$.

It is immediate that the terms of $\mathcal{F}$ are normal formulas of $\mathcal{G}$ of type $\mathcal{D}$ and that the formulas of $\mathcal{F}$ are normal formulas of $\mathcal{G}$ of type 2.

Note that a normal formula φ of type $\mathcal{D}$ or 2 in $\mathcal{G}$ cannot begin with a combinator. For let φ be

$$T\varphi_1 \ldots \varphi_n$$

where T is a constant and $n \geq 0$. If $T = K_{A,B}$ then $n < 2$, since φ is normal. But then φ is of type $A \Rightarrow (B \Rightarrow A)$ or of type $B \Rightarrow B$). A similar argument shows that T cannot be of the form $S_{A,B,C}$.

Proof of a. Let φ be a normal formula of $\mathcal{G}$ of type $\mathcal{D}$. We prove by induction on the number of occurrences of constants in it that it is a term of $\mathcal{F}$. Let φ be

$$T\varphi_1 \ldots \varphi_n$$

where T is a constant or variable.

- If T is a variable, then $n = 0$ and $\varphi = T$ is a term of $\mathcal{F}$.
- If T is a constant, then it is an n-ary function constant ($n \geq 0$) and $\varphi_1, \ldots, \varphi_n$ are normal terms of type $\mathcal{D}$. So by the induction hypothesis, they are terms of $\mathcal{F}$ and hence so is φ.

Proof of b. We assume now that $\varphi = T\varphi_1 \ldots \varphi_n$ is a normal formula of $\mathcal{G}$ of type 2, where T is a constant. We prove by induction on the number of nested occurrences of $\forall$ in φ and within that, by induction on the number of occurrences of symbols in it, that φ is a formula of $\mathcal{F}$.

- T cannot be a variable.
- T is a non-logical constant. Then it is an n-ary relation constant ($n \geq 0$) and, by the first part of the theorem, the φ_i are terms of $\mathcal{F}$. So $\varphi = \varphi^*$ is an atomic formula of $\mathcal{F}$.
- $T =\perp$. Then $n = 0$ and $\varphi =\perp$.
- $T =\rightarrow$. Then $n = 2$ and by the induction hypothesis, φ_1 and φ_2 are equivalent to formulas of $\mathcal{F}$. Hence so is φ.
- $T = \forall$. Then $n = 1$ and φ_1 is normal and of type $\mathcal{D} \Rightarrow 2$. By the induction hypothesis on the number of occurrences of $\forall$, we can assume that $\varphi_1 v$ is equivalent to a formula $\psi'(v)$ of $\mathcal{F}$. So $\varphi_1 \equiv \Lambda x\varphi_1 x \equiv \Lambda x\psi'(x)$, *where the first equivalence requires* η-conversion. It follows that φ is equivalent to $(\forall x\psi'(x))$.

The restricted use of η-conversion is necessary here, as the example $\forall R$, where R is a unary relation constant of $\mathcal{F}$, makes clear.

By an *n-predicate* I will mean a formula of $\mathcal{G}$ of type $\mathcal{D} \Rightarrow_n 2$ containing no free variables. Every formula of $\mathcal{F}$ is to within equivalence of the form $\varphi v_1 \ldots v_n$

for some $n \geq 0$, where φ is an n-predicate and the v_i are distinct variables of type 2. Namely, write the formula as $\varphi(v_1, \ldots, v_n)$, where the v_i include all the variables in the formula. Then it is $\equiv$

$$(\Lambda x_1 \ldots \Lambda x_n.\varphi(x_1, \ldots, x_n)v_1 \ldots v_n.$$

By an *n-term*, I mean a formula of $\mathcal{G}$ of type $\mathcal{D} \Rightarrow_n \mathcal{D}$ containing no free variables. Every term of $\mathcal{F}$ is to within equivalence of the form $tv_1 \ldots v_n$, where t is an n-term and the v_i include all the free variables in the term.

7. It will be useful to briefly describe Quine's method of eliminating bound variables in first-order formulas (Quine 1960, 1981a) and relate it to the present approach. Quine was somewhat more dismissive of the theory of combinators of Schönfinkel and Curry than an attentive peruser of (Curry and Feys 1958) might have been.

Remark 2 A year prior to Quine's paper, as he himself observed, Bernays published a paper (Bernays 1959) that he had delivered at a colloquium on constructive mathematics in 1957 in Amsterdam and in which he accomplishes essentially the same thing. Bernays abstracted from the von Neumann-Bernays operations for constructing the first-order definable classes of sets in axiomatic set theory to obtain operations for constructing the first-order definable subclasses of $\mathcal{D}^n$ $(n \geq 0)$ in our ambient model of $\mathcal{F}$. There is an interesting historical puzzle connected with Bernays' lecture: he refers to Curry's 'combinatory theory of functionality' in connection with his construction, but without any details or further reference to Curry. Chapter 9 of Curry-Feys is entitled "The Basic Theory of Functionality", and that is where type structure is introduced into combinator theory. If Bernays had just been referring to untyped combinatory logic, it would have been natural for him to refer also to Schönfinkel—after all, he had known him and certainly knew his paper on combinators. On the other hand, the publication date of Curry-Feys is 1958, a year after Bernays' lecture.

Quine's treatment is a bit easier to present than Bernays' and we can use bits of it. In fact, Quine makes a restriction on the first-order language $\mathcal{F}$ in that he assumes that there are no individual or function constants. Another difference from our treatment is that Quine takes $\wedge$, $\neg$, and $\exists$ as the logical constants, whereas we take $\rightarrow$, $\bot$ and $\forall$. But lets leave these differences aside and interpret Quine's term 'n-ary *predicate*, simply to mean an $n-$predicate in our sense. Let P^n denote the set of n-predicates. The *atomic* n-predicates are those containing no occurrences of $\rightarrow$ or $\forall$. (In the absence of individual or function constants, this would coincide with Quine's notion of an atomic n-ary predicate.) Quine introduces certain operations on these classes such all n-predicates can be built up from the atomic ones by means of these operations. The operations in question, which are all definable in terms of the combinators, are, for each $n \geq 0$:

- The operation

$$\rightarrow^n \in P^n \Rightarrow_2 P^n$$

defined by

$$\rightarrow^n \varphi\psi := \Lambda x_1 \dots x_n.[\varphi x_1 \dots x_n \rightarrow \psi x_1 \dots x_n].$$

- The operation

$$\forall^n \in P^{n+1} \Rightarrow P^n$$

defined by

$$\forall^n \varphi := \Lambda x_1 \dots x_n.\forall[\varphi x_1 \dots x_n]$$

.

- The operation

$$\uparrow^n \in P^n \Rightarrow P^{n+1}$$

defined by

$$\uparrow^n \varphi := \Lambda x_1 \dots x_{n+1}.\varphi x_1 \dots x_n.$$

So $\uparrow^n$ transforms an n-predicate into an $n+1$-predicate by adding a dummy last argument to it.

- The operation

$$\downarrow^n \in P^{n+1} \Rightarrow P^n$$

defined by

$$\downarrow^n \varphi := \Lambda x_1 \dots \Lambda x_n.\varphi x_1 \dots x_n x_n.$$

- For each permutation π of the set $\{1, \dots, n\}$, the operation

$$\pi \in P^n \Rightarrow P^n$$

defined by

$$\pi\varphi := \Lambda x_1 \dots x_n.\varphi x_{\pi 1} \dots x_{\pi n}.$$

Quine makes use of the fact that all the permutations of $\{1, \dots, n\}$ can be generated from just two, the transposition $(1, 2)$ and the cycle $(1, \dots, n)$, to restrict the operations π on P^n to just these two.

The superscripts n on $\rightarrow^n$, $\forall^n$, $\uparrow^n$ and $\downarrow^n$ are tedious and can be deduced from the nature of the predicates to which the operations in question are applied; so I will drop them when there can be no confusion. We will also have occasion to use the object

$$\perp^n := \Lambda x_1 \dots \Lambda x_n \perp$$

but, again, will decently refrain from adding the superscript unnecessarily.

When φ is an $n+1$-predicate, r is an n-term and t is a term of $\mathcal{F}$, we define $\varphi^* r$ and $\varphi * t$ by

$$\varphi * r := \Lambda x_1 \dots x_n.\varphi x_1 \dots x_n(r x_1 \dots x_n).$$

$$\varphi * t := \Lambda x_1 \ldots x_n . \varphi x_1 \ldots x_n t.$$

When φ is an n-predicate, its *universal closure* is the sentence

$$\varphi^+ = \forall \ldots \forall \varphi \; (= \forall^0 \ldots \forall^{n-1} \varphi).$$

16.3 Part II

8. We noted that variable binding in first-order logic doesn't stop with the explicit variable binding of the quantifiers: it occurs in deductions, too.

For example, when by the rule

$$\begin{array}{c} \vdots \\ \varphi(v) \\ \hline \forall x \varphi(x) \end{array}$$

of $\forall$-*Introduction* we pass from a deduction $p(v)$ of $\varphi(v)$ with free individual variable v to a deduction of $\forall x \varphi(x)$, we are essentially binding the variable in $p(v)$—Lets denote the result by

$$\Lambda x \in \mathcal{D}.p(x).$$

(It will become clear why this is an appropriate notation.) Remember the usual restriction: If the deduction $p(v)$ contains an hypothesis ψ and v occurs in ψ, then v is said to be *fettered* in $p(v)$. The required restriction is that v be unfettered in $p(v)$. Otherwise, for example, from the hypothesis $\varphi(v)$ itself we would be able to deduce $\forall x \in \mathcal{D}.\varphi(x)$.

Similarly, when by the rule

$$\begin{array}{c} [\varphi] \\ \vdots \\ \psi \\ \hline \varphi \to \psi \end{array}$$

of $\to$-*Introduction*, we pass from a deduction p of ψ from the hypothesis φ to a deduction of $\varphi \to \psi$, the hypothesis should be thought of as a free variable ranging over proofs of φ—so that p should really be written $p(v)$ again, where v is a free variable ranging over proofs of φ. In passing from the deduction of ψ from φ, we 'discharge zero or more occurrences of the hypothesis φ in the deduction: i.e. we are binding that variable v. That is the significance of the square brackets around φ.—Lets denote the result by

$$\Lambda x \in \varphi.p(x).$$

Here we don't need the restriction that v be unfettered in $p(v)$: it is always satisfied. For in first-order logic variables ranging over proofs of other formulas (if we don't count $\mathcal{D}$ as a formula) do not occur in formulas.[3]

But before discussing this further, we need to change our attitude towards the nature of sentences and deductions. If one thinks of deductions simply as arrays of symbols with a specified structure (as for example did Hilbert in his proof theory), then no semantics is involved. But if one takes the more interesting position that deductions denote something, namely proofs, and that proofs themselves have mathematical structure (and perhaps this position was first manifested by Brouwer in his argument for the Bar Theorem), then semantics is in business.

We may think of the class of proofs of a sentence as its meaning. In Part I we assigned formulas in $\mathcal{G}$ types: these are the types of the objects they *denote*, and sentences all have type 2. But when we are talking about proof, we are not talking about truth-values: 1 doesn't need a proof and 0 doesn't have one. A sentence not only has a truth-value, it has a *sense*: it expresses a 'thought' as Frege put it, or a proposition, as I would prefer to put it. We may regard the sense of a mathematical sentence to be given by the class of objects that count as proofs of it. In what follows we shall use the sentence to denote the class of its proofs; in Frege's words, we will use it with *oblique reference*—to denote what, in non-oblique contexts, is its sense.

If one accepts this 'proposition-as-type' ideology, then eliminating bound variables from deductions is an essential part of providing a compositional semantics for sentences. But this ideology aside, just assuming one takes seriously the idea of proofs as objects, the problem of providing deductions with a compositional semantics remains.

An n-predicate then denotes a Cartesian product

$$\Pi_{x_1 \dots x_n} Q(x_1, \dots, x_n)$$

where the x_i range over $\mathcal{D}$. We call this its *type*. So the type of a sentence in the present sense is a set Q—which we have agreed to simply denote by φ. In what follows, we will identify an n-predicate or n-term with its normal form. The connection between the type (viz. 2) of the sentence as an element of $\mathcal{G}$ in Part I and its type Q in the present sense is simply that it denotes 1 in the earlier sense just in case Q is non-empty.[4]

- We assume that the atomic n-ary predicates have already been assigned denotations.

  This is of course more of a demand on the given semantics of $\mathcal{F}$ than we assumed in Part I, where it was only assumed that a denotation in $\mathcal{D} \Rightarrow_n \mathcal{D}$ is assigned

[3] Of course, we could consider a richer system of propositions in which this is no longer the case. This would lead to the Curry-Howard theory of dependent types, which is more complicated. But there is no—or at least, insufficient—reason to go there now. Again, see Tait (1998) for a formalization without bound variables of Curry-Howard type theory.

[4] From the point of view of abandoning denotation in favor of sense, however, here is a retro element in our interpretation of $n+1$-predicates: $Q(x_1, \dots, x_{n+1})$ doesn't depend on the sense of (the assignments of values to) the x_i, only on their denotations. I will not go into this.

to each n-ary function constants and a denotation in $\mathcal{D} \Rightarrow_n 2$ is assigned to each n-ary relation constant.
Notice that the n-predicates include not just atomic relation symbols, but all expressions $\Lambda x_1 \ldots \Lambda x_n \varphi(x_1, \ldots x_n)$, where $\varphi(v_1, \ldots v_n)$ is an atomic formula of $\mathcal{F}$, possibly containing function symbols.

- Of course, we know what class of proofs $\perp$ denotes: it is the null class.
- What about $\varphi \rightarrow \psi$ or, in official notation, $\rightarrow \varphi\psi$? Its proofs should should enable us to move from a proof of φ to a proof of ψ: That is what the rule

$$\frac{\varphi \qquad \varphi \rightarrow \psi}{\psi}$$

of $\rightarrow$-*Elimination* or *Modus Ponens* tells us. This suggests that a proof of $\varphi \rightarrow \psi$ should be a function from proofs of φ to proofs of ψ—i.e $\varphi \rightarrow \psi$ denotes $\varphi \Rightarrow \psi$. We shall accept that suggestion. In view of this, we can formulate $\rightarrow$-Elimination more fully as the rule

$$\frac{p \in \varphi \qquad f \in \varphi \rightarrow \psi}{fp \in \psi}$$

of application of a function to an argument, in accordance with our aim of compositional semantics.
In particular, consider the proof of $\varphi \rightarrow \psi$ that we denoted by $\Lambda x \in \varphi.p(x)$ and that we obtain by $\rightarrow$-Introduction from the deduction $p(v)$ of ψ from the hypothesis v of φ. The notation is justified in that the proof is a function, namely the one defined by the lambda-conversion

$$[\Lambda x \in \varphi.p(x)]r \equiv p(r)$$

when r is a proof of φ.

- $\forall x \in \mathcal{D}.\varphi(x)$ is very like an implication: a proof should yield us, for every $d \in \mathcal{D}$, a proof of $\varphi(d)$. That is what the rule

$$\frac{\forall x \in \mathcal{D}.\varphi(x)}{\varphi(t)}$$

of $\forall$-*Elimination* or of *Instantiation* tells us. Thus, $\forall x \in \mathcal{D}.\varphi(x)$ should express the type of all functions f defined on $\mathcal{D}$ such that, for each $d \in \mathcal{D}$, $fd \in \varphi(d)$. In other words, $\forall x \in \mathcal{D}.\varphi(x)$ just expresses the Cartesian product $\Pi_{d \in \mathcal{D}}.\varphi(d)$. In view of this, $\forall$-Elimination is more fully expressed as the rule

$$\frac{t \in \mathcal{D} \qquad f \in \forall x.\varphi(x)}{ft \in \varphi(t)}$$

of application of a function to an argument.

Again, consider the proof $\Lambda v.p(v)$ that we obtain by $\forall$-Introduction from the deduction $p(v)$ of $\varphi(v)$ from the 'hypothesis' v of $\mathcal{D}$. That too denotes a function, namely the one defined by the lambda conversion

$$[\Lambda x.p(x)]r = p(r)$$

when $r \in \mathcal{D}$, i.e. when r is a 'proof' of $\mathcal{D}$.

In the richer environment of the Curry-Howard type theory, one can understand $\varphi \to \psi$ as just a special case of $\forall x \in \varphi.\psi(x)$, namely the case in which v does not occur in $\psi(v)$. But that unification isn't available to us, since we are not admitting quantification over types other than $\mathcal{D}$. (We may regard $\mathcal{D}$ as a type in the present sense, as opposed to the type structure associated with $\mathcal{G}$: it is the type of the objects in $\mathcal{D}$.)

Remark 3 The extension of the Curry-Feys analogy between the theory of functionality expressed by typed combinators and positive implicational logic to a conception of the formulas of first-order predicate logic as types was, as I mentioned above, due to Howard (1980).

9. Now we have made explicit the problem, namely of eliminating the variable-binding operation Λ from deductions. Our deductions are going to be built up by means of evaluation ultimately from given constant deductions of closed formulas of $\mathcal{F}$ and variable deductions of formulas of $\mathcal{F}$. So, again, every deduction will be *uniquely* of the form

$$p_0 p_1 \cdots p_n$$

where p_0 is either a constant or a variable and the remaining p_i are either deductions or terms of $\mathcal{F}$ ('deductions' of $\mathcal{D}$). The types of the constant deductions then are the *axioms* and we must, in each case,[5] give a definition of the constant by means of conversion rules that is in keeping with its type, the axiom. I will write

$$p \vdash \varphi$$

to mean that p is a deduction of φ.

Before proceeding, it is convenient to add a condition on $\mathcal{F}$, namely that it contain at least one individual constant, witnessing the fact that $\mathcal{D}$ is non-empty. (It is not really a restriction: we could set aside a variable to play the special role of the constant. Then, where we speak of a variable-free deductions, i.e. without assumptions. it should be understood to mean one in which there are no variables other than the distinguished one.)

The obvious way to eliminate bound variables is, of course, to consult Schönfinkel again: we need to define $\to$-Introduction $\Lambda x \in \varphi.p(x)$ and $\forall$-Introduction $\Lambda x \in \mathcal{D}.p(x)$ in terms of suitable combinators. These will be constant deductions whose types then are axioms.

[5] With the exception of the axioms for $\perp$, which we will discuss below.

- For all n-predicates φ, ψ and χ ($n \geq 0$), we introduce the constant deductions

$$K_{\chi,\varphi} \vdash [\varphi \to (\chi \to \varphi)]^+ \qquad K_\varphi \vdash [\varphi \to \forall \uparrow \varphi]^+$$

  Let $\bar{t}$ be the string of n terms of $\mathcal{F}$. Then

$$K_{\chi,\varphi}\bar{t} \vdash \varphi\bar{t} \to (\chi\bar{t} \to \varphi\bar{t}) \qquad K_\varphi\bar{t} \vdash \varphi\bar{t} \to \forall \uparrow (\varphi\bar{t}).$$

  In each case, K is defined by

$$K\bar{t}pc \ CONV \ p.$$

  In both cases, p is a deduction of $\varphi\bar{t}$. In the first case, c is a deduction of $\chi\bar{t}$. In the second case, c is a term of $\mathcal{F}$ (i.e. of type $\mathcal{D}$). $(\forall \uparrow (\varphi t))c \equiv \varphi\bar{t}$ and so this makes sense.
- For all n-predicates φ, ψ and χ ($n \geq 0$), we introduce the constant deduction

$$S_{\chi,\varphi,\psi} \vdash [\chi \to (\varphi \to \psi)] \to [(\chi \to \varphi) \to (\chi \to \psi))]^+$$

  and for each n and all $n+1$-predicates φ and ψ, the constant deduction

$$S_{\varphi,\psi} \vdash [\forall(\varphi \to \psi) \to (\forall\varphi \to \forall\psi)]^+.$$

  Again, for $\bar{t}$ a string of n terms of $\mathcal{F}$,

$$S_{\chi,\varphi,\psi} \in [\chi t \to (\varphi\bar{t} \to \psi\bar{t})] \to [(\chi\bar{t} \to \varphi\bar{t}) \to (\chi\bar{t} \to \psi\bar{t}))]$$

  and

$$S_{\varphi,\psi}\bar{t} \in [\forall(\varphi\bar{t} \to \psi\bar{t}) \to (\forall\varphi\bar{t} \to \forall\psi\bar{t})].$$

  In both cases then S is defined by

$$Stpqc \ CONV \ (pc)(qc).$$

  In the first case, $p \in \chi\bar{t} \to (\varphi\bar{t} \to \psi t)$, $q \in \chi\bar{t} \to \varphi\bar{t}$ and $c \in \chi\bar{t}$. In the second case, $p \in \forall(\varphi\bar{t} \to \psi\bar{t})$, $q \in \forall\varphi\bar{t}$ and c is a term of $\mathcal{F}$.

As we mentioned above, the two forms of the K combinators and the two forms of the S combinators collapse to one when we regard $\varphi \to \psi$ as the vacuous quantification $\forall x\varphi.\psi$.

We need three more kinds of constants: for each $n+1$-predicate φ, n-predicate χ and n-term r, we introduce the constants

$$C_{\varphi,r} \vdash [\forall\varphi \to \varphi * r]^+$$

and

$$D_\varphi \vdash [\forall\varphi \to \forall x(\varphi * x)]^+$$

and for φ an $n+2$-predicate we introduce the constant

$$E_\varphi \vdash [\forall\forall\varphi \to \forall \downarrow \varphi]^+$$

$C_{\varphi,r}$ is defined as follows: if $\bar{t}$ is a string of n terms of $\mathcal{F}$ and $p(\vdash \forall\varphi\bar{t})$, then $C_{\varphi,r}$ is defined by

$$C_{\varphi,r}\bar{t}p \; CONV \; p(r\bar{t}).$$

Of course, when $\mathcal{F}$ contains no individual or function constants, there are no n-terms r and so the constants $C_{\varphi,r}$ are not in play. $\mathcal{D}_\varphi$ is defined by

$$D_\varphi\bar{t}ps \; CONV \; ps.$$

and E_φ is defined by

$$E_\varphi\bar{t}ps \; CONV \; pss.$$

We define the notion of one deduction p reducing to another q, $p \succeq q$, as usual; however in this case we do not require η-conversion. Both the Church-Rosser Theorem and the Well-foundedness Theorem are proved essentially as they are in Part I, so that the relation of two deductions being equivalent, $p \equiv q$, is a well-defined and decidable equivalence relation.

We complete the definition of $\vdash$ by specifying that

$$p \equiv q, \; \varphi \equiv \psi, \; p \vdash \varphi \Rightarrow \; q \vdash \psi.$$

10. Before proving that $\to$-introduction and $\forall$-introduction can be eliminated (Explicit Definition Theorem III), we need some lemmas. The first lemma establishes propositional logic under the closure operation $^+$ and derives from Quine (1951, p. 90).

Lemma 16.1 *Let φ and ψ be n-predicates. If $p \vdash (\varphi \to \psi)^+$ and $q \vdash \varphi^+ \Rightarrow \vdash \psi^+$., then there is an $F \vdash \psi^+$ such that for t a string of variables of type $\mathcal{D}$,*

$$Ft \succeq pt(qt).$$

Proof by induction on n. When $n = 0$, $F = pq$. Assume $n > 0$. We have

$$p \vdash (\forall(\varphi \to \psi))^+$$

and (omitting type subscripts) the combinator

$$S \vdash [\forall(\varphi \to \psi) \to (\forall\varphi \to \forall\psi)]^+.$$

So by the induction hypothesis, we have $G \vdash (\forall\varphi \rightarrow \forall\psi)^+$, *where* $G\bar{u} \succeq Su(p\bar{u})$, *for a string* $\bar{u}$ *of* $n-1$ *variables of type* $\mathcal{D}$. $q \vdash (\forall\varphi)^+$ *and so by the inductive hypothesis again, there is an* $F \vdash \psi^+$ *with* $F\bar{u} \succeq G\bar{u}(q\bar{u}) \succeq S\bar{u}(p\bar{u})(q\bar{u})$ *and so with* v *a variable of type* $\mathcal{D}$

$$F\bar{u}v \succeq S\bar{u}(pu\bar{u})(q\bar{u})v \succeq p\bar{u}v(q\bar{u}v).$$

Lemma 16.2 *Let* χ *and* φ *be n-predicates and* $\bar{t}$ *a string of n variables of type* $\mathcal{D}$.

(i) If $p \vdash \varphi^+$, *then there is an* $F \vdash (\chi \rightarrow \varphi)^+$ *such that* $F\bar{t}q \succeq p\bar{t}$ *for* $q \vdash \chi\bar{t}$.
(ii) If $p \vdash (\chi \rightarrow (\varphi \rightarrow \psi)^+$, *then there is an* $F \vdash [(\chi \rightarrow \varphi) \rightarrow (\chi \rightarrow \psi)]^+$ *such that* $F\bar{t}qr \succeq p\bar{t}r(qr)$ *for* $q \vdash \chi\bar{t} \rightarrow \varphi t$ *and* $r \vdash \chi\bar{t}$.

By Lemma 16.1 using the combinators $K_{\chi,\varphi}$ *and* $S_{\chi,\varphi,\psi}$, *resp.* □

Lemma 16.3 *Let* χ *be an n-predicate,* φ *an* $n+1$*-predicate,* ψ *an* $n+2$*-predicate and r an n-term. Then there are variable-free*

$$C'_{\chi,\varphi,r} \vdash [(\chi \rightarrow \forall\varphi) \rightarrow (\chi \rightarrow \varphi * r)]^+$$

$$D'_{\chi,\varphi} \vdash [(\chi \rightarrow \forall\varphi) \rightarrow (\chi \rightarrow \forall x\varphi * x)]^+$$

such that, for $\bar{t}$ *a string of n variables of type* $\mathcal{D}$

$$C'_{\chi,\varphi,r}\bar{t}pq \succeq C_{\varphi,r}\bar{t}(pq), \qquad D'_{\chi,\varphi}\bar{t}pq \succeq D_{\varphi}\bar{t}(pq).$$

By Lemma 16.2 (i) we have $F \vdash [\chi \rightarrow (\forall\varphi \rightarrow \varphi * r)]^+$ *with* $F\bar{t}r \succeq C_{\varphi,r}\bar{t}$, *where* $r \vdash \chi\bar{t}$. *So by Lemma 16.2 (ii), we have* $C'_{\chi,\varphi,r}$ *such that*

$$C'_{\chi,\varphi,r}\bar{t}pq \succeq F\bar{t}q(pq) \succeq C_{\varphi.r}\bar{t}(pq).$$

Similarly for $D'_{\chi,\varphi}$. □

Lemma 16.4 *Let* $p \vdash \varphi$, *where* φ *is normal, and let v be a variable of type* $\mathcal{D}$. *If p does not contain v, then neither does* φ.

This is easily proved by induction on the complexity of p. Note that, if φ *is a formula of* $\mathcal{G}$ *that does not contain v, then its normal form does not contain v either.*

For φ an n-predicate and $\bar{t}$ a list of n variables of type $\mathcal{D}$, the identity function $I_{\varphi\bar{t}} \vdash \varphi\bar{t} \rightarrow \varphi\bar{t}$ is defined as usual:

$$I_{\varphi\bar{t}} = S_{\varphi,\varphi\rightarrow\varphi,\varphi}\bar{t}(K_{\varphi,\varphi\rightarrow\varphi}\bar{t})(K_{\varphi,\varphi}\bar{t}).$$

(The notation $I_{\varphi,\bar{t}}$ may be misleading, since it looks like $I_{\varphi\bar{t}}$ is a constant. It is constant only when $n = 0$.) So $I_{\varphi,\bar{t}}p \equiv p$ for $p \vdash \varphi\bar{t}$.

Explicit Definition Theorem III.

a. Let v be a variable of type χ and $p(v) \vdash \varphi$. There is a deduction $p' \vdash \chi \to \varphi$ built up from the combinators and constants or variables in $p(v)$ other than v, such that $p'v \equiv p(v)$.
b. Let v be a variable of type $\mathcal{D}$ and $p(v) \vdash \varphi(v)$, where v is unfettered $p(v)$. Then there is a $p' \vdash \forall\varphi' = \forall x\varphi(x)$ built up from the combinators and constants or variables in $p(v)$ other than v, such that $p'v \equiv p(v)$.

Proof of a. For all sufficiently large n, we have $\varphi \equiv \varphi_0\bar{t}$ and $\chi \equiv \chi_0\bar{t}$, where φ_0 and χ_0 are n-predicates and $\bar{t}$ is a list of n variables of type $\mathcal{D}$.

(i) $p(v) = v$. Then $p' = I_{\varphi_0,\bar{t}}$.
(ii) $p(v) = p$ does not contain v. Then $p' = K_{\chi_0,\varphi_0}\bar{t}p$.
(iii) $p(v) = q(v)r(v)$, where $q(v) \vdash \psi \to \varphi$ and $r(v) \vdash \psi$. By taking n large enough, we can assume that $\psi \equiv \psi_0\bar{t}$. Then $p' = S_{\chi_0,\psi_0,\varphi_0}\bar{t}q'r'$.
(iv) $p(v) = q(v)r$, where $q(v) \vdash \forall\psi$ and r is a term of $\mathcal{F}$. Then $q(v) \vdash \forall\psi$ and $q' \vdash \chi \to \forall\psi$. There are two subcases:

- r is a variable. Set $p' = D'_{\chi,\psi}\bar{t}q'r$.
- Otherwise we can assume that $r \equiv r_0\bar{t}$ where r_0 is an n-term. Set $p' = C'_{\chi_0,\psi_0,r_0}\bar{t}q'$.

Proof of b. For sufficiently large n, $\varphi \equiv \varphi_0\bar{t}$, where φ_0 is an $n+1$-predicate and $\bar{t}$ is a string of distinct variables of type $\mathcal{D}$

(i) $p(v) \neq v$.
(ii) $p(v) = p$ does not contain v. Then by Lemma 16.4, the normal form of φv doesn't either. The normal form is $\equiv \varphi_1\bar{t}$ for some n predicate φ_1. Set $p' = K_{\varphi_1}\bar{t}p$.
(iii) $p(v) = q(v)r(v)$ where $q(v) \vdash \chi(v) \to \varphi(v)$. Set $p' = S_{\chi_0,\varphi_0}\bar{t}q'r'$.
(iv) $p(v) = q(v)r(v)$ where $q(v) \vdash \forall\psi(v)$ and $r(v)$ is a term of $\mathcal{F}$. We can assume that $\psi(v) \equiv \psi_0\bar{t}v$ and that $r(v)$ is either a variable or is $\equiv r_0\bar{t}v$, where ψ_0 is an $n+2$-predicate and r_0 is an n-term. $q' \vdash \forall\forall\psi_0\bar{t}$. There are three subcases.

- $r(v) \equiv r_0\bar{t}v$, where r_0 is an n-term. $C\bar{t} = C_{\forall\psi_0,r_0}\bar{t} \vdash \forall\forall\psi_0\bar{t} \to \forall\psi\bar{t}*(r_0\bar{t})$ and $q' \vdash \forall\forall\psi_0\bar{t}$. So $p' = Cu(q_0\bar{t}) \vdash \forall\psi_0\bar{t} * (r_0\bar{t})$. Thus $p'v \vdash \psi_0\bar{t}v(r_0\bar{t}v) \equiv \varphi(v)$.
- $r(v) = r$ is a variable other than v. Set $p' = D_{\forall\psi_0}\bar{t}q'r$.
- $r = v$. Then set $p' = E_{\psi_0}\bar{t}q'r$.

□

Note that in part (b), *(ii)* – *(iv)* are the only possible cases only because v is unfettered in $p(v)$. For example , if $p(v)$ were a variable of type φv, it wouldn't fall under either of these cases.

11. To obtain a complete notion of logical deduction, we have only to introduce the axioms for negation $\bot$; $\bot$*-elimination*

$$N_\varphi \vdash [\bot^n \to \varphi]^+$$

and (for classical logic) *Double* $\bot$*-elimination*

$$DN_\varphi \vdash [((\varphi \to \bot^n) \to \bot^n) \to \varphi]^+$$

for each n-predicate φ. Unlike the other constant deductions, these have no definitions. We can think of these as theological axioms. (1) If there is a benevolent deity, there are no variable-free deductions of $\bot$ to which N_φ can be applied, and (2) the classical doctrine expressed by DN_φ is irreducible. God has made the rational universe simple: if we have established $\neg\neg\varphi$, then φ is indeed true, although we may not have a direct proof of it. Indeed, when φ is a sentence and $p \vdash (\varphi \to \bot) \to \bot$, that DNp cannot in general be computed, yielding a 'real' deduction of φ, is precisely what nonbelievers (i.e. constructive mathematicians) complain about.

So here are the axioms, i.e. the types of the constants, from which all the logical first-order truths in the language $\mathcal{F}$ can be derived using modus ponens:

(i) $[\varphi \to (\chi \to \varphi)]^+$
(ii) $[\chi \to (\varphi \to \psi)] \to [(\chi \to \varphi) \to (\chi \to \psi))]^+$
(iii) $[\varphi \to \forall x\varphi]^+$
(iv) $\forall(\varphi \to \psi) \to (\forall\varphi \to \forall\psi)]^+$
(v) $[\forall\forall\varphi \to \varphi * r]^+$ Here φ is an $n+1$-predicate and r an n-term.
(vi) $[\forall\varphi x \to \forall x\varphi * x]^+$
(vii) $[\forall\forall\varphi \to \forall \downarrow \varphi]^+$
(viii) $[\bot \to \varphi]^+$
(ix) $[(\varphi \to \bot) \to \bot]^+$

For $n = 0$, (i) and (ii) are the standard axioms for the theory of implication. Added to (viii) and (ix) for $n = 0$, they are complete for propositional logic. If we replace (i), (ii), (vii) and (viii) by the axiom schema

$$\varphi^+ \qquad \text{(for every tautology } \varphi \text{ in } \mathcal{F})$$

and drop (v), then this axiom system is easily seen (using Lemma 16.1) to be equivalent to that of Quine's *Mathematical Logic* (Quine 1951). Axiom (v) is rendered inoperative in Quine's system because the only terms are variables: there are no n-terms.

References

Bernays, P. (1959). Über eine natürliche Erweiterung des Relationkalküls. In A. Heyting (Ed.), *Proceedings of the Colloquium Held in Amsterdam, 1957* (pp. 1–14). Amsterdam: North-Holland Publishing Company.
Church, A. (1941). *The calculi of lambda-conversion*. Princeton: Princeton University Press.
Church, A., & Rosser, J. B. (1936). Some properties of conversion. *Transactions of the American Mathematical Society, 39*(3), 472–482.

Curry, H. & Feys, R. (1958). *Combinatory logic I: studies in logic and the foundations of mathematics* (2nd ed. 1968). Amsterdam: North-Holland Publishing Company.
Howard, W. (1980). The formula-as-types notion of construction. In J. P. Seldin & J. R. Hindley (Eds.), To H.B. Curry: *Essays on Combinatory Logic, Lambda Calculus and Formalism* (pp. 479–490). London: Academic Press.
Quine, W. (1951). *Mathematical logic: revised edition*. Cambridge: Harvard University Press.
Quine, W. (1960). Variables explained away. *Proceedings of the American Philosophical Society, 104*(3), 343–347.
Quine, W. (1966a). On Carnap's views on ontology. In *Selected Logical Papers* (pp. 127–134). New York: Random House.
Quine, W. (1981a). Predicates, terms and classes. In *Theories and Things* (pp. 164–172). Cambridge: Harvard University Press.
Schönfinkel, M. (1924). Über die Bausteine der mathematischen Logik. *Mathematische Annalen, 92*(3–4), 305–316.
Tait, W. W. (1963). A second order theory of functionals of higher type, with two appendices. appendix A: Intensional functionals. Appendix B: An interpretation of functionals by convertible terms. In Stanford Seminar Report, pp 171–206, unpublished. A published version is Tait (1967).
Tait, W. W. (1967). Intensional interpretations of functionals of finite type I. *Journal of Symbolic Logic, 32*(2), 198–212.
Tait, W. W. (1998). Variable-free formalization of the Curry-Howard type theory. In G. Gambin & J. Smith (Eds.), *Twenty-five years of constructive type theory*. Oxford: Oxford University Press.

Chapter 17
On Gentzen's Structural Completeness Proof

Neil Tennant

Abstract In his very first publication, Gentzen introduced the structural rules of thinning and cut on sequents. He did not consider rules for logical operators. Gentzen provided a most interesting 'structural completeness proof', which it is the concern of this study to explain and clarify. We provide an improved (because more detailed) proof of Gentzen's completeness result. Then we reflect on the self-imposed limitations of this, Gentzen's earliest sequent-setting, and explore how his approach might have been generalized, even in the absence of logical operators, so as to cover cases involving sequents with empty antecedent or succedent, and logical consequences of infinite sets of sequents.

Keywords Cut · Thinning · Structural completeness · Linear sequents · Gentzen-proof · Normal proof · Normal deducibility · Super-normal deducibility · Undermining · Confirming · Empty antecedents · Empty succedents

17.1 Introduction and Motivation

Gentzen's first publication was Gentzen (1932). It introduced the structural rules of thinning (*Verdünnung*) and cut (*Schnitt*) on sequents (unhelpfully called *Sätze*). To call sequents *Sätze* is unhelpful because the main application, subsequently, is to be one where the sequents are made up of a set of sentences on the left, and a sentence on the right. Since the usual reading of *Sätze* is 'sentences', this could lead to confusion.

Invited paper for Dag Prawitz on Proofs and Meaning, edited by Heinrich Wansing, in the Studia Logica series Trends in Logic. The author wishes to thank Curtis Franks for a pre-publication copy of his paper 'Cut as Consequence' (Franks 2010), which prompted closer consideration of Gentzen's methods and results in (Gentzen 1932). He is also grateful to Peter Schroeder-Heister for helpful comments in correspondence. It is a pleasure to acknowledge the considerable intellectual debt owed to Dag Prawitz, whose writings on proof theory and its philosophical importance significantly shaped the author's interests as a young scholar.

N. Tennant (✉)
Department of Philosophy, The Ohio State University, Columbus, OH 43210, USA
e-mail: tennant.9@osu.edu

H. Wansing (ed.), *Dag Prawitz on Proofs and Meaning*, Outstanding Contributions to Logic 7, DOI 10.1007/978-3-319-11041-7_17

Gentzen did not consider any other rules—in particular, he gave no rules for logical operators. He provided, however, a most interesting 'structural completeness proof', which it is the concern of this study to explain and clarify. In Sect. 17.2 we set out notation and provide an improved (because more detailed) proof of Gentzen's completeness result. In Sect. 17.3 we reflect on the self-imposed limitations of this, Gentzen's earliest sequent-setting, and explore how his approach might have been generalized even in the absence of logical operators.

Our aim here is to re-cast Gentzen's structural completeness proof into a form that allows for ready generalization to cover cases of completeness that Gentzen himself did not consider: the cases involving sequents with empty antecedent or succedent, and logical consequences of infinite sets of sequents. In order to accomplish this, we break the proof down into proofs of more lemmas than Gentzen himself cared to isolate for separate statement and proof. The advantage of doing this is that one comes to appreciate better how 'all the bits fit together', as it were; and which ones have to be tweaked, or re-ordered, in order to effect the generalizations that are sought in this study.

In order to regiment Gentzen's reasoning more rigorously, we have cast into the form of a proof by induction (to be found in the proof of our Lemma 17.2) a crucial passage of his reasoning, which he presents very briefly and intuitively, and which relies on the conviction that a certain procedure, iterated sufficiently many times, will produce a certain result because of the way it eventually exhausts a finite set of possibilities (see Footnote 7).[1]

For the average reader, this particular work of Gentzen is little known. Another service we try to render is to make Gentzen's definitions more perspicuous, by *parametrizing* them in a judicious way. Gentzen demanded a lot of his reader, by introducing arbitrary ciphers—single letters in unusual fonts—as cryptic abbreviations of concepts that contained a considerable amount of logical structure and involved more than one important parameter embedded within them. So we have tried

[1] As Peter Schroeder-Heister notes (Schroeder-Heister 2002, p. 261), here Gentzen, following Hertz (1928), uses 'the fixed point construction which is now standard in the theory of logic programming'. The interest of Schroeder-Heister's paper is twofold. First, he provides a detailed comparison of the work of Gentzen with that of his predecessor Hertz, whom Schroeder-Heister credits with the invention of proof-trees. Secondly, Schroeder-Heister is concerned to reveal how the theory of SLD resolution (part of the modern theory of logic programming), when understood proof-theoretically, is actually a part of structural proof theory.

Our aim here is orthogonal to that of Schroeder-Heister. We seek to provide more detail about the logical structure of Gentzen's completeness proof than Schroeder-Heister was able to provide, in his summary thereof, within the constraints of his wider comparative projects. The extra logical detail that we supply enables us to carry out the generalizations desired, to the case of sequents empty on the left or right, and the case of infinite premise-sets of sequents.

Perhaps by adapting an old metaphor we can make the contrast between Schroeder-Heister (2002) and the present study a little clearer. In Schroeder-Heister (2002), it is shown that modern logic-programming reinvents a certain proof-theoretical wheel. In the present study, we are concerned to supply some extra spokes for the original wheel, in order to attain a better understanding of how it worked.

to fashion slightly more expansive but still easily manipulable abbreviations that will obviate the need, as the exposition proceeds, to keep consulting earlier definitions of cryptic symbols.

17.2 Exposition of Gentzen's Completeness Results

We shall use here notational conventions preferred by the present author, which are more current in modern proof theory.[2]

Definition 17.1 Gentzen's *sequents* are of the form

$$\Delta : \psi,$$

where Δ is a *non-empty*, finite set of 'elements' (*Elemente*) of the same kind as ψ. Δ is called the *antecedent* and ψ is called the *succedent* of the sequent.

One can give the general form of a sequent as

$$\varphi_1, \ldots, \varphi_n : \psi,$$

on the understanding that the ordering of the elements on the left (i.e., before the colon) is of no consequence. Today, of course, we would think of the 'elements' as *sentences*, most probably of some formal language. As such, they could have internal logico-grammatical structure. Significant primitive expressions imparting that structure—such as logical connectives—would have sequent-rules governing them specifically. But in his paper Gentzen (1932), Gentzen was not at all concerned with sentential structure. His 'elements' were indeed elemental. He did not inquire after their internal structure.

Nor did Gentzen seek to read a sequent $\varphi_1, \ldots, \varphi_n : \psi$ as making only a claim of logical consequence. That would be but one permissible reading—'When the statements $\varphi_1, \ldots, \varphi_n$ are correct, so too is the statement ψ'. He gave examples of other possible readings for his sequents, such as 'Any domain of elements that contains $\varphi_1, \ldots, \varphi_n$ also contains ψ'. He also treated the relation of logical consequence as holding among *sequents*, not among sentences, as premises and conclusions.

Definition 17.2 A sequent $\varphi : \psi$ Gentzen called *linear*, and a sequent of the form $\psi : \psi$ he called *tautologous*. Any sequent $\Delta : \psi$ with ψ in Δ he called *trivial*.

The reason why Gentzen calls a sequent of the form $\varphi : \psi$ linear is that if one thinks of making a downward inference from the premises of a sequent (in its antecedent)

[2] Gentzen used '$\rightarrow$' where we use a colon, and preferred Latin variables to Greek ones.

to its conclusion (i.e., its succedent), then with multiple premises $\varphi_1, \ldots, \varphi_n$ there would be *branching*, whereas with but a single premise φ there would not be:[3]

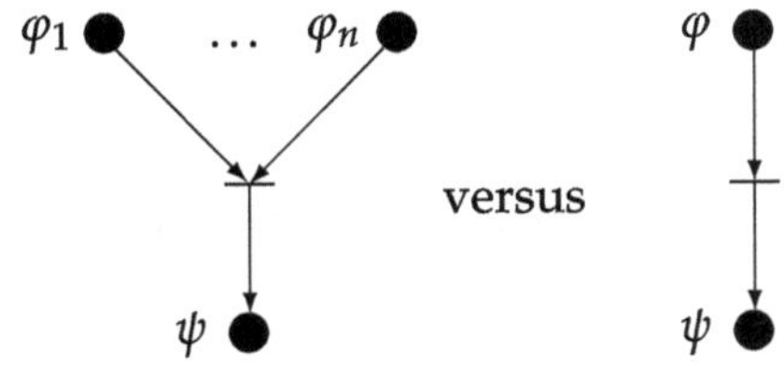

Gentzen considered two modes of inference involving sequents. The first, *thinning* (*Verdünnung*), allows one to put more elements on the left:

$$\text{THINNING} \qquad \frac{\Delta : \psi}{\Gamma, \Delta : \psi}.$$

(Read the comma as the sign for set union.) The second rule, which Gentzen called *cut* (*Schnitt*), allows one to avail oneself of the *transitivity* that is implicit in the two examples already given of how one might read a sequent:

$$\text{CUT} \qquad \frac{\Gamma : \varphi \qquad \Delta, \varphi : \psi}{\Gamma, \Delta : \psi}.$$

(Assume that φ is not in Δ, and bear in mind that Δ could be empty.)

In both rules, the sequents above the line are called *premises*; the ones below the line are called *conclusions*. So: premises and conclusions are not single sentences; rather, they are sequents. In the case of CUT, $\Gamma : \varphi$ is called the *left* premise, and $\Delta, \varphi : \psi$ is called the *right* premise.

In applications of CUT, if both premises are linear, then so too is the conclusion:

$$\frac{\theta : \varphi \qquad \varphi : \psi}{\theta : \psi}$$

What Gentzen calls *proofs* (*Beweise*) may be built up *in linear fashion* using finitely many *starting sequents*, so as to reach an *end sequent*. Each individual *step* within a proof is an application of THINNING or of CUT. There are only finitely many steps in a proof. Here is how Gentzen defines proofs:

> Unter einem *Beweis* eines Satzes $\mathfrak{q}$ aus den Sätzen $\mathfrak{p}_1, \ldots, \mathfrak{p}_\nu$ $(\nu \geqq 0)$ verstehen wir nunmehr *eine geordnete Anzahl von Schlüssen* (d.h. Verdünnungen und Schnitten[fn]), deren letzter $\mathfrak{q}$ als Konklusion besitzt, und in der jede Prämisse entweder zu den $\mathfrak{p}$ gehört oder tautologisch ist, oder mit einer vorangehenden Konklusion übereinstimmt. [Emphasis added—NT.]

Gentzen's phrase 'eine geordnete Anzahl von Schlüssen' is rather ambiguous in context. There are two interpretative possibilities.

[3] These will be the only diagrams we use in which tree-like arrays have their nodes labeled by *sentences*. In the rest of this discussion, we shall be considering only trees whose nodes are labeled by *sequents*.

1. The steps (Schlüssen) are to be represented in tree-like fashion. The sequent $\mathfrak{q}$ is at the root of the tree. Each upward furcation from a conclusion to one or to two premises corresponds to a distinct step (a thinning or a cut, respectively). Gentzen's definition of proof would imply that bifurcations could only ever take place when at least one premise-node is a *leaf* node of the proof-tree.
2. A proof is to be thought of as a Hilbert-like sequence of sequents (as opposed to sentences). That would impose the linearity that his phrase implies. But it would allow for one also to descry 'within the proof' bifurcations involving *two* complex subproofs. That is to say, *both* premises of an application of CUT could stand as conclusions of complex subproofs.

Of these two possibilities, (2) is the less plausible, since Gentzen speaks of 'eine geordnete Anzahl von *Schlüssen*' [emphasis added—NT] rather than of 'eine geordnete Anzahl von *Sätze*'. Moreover, he only ever depicts a *Schluss* as a fragment of a tree, with each node labeled by a sequent,[4] as, for example, in the statements of THINNING and of CUT above.

Although Gentzen did not state his definition of proof in an inductive form, it is helpful to have it as an inductive definition. In giving the following definition, we are seeking to capture interpretation (1) above of what Gentzen intended.

Let us use $\mathfrak{p}$ and $\mathfrak{q}$ (as Gentzen did), with or without numerical subscripts, as sortal variables ranging over sequents. Let us also use $\mathfrak{P}$ and $\mathfrak{Q}$ for finite sets of sequents. First we define what we mean by a *tree of sequents.*

1. Any sequent $\mathfrak{p}$ counts as a tree of sequents.
2. If Π and Σ are finite trees of sequents, then so is

$$\frac{\Pi \quad \Sigma}{\mathfrak{p}}$$

   (This is the finite tree with root-node $\mathfrak{p}$ and immediate sub-trees Π and Σ.)
3. (Closure) Every finite tree of sequents can be shown to be so by means of clauses (1) and (2).

We are now in a position to give our promised inductive definition of a *Gentzen-proof.* Clause (1) below is the basis clause, and clauses (2)–(4) are the inductive clauses. Clause (5) is the closure clause. Together these clauses define the ternary relation 'Π is a Gentzen-proof of the sequent $\mathfrak{q}$ from the (finite) set $\mathfrak{P}$ of sequents'.

Inductive Definition of Gentzen-Proof

1. Any non-tautologous sequent $\mathfrak{p}$ is a Gentzen-proof of $\mathfrak{p}$ from $\{\mathfrak{p}\}$; and any tautologous sequent $\mathfrak{p}$ is a Gentzen-proof of $\mathfrak{p}$ from $\emptyset$.
2. If Π is a Gentzen-proof of the sequent $\Delta : \psi$ from the set $\mathfrak{P}$ of sequents, and Γ is a finite set of elements, then

[4] If one is to be really careful in offering tree-representations of proofs, one might consider also treating each inference stroke as labelling a node lying above the conclusion-node, and below the relevant premise-nodes, of the inference in question.

$$\frac{\Pi}{\Gamma, \Delta : \psi}$$

is a Gentzen-proof of the sequent $\Gamma, \Delta : \psi$ from the set $\mathfrak{P}$ of sequents.

3. If Π is a Gentzen-proof of the sequent $\Gamma : \varphi$ from the set $\mathfrak{P}$ of sequents, and Δ is a finite set of elements other than φ, then

$$\frac{\Pi \qquad \Delta, \varphi : \psi}{\Gamma, \Delta : \psi}$$

is a Gentzen-proof of the sequent $\Gamma, \Delta : \psi$ from the set $\mathfrak{P} \cup \{\Delta, \varphi : \psi\}$ of sequents.

4. If Σ is a Gentzen-proof of the sequent $\Delta, \varphi : \psi$ $(\varphi \notin \Delta)$ from the set $\mathfrak{Q}$ of sequents, and Γ is a finite set of elements, then

$$\frac{\Gamma : \varphi \qquad \Sigma}{\Gamma, \Delta : \psi}$$

is a Gentzen-proof of the sequent $\Gamma, \Delta : \psi$ from the set $\mathfrak{Q} \cup \{\Gamma : \varphi\}$ of sequents.

5. (Closure) Every Gentzen-proof can be shown to be so by means of clauses (1)–(4).

Note that whenever bifurcation is involved within a Gentzen-proof (i.e. whenever CUT is applied), at most one of the sub-trees is a *complex* tree, i.e. something other than a sequent. That is why the two clauses (3) and (4) are devoted to covering the possible forms that can be taken by applications of CUT in building up a Gentzen-proof.

Definition 17.3 Gentzen called *normal* proofs of the forms

$$\frac{\mathfrak{s}_0}{\mathfrak{q}}, \quad \frac{\dfrac{\mathfrak{t}_0 \qquad \mathfrak{s}_0}{\mathfrak{s}_1}}{\mathfrak{q}}, \quad \frac{\dfrac{\mathfrak{t}_1 \qquad \dfrac{\mathfrak{t}_0 \qquad \mathfrak{s}_0}{\mathfrak{s}_1}}{\mathfrak{s}_2}}{\mathfrak{q}}, \quad \ldots, \quad \frac{\dfrac{\mathfrak{t}_{\rho-1} \qquad \begin{matrix} \dfrac{\mathfrak{t}_1 \qquad \dfrac{\mathfrak{t}_0 \qquad \mathfrak{s}_0}{\mathfrak{s}_1}}{} \\ \iddots \\ \mathfrak{s}_{\rho-1} \end{matrix}}{\mathfrak{s}_\rho}}{\mathfrak{q}}, \quad \ldots$$

Here the terminal single-premise steps are applications of THINNING, and the earlier double-premise steps are applications of CUT. The cut-element is always in the succedent of the $\mathfrak{t}$-sequent (hence in the antecedent of the corresponding $\mathfrak{s}$-sequent). Note that the $\mathfrak{s}$-sequents will all have the same succedent as $\mathfrak{q}$.

Observation 17.1 *For any trivial sequent $\mathfrak{s}$, there is a normal, one-step proof of $\mathfrak{s}$ from $\emptyset$.*

Proof Suppose $\mathfrak{s}$ is $\Delta : \varphi$. Since $\mathfrak{s}$ is trivial, we have $\varphi \in \Delta$. The proof

$$\frac{\varphi : \varphi}{\Delta : \varphi}$$

begins with the tautologous sequent $\varphi : \varphi$, has one step of THINNING, and proves $\Delta : \varphi$ from $\emptyset$ in accordance with clauses (1) and (2) of the inductive definition of Gentzen-proof. Moreover, the proof is normal, since its form is that of the first in the series of forms listed in Definition 17.3.

Definition 17.4 We go one further than Gentzen and call *super-normal* such proofs as are of the forms above except in so far as they do not contain the indicated terminal step of THINNING.

Definition 17.5 We say $\mathfrak{q}$ is *normal-deducible* from $\mathfrak{P} = \{\mathfrak{p}_1, \ldots, \mathfrak{p}_n\}$ (abbreviated: $\mathfrak{P} \vdash_N \mathfrak{q}$) if and only if there is a normal proof of $\mathfrak{q}$ from (some subset of) $\mathfrak{P}$.

Definition 17.6 We say $\mathfrak{q}$ is *super-normal-deducible* from $\mathfrak{P} = \{\mathfrak{p}_1, \ldots, \mathfrak{p}_n\}$ (abbreviated: $\mathfrak{P} \vdash_S \mathfrak{q}$) if and only if there is a super-normal proof of $\mathfrak{q}$ from (some subset of) $\mathfrak{P}$.

Lemma 17.1 *Suppose* $\Delta \subseteq \Gamma$ *and* $\mathfrak{P} \vdash_S \Delta : \psi$. *Then* $\mathfrak{P} \vdash_N \Gamma : \psi$.

Proof There is a super-normal proof of $\Delta : \psi$ from $\mathfrak{P}$. Call it Π. Then the normal proof

$$\begin{array}{c} \mathfrak{P} \\ \Pi \\ \Delta : \psi \\ \hline \Gamma : \psi \end{array}$$

establishes that $\mathfrak{P} \vdash_N \Gamma : \psi$.

Corollary 17.1 *Suppose* $\exists\Delta(\Delta \subseteq \Gamma \wedge \mathfrak{P} \vdash_S \Gamma : \psi)$. *Then* $\mathfrak{P} \vdash_N \Gamma : \psi$.

Proof Immediate from Lemma 17.1 by existential elimination.

Definition 17.7 We define $(\psi, \mathfrak{P})$*-sequents* to be sequents $\mathfrak{s}$ such that
ψ is the succedent of $\mathfrak{s}$ and there is a super-normal proof of $\mathfrak{s}$ from $\mathfrak{P}$.

Definition 17.8 Γ is $(\psi, \mathfrak{P})$*-weak* if and only if $\neg\exists\Delta(\Delta \subseteq \Gamma \wedge \mathfrak{P} \vdash_S \Gamma : \psi)$—that is, if and only if *no* $(\psi, \mathfrak{P})$-sequent has its antecedent included in Γ.

Definition 17.9 A (finite) set of elements is said to *undermine* a sequent $\Delta : \varphi$ just in case it contains all of Δ but does not contain φ.

Definition 17.10 A (finite) set of elements is said to *confirm* a sequent $\Delta : \varphi$ just in case it does not undermine $\Delta : \varphi$.[5]

Observation 17.2 *A (finite) set of elements confirms a sequent* $\Delta : \varphi$ *if and only if it either lacks some member of* Δ *or contains* φ.

---

[5] These are our terms of art, not Gentzen's. Gentzen used 'genügen' for 'confirm', and used its negation for 'undermine'.

Observation 17.3 *Suppose* $\Gamma \subseteq \Theta$ *and* Θ *confirms* $\Gamma : \psi$. *Then* $\psi \in \Theta$.

Observation 17.4 *Suppose* Γ *is* $(\psi, \mathfrak{P})$*-weak. Then* Γ *does not undermine any* $(\psi, \mathfrak{P})$*-sequent. Hence,* Γ *confirms each* $(\psi, \mathfrak{P})$*-sequent.*

Definition 17.11 We say that $\mathfrak{q}$ is a *consequence of* $\mathfrak{P} = \{\mathfrak{p}_1, \ldots, \mathfrak{p}_n\}$ (abbreviated: $\mathfrak{P} \models \mathfrak{q}$) just in case any set of elements involved in the sequents $\mathfrak{p}_1, \ldots, \mathfrak{p}_n, \mathfrak{q}$ that confirms $\mathfrak{p}_1, \ldots, \mathfrak{p}_n$ confirms $\mathfrak{q}$. Equivalently: ...just in case no set of elements involved in the sequents $\mathfrak{p}_1, \ldots, \mathfrak{p}_n, \mathfrak{q}$ confirms $\mathfrak{p}_1, \ldots, \mathfrak{p}_n$ but undermines $\mathfrak{q}$.

For the rest of this section, consider the $\mathfrak{p}_i$ ($= \Xi_i : \varphi_i$, say) given in some fixed order. Collectively, we shall call them $\vec{\mathfrak{P}}$.

Definition 17.12 The finite sequence

$$\Gamma = \Gamma_1 \subseteq \Gamma_2 \subseteq \ldots \subseteq \Gamma_{n+1} = \Gamma^{\vec{\mathfrak{P}}}$$

of sets of elements, whose last member $\Gamma^{\vec{\mathfrak{P}}}$ is called the $\vec{\mathfrak{P}}$*-completion of* Γ, is defined inductively as follows.[6] Let Γ_i be in hand. We define Γ_{i+1} by the dichotomous cases (α) and (β):

(α) For some k Γ_i undermines $\mathfrak{p}_k$. Let j be the least such k. Set $\Gamma_{i+1} = \Gamma_i \cup \{\varphi_j\}$.
(β) For no k does Γ_i undermine $\mathfrak{p}_k$. Set $\Gamma_{i+1} = \Gamma_i$.

Observation 17.5 $\Gamma \subseteq \Gamma^{\vec{\mathfrak{P}}}$.

Gentzen proves the following theorems.

Theorem 17.1 (Soundness) *If there is a Gentzen-proof of* $\mathfrak{q}$ *from* $\mathfrak{p}_1, \ldots, \mathfrak{p}_n$, *then* $\mathfrak{q}$ *is a consequence of* $\mathfrak{p}_1, \ldots, \mathfrak{p}_n$.

Theorem 17.2 (Completeness) *If a non-trivial sequent* $\mathfrak{q}$ *is a consequence of* $\mathfrak{p}_1, \ldots, \mathfrak{p}_n$, *then there is a normal proof of* $\mathfrak{q}$ *from* $\mathfrak{p}_1, \ldots, \mathfrak{p}_n$.

Corollary 17.2 (Normalizability) *If there is a Gentzen-proof of a non-trivial sequent* $\mathfrak{q}$ *from* $\mathfrak{p}_1, \ldots, \mathfrak{p}_n$, *then there is a normal proof of* $\mathfrak{q}$ *from* $\mathfrak{p}_1, \ldots, \mathfrak{p}_n$.

Lemma 17.2 *For each* $i \geq 1$, *there are at least* i *distinct sequents* $\mathfrak{p}_{r_1}, \ldots, \mathfrak{p}_{r_i}$ *among the* $\mathfrak{p}_1, \ldots, \mathfrak{p}_n$ *such that* Γ_{i+1} *confirms each of* $\mathfrak{p}_{r_1}, \ldots, \mathfrak{p}_{r_i}$; *and, if* Γ_{i+1} *undermines at least one of* $\mathfrak{p}_1, \ldots, \mathfrak{p}_n$, *then any set extending* Γ_{i+1} *by adding succedents of the sequents* $\mathfrak{p}_1, \ldots, \mathfrak{p}_n$ *confirms each of* $\mathfrak{p}_{r_1}, \ldots, \mathfrak{p}_{r_i}$.

---

[6] Here we supply a little more detail than Gentzen did, so as to enable a proper proof by mathematical induction of the desired properties of the set $\Gamma^{\vec{\mathfrak{P}}}$ being constructed. Our method has the advantage of being more readily generalizable to the infinite case.

For the reader interested in comparing our proof with that of Gentzen: he wrote u for our φ, v for our ψ, L for our Γ, M_i for our Γ_i, and N for our $\Gamma^{\vec{\mathfrak{P}}}$.

Proof By induction.

Basis ($i = 1$). We need to show that

> there is at least one sequent $\mathfrak{p}_{r_1}$ among the $\mathfrak{p}_1, \ldots, \mathfrak{p}_n$ such that Γ_2 confirms $\mathfrak{p}_{r_1}$; and, if Γ_2 undermines at least one of $\mathfrak{p}_1, \ldots, \mathfrak{p}_n$, then any set extending Γ_2 by adding succedents of the sequents $\mathfrak{p}_1, \ldots, \mathfrak{p}_n$ confirms $\mathfrak{p}_{r_1}$.

Now we reason by the dichotomous cases (α) and (β) for $i = 1$.

(α) For some k Γ_1 undermines $\mathfrak{p}_k$ ($= \Xi_k : \varphi_k$, say). Let m be the least such k. So $\Xi_m \subseteq \Gamma_1$ but $\varphi_m \notin \Gamma_1$. Also by definition $\Gamma_2 = \Gamma_1 \cup \{\varphi_m\}$. Hence Γ_2 (and any set extending Γ_2) confirms $\Xi_m : \varphi_m$. So Γ_2 confirms at least one of the sequents $\mathfrak{p}_1, \ldots, \mathfrak{p}_n$; and, if Γ_2 undermines at least one of $\mathfrak{p}_1, \ldots, \mathfrak{p}_n$, then any set extending Γ_2 by adding succedents of the sequents $\mathfrak{p}_1, \ldots, \mathfrak{p}_n$ confirms $\Xi_m : \varphi_m$.
(β) For no k does Γ_1 undermine $\mathfrak{p}_k$. So Γ_1 confirms $\Xi_j : \varphi_j$, $1 \leq j \leq n$. Also by definition $\Gamma_2 = \Gamma_1$. Hence Γ_2 confirms $\Xi_j : \varphi_j$, $1 \leq j \leq n$. So Γ_2 confirms at least one of the sequents $\mathfrak{p}_1, \ldots, \mathfrak{p}_n$; and, if Γ_2 undermines at least one of $\mathfrak{p}_1, \ldots, \mathfrak{p}_n$, then any set extending Γ_2 by adding succedents of the sequents $\mathfrak{p}_1, \ldots, \mathfrak{p}_n$ confirms that sequent. (The last conjunct is vacuously true, since its antecedent, *ex hypothesi*, is false.)

Inductive Hypothesis. There are at least i sequents $\mathfrak{p}_{r_1}, \ldots, \mathfrak{p}_{r_i}$ among the $\mathfrak{p}_1, \ldots, \mathfrak{p}_n$ such that Γ_{i+1} confirms each of $\mathfrak{p}_{r_1}, \ldots, \mathfrak{p}_{r_i}$; and, if Γ_{i+1} undermines at least one of $\mathfrak{p}_1, \ldots, \mathfrak{p}_n$, then any set extending Γ_{i+1} by adding succedents of the sequents $\mathfrak{p}_1, \ldots, \mathfrak{p}_n$ confirms each of $\mathfrak{p}_{r_1}, \ldots, \mathfrak{p}_{r_i}$.

Inductive Step. We need to show that

> There are at least $i + 1$ sequents $\mathfrak{p}_{r_1}, \ldots, \mathfrak{p}_{r_{i+1}}$ among the $\mathfrak{p}_1, \ldots, \mathfrak{p}_n$ such that Γ_{i+2} confirms each of $\mathfrak{p}_{r_1}, \ldots, \mathfrak{p}_{r_{i+1}}$; and, if Γ_{i+2} undermines at least one of $\mathfrak{p}_1, \ldots, \mathfrak{p}_n$, then any set extending Γ_{i+2} by adding succedents of the sequents $\mathfrak{p}_1, \ldots, \mathfrak{p}_n$ confirms each of $\mathfrak{p}_{r_1}, \ldots, \mathfrak{p}_{r_{i+1}}$.

Once again we reason by the dichotomous cases (α) and (β).

(α) For some k Γ_{i+1} undermines $\mathfrak{p}_k$ ($= \Xi_k : \varphi_k$, say). Let m be the least k such that Γ_{i+1} undermines $\mathfrak{p}_k$. So $\Xi_m \subseteq \Gamma_{i+1}$ but $\varphi_m \notin \Gamma_{i+1}$. Also by definition $\Gamma_{i+2} = \Gamma_{i+1} \cup \{\varphi_m\}$. So Γ_{i+2} confirms $\Xi_m : \varphi_m$.
Let $\mathfrak{p}_{r_1}, \ldots, \mathfrak{p}_{r_i}$ be as in IH. By IH, Γ_{i+2} confirms $\mathfrak{p}_{r_1}, \ldots, \mathfrak{p}_{r_i}$. But $\Xi_m : \varphi_m$ cannot be among these sequents. So Γ_{i+2} confirms the sequents $\mathfrak{p}_{r_1}, \ldots, \mathfrak{p}_{r_i}$, as well as the sequent $\Xi_m : \varphi_m$, which we can now take for $\mathfrak{p}_{r_{i+1}}$. Moreover, if Γ_{i+2} undermines at least one of $\mathfrak{p}_1, \ldots, \mathfrak{p}_n$, then by IH any set extending Γ_{i+2} (hence extending Γ_{i+1}) by adding succedents of the sequents $\mathfrak{p}_1, \ldots, \mathfrak{p}_n$ confirms all of $\mathfrak{p}_{r_1}, \ldots, \mathfrak{p}_{r_i}$; and also confirms $\mathfrak{p}_{r_{i+1}}$, because the succedent of this sequent is in Γ_{i+2}.

(β) For no k does Γ_{i+1} undermine $\mathfrak{p}_k$. So Γ_{i+1} confirms every one of the sequents $\mathfrak{p}_1, \ldots, \mathfrak{p}_n$. By definition $\Gamma_{i+2} = \Gamma_{i+1}$. Hence Γ_{i+2} confirms every one of the sequents $\mathfrak{p}_1, \ldots, \mathfrak{p}_n$. The result follows.

Corollary 17.3 *$\Gamma^{\vec{\mathfrak{P}}}$, which is Γ_{n+1}, confirms $\mathfrak{p}_j$, $1 \leq j \leq n$.*[7]

Lemma 17.3 *Suppose Γ is $(\psi, \mathfrak{P})$-weak. Then each Γ_k confirms every $(\psi, \mathfrak{P})$-sequent.*

Proof By induction.
Basis step. By Observation 17.4, Γ $(=\Gamma_1)$ confirms every $(\psi, \mathfrak{P})$-sequent.

Inductive Hypothesis. Suppose that Γ_i confirms every $(\psi, \mathfrak{P})$-sequent.

Inductive Step. Show that Γ_{i+1} confirms every $(\psi, \mathfrak{P})$-sequent.
If $\Gamma_{i+1} = \Gamma_i$, then by IH we are done.
If $\Gamma_{i+1} \neq \Gamma_i$, we argue as follows.
Suppose that in the construction of the Γ-sequence,

$\mathfrak{p}^i$ $(= \Xi^i : \varphi^i$, say) is the first sequent among
the sequents $\mathfrak{p}_1, \ldots, \mathfrak{p}_n$that is undermined by Γ_i

—so that

$$\Xi^i \subseteq \Gamma_i \tag{17.1}$$

and

$$\varphi^i \notin \Gamma_i, \tag{17.2}$$

—and in accordance with the definition we set

$$\Gamma_{i+1} = \Gamma_i \cup \{\varphi^i\}. \tag{17.3}$$

---

[7] We claim here some improvement on Gentzen's terse claim (p. 336)

> Offenbar tritt nach endlich vielen Hinzufügungen der Fall ein, daß der letzte Komplex $[\Gamma^{\vec{\mathfrak{P}}}]$ allen $\mathfrak{p}$ genügt. Denn die $\mathfrak{p}$ haben nur endlich viele Elemente, und der Komplex aller dieser Elemente genügt sicher jedem $\mathfrak{p}$. (Clearly after finitely many additions we reach a stage at which the last complex $[\Gamma^{\vec{\mathfrak{P}}}]$ confirms every $\mathfrak{p}$. For the $\mathfrak{p}$ have only finitely many elements, and the complex of all these elements certainly confirms each $\mathfrak{p}$.)

Gentzen's reasoning here is actually fallacious. This is because at each stage of construction of the Γ sets one can add only the *succedent* of some $\mathfrak{p}_i$. So there is no guarantee that one would ever use up *all* of the elements in the sequents $\mathfrak{p}_1, \ldots, \mathfrak{p}_n$. Gentzen should have limited his remark by saying that the set of all *succedents* of $\mathfrak{p}_1, \ldots, \mathfrak{p}_n$ confirms all of $\mathfrak{p}_1, \ldots, \mathfrak{p}_n$. He could have done this by replacing the first occurrence of 'Elemente' in the last quote with the plural of 'Sukzedens'.

Now suppose for *reductio* that Γ_{i+1} undermines some $(\psi, \mathfrak{P})$-sequent $\mathfrak{s}$, say. By the definition of $(\psi, \mathfrak{P})$-sequent, $\mathfrak{s}$ is of the form $\Delta : \psi$.

There are now two cases to consider: either

(i) $\varphi^i \in \Delta$, or
(ii) $\varphi^i \notin \Delta$.

Ad (i): $\varphi^i \in \Delta$. Let Δ' be $\Delta \backslash \varphi^i$. So $\mathfrak{s}$ is of the form $\Delta', \varphi^i : \psi$. Moreover, since $\mathfrak{s}$ is a $(\psi, \mathfrak{P})$-sequent, it has a super-normal proof, Π say, from (finitely many) members of $\mathfrak{P}$:

$$\begin{array}{c}\mathfrak{P}\\ \Pi\\ \Delta', \varphi^i : \psi\ (= \mathfrak{s})\end{array}$$

We are supposing for *reductio* that Γ_{i+1} undermines $\Delta', \varphi^i : \psi$. It follows from this supposition that $(\Delta', \varphi^i) \subseteq \Gamma_{i+1}$. By (17.3), $\Gamma_{i+1} = \Gamma^i \cup \{\varphi^i\}$, where by (17.2) $\varphi^i \notin \Gamma_i$. So we can conclude that

$$\Delta' \subseteq \Gamma_i. \tag{17.4}$$

Consider now the super-normal proof

$$\Sigma \quad = \quad \frac{\Xi^i : \varphi^i\ (= \mathfrak{p}^i) \qquad \begin{array}{c}\mathfrak{P}\\ \Pi\\ \Delta', \varphi^i : \psi\ (= \mathfrak{s})\end{array}}{\Xi^i, \Delta' : \psi}$$

Since $\mathfrak{p}^i \in \mathfrak{P}$, Σ is a super-normal proof of the sequent $\Xi^i, \Delta' : \psi$ from $\mathfrak{P}$. So $\Xi^i, \Delta' : \psi$ is a $(\psi, \mathfrak{P})$-sequent. By (17.1) we have $\Xi^i \subseteq \Gamma_i$; by (17.4) we have $\Delta' \subseteq \Gamma_i$; and we are assuming for this *reductio* that Γ_{i+1} undermines $\Delta', \varphi^i : \psi$, whence $\psi \notin \Gamma_{i+1}$, whence in turn $\psi \notin \Gamma_i$. Hence Γ_i undermines the $(\psi, \mathfrak{P})$-sequent $\Xi^i, \Delta' : \psi$. This contradicts IH.

The picture in case (i) is this:

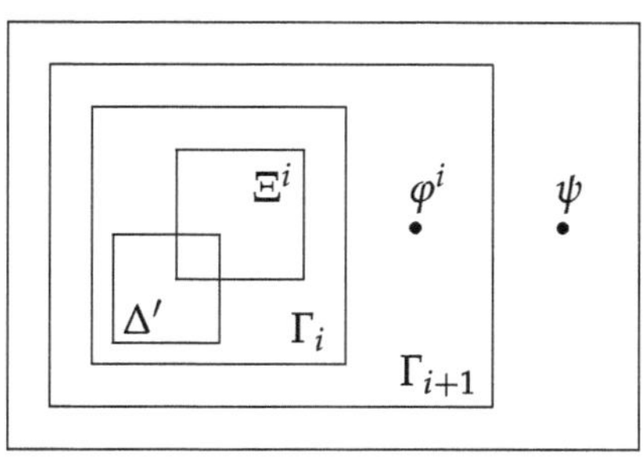

$\psi \notin \Gamma_{i+1}$
$\Delta = \Delta' \cup \{\varphi^i\}$
$\Gamma_{i+1} = \Gamma_i \cup \{\varphi^i\}$

Ad (ii): $\varphi^i \notin \Delta$. We are supposing that Γ_{i+1} undermines $\Delta : \psi$. It follows from this supposition that $\Delta \subseteq \Gamma_{i+1}$ and $\psi \notin \Gamma_{i+1}$. By (17.3), $\Gamma_{i+1} = \Gamma_i \cup \{\varphi^i\}$, where $\varphi^i \notin \Gamma_i$. So $\Delta \subseteq \Gamma_i \cup \{\varphi^i\}$. But we are supposing that $\varphi^i \notin \Delta$. Thus $\Delta \subseteq \Gamma_i$. By (17.1) we have $\Xi^i \subseteq \Gamma_i$. So Γ_i undermines the $(\psi, \mathfrak{P})$-sequent $\Xi^i, \Delta : \psi$. Once again this contradicts IH.

The picture in case (ii) is this:

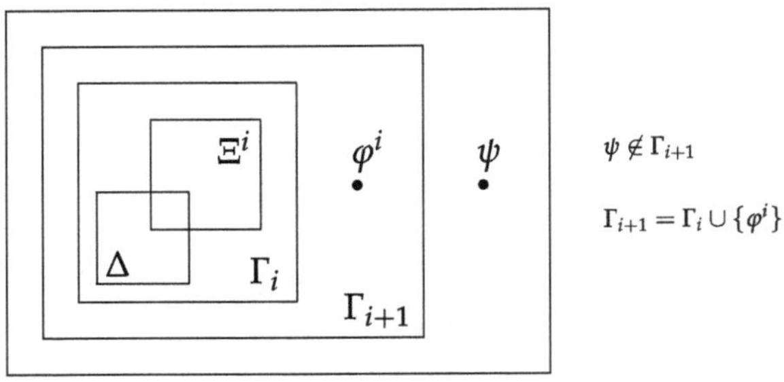

We have now reduced to absurdity the assumption that Γ_{i+1} undermines some $(\psi, \mathfrak{P})$-sequent. We conclude that Γ_{i+1} confirms every $(\psi, \mathfrak{P})$-sequent.

Lemma 17.4 *Suppose* Γ *is* $(\psi, \mathfrak{P})$*-weak. Then each* Γ_k *does not contain* ψ.

Proof By induction.
Basis step. By initial supposition $\mathfrak{q}$ is non-trivial, that is, Γ does not contain ψ. So Γ_1 does not contain ψ.
Inductive Hypothesis. Suppose that $\psi \notin \Gamma_i$.
Inductive Step. Show that $\psi \notin \Gamma_{i+1}$.
If $\Gamma_{i+1} = \Gamma_i$, then by IH we are done.
If $\Gamma_{i+1} \neq \Gamma_i$, we argue as follows.
Suppose that in the construction of the Γ-sequence,

$\mathfrak{p}^i$ $(= \Xi^i : \varphi^i$, say) is the first sequent among the sequents $\mathfrak{p}_1, \ldots, \mathfrak{p}_n$ that is undermined by Γ_i

—so that

$$\Xi^i \subseteq \Gamma_i \tag{17.5}$$

—and in accordance with the definition we set

$$\Gamma_{i+1} = \Gamma_i \cup \{\varphi^i\}. \tag{17.6}$$

By IH, $\psi \notin \Gamma_i$. Hence, given (17.5), we have that

$$\Gamma_i \text{ undermines } \Xi^i : \psi. \tag{17.7}$$

Now suppose for *reductio* that $\psi \in \Gamma_{i+1}$. By (17.6) either $\psi \in \Gamma_i$ or $\varphi^i = \psi$. By IH, $\psi \notin \Gamma_i$. So $\varphi^i = \psi$. Substituting φ^i for ψ in (17.7), we infer that

$$\Gamma_i \text{ undermines } \Xi^i : \varphi^i.$$

But $\Xi^i : \varphi^i$ is $\mathfrak{p}^i$, which, since $\varphi^i = \psi$, is a $(\psi, \mathfrak{P})$-sequent. So Γ_i undermines a $(\psi, \mathfrak{P})$-sequent.

This contradicts Lemma 17.3. So $\psi \notin \Gamma_{i+1}$.

Corollary 17.4 *Suppose* Γ *is* $(\psi, \mathfrak{P})$*-weak. Then* $\psi \notin \Gamma^{\vec{\mathfrak{P}}}$.

Proof Immediate by Lemma 17.4, since $\Gamma^{\vec{\mathfrak{P}}}$ is Γ_{n+1}.

Proof of Theorem 17.2.
Suppose that $\mathfrak{P} \models \Gamma : \psi$. By Corollary 17.3, $\Gamma^{\vec{\mathfrak{P}}}$ confirms $\mathfrak{P}$. Hence $\Gamma^{\vec{\mathfrak{P}}}$ confirms $\Gamma : \psi$. By Observation 17.5, $\Gamma \subseteq \Gamma^{\vec{\mathfrak{P}}}$. Hence $\psi \in \Gamma^{\vec{\mathfrak{P}}}$.

Suppose for *reductio* that Γ is $(\psi, \mathfrak{P})$-weak, i.e. $\neg\exists\Delta(\Delta \subseteq \Gamma \wedge \mathfrak{P} \vdash_S \Delta : \psi)$. By Corollary 17.4, $\psi \notin \Gamma^{\vec{\mathfrak{P}}}$. Contradiction.

So, by *classical reductio*, $\exists\Delta(\Delta \subseteq \Gamma \wedge \mathfrak{P} \vdash_S \Delta : \psi)$. By Corollary 17.1, $\mathfrak{P} \vdash_N \Gamma : \psi$. □

This argument has the formal structure

$$\dfrac{\dfrac{\dfrac{\overset{\text{O5:}}{\Gamma \subseteq \Gamma^{\vec{\mathfrak{P}}}} \qquad \dfrac{\overset{\text{C3:}}{\Gamma^{\vec{\mathfrak{P}}} \text{ confirms } \mathfrak{P}} \qquad \mathfrak{P} \models \Gamma : \psi}{\Gamma^{\vec{\mathfrak{P}}} \text{ confirms } \Gamma : \psi}}{\psi \in \Gamma^{\vec{\mathfrak{P}}}}\,(\text{O3}) \qquad \dfrac{\dfrac{}{\neg\exists\Delta(\Delta \subseteq \Gamma \wedge \mathfrak{P} \vdash_S \Delta : \psi)}\,(1)}{\psi \notin \Gamma^{\vec{\mathfrak{P}}}}\,(\text{C4})}{\dfrac{\dfrac{\bot}{\exists\Delta(\Delta \subseteq \Gamma \wedge \mathfrak{P} \vdash_S \Delta : \psi)}\,(1)}{\mathfrak{P} \vdash_N \Gamma : \psi}\,(\text{C1})}$$

17.3 Generalizing: Sequents Empty on the Left or Right

Observation 17.1 and Theorem 17.2 together yield the fuller completeness result

Theorem 17.3 *If a sequent* $\mathfrak{q}$ *is a consequence of* $\mathfrak{p}_1, \ldots, \mathfrak{p}_n$*, then there is a normal proof of* $\mathfrak{q}$ *from* $\mathfrak{p}_1, \ldots, \mathfrak{p}_n$.

In this statement of completeness, $\mathfrak{q}$ is not required to be non-trivial.

The real interest, however, lies in the case where $\mathfrak{q}$ *is* non-trivial, since (by Observation 17.1) trivial $\mathfrak{q}$ has such trivial proof! Note that even in the statement of Theorem 17.2, with its restriction to non-trivial $\mathfrak{q}$, there is no mention of any corresponding requirement on the premise-sequents $\mathfrak{p}_1, \ldots, \mathfrak{p}_n$. This invites further reflection. Did Gentzen miss some opportunity here? Could he have stated and proved a sharper, more informative result?

In Gentzen's normal proofs, the rule of THINNING is applied only once, if at all, and then only at the terminal step. He *does* leave open, however, the possibility that among the premise-sequents $\mathfrak{p}_1, \ldots, \mathfrak{p}_n$ could be some trivial ones. (His formal definition of proof rules out mention only of *tautologous* sequents, but not trivial ones. And this is true also of normal proofs.)

Yet a trivial premise-sequent really makes no contribution at all to the result supposedly being proved. A trivial sequent is confirmed by *every* set of elements, and undermined by *none.* So, if $\mathfrak{Q}$ is a set of non-trivial sequents, $\mathfrak{P}$ a set of trivial sequents, and $\mathfrak{q}$ a non-trivial sequent, we have

$$\mathfrak{q} \text{ is a consequence of } \mathfrak{P} \cup \mathfrak{Q} \text{ if and only if } \mathfrak{q} \text{ is a consequence of } \mathfrak{Q}.$$

Clearly, in light of this, the theoretical focus should be only on consequences involving sets $\mathfrak{Q}$ of *non-trivial* sequents.

Further reflection on the proof-theoretical side backs this up. There really is no point in ever beginning a proof with an application of a rule of inference to a trivial sequent. This is borne out by the following lemmas. Their combined effect is to guarantee that even without increasing the number of steps in a proof, one never needs a non-tautologous but trivial sequent at a leaf-node of any Gentzen-proof tree.

Lemma 17.5 *Any thinning of a trivial premise can be replaced by a thinning of a tautologous premise:*

$$(\textit{where } \varphi \in \Delta\ :) \qquad \frac{\Delta : \varphi}{\Delta, \Gamma : \varphi} \qquad \mapsto \qquad \frac{\varphi : \varphi}{\Delta, \Gamma : \varphi}$$

Lemma 17.6 *Any cut with trivial left premise can be replaced by a thinning of its right premise:*

$$(\textit{where } \varphi \in \Delta\ :) \qquad \frac{\Delta : \varphi \quad \Gamma, \varphi : \psi}{\Delta, \Gamma : \psi} \qquad \mapsto \qquad \frac{\Gamma, \varphi : \psi}{\Delta, \Gamma : \psi}$$

Lemma 17.7 *Any cut whose right premise is trivial and has the cut element as succedent can be replaced by a thinning of its left premise:*

$$\frac{\Delta : \psi \quad \Gamma, \psi : \psi}{\Delta, \Gamma : \psi} \qquad \mapsto \qquad \frac{\Delta : \psi}{\Delta, \Gamma : \psi}$$

Lemma 17.8 *Any cut whose right premise is trivial and does not have the cut element as succedent has a trivial conclusion, which could just as well have been derived from a tautologous sequent by a thinning:*

$$(where\ \psi \in \Gamma\ :) \quad \frac{\Delta : \varphi \quad \Gamma, \varphi : \psi}{\Delta, \Gamma : \psi} \quad \mapsto \quad \frac{\Gamma, \varphi : \psi}{\Delta, \Gamma : \psi} \quad \mapsto \quad \frac{\psi : \psi}{\Delta, \Gamma : \psi}$$

The upshot is clear: we can limit our attention to Gentzen-proofs of non-trivial sequents from non-trivial sequents (just as we could for the semantic relation of consequence among sequents).

Theorem 17.2 of course still holds under this limitation to non-trivial premise-sequences. Now, what is most remarkable about Theorem 17.2 is the very constrained form of the normal proofs that it affords for consequences among sequents. As noted earlier, in a normal proof the rule of THINNING is applied only once, if at all, and then only at the terminal step.

Hence the *penultimate* sequent of a normal proof is the strongest statement (of consequence) that one can glean from the proof. And it can be obtained (perhaps not surprisingly, given that THINNING can only weaken a claim of consequence) by means of cuts alone, making up what we called a *super-normal* proof.

CUT is the crucible in which optimally strong statements of consequence among (non-trivial) sequents can be forged. In the limited context of this investigation by Gentzen, however, the *elements* of these (non-trivial) sequents behave like propositional variables. They are assumed to be able to take their semantic values quite independently of each other. The underlying thought appears to be that logical relations among elements of sequents (such as, say, contrariety or mutual inconsistency) would need to be considered only when logical operators (such as negation) are introduced. At that point, sequents would no longer consist of unstructured *elements* on the right and on the left, but would consist, rather, of *sentences* in the new formal language that provides not only for atomic sentences but also for logically structured ones.

This postponement to complex languages of possible consideration of logical relations is, however, theoretically short-sighted. We need to examine what might happen with the structural relations of consequence and deducibility among sequents if we allow for the possibility that logical relations are entered into, and logical properties enjoyed, by even the *unstructured elements* that we have been considering thus far. Naturally this calls for a conception of the elements as corresponding more to propositional constants than to propositional variables. They will be behaving more like *propositional variables under an interpretation*. But—and this is the crucial feature—such behavior will not be the result of such logical form as would be bestowed by logical operators occurring within them. For, ex hypothesi, they will be operator-free.

We are arguing here, in effect, for consideration of what modern proof-theorists call *atomic rules of inference*. Take, for example, the two atomic sentences 'this is red' (ρ) and 'this is colored' (γ). The meaning-connection between them is that the former entails the latter. This can be registered by the sequent

$$\rho : \gamma.$$

One might think *'Well and good; so, let's just allow this sequent $\rho : \gamma$ always to be available for use as a (non-trivial) premise-sequent within Gentzen-proofs. What's the problem?'*

The problem becomes apparent only when we consider two atomic sentences such as 'this is red' (ρ) and 'this is blue' (β). These are *contraries*. They cannot be true together. Modern proof-theorists have at their disposal the following sequent-expression of contrariety:

$$\rho, \beta : \emptyset,$$

or, more simply,

$$\rho, \beta : \quad .$$

This is a sequent with *empty succedent*. We may find it convenient to write

$$\rho, \beta : \quad \bot,$$

in order to have a symbol that will both emphasize the fact that the succedent is empty, and remind one of the semantical significance of that fact.

Gentzen, however, in his 1932 study, *did not allow for sequents with empty succedent*, i.e. for explicit statements to the effect that the antecedent in question is unsatisfiable (or inconsistent).[8]

The reader will see also that Gentzen *did not allow for sequents with empty antecedent*.[9] Thus he would have been unable to express the fact that certain atomic sentences are *logically true*, such as '$0 = 0$'. The sequent that expresses this is

$$: 0 = 0.$$

---

[8] See Definition 17.1 above, which is faithful to Gentzen's text at p. 330:

Ein *Satz* hat die Form

$$u_1 u_2 \ldots u_\nu \rightarrow v \qquad (\nu \geqq 1).$$

Die u und v heißen *Elemente*.

(A *sequent* has the form

$$u_1 u_2 \ldots u_\nu \rightarrow v \qquad (\nu \geqq 1).$$

The u and v are called *elements*.)

[9] The closest that Gentzen could have come to expressing that λ is logically true would be to have available as a premise each and every non-trivial sequent of the form $\Delta : \lambda$. If, however, there are only finitely many sets Δ, this falls short of saying that λ is logically true. Indeed, even if there are *infinitely* many sets Δ, this would still fall short of saying that λ is logically true. For the language could be extended by new elements not yet involved in any of these (infinitely many) sets Δ.

One might legitimately wish to express the polar opposite, by being able to say that the atomic sentence '$0=1$' is *logically false*. The sequent that expresses this is

$$0 = 1 : \quad .$$

The temptation is clear, and irresistible: the 'elemental sequent-theorist' should be able to provide a treatment that allows for these expressive possibilities involving 'logically unstructured' sentences, even without considering any logical operators such as connectives and quantifiers.

Suppose then that we modify Definition 17.1 as follows.

Definition 17.13 *Extended sequents* are of the form $\Delta : \psi$ or $\Delta : \bot$, where Δ is a (possibly empty) finite set of elements and ψ is an element.

If we now had to consider *extended* sequents instead of Gentzen's original sequents (i.e., those of just the first of the three permitted forms above), what would happen to Gentzen's main results proved above?

First we would have to inquire after the forms that might now be taken by the rules of THINNING and CUT. The temptation would be strong to have THINNING cover inferences of the following form:

$$\frac{\Delta :}{\Delta : \psi}$$

Likewise, one would be tempted to allow CUT to apply when the right premise has empty succedent so as to yield a conclusion with empty succedent[10]:

$$\frac{\Gamma : \varphi \quad \Delta, \varphi :}{\Gamma, \Delta :}$$

We now seek to go beyond Gentzen's results in his 1932 paper. Fortunately we can get by with minor adaptations of our exposition of his completeness proof above, by yielding to these two temptations.

Theorem 17.4 *If a non-trivial extended sequent* $\mathfrak{q}$ *is a consequence of non-trivial extended sequents* $\mathfrak{p}_1, \ldots, \mathfrak{p}_n$, *then there is a normal proof of* $\mathfrak{q}$ *from* $\mathfrak{p}_1, \ldots, \mathfrak{p}_n$.

Definition 17.14 We liberalize the definition of *super-normal* proofs so as to allow them to have a terminal step of THINNING *on the right*.

We liberalize also the notion of $(\sigma, \mathfrak{P})$-sequent, so that it is understood by reference to the preceding more liberal notion of super-normal proof.

Definition 17.15 We define $(\sigma, \mathfrak{P})$*-sequents* (where σ is either $\bot$ or ψ) to be sequents $\mathfrak{s}$ such that

[10] As Peter Schroeder-Heister has pointed out (personal correspondence), this cut rule with empty succedent is the SLD resolution rule used in logic programming (in the propositional case, i.e. without substitution).

σ is the succedent of $\mathfrak{s}$ and there is a super-normal proof of $\mathfrak{s}$ from sequents in the set $\mathfrak{P}$.

Our proposed generalization to extended sequents brings with it a need to revisit the notion of a set of elements *undermining* or *confirming* a sequent. This is because we can now have empty antecedents or succedents. (Remember that we use $\perp$ to indicate an empty succedent.)

Our earlier definition (on Gentzen's behalf) of the notion of *undermining* (Definition 17.9), though speaking blandly of membership of an arbitrary set (the set that does the undermining), really turned on the idea that undermining is a matter of *making all the elements on the left* **true**, while *making (all) the element(s) on the right* **false**. If we had thought of the succedent as a *set*, it would have been a singleton (since, as we have pointed out, Gentzen did not allow for empty succedents). So, even at that early stage, one could have symmetrized the expression of the definition of undermining, so that it could have read

> Θ undermines $\Delta : \sigma$ if and only if Θ contains every element of Δ and Θ lacks every element of σ.

(Clearly, the motivating idea here is that Θ is the set of truths on some interpretation.) With extended sequents, we can now have empty succedents σ. This symmetrized definition of undermining will now serve our purposes perfectly.

Definition 17.16 Suppose $\Delta : \sigma$ is an extended sequent. Then Θ undermines $\Delta : \sigma$ if and only if $\Delta \subseteq \Theta$ and $\Theta \cap \sigma = \emptyset$.

Observation 17.6 *A (finite) set Θ of elements undermines an extended sequent $\mathfrak{s}$ if and only if:*

1. *$\mathfrak{s}$ is of the form $\Delta : \varphi$, and $\Delta \subseteq \Theta$ but $\varphi \notin \Theta$; or*
2. *$\mathfrak{s}$ is of the form $\Delta : \perp$, and $\Delta \subseteq \Theta$.*

Definition 17.17 A (finite) set of elements is said to *confirm* a sequent $\Delta : \varphi$ just in case it does not undermine $\Delta : \varphi$.

Observation 17.7 *Suppose $\Gamma \subseteq \Theta$ and Θ confirms $\Gamma : \psi$. Then $\psi \in \Theta$.*

Three new possibilities arise for extended sequents, which did not obtain for Gentzen's sequents.

First new possibility. Given certain set $\mathfrak{P}$ of extended sequents as premises, it can be shown that a set Θ of elements is incoherent, in the sense that there is a (normal) proof, from $\mathfrak{P}$, of the extended sequent $\Theta : \perp$. By way of illustration, take for $\mathfrak{P}$ the set consisting of just the extended sequent $\theta : \perp$, and for the proof take just that sequent on its own! This example is of course very degenerate, but serves our purposes. A less degenerate example would be the following.

Example Consider the extended sequents

$$a : b \ ; \ a : c \ ; \ b, c : \bot .$$

The (normal) proof

$$\dfrac{a : c \qquad \dfrac{a : b \quad b, c : \bot}{a, c : \bot}}{a : \bot}$$

shows that $\{a\}$ is inconsistent.

Second new possibility. Definition 17.13 of extended sequents allows the *empty* sequent $\emptyset : \emptyset$ (or $\emptyset : \bot$) to count as an extended sequent. The empty sequent is undermined by every set of elements, hence confirmed by none. Some collections of extended sequents allow one to construct a (normal) proof of the empty sequent. The simplest example of this would be the premise-collection

$$\emptyset : a \ ; \ a : \bot ,$$

used in the proof

$$\dfrac{\emptyset : a \qquad a : \bot}{\emptyset : \bot}$$

Such a proof shows that the premise-collection (of *sequents*) is *incoherent*. It is impossible for any set of elements to confirm both its members.

In the restricted context of Gentzen's sequents, which could not be empty on the left or on the right, no premise-set of sequents allows one to deduce the empty sequent as a conclusion. For every sequent at a leaf-node in a Gentzen proof-tree (even if tautologous) has at least one element on the left, and at least one element on the right. And each rule (THINNING or CUT) preserves that property from its premise(s) to its conclusion. So, by induction, the conclusion of any Gentzen sequent-proof has at least one element on the left, and at least one element on the right.

Not so, however, for sequent-proofs involving extended sequents, as is seen from our last example.

It is important that one's sequent calculus be able to reveal the incoherence of a set of premise-sequents that indeed cannot have all its members confirmed by any set of elements. This in effect becomes a new requirement of completeness on a sequent calculus.

Our first new possibility above concerned the 'incoherence' of certain sets Θ of elements being demonstrable on the basis of a set $\mathfrak{P}$ of extended sequents. The demonstration consists in a deduction of the sequent $\Theta : \bot$ from $\mathfrak{P}$.

Third new possibility. The third new possibility that arises for extended sequents concerns the other logical extreme. It is now possible to demonstrate, on the basis of a set $\mathfrak{P}$ of extended sequents, that a certain element θ 'must be true'. Such a demonstration consists in a deduction of the sequent $\emptyset : \theta$ from $\mathfrak{P}$. By way of illustration, take for $\mathfrak{P}$ the set consisting of just the extended sequent $\emptyset : \theta$, and for the proof take just that sequent on its own! This example is of course very degenerate, but serves our purposes. A less degenerate example would be the following.

Example $\mathfrak{P}$ is the set consisting of just the extended sequents

$$\emptyset : a \quad ; \quad \emptyset : b \quad ; \quad a, b : \theta$$

and the (normal) proof showing that θ 'must be true' (given the foregoing sequents) is

$$\frac{\emptyset : b \qquad \dfrac{\emptyset : a \quad a, b : \theta}{b : \theta}}{\emptyset : \theta}$$

Observation 17.8 *Suppose* Γ *is* $(\psi, \mathfrak{P})$*-weak. Then* Γ *does not undermine any* $(\psi, \mathfrak{P})$*-sequent. Hence,* Γ *confirms each* $(\psi, \mathfrak{P})$*-sequent.*

Definition 17.18 We say that $\mathfrak{q}$ is a *consequence of* $\mathfrak{P} = \{\mathfrak{p}_1, \ldots, \mathfrak{p}_n\}$ (abbreviated: $\mathfrak{P} \models \mathfrak{q}$), where all the sequents involved are *extended* sequents, just in case any set of elements involved in the sequents $\mathfrak{p}_1, \ldots, \mathfrak{p}_n, \mathfrak{q}$ that confirms $\mathfrak{p}_1, \ldots, \mathfrak{p}_n$ confirms $\mathfrak{q}$. Equivalently: ...just in case no set of elements involved in the sequents $\mathfrak{p}_1, \ldots, \mathfrak{p}_n, \mathfrak{q}$ confirms $\mathfrak{p}_1, \ldots, \mathfrak{p}_n$ but undermines $\mathfrak{q}$.

We are considering the $\mathfrak{p}_i$ ($= \Xi_i : \sigma_i$, say) given in some fixed order. We are calling the collection $\vec{\mathfrak{P}}$, in order to emphasize the underlying ordering. In each case the succedent σ_i is either $\bot$ or some element φ_i.

Definition 17.19 The finite sequence

$$\Gamma = \Gamma_1 \subseteq \Gamma_2 \subseteq \ldots \subseteq \Gamma_{n+1} = \Gamma^{\vec{\mathfrak{P}}}$$

of sets of elements, whose last member $\Gamma^{\vec{\mathfrak{P}}}$ is called the $\vec{\mathfrak{P}}$*-completion of* Γ, is defined inductively. Let Γ_i be in hand. Γ_{i+1} is then constructed in the trichotomous cases $(\alpha.1)$, $(\alpha.2)$ and (β) as follows.

(α) For some k Γ_i undermines $\mathfrak{p}_k$. Let j be the least such k.

(1) $\mathfrak{p}_j$ is of the form $\Xi_j : \varphi_j$. Set $\Gamma_{i+1} = \Gamma_i \cup \{\varphi_j\}$.
(2) $\mathfrak{p}_j$ is of the form $\Xi_j : \bot$. Set $\Gamma_{i+1} = \Gamma_i$.

(β) For no k does Γ_i undermine $\mathfrak{p}_k$. Set $\Gamma_{i+1} = \Gamma_i$.

Observation 17.9 $\Gamma \subseteq \Gamma^{\mathfrak{P}}$.

Lemma 17.9 *(for extended sequents). Suppose* Γ *is* $(\psi, \mathfrak{P})$*-weak. Then each* Γ_k *confirms every* $(\psi, \mathfrak{P})$*-sequent.*

Proof Exactly as for Lemma 17.3, except that in the case where $\Gamma_{i+1} \neq \Gamma_i$, the premise-sequent $\mathfrak{p}^i$ ($= \Xi^i : \varphi^i$, say) is chosen to be the first sequent among the sequents $\mathfrak{p}_1, \ldots, \mathfrak{p}_n$ *that does not have succedent* $\perp$ and is undermined by Γ_i. From that point on, the proof goes through unchanged.

Lemma 17.10 *(for extended sequents). Suppose* Γ *is* $(\perp, \mathfrak{P})$*-weak. Then each* Γ_k *confirms every* $(\perp, \mathfrak{P})$*-sequent.*

Proof By induction.
Basis step. By Observation 17.8, Γ ($= \Gamma_1$) confirms every $(\perp, \mathfrak{P})$-sequent.
Inductive Hypothesis. Suppose that Γ_i confirms every $(\perp, \mathfrak{P})$-sequent.
Inductive Step. Show that Γ_{i+1} confirms every $(\perp, \mathfrak{P})$-sequent.
If $\Gamma_{i+1} = \Gamma_i$, then by IH we are done.
If $\Gamma_{i+1} \neq \Gamma_i$, we argue as follows.
Suppose that in the construction of the Γ-sequence,

> $\mathfrak{p}^i$ ($= \Xi^i : \varphi^i$, say) is the first sequent among the sequents $\mathfrak{p}_1, \ldots, \mathfrak{p}_n$ that does not have succedent $\perp$ and is undermined by Γ_i

—so that

$$\Xi^i \subseteq \Gamma_i \tag{17.8}$$

and

$$\varphi^i \notin \Gamma_i, \tag{17.9}$$

—and in accordance with the definition we set

$$\Gamma_{i+1} = \Gamma_i \cup \{\varphi^i\}. \tag{17.10}$$

Now suppose for *reductio* that Γ_{i+1} undermines some $(\perp, \mathfrak{P})$-sequent $\mathfrak{s}$, say. There are now two cases to consider:

(i) $\varphi^i \in \Delta$; or
(ii) $\varphi^i \notin \Delta$.

Ad (i): $\varphi^i \in \Delta$. Let Δ' be $\Delta \setminus \varphi^i$. Since $\mathfrak{s}$ is a $(\perp, \mathfrak{P})$-sequent, it has a super-normal proof, Π say, from (finitely many) members of $\mathfrak{P}$:

$$\begin{array}{c} \mathfrak{P} \\ \Pi \\ \Delta', \varphi^i : \perp \; (= \mathfrak{s}) \end{array}$$

We are supposing for *reductio* that Γ_{i+1} undermines $\Delta', \varphi^i : \perp$. It follows from this supposition that $(\Delta', \varphi^i) \subseteq \Gamma_{i+1}$. By (17.10), $\Gamma_{i+1} = \Gamma^i \cup \{\varphi^i\}$, where by (17.9) $\varphi^i \notin \Gamma_i$. So we can conclude that

$$\Delta' \subseteq \Gamma_i. \tag{17.11}$$

Consider now the super-normal proof

$$\Sigma \quad = \quad \cfrac{\Xi^i : \varphi^i \;(= \mathfrak{p}^i) \qquad \begin{array}{c}\mathfrak{P}\\ \Pi\\ \Delta', \varphi^i : \bot \;(= \mathfrak{s})\end{array}}{\Xi^i, \Delta' : \bot}$$

Since $\mathfrak{p}^i \in \mathfrak{P}$, Σ is a super-normal proof of the sequent $\Xi^i, \Delta' : \bot$ from $\mathfrak{p}_1, \ldots, \mathfrak{p}_n$. So $\Xi^i, \Delta' : \bot$ is a $(\bot, \mathfrak{P})$-sequent. By (17.8) we have $\Xi^i \subseteq \Gamma_i$; by (17.11) we have $\Delta' \subseteq \Gamma_i$. Hence Γ_i undermines the $(\bot, \mathfrak{P})$-sequent $\Xi^i, \Delta' : \bot$. This contradicts IH.

The picture in case (i) is this:

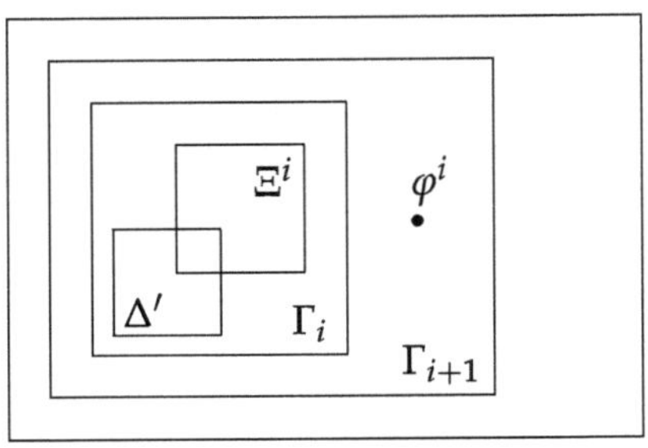

$\Delta = \Delta' \cup \{\varphi^i\}$
$\Gamma_{i+1} = \Gamma_i \cup \{\varphi^i\}$

Ad (ii): $\varphi^i \notin \Delta$. We are supposing that Γ_{i+1} undermines $\Delta : \bot$. It follows from this supposition that $\Delta \subseteq \Gamma_{i+1}$. By (17.10), $\Gamma_{i+1} = \Gamma_i \cup \{\varphi^i\}$, where $\varphi^i \notin \Gamma_i$. So $\Delta \subseteq \Gamma_i \cup \{\varphi^i\}$. But we are supposing that $\varphi^i \notin \Delta$. Thus $\Delta \subseteq \Gamma_i$. By (17.8) we have $\Xi^i \subseteq \Gamma_i$. So Γ_i undermines the $(\bot, \mathfrak{P})$-sequent $\Xi^i, \Delta : \bot$. Once again this contradicts IH.

The picture in case (ii) is this:

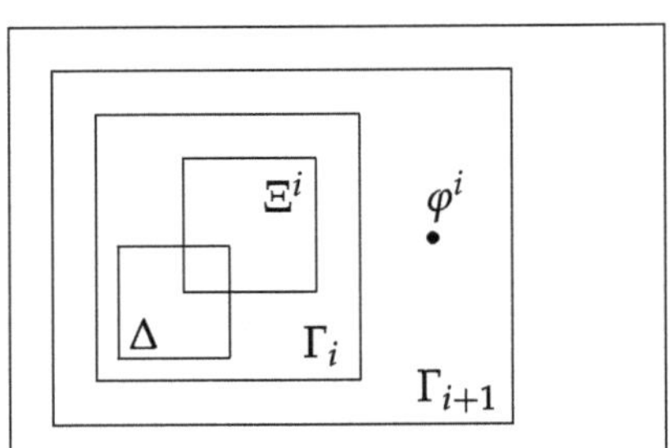

$\Gamma_{i+1} = \Gamma_i \cup \{\varphi^i\}$

We have now reduced to absurdity the assumption that Γ_{i+1} undermines some $(\bot, \mathfrak{P})$-sequent. We conclude that Γ_{i+1} confirms every $(\bot, \mathfrak{P})$-sequent.

Lemma 17.11 *Suppose Γ is $(\sigma, \mathfrak{P})$-weak. Then each Γ_k does not contain ψ, if $\sigma = \psi$.*

Proof By induction.

Basis step. Suppose $\sigma = \psi$. By initial supposition $\mathfrak{q}$ is non-trivial, that is, Γ does not contain ψ. So Γ_1 does not contain ψ, if $\sigma = \psi$.
Inductive Hypothesis. Suppose that $\psi \notin \Gamma_i$, if $\sigma = \psi$.
Inductive Step. Show that $\psi \notin \Gamma_{i+1}$, if $\sigma = \psi$.
If $\Gamma_{i+1} = \Gamma_i$, then by IH we are done.
If $\Gamma_{i+1} \neq \Gamma_i$, we argue as follows.
Suppose that in the construction of the Γ-sequence,

> $\mathfrak{p}^i$ ($= \Xi^i : \varphi^i$, say) is the first sequent among the sequents $\mathfrak{p}_1, \ldots, \mathfrak{p}_n$ that does not have succedent $\perp$ and is undermined by Γ_i

—so that

$$\Xi^i \subseteq \Gamma_i \tag{17.12}$$

—and in accordance with the definition we set

$$\Gamma_{i+1} = \Gamma_i \cup \{\varphi^i\}. \tag{17.13}$$

By IH, $\psi \notin \Gamma_i$, if $\sigma = \psi$. Hence, given (17.12), we have that

$$\Gamma_i \text{ undermines } \Xi^i : \sigma. \tag{17.14}$$

Suppose $\sigma = \psi$. Now suppose for *reductio* that $\psi \in \Gamma_{i+1}$. By (17.13) either $\psi \in \Gamma_i$ or $\varphi^i = \psi$. By IH, $\psi \notin \Gamma_i$. So $\varphi^i = \psi$. Hence $\varphi^i = \sigma$. Substituting φ^i for σ in (17.14), we infer that

$$\Gamma_i \text{ undermines } \Xi^i : \varphi^i.$$

But $\Xi^i : \varphi^i$ is $\mathfrak{p}^i$, which, since $\varphi^i = \psi$, is a $(\psi, \mathfrak{P})$-sequent. So Γ_i undermines a $(\psi, \mathfrak{P})$-sequent.
This contradicts Lemma 17.9. So $\psi \notin \Gamma_{i+1}$.
All this was on the supposition that $\sigma = \psi$. So $\psi \notin \Gamma_{i+1}$ if $\sigma = \psi$.

Lemma 17.12 *Suppose Γ is $(\sigma, \mathfrak{P})$-weak. Then for each $i \geq 1$, there are at least i distinct sequents $\mathfrak{p}_{r_1}, \ldots, \mathfrak{p}_{r_i}$ among the $\mathfrak{p}_1, \ldots, \mathfrak{p}_n$ such that Γ_{i+1} confirms each of $\mathfrak{p}_{r_1}, \ldots, \mathfrak{p}_{r_i}$; and, if Γ_{i+1} undermines at least one of $\mathfrak{p}_1, \ldots, \mathfrak{p}_n$, then any set extending Γ_{i+1} by adding succedents of the sequents $\mathfrak{p}_1, \ldots, \mathfrak{p}_n$ confirms each of $\mathfrak{p}_{r_1}, \ldots, \mathfrak{p}_{r_i}$.*

Basis ($i = 1$). We need to show that

> there is at least one sequent $\mathfrak{p}_{r_1}$ among the $\mathfrak{p}_1, \ldots, \mathfrak{p}_n$ such that Γ_2 confirms $\mathfrak{p}_{r_1}$; and, if Γ_2 undermines at least one of $\mathfrak{p}_1, \ldots, \mathfrak{p}_n$, then any set extending Γ_2 by adding succedents of the sequents $\mathfrak{p}_1, \ldots, \mathfrak{p}_n$ confirms $\mathfrak{p}_{r_1}$.

Now we reason by the trichotomous cases (α.1), (α.2) and (β) for $i = 1$.

(α.1) For some k Γ_1 undermines $\mathfrak{p}_k$; j is the least such k; and $\mathfrak{p}_j$ is of the form $\Xi_j : \varphi_j$.

By definition $\Gamma_2 = \Gamma_1 \cup \{\varphi_j\}$. Hence Γ_2 (and any set extending Γ_2) confirms $\Xi_j : \varphi_j$. So Γ_2 confirms at least one of the sequents $\mathfrak{p}_1, \ldots, \mathfrak{p}_n$; and, if Γ_2 undermines at least one of $\mathfrak{p}_1, \ldots, \mathfrak{p}_n$, then any set extending Γ_2 by adding succedents of the sequents $\mathfrak{p}_1, \ldots, \mathfrak{p}_n$ confirms $\Xi_j : \varphi_j$.

(α.2) For some k Γ_1 undermines $\mathfrak{p}_k$; j is the least such k; and $\mathfrak{p}_j$ is of the form $\Xi_j : \bot$. Remember that $\Gamma_1 = \Gamma$. So Γ undermines $\mathfrak{p}_j$ ($= \Xi_j : \bot$). Hence $\Xi_j \subseteq \Gamma$. Moreover $\Xi_j : \sigma$ is a $(\sigma, \mathfrak{P})$-sequent. This is immediate if $\sigma = \bot$. But if $\sigma = \psi$, the proof

$$\frac{\Xi_j : \bot}{\Xi_j : \psi},$$

which is super-normal (by Definition 17.14), shows that $\Xi_j : \sigma$ is a $(\sigma, \mathfrak{P})$-sequent. This contradicts the main supposition that Γ is $(\sigma, \mathfrak{P})$-weak—i.e., that *no* $(\sigma, \mathfrak{P})$-sequent has its antecedent included in Γ. So this case is impossible.

(β) For no k does Γ_1 undermine $\mathfrak{p}_k$. So Γ_1 confirms $\mathfrak{p}_k$ for every k. Also by definition $\Gamma_2 = \Gamma_1$. Hence Γ_2 confirms $\mathfrak{p}_k$ for every k. So Γ_2 confirms at least one of the sequents $\mathfrak{p}_1, \ldots, \mathfrak{p}_n$; and, if Γ_2 undermines at least one of $\mathfrak{p}_1, \ldots, \mathfrak{p}_n$, then any set extending Γ_2 by adding succedents of the sequents $\mathfrak{p}_1, \ldots, \mathfrak{p}_n$ confirms that sequent. (The last conjunct is vacuously true, since its antecedent, ex hypothesi, is false.)

Inductive Hypothesis. There are at least i sequents $\mathfrak{p}_{r_1}, \ldots, \mathfrak{p}_{r_i}$ among the $\mathfrak{p}_1, \ldots, \mathfrak{p}_n$ such that Γ_{i+1} confirms each of $\mathfrak{p}_{r_1}, \ldots, \mathfrak{p}_{r_i}$; and, if Γ_{i+1} undermines at least one of $\mathfrak{p}_1, \ldots, \mathfrak{p}_n$, then any set extending Γ_{i+1} by adding succedents of the sequents $\mathfrak{p}_1, \ldots, \mathfrak{p}_n$ confirms each of $\mathfrak{p}_{r_1}, \ldots, \mathfrak{p}_{r_i}$.

Inductive Step. We need to show that

> there are at least $i + 1$ sequents $\mathfrak{p}_{r_1}, \ldots, \mathfrak{p}_{r_{i+1}}$ among the $\mathfrak{p}_1, \ldots, \mathfrak{p}_n$ such that Γ_{i+2} confirms each of $\mathfrak{p}_{r_1}, \ldots, \mathfrak{p}_{r_{i+1}}$; and, if Γ_{i+2} undermines at least one of $\mathfrak{p}_1, \ldots, \mathfrak{p}_n$, then any set extending Γ_{i+2} by adding succedents of the sequents $\mathfrak{p}_1, \ldots, \mathfrak{p}_n$ confirms each of $\mathfrak{p}_{r_1}, \ldots, \mathfrak{p}_{r_{i+1}}$.

We reason again by the trichotomous cases (α.1), (α.2) and (β).

(α.1) For some k Γ_{i+1} undermines $\mathfrak{p}_k$; j is the least such k; and $\mathfrak{p}_j$ is of the form $\Xi_j : \varphi_j$.

By definition $\Gamma_{i+2} = \Gamma_{i+1} \cup \{\varphi_j\}$. So Γ_{i+2} confirms $\Xi_j : \varphi_j$. Let $\mathfrak{p}_{r_1}, \ldots, \mathfrak{p}_{r_i}$ be as in IH. By IH, Γ_{i+2} confirms the sequents $\mathfrak{p}_{r_1}, \ldots, \mathfrak{p}_{r_i}$ that Γ_{i+1} confirms. But $\Xi_j : \varphi_j$ cannot be among these. So Γ_{i+2} confirms the sequents $\mathfrak{p}_{r_1}, \ldots, \mathfrak{p}_{r_i}$, as well as the sequent $\Xi_j : \varphi_j$, which we can now take for $\mathfrak{p}_{r_{i+1}}$.

Moreover, if Γ_{i+2} undermines at least one of $\mathfrak{p}_1, \ldots, \mathfrak{p}_n$, then by IH any set extending Γ_{i+2} (hence extending Γ_{i+1}) by adding succedents of the sequents $\mathfrak{p}_1, \ldots, \mathfrak{p}_n$ confirms all of $\mathfrak{p}_{r_1}, \ldots, \mathfrak{p}_{r_i}$; and also confirms $\mathfrak{p}_{r_{i+1}}$, because the succedent of this sequent is in Γ_{i+2}.

(α.2) For some k, Γ_{i+1} undermines $\mathfrak{p}_k$; j is the least such k ; and $\mathfrak{p}_j$ is of the form $\Xi_j : \bot$.

Suppose that σ is $\bot$, so that our main supposition is to the effect that Γ is $(\bot, \mathfrak{P})$-weak. Then $\mathfrak{p}_j$ $(= \Xi_j : \bot)$ is a $(\bot, \mathfrak{P})$-sequent that is undermined by Γ_{i+1}. This contradicts Lemma 17.10.

Now suppose that σ is ψ, so that our main supposition is to the effect that Γ is $(\psi, \mathfrak{P})$-weak. Then the super-normal proof

$$\frac{\Xi_j : \bot}{\Xi_j : \psi}$$

shows that $\mathfrak{p}_j$ $(= \Xi_j : \psi)$ is a $(\psi, \mathfrak{P})$-sequent. Moreover, by Lemma 17.11, Γ_{i+1} does not contain ψ. So Γ_{i+1} undermines $\Xi_j : \psi$. This contradicts Lemma 17.9. So this case is impossible.

(β) For no k does Γ_{i+1} undermine $\mathfrak{p}_k$.

In this case Γ_{i+1} confirms every one of the sequents $\mathfrak{p}_1, \ldots, \mathfrak{p}_n$. Also by definition $\Gamma_{i+2} = \Gamma_{i+1}$. Hence Γ_{i+2} confirms every one of the sequents $\mathfrak{p}_1, \ldots, \mathfrak{p}_n$. The result follows.

Comment. Note that we have appealed to Lemmas 17.9 and 17.10 in proving Lemma 17.12. This is unlike the situation with Gentzen's original results. When the sequents are not allowed to be empty on the left or the right, the proof of Lemma 17.2 (the analogue of Lemma 17.12) did *not* appeal to Lemma 17.3 (which has split into the two analogues Lemma 17.9 and 17.10 for extended sequents).

Corollary 17.5 *Suppose* Γ *is* $(\sigma, \mathfrak{P})$*-weak. Then* $\Gamma^{\vec{\mathfrak{P}}}$ *confirms* $\mathfrak{p}_j$, $1 \leq j \leq n$.

Proof Immediate by Lemma 17.12, since $\Gamma^{\vec{\mathfrak{P}}}$ is Γ_{n+1}.

Corollary 17.6 *Suppose* Γ *is* $(\psi, \mathfrak{P})$*-weak. Then* $\psi \notin \Gamma^{\vec{\mathfrak{P}}}$.

Proof Immediate by Lemma 17.11, since $\Gamma^{\vec{\mathfrak{P}}}$ is Γ_{n+1}.

Proof of Theorem 17.4.
Suppose that $\mathfrak{P} \models \Gamma : \psi$. By Corollary 17.5, $\Gamma^{\vec{\mathfrak{P}}}$ confirms $\mathfrak{P}$. Hence $\Gamma^{\vec{\mathfrak{P}}}$ confirms $\Gamma : \psi$. By Observation 17.9, $\Gamma \subseteq \Gamma^{\vec{\mathfrak{P}}}$. Hence $\psi \in \Gamma^{\vec{\mathfrak{P}}}$.

Suppose for *reductio* that Γ is $(\psi, \mathfrak{P})$-weak, i.e. $\neg\exists\Delta(\Delta \subseteq \Gamma \wedge \mathfrak{P} \vdash_S \Delta : \psi)$. By Corollary 11, $\psi \notin \Gamma^{\vec{\mathfrak{P}}}$. Contradiction.

So, by *classical reductio*, $\exists\Delta(\Delta \subseteq \Gamma \wedge \mathfrak{P} \vdash_S \Delta : \psi)$. By Corollary 17.1, $\mathfrak{P} \vdash_N \Gamma : \psi$. □

This argument has the formal structure

$$\dfrac{\dfrac{\text{O9: } \Gamma \subseteq \Gamma^{\vec{\mathfrak{P}}} \qquad \dfrac{\dfrac{\dfrac{}{\neg\exists\Delta(\Delta \subseteq \Gamma \wedge \mathfrak{P} \vdash_S \Delta : \psi)}\,(1)}{\Gamma^{\vec{\mathfrak{P}}} \text{ confirms } \mathfrak{P}}\,(\text{C5}) \qquad \mathfrak{P} \models \Gamma : \psi}{\Gamma^{\vec{\mathfrak{P}}} \text{ confirms } \Gamma : \psi}}{\psi \in \Gamma^{\vec{\mathfrak{P}}}}\,(\text{O7}) \qquad \dfrac{\dfrac{}{\neg\exists\Delta(\Delta \subseteq \Gamma \wedge \mathfrak{P} \vdash_S \Delta : \psi)}\,(1)}{\psi \notin \vec{\mathfrak{P}}}\,(\text{C6})}{\dfrac{\dfrac{\bot}{\exists\Delta(\Delta \subseteq \Gamma \wedge \mathfrak{P} \vdash_S \Delta : \psi)}\,(1)}{\mathfrak{P} \vdash_N \Gamma : \psi}\,(\text{C1})}$$

Note here how Corollary 17.5 has as its hypothesis the *reductio* assumption $\neg\exists\Delta(\Delta \subseteq \Gamma \wedge \mathfrak{P} \vdash_S \Delta : \psi)$. This has been occasioned by the need to accommodate extended sequents. In our regimentation of Gentzen's simpler completeness result in Sect. 17.2, the analogue of Corollary 17.5, namely Corollary 17.3, did not need $\neg\exists\Delta(\Delta \subseteq \Gamma \wedge \mathfrak{P} \vdash_S \Delta : \psi)$ as an hypothesis.

17.4 Generalizing: The Infinite Case

Thus far we have been considering only finite sets of premise-sequents. But the infinite case merits attention.[11] We shall consider infinitely many premise-sequents $\mathfrak{p}_i$ $(= \Xi_i : \sigma_i$, say) given in some fixed order. We shall call the collection $\vec{\mathfrak{P}}$, so as to emphasize its underlying ordering. In each $\mathfrak{p}_i$ the succedent σ_i is either $\bot$ or some element φ_i. Moreover, each antecedent Ξ_i is still finite. The conclusion-sequent $\mathfrak{q}$ likewise has a finite antecedent, which is called Γ.

Definition 17.20 Let $\mathfrak{P}$ be any set of extended sequents, possibly infinite. We say that an extended sequent $\mathfrak{q}$ is a *consequence of* $\mathfrak{P}$ just in case any set of elements involved in sequents in $\mathfrak{P}$ or involved in $\mathfrak{q}$ that confirms every sequent in $\mathfrak{P}$ confirms $\mathfrak{q}$. Equivalently: …just in case no such set of elements confirms every sequent in $\mathfrak{P}$ but undermines $\mathfrak{q}$.

The construction of $\Gamma^{\vec{\mathfrak{P}}}$ requires an extra degree of delicacy in the infinite case, so as to ensure the desired result that $\Gamma^{\vec{\mathfrak{P}}}$ confirms every sequent in $\vec{\mathfrak{P}}$ (see Lemma 17.17 below). $\vec{\mathfrak{P}}$ is an infinite, ordered set $\{\mathfrak{p}_1, \mathfrak{p}_2, \ldots, \mathfrak{p}_n, \ldots\}$. We shall denote by $\mathfrak{P}_n$ its 'initial segment' $\{\mathfrak{p}_1, \mathfrak{p}_2, \ldots, \mathfrak{p}_n\}$. $\mathfrak{P}_0$ is accordingly $\emptyset$; whence $\Gamma^{\mathfrak{P}_0}$ is $\Gamma^{\emptyset}$. And $\Gamma^{\emptyset}$ is of course Γ itself.

[11] Schroeder-Heister remarks (Schroeder-Heister 2002, p. 247), that 'Structural rules may be looked upon as axiomatizing a consequence relation in Tarski's sense (for the *finite case*, of course).' [Emphasis added—NT.] Here we are concerned to cancel any implicature that this remark may carry to the effect that a consequence relation in Tarski's sense that is defined on *infinite* sets of premise-sequents could not be axiomatized in exactly the same way.

Definition 17.21 (*the $\vec{\mathfrak{P}}$-completion of* Γ, *where* $\vec{\mathfrak{P}}$ *is infinite*) We set

$$\begin{aligned}
\Gamma^0 &= \Gamma^{\mathfrak{P}_0};\\
\Gamma^1 &= (\Gamma^0)^{\mathfrak{P}_1};\\
\Gamma^2 &= (\Gamma^1)^{\mathfrak{P}_2};\\
&\vdots\\
\Gamma^{i+1} &= (\Gamma^i)^{\mathfrak{P}_{i+1}};\\
&\vdots
\end{aligned}$$

Finally, we set

$$\bigcup_i \Gamma^i = \Gamma^{\vec{\mathfrak{P}}}$$

Note that in order to determine Γ^{n+1} from Γ^n, one takes Γ^n as one's 'initial set' for the *earlier* kind of completion-construction, and one carries out the earlier method of construction with respect to the finite ordered set $\{\mathfrak{p}_1, \ldots, \mathfrak{p}_{n+1}\}$, so as to obtain the $\{\mathfrak{p}_1, \ldots, \mathfrak{p}_{n+1}\}$-completion of Γ^n, denoted more succinctly as $(\Gamma^n)^{\mathfrak{P}_{n+1}}$.

This means that the construction at each finite stage is as it was earlier. For given n, and for $1 \le i \le n$, let $(\Gamma^n)_i$ be in hand. $(\Gamma^n)_{i+1}$ is then constructed in the trichotomous cases $(\alpha.1)$, $(\alpha.2)$ and (β) as follows.

(α) $(\Gamma^n)_i$ undermines at least one of $\mathfrak{p}_1, \ldots, \mathfrak{p}_{n+1}$. Let j be the index of the first one.

(1) $\mathfrak{p}_j$ is of the form $\Xi_j : \varphi_j$. Set $(\Gamma^n)_{i+1} = (\Gamma^n)_i \cup \{\varphi_j\}$.
(2) $\mathfrak{p}_j$ is of the form $\Xi_j : \bot$. Set $(\Gamma^n)_{i+1} = (\Gamma^n)_i$.

(β) $(\Gamma^n)_i$ undermines none of $\mathfrak{p}_1, \ldots, \mathfrak{p}_{n+1}$. Set $(\Gamma^n)_{i+1} = (\Gamma^n)_i$.

Observation 17.10

$\Gamma =$

$\Gamma^0 =$

$(\Gamma^0)_1 \subseteq (\Gamma^0)_2 = (\Gamma^0)^{\mathfrak{P}_1} =$

$\Gamma^1 =$

$(\Gamma^1)_1 \subseteq (\Gamma^1)_2 \subseteq (\Gamma^1)_3 = (\Gamma^1)^{\mathfrak{P}_2} =$

Γ^2

$\vdots$

$$\Gamma^n =$$

$$(\Gamma^n)_1 \subseteq (\Gamma^n)_2 \subseteq (\Gamma^n)_3 \subseteq \ldots \subseteq (\Gamma^n)_{n+2} = (\Gamma^n)^{\mathfrak{P}_{n+1}} =$$

$$\Gamma^{n+1}$$
$$\vdots$$
$$\subseteq \Gamma^{\vec{\mathfrak{P}}}.$$

Each subset-chain in Observation 17.10 is produced in accordance with our earlier Definition 17.19, with respect to the initial segment $\{\mathfrak{p}_1, \ldots, \mathfrak{p}_{n+1}\}$ of $\vec{\mathfrak{P}}$.

Lemma 17.13 (for infinitely many extended sequents) *Suppose* Γ *is* $(\psi, \mathfrak{P})$*-weak. Then each* Γ^k *confirms every* $(\psi, \mathfrak{P})$*-sequent.*

Proof Exactly as for Lemma 17.9.

Corollary 17.7 (for infinitely many extended sequents) *Suppose* Γ *is* $(\psi, \mathfrak{P})$*-weak. Then* $\Gamma^{\vec{\mathfrak{P}}}$ *confirms every* $(\psi, \mathfrak{P})$*-sequent.*

Proof Immediate by Lemma 17.13, since $\Gamma^{\vec{\mathfrak{P}}}$ is $\bigcup_i \Gamma^i$.

Lemma 17.14 (for infinitely many extended sequents) *Suppose* Γ *is* $(\bot, \mathfrak{P})$*-weak. Then each* Γ^k *confirms every* $(\bot, \mathfrak{P})$*-sequent.*

Proof Exactly as for Lemma 17.10.

Lemma 17.15 (for infinitely many extended sequents) *Let* σ *be either* $\bot$ *or* ψ*. Suppose* Γ *is* $(\sigma, \mathfrak{P})$*-weak. Then* $\Gamma^{\vec{\mathfrak{P}}}$ *confirms every* $(\sigma, \mathfrak{P})$*-sequent.*

Proof Let $\mathfrak{s}$ be any $(\sigma, \mathfrak{P})$-sequent. Suppose for *reductio* that $\Gamma^{\vec{\mathfrak{P}}}$ (which we defined to be $\bigcup_i \Gamma^i$) undermines $\mathfrak{s}$. Then for some k, Γ^k undermines $\mathfrak{s}$. This contradicts Lemma 17.14 if σ is $\bot$, and contradicts Lemma 17.13 if σ is ψ. Hence $\Gamma^{\vec{\mathfrak{P}}}$ confirms $\mathfrak{s}$. But $\mathfrak{s}$ was an arbitrary $(\sigma, \mathfrak{P})$-sequent. Hence $\Gamma^{\vec{\mathfrak{P}}}$ confirms every $(\sigma, \mathfrak{P})$-sequent.

Lemma 17.16 *Suppose* Γ *is* $(\sigma, \mathfrak{P})$*-weak. Then each* Γ^k *does not contain* ψ*, if* $\sigma = \psi$.

Proof Exactly as for Lemma 17.11.

Corollary 17.8 (for infinitely many extended sequents) *Suppose* Γ *is* $(\psi, \mathfrak{P})$*-weak. Then* $\psi \notin \Gamma^{\vec{\mathfrak{P}}}$.

Proof Suppose for *reductio* that ψ is in $\Gamma^{\vec{\mathfrak{P}}}$. $\Gamma^{\vec{\mathfrak{P}}}$ is $\bigcup_i \Gamma^i$. Thus for some i, Γ^i contains ψ. This contradicts Lemma 17.16. Thus $\psi \notin \Gamma^{\vec{\mathfrak{P}}}$.

Lemma 17.17 *Suppose* Γ *is* $(\sigma, \mathfrak{P})$*-weak. Then for each* $n \geq 0$, *and for* $1 \leq i \leq (n+1)$, *there are at least i distinct sequents* $\mathfrak{p}_{r_1}, \ldots, \mathfrak{p}_{r_i}$ *among* $\mathfrak{p}_1, \ldots, \mathfrak{p}_{n+1}$ *such that* $(\Gamma^n)_{i+1}$ *confirms each of* $\mathfrak{p}_{r_1}, \ldots, \mathfrak{p}_{r_i}$; *and, if* $(\Gamma^n)_{i+1}$ *undermines at least one of* $\mathfrak{p}_1, \ldots, \mathfrak{p}_{n+1}$, *then any set extending* $(\Gamma^n)_{i+1}$ *by adding succedents of the sequents* $\mathfrak{p}_1, \ldots, \mathfrak{p}_{n+1}$ *confirms each of* $\mathfrak{p}_{r_1}, \ldots, \mathfrak{p}_{r_i}$.

Proof For fixed n, the induction on i is as in the proof of Lemma 17.12. The very first basis step, for $n = 0$ and $i = 1$, is obvious. By Corollary 17.5, we thereby obtain, for successive values of n, the result that $(\Gamma^n)^{\vec{\mathfrak{P}}_{n+1}}$ confirms $\mathfrak{p}_j$, $1 \leq j \leq n+1$.

Corollary 17.9 *Suppose* Γ *is* $(\sigma, \mathfrak{P})$*-weak. Then* $\Gamma^{\vec{\mathfrak{P}}}$ *confirms every member of* $\mathfrak{P}$.

Proof Immediate by Lemma 17.17, since $\Gamma^{\vec{\mathfrak{P}}}$ is $\bigcup_i \Gamma^i$.

Theorem 17.5 *If a non-trivial extended sequent* $\mathfrak{q}$ *is a consequence of the set* $\mathfrak{P}$ *of non-trivial extended sequents, then there is a normal proof of* $\mathfrak{q}$ *from sequents in* $\mathfrak{P}$.

Proof of Theorem 17.5.
Suppose that $\mathfrak{P} \models \Gamma : \psi$. By Corollary 17.9, $\Gamma^{\vec{\mathfrak{P}}}$ confirms $\mathfrak{P}$. Hence $\Gamma^{\vec{\mathfrak{P}}}$ confirms $\Gamma : \psi$. By Observation 17.10, $\Gamma \subseteq \Gamma^{\vec{\mathfrak{P}}}$. Hence $\psi \in \Gamma^{\vec{\mathfrak{P}}}$.

Suppose for *reductio* that Γ is $(\psi, \mathfrak{P})$-weak, i.e. $\neg\exists\Delta(\Delta \subseteq \Gamma \wedge \mathfrak{P} \vdash_S \Delta : \psi)$. By Corollary 17.8, $\psi \notin \Gamma^{\vec{\mathfrak{P}}}$. Contradiction.

So, by *classical reductio*, $\exists\Delta(\Delta \subseteq \Gamma \wedge \mathfrak{P} \vdash_S \Delta : \psi)$. By Corollary 17.1, $\mathfrak{P} \vdash_N \Gamma : \psi$. □

This argument has the formal structure

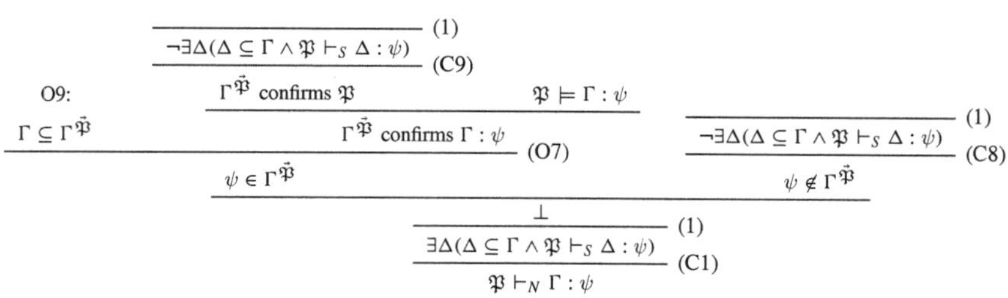

Corollary 17.10 *Logical consequence among non-trivial extended sequents is compact.*

References

Franks, C. (2010). Cut as consequence. *History and Philosophy of Logic*, *31*(4), 349–379.

Gentzen, G. (1932). Über die Existenz unabhängiger Axiomensysteme zu unendlichen Satzsystemen. *Mathematische Annalen*, *107*(1), 329–350.

Hertz, P. (1928). Reichen die üblichen syllogistischen Regeln für das Schließen in der positiven Logik elementarer Sätze aus? *Annalen der Philosophie und Philosophischen Kritik*, *7*(1), 272–277.

Schroeder-Heister, P. (2002). Resolution and the origins of structural reasoning: early proof-theoretic ideas of Hertz and Gentzen. *Bulletin of Symbolic Logic*, *8*(2), 246–265.

Chapter 18
A Notion of *C*-Justification for Empirical Statements

Gabriele Usberti

Abstract After having argued that Prawitz's notion of ground for A is not epistemically transparent when A is an empirical statement (Sect. 18.2), a non-factive and defeasible notion of C-justification ("C" for "computational") for empirical statements is defined and proposed as the key notion of the theory of meaning. C-justifications for A are conceived as cognitive states (Sect. 18.3), and are defined by recursion on the logical complexity of A. In the atomic case (Sect. 18.4) they are defined in terms of two other concepts: the one of C-authorization to use a name to refer to a given entity, and the one of C-authorization to use a predicate in order to apply an accessible concept to objects. In the logically complex cases (Sect. 18.5) the meaning of the logical constants as applied to empirical statements is discussed, and the necessity is shown of Nelson's strong negation besides intuitionistic negation. In the conclusion (Sect. 18.6) it is argued that the notion defined is epistemically transparent and not exposed to traditional objections.

Keywords Theory of meaning · Justificationist semantics · Anti-realism · Dag Prawitz · Justification · Name · Predicate · Concepts

18.1 Introduction: Prawitz on Inference and Knowledge

In recent years Dag Prawitz, posing the problem of the relation between inference and knowledge at the center of his philosophical reflection, proposed the notion of *ground* for (the assertion of) a statement as the key notion in terms of which the

[1] See Prawitz (2009, 2012); the idea was anticipated in Prawitz (1970).

I am most grateful to Beatrice Bernazzi, Luca Tranchini and Giuseppe Varnier for helpful comments to earlier versions of this paper. This work was partially supported by the MIUR fund n. 20107738C5_002.

G. Usberti (✉)
Dipartimento di Scienze Sociali, Politiche e Cognitive,
Via Roma 56, 53100 Siena, Italy
e-mail: gabriele.usberti@unisi.it

H. Wansing (ed.), *Dag Prawitz on Proofs and Meaning*, Outstanding Contributions to Logic 7, DOI 10.1007/978-3-319-11041-7_18

meaning of mathematical and empirical statements is to be explained.[1] This amounts to a substantial modification of his former theory of meaning, according to which meaning was explained in terms of the notion of canonical argument.[2] What is the rationale for this replacement?

Both notions are intended to turn the intuitive notion of conclusive evidence for (the assertion of) a statement into a precise one, but along different paths. Intuitively, what confers conclusive evidence to a mathematical statement is a proof of it; but what is a proof? There are two basic ways of providing an answer.

The first one consists in asking how proofs are linguistically given to us; then a natural answer is: through valid arguments, where an argument is essentially a chain of elementary inferential steps. This idea goes hand in hand with Gentzen's formalization of logic in natural deduction, in which the application of a rule corresponds to a single inferential step; a proof, from this point of view, is what is represented by a valid argument.[3]

Gentzen had also another important idea about the rules of a natural deduction system: that introductions define the meaning of the logical constants, and that eliminations must be, in some sense, in harmony with introductions. This idea has become the keystone of neo-verificationist theories of meaning; according to such theories valid arguments, more precisely canonical valid arguments, play therefore a *double* explanatory role: on the one hand, they are the notion in terms of which the meaning of the logical constants is explained, and which is therefore proposed as the *explicans* of the intuitive notion of evidence; on the other hand, they are linguistic presentations of proofs, and are essential ingredients in the explanation of assertion and inference.

However, the notion of canonical valid argument has a substantial shortcoming when it is charged with the role of explaining meaning. The *explicans* of the notion of evidence—let me call it "E" —must have a property I shall call *epistemic transparency*, which can be characterized as follows:

(1) E is epistemically transparent if and only if an idealized subject who is in possession of E is in a position to know that he is.[4]

According to Dummett, if evidence had not this property it would be impossible, for knowledge of the meaning of a sentence (as defined in terms of evidence), to manifest itself; and Prawitz agrees about the necessity of transparency:

> to know the meaning of a sentence is thus to know what counts as [evidence] for the assertion of the sentence. As Dummett as argued forcefully, such knowledge has at the end

---

[2] See for instance Prawitz (1973, 1977).

[3] More precisely, by each member of a class of arguments having a common normal form.

[4] I use here "is in a position to know" in the sense defined by Williamson:

> To be in a position to know p, it is neither necessary to know p nor sufficient to be physically and psychologically capable of knowing p. No obstacle must block one's path to knowing p. If one is in a position to know p, and one has done what one is in a position to do to decide whether p is true, then one does know p (Williamson 2000, p. 95).

> to be implicit, and consists then in the ability to recognize whether something constitutes [evidence] for the assertion of the sentence. (Prawitz 2011, p. 22)[5]

Unfortunately, the notion of canonical valid argument[6] cannot be plausibly proposed as epistemically transparent. Consider for instance a J-canonical argument Π for $A \supset B$, i.e. an argument whose last inference is of the form

$$\frac{\begin{array}{c}[A]\\ \Pi'\\ B\end{array}}{A \supset B} \tag{18.1}$$

where Π' is a J-valid open argument for B (from A as assumption) and J is a set of justifiying procedures. There are two distinct reasons that make highly implausible the assumption that the J-canonicity of Π is transparent. Firstly, Π' is J-valid if and only if all its closed instances are J'-valid, for all the extensions J' of J; and it is not clear how the totality of the extensions of J could be regimented. One might think about modifying the explanation of implication by requiring, as a second clause, that a J-canonical argument for $A \supset B$ contains also a J-valid argument Π'' for the statement asserting that Π' is J-valid, as suggested by Kreisel. But the transparency of the J-validity of Π'' could not be warranted without starting an infinite regress.

Secondly,[7] consider a closed instance of Π': it is valid if and only if the composition of the procedures in J yields an effective method to transform every J-canonical argument for A into a J-canonical argument for B; but we know from Gödel's theorem that the totality of effective methods cannot be generated by any formal system, so the assumption that an effective method with the required property is transparent seems hardly plausible, as Prawitz remarks in Prawitz (1977).

Owing to its lack of epistemic transparency, the notion of canonical argument cannot be proposed as the key notion of the theory of meaning. It is therefore necessary to resort to the other way to answer the question "What is a proof?": the answer given by Heyting with his characterization of the notion of proofs of A by recursion on the logical complexity of A, the basis of the so-called BHK-explanation of the meaning of the logical constants. The notion of proof defined by Heyting can, on the one hand, be plausibly assumed to be epistemically transparent, and therefore can play the role of the key notion of the theory of meaning; on the other hand, Heyting's proofs cannot be seen as proofs in the first sense defined above, namely as chains of valid inferential acts. One reason for this is that Heyting requires that a proof of $A \vee B$ is either a proof of A or a proof of B, whereas there are perfectly acceptable (even by intuitionists) proofs-as-chains of $A \vee B$ that are neither Heyting-proofs of A nor Heyting-proofs of B. This objection might be (and has been) obviated by introducing

[5] I have replaced "ground" with "evidence", since grounds—as explained below in the text—are precisely the formal counterpart of what is intuitively called "evidence".

[6] As defined for instance in Prawitz (1973). For the definition of canonicity I use below see Usberti (1995, [III.5]).

[7] Cp. Prawitz (1977, [Sect. III.3]).

a distinction between canonical and non-canonical proofs, and taking Heyting-proofs as *canonical* proofs. But this maneuver would be insufficient: in certain cases (as implication, for instance) "[A] canonical proof of a compound sentence must [...] be defined in terms of what counts as non-canonical proofs of the constituents" (Prawitz 2012, p. 12); in the terms just introduced, this means that Heyting-proofs of an implication should be defined in terms of proofs-as-chains of the constituents; but this is just what cannot be done, on pain of being exposed to an objection of circularity. Such objection was formulated by Gentzen with the following words:

> In interpreting $A \supset B$ in this way, I have presupposed that the available proof of B from the assumption A contains merely inferences *already recognized as permissible*. On the other hand, such a proof may contain other $\supset$-inferences and then our interpretation *breaks down*. For it is *circular* to justify the $\supset$-inferences on the basis of a $\supset$-interpretation which itself already involves the presupposition of the admissibility of *the same* form of inference. (Gentzen 1936, p. 167)

This difficulty can be avoided by dropping the assumption that the available proof of B from A contains merely inferences *already recognized as permissible* (as Prawitz has shown with his definition of valid argument); but this amounts precisely to giving up the idea that Heyting-proofs of $A \supset B$ can be defined in terms of proofs-as-chains of the constituents.

If my reconstruction is correct, we can see the reasons why Prawitz makes now reference to two different notions, assigning them different roles: to the notion of ground the role of the key notion of the theory of meaning, to the notion of deduction[8] the role of (linguistic presentation of a) proof, and therefore the role of explaining the intuitive concept of inference. The crucial difference between the two notions is that grounds for A are defined by recursion on the complexity of A, while a deduction of A is defined by induction on the number of its steps. So, in the case of grounds for $A \supset B$, it is legitimate to abstract from the intrinsic complexity of the function that transforms each ground for A into a ground for B, and, provided the knowing subject is presupposed to be capable to recognize general methods and their properties,[9] it becomes plausible to postulate that grounds are epistemically transparent. On the other hand, a deduction of A can now be defined as a chain of valid inferential acts, where an inferential act is valid if its result is a ground. This is a much more intuitive approach to validity than the one that was necessary when the key notion was that of valid canonical argument; in that case the validity of a single inferential step was explained in terms of the validity of the whole argument it belongs to—a clear reversal of the natural order of explanation.

If the proposed analysis is correct, still a central question remains open: is it legitimate to *assume* that the notion of ground is epistemically transparent? It is *an assumption*, because the notion of ground is, like Heyting's notion of proof, a semi-intuitive one; however, an assumption may be more or less plausible. It is a central question too, because, if grounds were not transparent, knowledge of the meaning

[8] See for example Prawitz (2011, [p. 18]).

[9] Notice that in Heyting (1974) the notion of general method of construction is explicitly mentioned as one of the primitive notions of intuitionistic mathematics.

of A could not be characterized as the capacity to recognize the grounds for A. In the next Sect. I will argue that Prawitz's notion of ground cannot plausibly be assumed to be transparent.

18.2 Is Prawitz's Notion of Ground Epistemically Transparent?

There are atomic statements whose grounds are not obtained by inferences:

> all grounds cannot be obtained by inferences. There must in other words be some propositions like $t = t$ or '0 is a natural number' for which it is constitutive that there are specific grounds for them that are not derived or built up from something else. (Prawitz 2009, p. 186)

In the case of empirical atomic statements, what do their grounds look like? Prawitz is very sparing in remarks about this question, limiting himself to some paradigmatic examples:

> For instance, a ground for a proposition 'it is raining' is taken to consist in seeing that it rains; taking "seeing" in a veridical sense, it constitutes a conclusive ground.
>
> (Prawitz 2009, p. 186)

A preliminary remark. "Conclusive" is in fact ambiguous, meaning sometimes indefeasible, sometimes factive:

> Define a way of having warrant to assert *p* to be *defeasible* just in case one can have warrant to assert *p* in that way and then cease to have warrant to assert *p* merely in virtue of gaining new evidence. A way of having warrant to assert *p* is indefeasible just in case it is not defeasible. [...] A way of having warrant to assert p is factive just in case a necessary condition of having warrant to assert p in that way is that p is true. (Williamson 2000, p. 265)

If we take "seeing" in a veridical sense, then seeing that A is a factive way of having warrant to assert that A; I will therefore assume that, in the passage quoted above, Prawitz is using "conclusive" in the sense of "factive". Now, if a ground for "It is raining" consists in seeing that it rains, and seeing is a factive way of having warrant to assert "It is raining", then it is illegitimate to assume that having such a ground is epistemically transparent, since, in general, the evidence actually available to a subject is not sufficient for him to discriminate between seeing (factively) that it rains and merely having the impression to see that it rains: the experience of the subject is the same, whether he is seeing or merely having the impression to see; it may therefore happen that the subject is in possession of a ground for "It is raining" without being in a position to know that he is.

Moreover, if "seeing" is taken in a veridical sense, then, when a subject believes that it is raining because he has the visual experience of presently falling rain, while, instead, he is actually exposed to a cortical stimulation, then the subject is making a mistake:

> In mathematics we operate with the notion of conclusive proofs although of course we may also make mistakes there, in which case we say that what we thought was a proof was not

> really a proof. Similarly I would say that we may think to have seen something but that it later turns out that we did not see it. (Prawitz 2002, pp. 90–91)

Contrary to this assumption, I find that, in this respect, there is a deep and significant difference between the empirical and the mathematical case: a subject who takes for a mathematical proof of *A* an argument containing a flaw is not intuitively justified in asserting *A*, because he *has* actually made a mistake: he has neglected, or has not paid attention to, some information he *should* have taken into consideration; whereas the subject who has the visual experience of falling rain is intuitively justified in asserting "It is raining", and has made no mistake, since there is no information he *should* have taken into consideration.

By the way, it may be interesting to pause a while and reflect upon the reasons of this disparity between mathematical and empirical domains—a disparity that some years ago Prawitz himself acknowledged:

> Outside mathematics the conditions for correct assertions may allow assertion of sentences that in fact are not true. If a sentence is asserted in mathematics on the basis of what one thinks is a proof of it and it later turns out that the sentence is false, one would ordinarily say that the alleged proof was not a proof and that therefore the sentence was incorrectly asserted. But outside mathematics, one may want to say that a sentence was correctly asserted (on sufficient ground) although it later turned out that the sentence was false, *i.e.*, the grounds on which the sentence was asserted are still regarded as having been sufficient in the situation in question (although they are not so anymore). (Prawitz 1980, p. 8)

Consider the subject *s* who takes for a mathematical proof of *A* an argument containing a flaw: we feel that there must be some *intrinsic* feature of the argument such that, had *s* noticed it, he would have realized that the argument is not a proof. The reason why we judge that *s should* have taken into consideration that feature is not simply that it was available to him and relevant to the question, but also that is was *intrinsic*, in the sense that it was a property of the argument he could have known through a careful analysis of the argument itself, without resorting to other, external pieces of information: that feature was not *immediately* available, but it was available, and he did not notice it because of some lack of attention or memory, or because he unduly trusted some of his informants, etc.; he therefore failed to do his epistemic duty, according to which he ought to have refrained from believing that a proof of *A* had been presented to him until he had carefully analysed all the available data. Consider now the subject who has the visual experience of falling rain: there seems to be no *intrinsic* feature of his experience such that, had *s* noticed it, he would have realized that his experience was not an experience of (veridical) seeing; the only way for him to realize it would consist in acquiring new relevant information, which is *not* available in his present cognitive state. Therefore, if the subject believes that it is raining, he has not infringed the epistemic duty mentioned above, and his belief is justified.

Concluding, if we look for a transparent notion of having empirical warrant to assert, we cannot look for a factive notion. But then we cannot look for an indefeasible notion either, for non-factivity entails defeasibility[10]: if one has warrant to assert *A*

[10] This is an antirealist thesis, as is clear from the following argument.

and *A* is not true, there must be some reason why *A* is not true; so, if one comes to know that reason, one has no longer warrant to assert *A*.

The reason why Prawitz takes "seeing" (and of course many other perceptual verbs) in a veridical sense is that he looks for conclusive grounds even for empirical statements, and the reason for *this* is presumably that he holds there are insuperable difficulties in developing a theory of meaning based on a non-conclusive key notion like justification. But, as we have just seen, the adoption of a factive key notion like ground or verification entails the non-transparency of grounds/verifications. So we face a dilemma here: either we choose a factive key notion, at the price of its non-transparency,[11] or we choose a non-factive and defeasible (and hopefully transparent) key notion, at the price of serious difficulties. In the following section I will consider these difficulties, and argue that there is a way out of them.

18.3 Justifications as Cognitive States

The crucial problem Prawitz sees in developing a theory of meaning based on a non-conclusive key notion, like justification, is stated in the following passage:

> It is not possible to explain the content of such sentences [sentences for which the notion of conclusive verification does not apply (G.U.)] in terms of possible justifications. The situation can easily be illustrated by considering ordinary universal sentences. To give an account of when the assertion of such a sentence is justified we may have to deal with all the problems of induction. However, even if we solved all problems of that kind and gave a completely accurate account of all the tests that have to be performed to be justified in asserting a universal sentence, it would be of no help for saying what the content of a universal sentence is. As long as the justifications in question are not conclusive, there is always the possibility that the sentence is refuted by a counterinstance although the justification of the assertion at an earlier time was completely in order. The assertion has then to be withdrawn. But there would be no reason to do so, if the content of the sentence were explained in terms of the possibility for a suitably placed person to justify the assertion of the sentence; *the existence of such a justification would be all that had been asserted, and the speaker would be right about that.* This clearly shows that the proposed understanding of such sentences in terms of possible justifications does not square with our linguistic practice with respect to these sentences (Prawitz 2002, p. 91). (Italics mine.)

The crucial passage is the italicized remark, which shows also the presupposition behind the whole argument, namely that the content of the assertion of *A* is that there is a verification of *A*. I have argued elsewhere against this presupposition (Usberti 2012, Sect. 4); the content of the assertion of *A* – I claim—is simply the proposition that *A*. If we give up Prawitz's assumption, his argument is blocked: the content of the assertion that all ravens are black is not that we actually have, nor that there is, a justification for the statement "All ravens are black", but simply (the proposition)

[11] Another price is put into evidence by Cozzo (2015, Sect. 5): as a matter of fact, we often infer conclusions from wrong premises, for which we have no grounds; since, according to Prawitz's definition of inference, an inference act must involve (conclusive) grounds for the premises, our acts are not inferences.

that all ravens are black, and the practice of withdrawing that assertion, when a white raven is found, can be accounted for in a natural way: since having a justification for A is a defeasible way of having warrant to assert A, it may happen that at t one has a justification for A and at a later time t' one has no justification for A, and therefore that the assertion made at t is withdrawn at t'.

Another objection has been raised against the idea of basing a theory of meaning on a *defeasible* key notion - an objection formulated by Casalegno (2002) but explicitly endorsed by Prawitz (2002, footnote 1).[12] Assume that a justification for "It is raining" consists in seeing that it rains, but that "seeing" is now taken in a non-veridical sense, as meaning approximately having the visual experience that it rains. Now suppose that, at time t, the subject s has a visual experience of rain falling right where he stands; then s has, at t, a justification to assert "It is raining". At a later time t' s is told that, as a matter of fact, at t it was not raining, but that, unbeknownst to him, a cortical stimulation had produced his visual experience; s believes what he is told and as a consequence he withdraws the assertion made at t, even though he continues to hold that, at t, he was entitled to make that assertion. So, is his visual experience a justification for him to assert that it rains or not? To say that it is such a justification at t, and it is not at t', doesn't help: for the piece of information that, in the situation just described, s acquires at t' could have been available to him already at t; and if this had been the case, his visual experience would not have entitled him to make that assertion at t. The conclusion is that the notion of defeasible justification is ultimately inconsistent.

The objection is serious, but it seems to me that there is a way out. Observe that the crucial step of the preceding argument is that the very same 'thing' —the visual experience of the falling rain—is a justification for "It is raining" in one situation and not in the other; so, if we choose, as a justification for the assertion of that statement, something that cannot remain the same as the situation changes, the argument is blocked. Well: what certainly changes in passing from one situation to another is the situation itself. But what, exactly, makes the difference between the two situations, in the case under consideration? The piece of information that the visual experience of s had been produced by a cortical stimulation. More generally, when questions about a subject's having or not having a justification for a statement in a given situation are at stake, what is relevant is all information available to the subject about that situation—in other terms, the cognitive state of the subject. Therefore, the way out consists in conceiving of justifications as global cognitive states. In the present case, the justification the subject 'has' for "It is raining" is not the visual experience of rain falling right where he stands, but the global cognitive state he is in at t, in which the hypothesis that it is raining is the best explanation of the data available to him; at a later time t', when the best explanation of the available data is that s' visual experience has been produced by a cortical stimulation, s attains a new cognitive state c'.[13] It would be senseless to say, of the former state, that *it* has lost its status of justification at t': a cognitive state gives rise to another cognitive state as new information is added,

[12] For a more detailed discussion of the following argument see Usberti (2004).

[13] The role of the notion of best explanation will be clarified below.

and in general there is no question of one and the same cognitive state undergoing a transformation. The argument is therefore blocked.

Someone might object that the notion of justification I am proposing is, again, an indefeasible notion: according to the picture I have proposed, after all, a cognitive state that is a justification for A does not loose its status of justification for A, but is simply replaced by another cognitive state which is not a justification for A. This remark is correct, but it misses the point. What may be defeasible or not is not the notion of justification or verification itself, but the notion of *having* a justification/verification; once justifications are conceived as cognitive states, *having* a justification for A amounts to *being in* a cognitive state which is a justification for A; and being in a specific cognitive state is defeasible.[14] As for factivity, the answer depends on how one answers the question: when does one have warrant to assert A? I will not enter into this question in this paper.

Coming back to the dilemma described above, the way is open to choose the second alternative, which consists in trying to define a defeasible and transparent notion of justification for A, to be adopted as the key notion of a theory of meaning of empirical statements.

18.4 Justification for Atomic Sentences

Even in my approach the epistemic transparency of justification can be at best assumed, since the notion I will define is a semi-intuitive one. In order to make such an assumption plausible I will define justifications as cognitive states having certain

[14] According to Williamson,

> even grasping a proof of a mathematical proposition is a defeasible way of having warrant to assert it. One can have warrant to assert a mathematical proposition by grasping a proof of it, and then cease to have warrant to assert it merely in virtue of gaining new evidence about expert mathematicians, utterances, without forgetting anything.
>
> (Williamson 2000, p. 265)

This is plausible if "proof" is understood as meaning a written proof or something similar; if a proof is conceived as a whole cognitive state, and having that proof is equated to being in that state, Williamson's view becomes much less plausible. If the subject s of Williamson's example has been in a mental state σ_1 that is a proof of A and later is in a mental state σ_2 that is not a proof of A, then either s has forgotten σ_1 when he is in σ_2, or he has chosen to trust the mathematicians instead of his own memory; the first case can be neglected: subjects are necessarily idealized, and idealized subjects have no limitation of memory, attention, etc.; in the second case s has made a mistake, hence he is *not* justified in believing that he has not warrant to assert A.

computational properties,[15] in such a way that the process of recognizing a cognitive state as a justification for a statement A can be seen a process of computation.

The definition of computational justifications for A (C-justifications for A, for short) will be by recursion on the logical complexity of A; in the atomic case, the idea is to characterize a C-justification for an atomic statement $P(n_1, \ldots, n_k)$ in terms of two other notions: the notion of C-authorization *to use the names* $n_1, \ldots, n_k$ to refer to k given entities, and the notion of C-authorization *to use the predicate* P to apply an accessible concept to objects.[16] These two kinds of cognitive authorizations are to be defined themselves in terms of cognitive states—more precisely in terms of *atomic* cognitive states.

18.4.1 Atomic Cognitive States

Assuming a computational view of mind, I will consider an atomic cognitive state as completely specified when two factors are specified: information available to the subject at a certain time, and the subject's cognitive structure; in other—more familiar—terms, an atomic cognitive state is specified by the data available at time t and by specific algorithms implemented by the subject's cognitive apparatus. Here is the definition:

Definition 18.1 An *atomic cognitive state* is a 7-tuple
$\sigma = \langle \mathrm{i}, ec, \mathrm{at}, class, inf, \langle p_{c1}, \ldots, p_{cn}\rangle, \mathrm{R}\rangle$, where

- i is a time;
- ec is a function associating to every name n and to every predicate P an epistemic content ec_n and ec_P, respectively;

[15] The problem is evidently a vast one, because the variety of atomic statements of a natural language is immense. I will be concerned only with a restricted number of cases, but I will select them in such a way that they are representative of a fairly large class of cases. In particular I will keep present, besides mathemathical ones, observational statements and several other empirical statements in the present tense and in the third singular person.

[16] This idea could be motivated with considerations analogous to Gareth Evans' ones leading to his Generality Constraint:

> For we cannot avoid thinking of a thought about an individual object x, to the effect that it is F, as the exercise of two separable capacities; one being the capacity to think of x, which could be equally exercised in thoughts about x to the effect that it is G or H; and the other being a conception of what it is to be F, which could be equally exercised in thoughts about other individuals, to the effect that they are F. (Evans 1982, p. 75)

Analogously, we cannot avoid thinking of a justification for $P(n)$ as the result of two separable components; one being a C-authorization to use n to refer to a given entity, which could equally be a component of a justification for $G(n)$ or $H(n)$; and the other being a C-authorization to use P to apply an accessible concept to objects, which could equally be a component of a justification for $P(m)$ or $P(k)$.

- at is a k-tuple of terms of **IRS** activated at time i[17];
- *class* is a function associating to every term t of **IRS** a classification criterion $class_t$;
- *inf* is a function associating to every term t of **IRS** a certain amount of *information* inf_t, and to every primitive predicate P a certain amount of (supplementary) information inf_P;
- $\langle p_{C1}, \ldots, p_{Cn} \rangle$ is a finite collection of feature-checking algorithms;
- R is a relevance relation between statements and pairs $\langle P_k, X \rangle$, where $X = \{P_1, \ldots, P_k, \ldots\}$.

This definition will be gradually explained in Sects. 4.2–4.4.

18.4.2 C-Authorizations for Names

What sort of cognitive state is a *C-authorization* to use a name to refer to a given entity? In more intuitive terms: in what conditions is a subject s cognitively authorized to use n to refer to a given entity?

18.4.2.1 A Given Entity

First of all, what do we mean when we say that a specific entity, for instance the Mount Everest, is *given* to a subject s? We can mean two very different things: either that what is *in fact* the Mount Everest is given to s, independently of his actually recognizing that object as the Mount Everest; or that s actually *recognizes* something *as* the Mount Everest, independently of its being in fact the Mount Everest. I hold that the two meanings are clearly distinct, and that we should try to answer to our question in *both* its meanings; however, I hold that the computational significance of the question is immediately *clear* only when it is understood in the second sense, so that it will be convenient to start from this interpretation.[18] Let me explain why. From the point of view of the subject, there is in principle no difference between being presented with the Mount Everest itself and being presented with an hologram of the Mount, or even with being presented with nothing at all, and being only stimulated with an electrode at some point of the cerebral cortex in such a way as to produce a mental image of the Mount Everest; when I say that there is no difference from the point of view of the subject, I mean that if our aim is to characterize the mental state of the subject and his mental operations or computations, the subject's mental state is exactly the same in the two situations: the two situations are very

[17] The choice of k varies according to the nature of the statement for which the cognitive state is a justification. **IRS** will be defined in Sect. 4.2.1.

[18] As a matter of fact, in this paper I will be concerned only with the second sense. I will take into consideration the first in a sequel to the present paper, which will be devoted to the notion of empirical warrant or truth-ground.

different, but the amount and the quality of information available to the subject is the same, and therefore his mental computations will be the same; in other terms, what makes the difference between the two situations is entirely irrelevant. What is relevant is only available information and the structure of the subject's computing apparatus; and these factors are presumably also the only ones responsible for the subject's perceptual experience, in particular for his *recognizing* something *as* the Mount Everest.

So: what does it mean, for a subject conceived as a computational device, that it recognizes something as a specific object? As I have argued right now, it is perfectly possible that there is absolutely *nothing*, in the surrounding environment, that is presented to the subject, even though it perceives something, for instance the Mount Everest. Well: it is possible that there is nothing *in the environment*, but there is surely something *in the subject's mind*. What? A mental representation,I suggest. I do not use "representation" in a relational sense: a representation is not a representation *of* an object in the real world, but simply a structured symbol, a *term* of some internal representational system (**IRS**).[19]

Does this mean that I am identifying objects with terms of **IRS**? Not exactly. Recall that we are looking for conditions at which a subject *recognizes* something *as* the Mount Everest. My hypothesis is that recognition should be computationally analyzed as the act of inserting the newly activated term either into a previously constituted class of terms or into a newly constituted class. Consider for instance two situations. In the first, the subject *s* is in front of Mount Everest, and he has already been presented with that mountain on other occasions, so that *s* already has in his cognitive space (in his memory) a previously constituted class of terms: his computational task is to scan the classes of terms present in his memory to verify whether there is *one and only one* class into which the actually activated term/representation can be inserted (on the basis of criteria to be specified), and to insert it if the answer is affirmative. In the second situation, *s* is in front of Mount Everest, and he has never been presented with that mountain on any other occasion: his computational task is the same, but the answer is negative; in this case he must perform a mental act, i.e. create a new class of terms and put the newly derived term into it. Under this hypothesis concerning recognition, objects are best equated to classes of terms of **IRS**.

Summing up, an entity is given to a subject when a term of **IRS** is activated and this term belongs to a definite class, either previously or newly constituted. Since also animals seem to be capable of perceiving and recognizing objects, it is natural to assume that the human cognitive apparatus is organized in such a way as to permit the constitution of objects, i.e. of classes of terms, *before* language comes in. For example, the visual descriptions[20] derived by a subject looking at a moving cat are

---

[19] This technical sense of "representation" is common in cognitive psychology; for instance, Chomsky (2000, p. 173) introduces it by saying that "there is nothing 'represented' in the sense of representative theories of ideas, for example."

[20] Following common use in cognitive psychology, I use here "description" as a synonymous of "term" and "representation".

related to one another by transformations belonging to a group (the rigid movements in 3-dimensional space) that leave properties of shape and size invariant; and such visual descriptions are put into one class, thereby constituting a single visual object.

Let me therefore introduce some general assumptions. Cognitive states, as defined above, are conceived as the states of a component of mind which might be called, following Chomsky, *conceptual-intentional system* (**C-IS**), a component of our cognitive apparatus that can be accessed, on the one hand, by several other components such as vision, audition, memory, imagination, and, on the other hand, by language, more precisely by syntax, in the sense that the syntactic structures generated by syntax are inputs of **C-IS**, which is dedicated to their (semantic) interpretation. Every component has its own representational systems, permitting the derivation of *descriptions*, i.e. structured arrays of primitive terms/symbols. Examples of descriptions derived in the visual representational system are Marr's 3-D model descriptions; examples of descriptions derived in the linguistic representational system are logical forms. I will assume, for the sake of simplicity, that the **IRS** of **C-IS** is unique, and that the descriptions derived in the systems to which **C-IS** is accessible are 'translated' into it.

Secondly, I assume that the terms belonging to **IRS** are organized in classes (C-objects), constituted by the components that have access to **C-IS**, except syntax. Each cognitive state is therefore inhabited by C-objects, which are constituted independently of language. When, in a cognitive state σ, a C-object is constituted and a term t is activated, σ can be assumed to be equipped with an appropriate *classification criterion* associated with t, i.e. with a function $class_t$ such that, for every term t', $class_t(t') = 1$ if and only if t and t' belong to the same class; intuitively t and t' will be treated by the subject as representations *of the same object*. Given a cognitive state σ and a term t of **IRS**, I will denote with "$/t/_\sigma$" the C-object t belongs to in σ.

18.4.2.2 Linguistic Recognition

Till now I have postulated that **C-IS** has some internal structure, in particular that there are principles according to which C-objects are constituted, but I have not introduced any specific principle. It is likely that some principles are inherited by the cognitive systems having access to **C-IS** (like vision, memory, etc.); in any case I will not deal with them; I will instead introduce a principle that, in my opinion, is operating at the intersection of **C-IS** with syntax.

When language comes into play the available terms of **IRS** become much more numerous; it might be thought that, consequently, the modes of constitution of classes of terms become much more varied. I think, on the contrary, that what happens is the opposite: the modes of constitution become more abstract, and their diversity drastically reduces. Let me illustrate this point by explaining how cognitive psychology explains recognition. According to Marr (1982) recognition involves four things: (i) a collection (or better a *catalogue*) of stored 3-D models, (ii) a device to derive new descriptions, (iii) a device K to establish whether a newly derived description matches a stored one, and (iv) a newly derived description. For example, the recognition by a subject of a disk in front of him can be sketched in the following terms: if d is the newly derived description and m is a description stored in the catalogue of 3-D models under the head DISK, the procedure K is applied to $\langle m, d \rangle$ and it yields answer 1 if d matches m, 0 otherwise.

Marr is chiefly interested in perceptual recognition, in which information stored in memory is compared with input information coming from perceptual modules, in particular from vision; but there is recognition also when stored information is compared with inputs from imagery, or from language; and this last case, 'linguistic' recognition, is especially interesting from our point of view. In order to give an account of the notion of authorization to use a name it is therefore useful to generalize and adapt Marr's conceptual framework.

From an abstract point of view the essential ingredients are: stored information, new information, and an operation of comparison or 'matching' between the two. Let us start from stored information. I assume that to every name n a certain stock of information is attached and stored in memory, which I call the *epistemic content* ec_n associated to the name. I make two further assumptions: (i) that the epistemic content associated to a name is articulated into a *lexical content* lc_n, constituted by information coming from the lexicon, and a situational content sc_n, constituted by information coming from perception, memory, imagery, the belief system, etc.[21]; (ii) that lc_n remains invariant in passing from a cognitive state to another, while sc_n is not subject to this restriction. Intuitively, information contained in lc_n is information without which a subject cannot be said to know the meaning of n, or to have semantic competence about it. Which pieces of information are to be put into lc_n and which in sc_n is partly an empirical question pertaining to lexical semantics, partly a question depending on our intuitions about synonymy and related notions. From the conceptual point of view the important thing is that *some* divide between lexical and situational content is acknowledged, in order to avoid Quine's inextricability thesis and its holistic consequences.

I shall not deal with the problem of how information contained in the epistemic content is organized, represented, or accessed; I will simply assume that it is computationally tractable, as it happens in the case of information encoded into lexical entries through (systems of arrays of) *features*. But features are surely not the only computationally tractable way of codifying information; Marr's 3-D model descriptions are another example.

Let us consider new information. In Marr's framework it is encoded into (i.e., explicitly represented in) the newly derived description; but if we take into consideration kinds of recognition different from the visual one, we cannot assume that in general relevant information is encoded directly into the activated term: it will simply be *associated* to the activated term in some way or other.[22] For example, consider a situation in which I hear someone speaking of my friend Paul, or in which a mathematician tells me "Consider the square root of 2": also in such cases it is natural to assume that I am recognizing an entity, and therefore that a mental representation,

[21] An analogous distinction is introduced and motivated in Bierwisch (1992, pp. 30–32).

[22] The need of associated information has also other reasons, which will be explained below.

or a term, is activated from my memory or from my imagery. Again, the associated information may be of very different kinds and formats; in our examples, it may be the face of my friend Paul, or the linguistic description "the number he is speaking about", for instance.

Consider finally the operation of matching. The name "matching" is metaphoric, where the metaphor tries to suggest in some way what is common to a family of operations, which may indeed be very different from one another, according to the format, and nature of 'matched' information. For instance, there may be matching between a mental model stored in memory and a visual or an acoustic description, between the features of a lexical item and information associated to a term stored in memory. I shall not enter into these questions, which belong to the second of the three levels at which, according to Marr (1982, pp. 22–25), any information-processing task must be understood; I shall be exclusively concerned with the first level, the level of the computational theory. From this point of view I shall postulate the existence, as an essential part of the computational device of human mind, of a single *matching* function *match* taking as arguments pairs $\langle k, t\rangle$—where k is a stock of information and t is a term of **IRS**, and whose value is 1 if there is matching, in the appropriate sense, between k and information associated to t, 0 otherwise.

18.4.2.3 The Denotation of Names

If a cognitive state σ is specified, then for every name n the following question has a definite computational meaning: "Does ec_n authorize an idealized subject to use n to refer to the entity given by at, in presence of inf_{at}?" It has a definite computational meaning in the sense that the answer does not depend on any other hidden feature of the context; it exclusively depends on the following two questions:

(3) *The matching question for the name n:*
Does $match(ec_n, inf_{at}) = 1$? I.e., is there an appropriate matching between ec_n and inf_{at}?

(4) *The Uniqueness question for the name n:*
For every term t', such that $match(ec_n, inf_{t'}) = 1$, is $/t'/_\sigma = /\mathrm{at}/_\sigma$?

If the answers to both questions are affirmative, in the cognitive state σ one is authorized to use n to refer to the entity given by at, otherwise one is not. We can therefore give the following definition:

Definition 18.2 A *C-authorization* to use n to refer to the entity given by at is an atomic cognitive state σ such that both the answer to the Matching Question for n and the answer to the Uniqueness Question for n is 1.

Let me explain the necessity of the uniqueness condition by means of an example. Suppose that a subject s associates to the name "Chomsky" an epistemic content including the features LINGUIST and TEACHES AT M.I.T., and imagine that on a certain occasion someone tells him something about *two* distinct linguists t_1 and t_2 who teach at M.I.T.. Intuitively s is not authorized to use "Chomsky" to refer to both t_1 and t_2, or to refer to whichever of them; the reason is plausibly that there is a general principle (tacitly known to every subject) according to which with a proper name one cannot refer to more than one object in a given context. According to the present approach this situation can be analyzed in the following way. The cognitive state σ modeling the situation is defined by choosing one of the two terms t_1 and t_2 as activated; for definiteness, let us say that at is t_1. The answer to the matching question is 1, since t_1 matches $ec_{Chomsky}$; but the answer to the uniqueness question is 0, since there is a term, t_2, such that $match(ec_{Chomsky}, inf_{t_2}) = 1$, and $/t_1/_\sigma \neq /t_2/_\sigma$; therefore σ is not an authorization to use "Chomsky" to refer to the object given by at.

Given a cognitive state σ, for every name n the following relation EQ_n can be defined on the set of terms of **IRS**:

Definition 18.3 $tEQ_nt' =_{def} match(ec_n, inf_t) = 1$ if and only if $match(ec_n, inf_{t'}) = 1$.

EQ_n is an equivalence relation, so it induces a partition on the set of terms of **IRS**.[23] I will denote by "$/t/_{\sigma,n}$" the equivalence class of the term t; the notation is motivated by the fact that $/t/_{\sigma,n}$ can be seen as a sort of *linguistic updating* of $/t/_\sigma$, the set associated to the classification criterion $class_t$. For example, suppose I see a person in front of me, that someone informs me that person is John, and that I previously associated to the name "John" the epistemic content MARY'S BROTHER; in this cognitive state at, the activated term, is the visual description of a man, $/\mathrm{at}/_\sigma$ is a certain class of representations coming from perception and memory, and there is a class $/t'/_\sigma$ of terms which match the epistemic content MARY'S BROTHER; through the piece of information that $/\mathrm{at}/_\sigma$ is John I learn that all the terms belonging to $/\mathrm{at}/_\sigma$ match the epistemic content MARY'S BROTHER and therefore stand to all the terms belonging to $/t'/_\sigma$ in the relation EQ_{John}; hence my computational apparatus performs an updating of $/\mathrm{at}/_\sigma$ and $/t'/_\sigma$ consisting in a 'fusion' of the two classes; in other terms, two formerly distinct C-objects are fused into one in the new cognitive state. Analogously, one C-object can split into two different C-objects in passing from a cognitive state to another.

When σ is a cognitive authorization to use n to refer to the entity given by at, an idealized subject s in σ is intuitively authorized to use n to refer to $/at/_{\sigma,n}$; so it is natural to identify $/\mathrm{at}/_{\sigma,n}$ with *the denotation* of n in the state σ, which I shall denote with the symbol "$|n|_\sigma$":

Definition 18.4 The *denotation of* the name n in the cognitive state σ (in symbols $|n|_\sigma$) is $/at/_{\sigma,n}$ if and only if σ is a C-authorization to use n to refer to the entity given by at.

---

[23] The classes belonging to this partition should not be confused with the 'pre-linguistic' ones corresponding to the criteria of identification associated to activated terms: two terms of **IRS** may be in the same class independently of their matching the epistemic content associated to any name; conversely, two terms of **IRS** may both match the epistemic content associated to the name n independently of their being in the same 'pre-linguistic' class.

18.4.3 C-Authorizations for Predicates

I will now consider *C*-authorizations to use a predicate in order to apply an accessible concept to objects. Let me first explain why I use this involved and somewhat abstruse expression instead of the much more simple and intuitive "authorization to concatenate a predicate with a name". I speak of application of a *concept* (let me skip "accessible" for a while) to an *object*, instead of concatenation of a *predicate* with a *name*, because I think that Frege was correct in conceiving predication as an operation involving concepts and objects, i.e. the entities *denoted* by predicates and names, and not directly predicates and names. This is correct, I hold, even for a computational approach like mine (which involves an internalistic notion of denotation, as we have seen); the difference from the Fregean view concerns the nature of the entities denoted by names and predicates, not the fact that such entities are *distinct* from linguistic entities. The main reason for this is that an important aspect of the cognitive preconditions for the use of predicates is that, if a subject is authorized to concatenate a predicate with a name, then he is authorized to concatenate it with any other name of the same object, provided he is authorized to believe that it is a name of the same object. For example, if I am justified to assert that the boy in front of me is running, then I am thereby justified to assert that Matthew is running, and that the elder son of my brother is running, provided I am justified to believe that the boy in front of me is Matthew, the elder son of my brother. In other terms, we might say that predication, the operation of concatenating a predicate with a name, has an implicit modal aspect, in the sense that we do not simply ask ourselves whether we are authorized to concatenate a predicate with a given name, but with any other name we *could* use to refer to *the same object*. This seems to be the main reason why proper names cannot simply pick out terms of **IRS**, but must be used to refer to *objects*.

An immediate consequence of this is that what applies to objects are *concepts*, in the Fregean sense of entities having the nature of functions. It should be noticed, however, that Frege never speaks of *concepts* as applying to objects, but directly of predicates. As Dummett (1973, p. 246) observes, for Frege "the crucial notion for the explanation of the sense of a predicate is that of its being true of an object [...]". As a consequence, "the relation between [a predicate] and its referent [i.e., a concept] does not have to be invoked" (Dummett 1973, p. 246); nor *could* it be invoked – I add—because "we can make no suggestion for what it would be to be given a concept" (Dummett 1973, p. 241).[24] An almost immediate consequence of this view is that "The only way we can gain an idea of [a concept] is as the referent of a predicate, [...] we approach it —apprehend it —via language" (Dummett 1973, p. 202); and a consequence of this thesis is that a human being has no concepts before the acquisition of a language, nor does any non-human animal have access to concepts . I find this conclusion unacceptable for many reasons; for one, it is incompatible with the idea that human beings are endowed with a rich innate conceptual structure—an

[24] See also Dummet (1973, p. 402): "the notion of identifying a concept [...] seems quite inappropriate".

idea strongly sustained by poverty of the stimulus arguments. For this reason, I think that the mention of *an accessible concept* is essential in the statement of the starting question concerning predicates; I don't say that they are *given*, like objects, but that they are *accessible*, and that we have access to them before language comes in.

18.4.3.1 An Accessible Concept

What does it mean that a concept is pre-linguistically accessible? And even before: what *is* a concept, from a computational point of view? As we have seen, the essence of Fregean concepts is that they are functions, as opposed to objects. But Fregean concepts take as arguments objects of the external world, whereas C-objects are sets of representations. Therefore C-concepts should take C-objects as arguments, and give as values 1 and 0, which will no longer be understood as truth-values, but as *Yes* or *No* answers the computational apparatus of **C-IS** associates to those inputs.

While it is intuitively clear what it is for an object of the external world to be a horse, it seems less clear what it is for a set of representations to have such property. But this is not the relevant question. The computational apparatus is intended to answer 1 not when a C-object is a horse, but when it is *evident* that it is; so the intuitive relevant question is: at which conditions is it *evident* that a given object is a horse, or that a man pursues a dog? And the natural answer is: when an appropriate *feature* is present in the actually derived description.

Let me start from the horse case. The subject is given an object he recognizes as a horse; this means that the subject is in a cognitive state σ in which a term at is activated which belongs to a C-object $/\mathrm{at}/_{\sigma}$, and there is matching between at and a stored description having the label HORSE. What is needed, for it to be evident to the subject *that* the given object is a horse, is an algorithm checking the presence, in at, of the feature configuration HORSE.

Consider now the case of the man pursuing a dog. Here the situation is more complex, with two activated terms, but the task the subject's computational apparatus is confronted with has an essential aspect in common with the preceding case: what is required is to check the presence, in the activated terms, of a certain feature configuration. More specifically, the activated terms belong to two distinct C-objects labeled MAN and DOG, respectively, by two appropriate algorithms; what is needed, for it to be evident to the subject that the first given object pursues the second given object, is an algorithm checking the presence, in the situation involving the two C-objects, of a feature configuration having approximately the following structure:

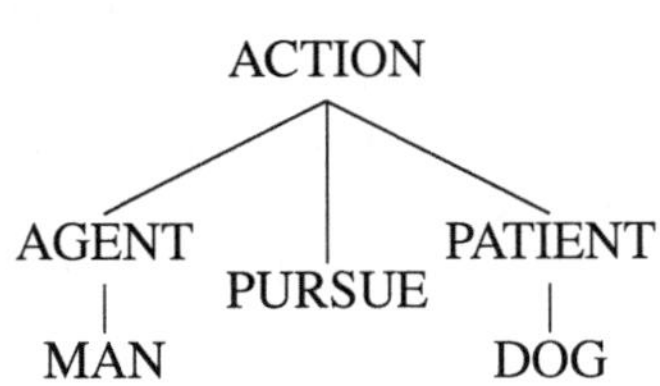

(18.2)

As a consequence, the algorithm I am postulating presupposes the execution of several subroutines: articulating the derived description into (the descriptions of) two objects and an ACTION; verifying the presence in (the descriptions of) the two objects of the features MAN and DOG, respectively; assigning them the roles AGENT and PATIENT, respectively; verifying the presence in (the descriptions of) the action of the feature PURSUE.[25]

I therefore postulate that an atomic cognitive state is characterized by a new component: a set of feature-checking algorithms $p_{C1}, p_{C2}, \ldots$, where each p_{Ci} verifies the presence of the feature configuration C_i. Each of these algorithms computes a k-ary function f_{Ci} taking as arguments k-tuples of C-objects and giving as values 1 or 0 according as the arguments have or not the feature configuration C_i. Under these assumptions, the intuitive expression "the C-concept C_i" is systematically ambiguous between the feature configuration C_i and the function f_{Ci}. Given a cognitive state σ, I will say that the C-concept C_i is accessible in σ if an algorithm p_{Ci} computing f_{Ci} is available in σ.

18.4.3.2 The Denotation of Predicates

Accessibility of C-concepts is already granted at the pre-linguistic level. If we now introduce predicates, and assume that an epistemic content is associated to them too, it is not difficult to give an account of how they can denote concepts that are pre-linguistically accessible. The same restrictions imposed upon the epistemic contents associated to names are imposed upon the ones associated to predicates. Now suppose that in the cognitive state σ an epistemic content is associated to the predicate P in which the feature configuration C is specified, and that there is, among the algorithms accessible in σ, a feature-checking algorithm computing the function f_C: then an idealized subject in σ is authorized to use P to apply the concept C to objects, and f_C is the obvious candidate to be the denotation of P in σ.

While pre-linguistically accessible concepts are already present in cognitive states, nothing prevents a predicate from denoting a 'linguistically constituted' C-concept. For example, consider the predicate "bachelor", and suppose that the associated epistemic content includes the features configuration $ADULT \wedge MALE \wedge \neg MARRIED$; if algorithms computing $f_{adult}, f_{male}, F_{married}, f_{\neg}$ and $f_{\wedge}$ are accessible in σ, and if the computational component of **C-IS** is equipped with some logical machinery,[26] also an algorithm checking the presence of the complex feature $ADULT \wedge MALE \wedge \neg MARRIED$ will be defined, hence $f_{adult \wedge male \wedge \neg married}$—i.e. $F_{bachelor}$—will be accessible.

The following definitions sum up the preceding discussion:

---

[25] It should be admitted that, while we have some hints about what recognizing a man or a dog amounts to in computational terms, more difficult is to give a computational analysis of actions and of the assignment of roles. On the other hand, it is a methodological assumption of computational psychology that it is *possible* to do it, and I see no *a priori* argument to the contrary.

[26] As it would be plausible to assume.

Definition 18.5 A *C-authorization to use P to apply the concept C to C-objects* is a cognitive state σ such that ec_P contains the feature configuration C and C is *accessible* in σ.

Definition 18.6 The *denotation* of the predicate P *in* the cognitive state σ (in symbols $|P|_\sigma$) is the function f_C if and only if σ is a C-authorization to use P to apply the concept C to C-objects.

18.4.4 The Application Question and the Problem of Relevance

Once a cognitive state σ has been specified, the following question has a definite computational meaning, for every atomic statement of the form $P(n_1, \ldots, n_k)$:

(6) *The Application Question for* $P(n_1, \ldots, n_k)$ *in* σ:
Does $|P|_\sigma$ apply to $\langle |n_1|_\sigma \ldots \langle |n_k|_\sigma \rangle$?

It has a definite computational meaning in the sense that the answer depends exclusively on information available in the given cognitive state; but the procedure through which the answer is obtained still has to be specified. The examples considered so far[27] suggest that in order to get the answer it is sufficient to take into consideration, as representatives of $|n_1|_\sigma, \ldots, |n_k|_\sigma$, the activated terms $at_1, \ldots, at_k$. But there are more intricate cases.

Consider the following situation: a subject s, sitting in position p_1, looks at a disk d placed on a table, and sees d as round. It seems intuitively correct to say that in the situation described—let's call it the cognitive state σ_1—s is authorized to apply the concept $|x$ *is ROUND*$|_{\sigma 1}$ to the C-object $|d|_{\sigma 1}$. Imagine now that at a subsequent time s moves to a position p_2, from which he see the disk as elliptical—in other terms, in the new cognitive state σ_2 the activated term is the representation of an elliptical disk. It is not intuitively correct to say that in σ_2 s is authorized to apply both the concept $|x$ *is ROUND*$|_{\sigma 1}$ and the concept $|x$ *is ELLIPTICAL*$|_{\sigma 1}$ to the C-object $|d|_{\sigma 1}$; the subject will probably be uncertain about the shape of the disk, and in normal conditions he will try to acquire new relevant information, for example by touching the disk, or by changing again his position, or some other way—a clear indication of the fact that he feels not authorized to apply either of the two concepts to the object in σ_2. It seems plausible to say that, in order to arrive at a cognitive state in which he is again authorized to apply one of the concepts to the object, s engages in a process whose goal is the selection of one representation of that disk, among the ones to which he has access through perception, memory, attention and so on, as *the best* one. For instance, he will select the visual representation that is 'in accord' with the tactile representation; he will select the representation that, together with some general laws, permits him to account for the others; and so on.

The sense in which a representation is better than another is relatively clear in specific cases, although it is not yet clear whether there is a single point of view

[27] I mean the horse and the man-pursuing-dog examples.

from which it can be characterized/defined. In any case, the example suggests that, in order to answer the Application Question, it is not always sufficient to consider the activated terms: sometimes it is necessary to consider the best terms belonging to the C-objects of the activated terms, for some sense of "best" to be explained.

The cases considered so far are all *direct*, in the sense that the Application Question concerned cognitive states σ in which the denotations of $n_1, \ldots, n_k$ were directly given to the subject; but in many other cases—which may be called *indirect*—a subject is intuitively authorized to apply $|P|_\sigma$ to $\langle |n_1|_\sigma \ldots \langle |n_k|_\sigma \rangle$ even if he is not 'presented' with those C-objects, in the sense that some or all the C-objects $|n_1|_\sigma, \ldots, |n_k|_\sigma$ have no activated term in σ. Here is an example. Suppose a subject s hears some noises in the room nearby where, as a matter of fact, Jack is running. If he had no other information, s would not be justified to believe

(7) Jack is running in the room nearby;

but suppose he has at his disposal the following supplementary pieces of information: (i) that in the room nearby there is only Jack, and (ii) that a person running in the room nearby produces noises similar to the ones he is hearing. In this cognitive state s is again intuitively justified to believe that Jack is running, but his justification is much more 'indirect' than before. In particular, relevant information is in no way limited by the syntactic structure of the sentence[28]; for instance, it is not sufficient to make reference to the meaning of "run" in order to know whether the pieces of information (i) and (ii) are relevant to a justification of "Jack is running".

The problems are even more involved. Suppose, as a second example, that when he wakes up John hears at the radio that the evening before a demonstration was held, and that the police used fireplugs; he wants to know whether the demonstration passed through a certain street, he goes there and he sees puddles in the street. In this case we would intuitively say that he has a justification for something like "The demonstration passed through the street"; but if John had a different question in mind—for instance "Which shoes should I put?"—it would have been correct to say that his seeing puddles in the street gave him a justification for something different, maybe for the belief that a certain pair of shoes is not good. How to account for this 'interest-relativity' of the notion of justification?

Finally, let us consider an example involving testimony. Smith has been informed that John is Louisa's husband by a friend of his; since Smith believes that his friend is well-informed and trustworthy, he is intuitively justified to believe that John is married. Smith's justification is obtained through some kind of inference; how can inference provide justification?

---

[28] As it might be thought if only cases similar to our first example were taken into consideration: in that case it might be suggested that the representation of a round disk is relevant because it is in some sense similar to the activated term, which is relevant because it is activated at the time of the cognitive state.

18.4.4.1 Justification and Explanations as Answers

It seems to me that a very natural answer to the preceding questions emerges if we look at the problem from the viewpoint of the theory of explanation. Let us come back to the round/elliptic disk example, and consider the state σ_2; if s associates to the predicates "is ROUND" and "is ELLIPTICAL" the epistemic contents usually associated to them (say, stored models of the two geometrical forms), and if we conceive the activated term (the representation of an elliptical disk) *and* the term activated at σ_1 (the representation of a round disk), together with further information associated to them, as the *data* available at σ_2, the problem of selecting one of them can most naturally be conceived as the problem of *explaining the data*: the selected representation is the one that explains the data better than the other, on the background of a theory consisting in the epistemic contents associated to the predicates plus several (unanalysed) explanatory principles. In this way the sense in which a representation is better than another is elucidated: it is preferable in the sense that it offers a better explanation of the data.

There are many theories of explanation. For a number of reasons, which will become clear presently, I will make reference to van Fraassen's (1980, Chap. 5) theory. According to it, explanations are answers to why-questions, and why-questions have a *contrastive* nature, in the sense that their logical form is not simply "Why P?" but "Why P *in contrast to* X?", where X is a set of alternatives. From this point of view "The demonstration passed through this street" and "It has been raining" can be seen as answers to two quite different questions—say "Why are there puddles in this street *in contrast with* there not being any in that one?" and "Why are there puddles in the streets *in contrast with* there not being any?", respectively. In this way, the subject's interest, which was intuitively seen as a disturbing subjective factor, is now transformed into an aspect of the objective situation; as a consequence, there *is* now some objective factor in terms of which a justification for one of the two statements can be differentiated from a justification for the other.

However, the interest-dependence of justifications, i.e. of answers to why-questions, cannot be explained away exclusively by reference to the contrastive interpretation of why-questions. For example [van Fraassen (1980, p. 142)], the question "Why does the blood circulate through the body?" can be answered in different ways—for instance "Because the heart pumps the blood through the arteries" or "To bring oxygen to every part of the body tissue" —independently of the contrasting class of alternatives, and depending on the kind of reason requested—a cause or a function, respectively. It seems natural to say that here a relation of *relevance* comes into play: in one case a causal reason is relevant, in the other a functional reason.

The importance of relevance is in fact much more vast than the preceding example suggests. Let us come back to our second example: I observed that it is not sufficient to make reference to the meaning of "run" in order to know whether the pieces of information (i) and (ii) are relevant to a justification for "Jack is running". This is true in most cases in which our justifications are 'indirect'. Looking at the example from the point of view of the theory of explanation, it is natural to suggest that the cognitive system of the subject is involved in a process of explanation, and that this process can be approximately characterized as follows: (i) it generates several potential explanations of the available data; (ii) it selects one of them as the best, on the basis of some selection criterion. How can the class of potential explanations be characterized, or at least conceptually circumscribed? Again, an appeal to relevance seems to be necessary in this connection: potential explanations are the relevant answers to the question "Why are there such and such noises in the room nearby, in contrast to their being silence in it?"

To sum up, a why-question Q expressed, in a given context, by an interrogative sentence may be identified with a triple $\langle P_k, X, R\rangle$, where P_k is the *topic*, $X = \{P_1, \ldots, P_k, \ldots\}$ is the *contrast-class*, and R is a *relevance relation* between propositions and pairs $\langle P_k, X\rangle$. An answer to a why-question Q is expressed by a sentence of the form

(8) P_k in contrast to (the rest of) X because A;

(8) is assumed to claim that P_k and A are true, that the other members of X are not true, and that A is a *reason*, i.e. that A bears relation R to $\langle P_k, X\rangle$ (or, equivalently, that A is relevant to the question Q). A proposition B is a *direct answer* to a question $Q = \langle P_k, X, R\rangle$ if and only if there is a proposition A (the core of answer B) such that A bears relation R to $\langle P_k, X\rangle$ and B is true if and only if all the members of the class $\{P_k;\ \text{for all}\ i \neq k, \neg P_i; A\}$ are true. A *presupposition* of a question Q is any proposition which is implied by all direct answers to Q. As a consequence, a why-question presupposes exactly (i) that its topic is true, (ii) that the other members of its contrast-class are not true, and (iii) that at least one of the propositions that is relevant to it is true; the conjunction of (i) and (ii) is called *the central presupposition* of the question.

In these terms we can settle a problem that is crucial for the present approach. My intuitive idea is that a justification for "It rained" is a cognitive state σ in which "It rained" is the best answer to a why-question arising in σ; but what does it mean that a why-question arises in a cognitive state? van Fraassen remarks that

> In the context in which the question is posed, there is a certain body K of accepted background theory and factual information. This is a factor in the context, since it depends on who the questioner and audience are. It is this background which determines whether or not the question arises; hence a question may arise (or conversely, be rightly rejected) in one context and not in another. (van Fraassen 1980, p. 145)

He therefore proposes that the phrase "The question Q arises in the context C" be taken to mean that K—the background knowledge available in C—implies the central presupposition of Q and does not imply the denial of any presupposition of Q.[29]

I skip other important questions. It is enough to mention that there will be some criteria according to which an answer to a why-question may be classified as a good answer, and one answer is selected as the best one among several possible ones.

---

[29] Van Fraassen remarks that the requirement that K does not imply the denial of any presupposition of Q is very different from the requirement that all the presuppositions of Q are true: "K may not tell us which of the possible answers is true, but this lacuna in K clearly does not eliminate the question." (van Fraassen 1980, p. 146)

18.4.4.2 Explanation and Computation

Van Fraassen's theory is sufficiently articulated and flexible to make explicit all the variables that are implicit in the intuitive relation of explanation, thereby making a computational treatment of explanation possible. The first step in this direction is the remark that the notion of context, which is fundamental in van Fraassen's approach but is left by him unanalysed, can be analysed, at least partially, in terms of the notion of cognitive state introduced above. Van Fraassen (1980, p. 135) conceives of a context of use in the usual way, i.e. as "an actual occasion, which happened at a definite time and place, and in which are identified the speaker [...], addressee [...], and so on." An important aspect of the intuitive notion, as it results from the passage quoted at the end of the preceding section, is that both "a certain body K of accepted background theory and factual information" is available in a given context; but van Fraassen does not analyze K, apart from saying that "it depends on who the questioner and audience are". Well: if we restrict ourseves to the contexts in which the speaker and the addressee are one and the same subject, it is not difficult to see how contexts can be defined in terms of cognitive states. Given a cognitive state σ, a context c can be defined in the following way: the subject is defined as the one to whom the state σ belongs (subjects include temporal sequences of cognitive states); the background theory is implicitly specified through the epistemic contents associated in σ to names and predicates; factual information is information encoded into, or associated to, the activated terms.

Secondly, a merit of van Fraassen's theory, from the point of view of a computational approach, is that the relation of relevance is taken as primitive: instead of explaining it in terms of other notions, van Fraassen explains other notions, in particular the notion of reason, in terms of it. Consider the case of the subject who sees puddles of water in the streets. Intuitively, we say that there is a relation of relevance between puddles of water in a region r at time t and rain in that region at a preceding time t'. From a realist point of view this relation is a causal one: rain has caused the puddles. From the computational point of view I adopt here it is an evidential or computational relation, i.e. a relation between the cognitive state in which it is evident that there are puddles in the streets and the cognitive state in which it is evident that it rains; and this relation is constitutive of the structure of our C-I system, not of the structure of external reality.[30] To say that rain is a *reason* for the puddles is to say nothing more than that there is such a relation, and this relation cannot in turn be explained in terms of other, more fundamental relations. This is why, in Definition 18.1, a relevance relation has been introduced as constitutive of the notion of atomic cognitive state.

In this way all the notions of van Fraassen's theory of explanation can be defined in terms of the notion of cognitive state, and the notion of explanation (i.e., of

[30] Nor is it constitutive of the meaning of "rain" or of "puddle": it is a fact concerning the structure of our **C-IS** that there is a relation of relevance between the concepts denoted by "puddle" and "rain", without this relation being constitutive of the two concepts. Of course, the existence of this relation can be seen as the result of an adaptation of our **C-IS** to the external environment; but this hypothesis plays no explanatory role in the theory of the structure of our **C-IS**.

answer) can be legitimately used in the computational characterization of the procedure yielding an answer to the Application Question (6):

$$|P|_\sigma(\langle |n_1|_\sigma \ldots \langle |n_k|_\sigma\rangle) = \begin{cases} 1 & \text{if there are terms } t_1, \ldots, t_k, \text{ belonging respectively to } |n_1|_\sigma, \ldots, |n_k|_\sigma, \text{ and a why-question } Q \text{ such that } Q \text{ arises in } \sigma \text{ and the hypothesis that } |P|_\sigma(\langle t_1, \ldots, t_k\rangle) = 1 \text{ is the best answer to } Q \text{ or is entailed by the best answer to } Q. \\ 0 & \text{if there are terms } t_1, \ldots, t_k, \text{ belonging respectively to } |n_1|_\sigma, \ldots, |n_k|_\sigma, \text{ and a why-question } Q \text{ such that } Q \text{ arises in } \sigma \text{ and the hypothesis that } |P|_\sigma(\langle t_1, \ldots, t_k\rangle) = 0 \text{ is the best answer to } Q \text{ or is entailed by the best answer to } Q. \end{cases} \quad (18.3)$$

Let us see how the preceding examples can be dealt with. In the first (round VS elliptical disk) the why-question is something like "Why in p_1 does that disk look round and in p_2 elliptical (in contrast to looking round/elliptical in both positions)?", and the term is the visual representation the subject has in p_1.

In the second example ("Jack is running"), the cognitive state σ of the subject is characterized by the following facts: (i) that sc_{Jack}, the situational component of the epistemic content associated to "Jack", contains the piece of information that Jack, and no other, is in the room nearby (at time *i*), (ii) that sc_{RUN} contains the piece of information that a person running in a room produces such and such noises; (iii) that the activated term is a representation of such and such noises; (iv) that in σ the question expressed by "Why are there such noises in the room nearby (in contrast to there not being noises)?" arises; (v) that the topic "There are such and such noises" bears relation *R* to such sentences as "John is running in the room nearby", "Jack is running in the room nearby", "Someone is running in the room nearby", etc. The sentence "Jack is running in the room nearby" belongs therefore to the class of potential answers to a question arising in σ; if a further computation selects it as the best answer to that question, then, according to (9), the answer to the Application Question is 1.

In the third example (the puddles in the street), the cognitive state σ of the subject is characterized by the following facts, among many others: (i) that $sc_{the\ demonstration}$ contains the piece of information that the evening before a demonstration was held, and that the police used fireplugs; (ii) that $lc_{fireplug}$ contains the piece of information that the use of fireplugs leaves water-traces (puddles, for the sake of simplification); (iii) that the question expressed by "Why are there puddles in this street in contrast with there not being in that one?" arises; (iv) that the topic "There are puddles in this street" bears relation *R* to such sentences as "The demonstration passed through this street", "It rained in this street", "Someone poured water in this street", etc. The sentence "The demonstration passed through this street" belongs therefore to the class of potential answers to a question arising in σ; if a further computation selects it as the best answer to that question, then, according to (9), the answer to the Application Question is 1.

In the case of Louisa's husband, Smith's cognitive state σ is characterized by the following facts: (i) that sc_{John} contains the piece of information that Smith's friend asserted that John is Louisa's husband; (ii) that $sc_{Smith's\ friend}$ contains the piece of information that Smith's friend is well informed and trustworthy; (iii) that the question expressed by "Why did Smith's friend assert that John is Louisa's husband in contrast with not asserting that?" arises; (iv) that the topic "Smith's friend asserted that John is Louisa's husband" bears relation R to such sentences as "Smith's friend knows that John is Louisa's husband", "Smith's friend believes that John is Louisa's husband", "Smith's friend was joking", etc. The sentence "Smith's friend knows that John is Louisa's husband" belongs therefore to the class of potential answers to a question arising in σ; this fact and (ii) may trigger a further computation ending with that sentence as the best answer to the question. Since lc_{KNOW} contains the piece of information that "s knows that A" entails that A, Smith infers that John is Louisa's husband, and then, by an analogous inference, that John is married. "A entails B" means here that whenever a subject s has a justification for A, s has a justification for B; and the act of inference can be seen as guided by the computation of justifications, along the lines of Prawitz's theory of inference.

18.4.5 C-Justifications for Atomic Statements

The following definition is the natural outcome of the preceding analysis:

Definition 18.7 A *C-justification for an atomic statement* of the form

- $P(n_1, \ldots, n_k)$ is an atomic cognitive state σ of S such that the answer to the Application Question for $P(n_1, \ldots, n_k)$ in σ is 1;
- $n_1 = n_2$ is an atomic cognitive state σ in which a why-question Q arises such that the hypothesis that $|n_1|_\sigma = |n_2|_\sigma$ is the best answer to Q or is entailed by the best answer to Q.

18.5 Cognitive States and C-Justifications for Logically Complex Statements

The notion of atomic cognitive state has been introduced in order to define the notion of justification for atomic statements. If we want to extend the definition to the whole class of statements of a first order language we need introduce, so to say, the logical concepts into cognitive states. More precisely, we must define cognitive states as atomic cognitive states in which the logical concepts are accessible. Intuitively, the conditions at which a logical concept is accessibile are analogous to the ones at which an extra-logical concepts is, namely when an algorithm is available checking the presence of the appropriate feature.

18.5.1 Cognitive States

18.5.1.1 Conjunction

If we keep present Heyting's characterization of a proof of $A \wedge B$ as a pair of a proof of A and a proof of B, we can say that the concept of conjunction is accessible when we can recognize the pairs of cognitive states such that the first is a justification for A and the second is a justification for B.[31] According to the explanation of concept accessibility given above, this means that an algorithm is available that computes the function $f_{\wedge}$ from pairs of cognitive states to {1,0}, such that $f_{\wedge}(\langle\sigma_1, \sigma_2\rangle) = 1$ if and only if $\sigma_1(A) = \sigma_2(B) = 1$.

18.5.1.2 Disjunction

According to Heyting, a proof of $A \vee B$ is either a proof of A or a proof of B. However, even in the mathematical domain this definition seems to be too restrictive: "$PRIME(n) \vee \neg PRIME(n)$", where n is some very large number, is intuitionistically assertible even if neither "$PRIME(n)$" nor "$\neg PRIME(n)$" is. I will therefore take as starting point the following emendation of Heyting's definition: A proof of $A \vee B$ is a procedure such that its execution yields,[32] after a finite time, either a proof of A or a proof of B.[33] Accordingly, the concept of disjunction is accessible when an algorithm is available that computes the function $f_{\vee}$ from procedures to {1,0} such that, for every procedure p, $f_{\vee}(p) = 1$ if and only if the execution of p yields, after a finite time, either a cognitive state σ_1 such that $\sigma_1(A) = 1$ or a cognitive state σ_2 such that $\sigma_2(B) = 1$.

18.5.1.3 Implication

According to Heyting, a proof of $A \supset B$ is a function that associates to each proof of A a proof of B. Accordingly, the concept of implication is accessible when an algorithm is available that computes the function $f_{\supset}$ such that, for every function g

---

[31] Hereafter I will write "$\sigma(A) = 1$" to mean that σ is a justification for A. The property of being a pair of cs' σ_1 and σ_2 such that $\sigma_1(A) = 1$ and $\sigma_2(B) = 1$ is, in this case, the feature to be checked.

[32] "Yields" is to be understood as equivalent to "is known to yield".

[33] For instance, a proof of "$PRIME(n) \vee \neg PRIME(n)$" is a primality test for n, i.e. an algorithm for determining whether n is prime. Such a test should not be confused with a general method consisting in applying to every number x a test for x (this general method is a proof of "$\forall x(PRIME(x) \vee \neg PRIME(x))$").

from cognitive states to cognitive states, $f_{\supset}(g) = 1$ if and only if, for every cognitive state σ, if $\sigma(A) = 1$, then $g(\sigma)(B) = 1$.

18.5.1.4 Universal Quantification

According to Heyting, a proof of $\forall xA$ is a function that associates to each c belonging[34] to a domain D a proof of $A(\underline{c})$. Accordingly, the concept of universal quantification is accessible when an algorithm is available that computes the function $f_{\forall}(c)$ such that, for every function g from cognitive domains to cognitive states, $f_{\forall}(g) = 1$ if and only if, for every C-object d, if $d \in D$,[35] then $g(d)(A[\underline{d}/x]) = 1$.

18.5.1.5 Existential Quantification

According to Heyting, a proof of $\exists xA$ is a pair $\langle c, p\rangle$, where c belongs to D and p is a proof of $A[\underline{c}/x]$. Against this characterization an objection similar to the one against Heyting's definition of disjunction can be raised: in order to prove $\exists xA$ an intuitionist mathematician need not actually have a pair $\langle c, p\rangle$: it is sufficient that he has a procedure whose execution yields one (after a finite time). I will therefore take as starting point the following emendation of Heyting's definition: A proof of $\exists xA$ is a procedure such that its execution yields, after a finite time, a pair $\langle c, p\rangle$, where c belongs to D and p is a proof of $A[\underline{c}/x]$. Accordingly, the concept of existential quantification is accessible when an algorithm is available that computes the function $f_{\exists}$ from procedures to $\{1,0\}$ such that, for every procedure p, $f_{\exists}(p) = 1$ if and only if the execution of p yields, after a finite time, a pair $\langle c, \sigma\rangle$, where $d \in D$[36] and σ is a cognitive state such that $\sigma(A[\underline{d}/x]) = 1$.

18.5.1.6 Negation

According to Heyting (1974, p. 82), a proof of $\sim A$ is a general method that associates to each proof of A a contradiction. Accordingly, the concept of intuitionistic negation is accessible when an algorithm is available that computes the function $f_{\sim}$ such that, for every function g from cognitive states to cognitive states, $f_{\sim}(g) = 1$ if and only if, for every cognitive state σ, if $\sigma(A) = 1$, then $g(\sigma)$ is a cognitive state in which a contradiction is observed.

However, there are important classes of empirical atomic statements whose negations cannot be conceived as intuitionistic. Let us come back to the example of the subject s, looking at a round disk from different positions (at the beginning of

---

[34] "Belongs" is to be understood as equivalent to "is known to belong".

[35] D is the cognitive domain (non-empty set of cognitive objects) over which x ranges.

[36] See footnote 35.

Sect. 3.4); when he sits in position p_1, and sees the disk as round, *s* has, besides an intuitive justification for the statement

(10) That disk is round,

also an intuitive justification for

(11) That disk is not elliptical;

what does this justification consist in? Suppose it is a method *m* that associates to each justification for

(12) The disk is elliptical

a contradiction, as Heyting's explanation requires; then, when *s* moves to a position p_2 (from which he sees the disk as elliptical), he would have no reason to be uncertain about the shape of the disk, since *m* would enable him to associate a contradiction to *all* the potential justifications for "That disk is elliptical", among which there is the visual experience from position p_2. But this is not what happens in fact: the subject hesitates and tries to acquire new relevant information—a clear indication that the two visual experiences are for him on a par as potential justifications for (11) and for (12), respectively.

The intuitionistic explanation of negation must be abandoned in the case of atomic empirical statements of this sort. An alternative is suggested by the following passage from Nelson (1959, p. 208):

> [I]t might be maintained that every significant observation must be an observation of some property, and further that the absence of a property P if it may be established empirically at all, must be established by the observation of (another) property N which is taken as a token for the absence of P.

In our example, the token for the absence of ELLIPTICAL could be the presence of ROUND; and a way to substantiate this suggestion would be to define a justification for (11) as a pair $\langle j_1, j_2 \rangle$, where j_1 is a justification for (10) and j_2 is a justification for something like "ELLIPTICAL and ROUND are incompatible". The problem of this solution is that the choice of j_1 as the first component of a justification for (11) is not more motivated than the choice of a justification for any other sentence of the form "That disk is *P*", where *P* is a feature incompatible with being ELLIPTICAL. Of course, given a specific justification *j* for "That disk is *P*", it would be natural to choose just *j* as the first component; but how should the first component be defined in general? The only possible strategy would consist in following some sort of rule like: "Choose a feature *P* such that it is incompatible with ELLIPTICAL and the answer to the Application Question "Does $|P|_\sigma$ apply to $|that\ disk|_\sigma$?" is 1". However, since there seems to be no way of regimenting the class of features incompatible with being elliptical, and since no effective rule for choosing *P* can apparently be given, the adoption of such a definition of the first component would entail the loss of the epistemic transparency of the relation "*x* is a justification for the negation of *A*", even when *A* is atomic.

The solution I suggest is to define a justification for the empirical negation $\neg A$ of an empirical atomic statement *A* as a cognitive state σ such that the answer to the

Application Question for A is 0. An immediate consequence of the definition is that a justification for $\neg\neg A$ is a justification for A, when A is atomic.

What about the negations of logically complex statements? Once the intuitionistic strategy has been abandoned and the idea has been accepted that the negation of an empirical sentence is in many cases[37] an operation deeply different from implication, a natural answer comes to mind: to define the notion of justification for negative statements by recursion on their logical complexity. This idea is supported by the observation that, in empirical contexts, the most natural way of justifying the negation of a sentence is to exhibit a counterexample to the sentence. For example, the most natural way to justify "Not all men are good" is to exhibit a bad man; by the way, to bring to contradiction the assumption that all men are good would be much more complicated, in spite of the fact that the justified proposition would be, in a sense, weaker. Analogously, the most natural way to justify "It is not the case that John is away and Mary is at home" is to justify either "John is not away" or "Mary is not at home"; and so on.

These examples suggest the idea of defining the justifications for $\neg A$, for an arbitrary A, by recursion on the logical complexity of A, so that the following principles are validated:

$$\begin{aligned}
&\neg(A \wedge B) \Leftrightarrow (\neg A \vee \neg B);\\
&\neg(A \vee B) \Leftrightarrow (\neg A \wedge \neg B);\\
&\neg(A \supset B) \Leftrightarrow (A \wedge \neg B);\\
&\neg \sim A \Leftrightarrow A;\\
&\neg\neg A \Leftrightarrow A;\\
&\neg\exists x A \Leftrightarrow \forall x \neg A;\\
&\neg\forall x A \Leftrightarrow \exists x \neg A.
\end{aligned} \tag{18.4}$$

In this way empirical negation turns out to coincide with strong negation.[38]

To sum up, my suggestion is to distinguish two kinds of negation: intuitionistic negation "$\sim$", to be applied to mathematical and to certain classes of empirical statements, and strong negation "$\neg$", to be applied to the vast majority of empirical statements.[39] Looking at principles (18.4) it is easy to realize that strong negation

---

[37] Not in all. In the case of many empirical sentences, their negations cannot be construed but as intuitionistic. Consider the following example (due to Paolo Casalegno):

(13) Not all prehistoric men were black-eyed;

probably we will never be able to say, of a specific prehistoric man, that he was black-eyed; at the same time, it is quite plausible to say that we have justifications to believe that (13) is true; and if we reflect on the nature of these justifications, we realize that each of them can be verbalized as a reductio ad absurdum of the assumption that all prehistoric men were black-eyed, as the intuitionistic explanation of negation requires.

[38] Cp. Nelson (1949).

[39] It may be a question of discussion which negation is involved in a given natural language statement.

does not require the accessibility of more logical concepts than the ones already introduced.

We can now define the notion of cognitive state. The idea is that cognitive states are atomic cognitive states in which the logical concepts are accessible and the logical constants denote them.

Definition 18.8 A *cognitive state* is an atomic cognitive state σ such that

- The concept of conjunction is accessible in σ and $| \wedge |_\sigma = f_\wedge$.
- The concept of implication is accessible in σ and $| \supset |_\sigma = f_\supset$.
- The concept of disjunction is accessible in σ and $| \vee |_\sigma = f_\vee$.
- The concept of universal quantification is accessible in σ and $|\forall|_\sigma = f_\forall$.
- The concept of existential quantification is accessible in σ and $|\exists|_\sigma = f_\exists$.
- The concept of intuitionistic negation is accessible in σ and $| \sim |_\sigma = f_\sim$.

18.5.2 C-Justifications for Logically Complex Statements

Definition 18.9 A *direct C-justification* for

- $\neg P(n_1, \ldots, n_k)$ is an atomic cs[40] σ such that the answer to the Application Question for $P(n_1, \ldots, n_k)$ in σ is 0;
- $\neg(n_1 = n_2)$ is an atomic cs σ in which $|n_1|_\sigma \neq |n_2|_\sigma$;
- $\bot$ is a cs σ in which a contradiction is observed;
- $\neg\bot$ is a cs σ in which no contradiction is observed;
- $A \wedge B$ is a cs σ in which two cs σ_1 and σ_2 are remembered such that $\sigma_1(A) = \sigma_2(B) = 1$;
- $\neg(A \wedge B)$ is a C-justification for $\neg A \vee \neg B$;
- $A \vee B$ is a cs σ in which a procedure p is available whose execution gives rise in a finite time either to a cs σ_1 such that $\sigma_1(A) = 1$ or to a cs σ_2 such that $\sigma_2(B) = 1$;
- $\neg(A \vee B)$ is a C-justification for $\neg A \wedge \neg B$;
- $A \supset B$ is a cs σ in which a function g from cs to cs is available such that, for every cs σ', if $\sigma'(A) = 1$, then $g(\sigma'(B)) = 1$;
- $\neg(A \supset B)$ is a C-justification for $A \wedge \neg B$;
- $\forall xA$ is a cs σ in which a function g is available from cognitive domains to cs such that, for every C-object d, if $d \in D$, then $g(d)(A[\underline{d}/x]) = 1$;
- $\neg\forall xA$ is C-justification for $\exists x\neg A$;
- $\exists xA$ is a cs σ in which a procedure p is available whose execution gives rise in a finite time to a pair $\langle c, \sigma' \rangle$, where $d \in D$ and σ' is a cs such that $\sigma'(A[\underline{d}/x]) = 1$;
- $\neg\exists xA$ is a C-justification for $\forall x\neg A$.

A *C-justification* for A is a cs σ in which the hypothesis that there is a direct C-justification for A is the best answer to a question arising in σ.

---

40 "cs" abbreviates "cognitive state".

Direct C-justifications are a subclass of C-justifications: a cs σ that is a direct C-justification for A is a cs in which the hypothesis that there is a direct C-justification for A is the best answer to a question arising in σ.

In the passage quoted at the beginning of Sect. 18.2 Prawitz rightly mentions "all the problems of induction". Consider the statement

(14) All ravens are black;

why can the repeated observation of black ravens constitute a justification to believe it? This is the problem, or better the bunch of problems, of induction. I shall not enter these problems here. I only remark that, if we look at them from the standpoint of the theory of explanation, an extension to empirical statements of Heyting's explanation of universal quantification becomes plausible; for the best explanation of the repeated observation of black ravens is the hypothesis that there is a law (perhaps connected to certain genotypic features of the species Corvidae), and this law can be stated as a general method, a function, f associating to every object c belonging to that species a justification $f(c)$ for $A[\underline{c}/x]$. It might be objected that such a general method is in fact an indefeasible justification for (14): if f is such a method at t, it must be such a method also at t', for every $t' \geq t$; true, but that is not the candidate I am proposing: a justification for $\forall xA$ – I am suggesting—is a cognitive state σ in which the existence of such a method f is the best answer to a question arising in σ; and it is clearly possible that a cognitive state σ' in which more information is available is no longer such that the existence of f is the best answer to a question arising in σ'.

18.6 Conclusion

I have defined a defeasible and non-factive notion of C-justification for A, which in my opinion could be adopted as the key notion of a theory of meaning for empirical statements. It seems to me preferable to Prawitz's notion of ground because, being non-factive, it is not exposed to the charge of non-transparency I addressed to grounds for atomic empirical statements. On the other hand, I tried to define it in such a way that the assumption of transparency becomes plausible. The leading idea has been to define justifications as mental states resulting from the execution of certain computations made possible by the availability of specific algorithms and data; in this way an answer has been suggested also to one of the questions Prawitz (rightly) considers crucial for an adequate explanation of inference:

> Could there be something like recognizing the validity of an inference, understood as less demanding than knowing but as something of sufficient substance to imply that one is justified in holding the conclusion true? (Prawitz 2010, p. 16).

The answer I propose is that one recognizes the validity of an inference whose conclusion is A when one is in a position to effect, on the basis of one's knowledge of the meaning of A, the computations required to recognize that one is in a cognitive state that is a justification for A. Of course, this answer is significant only if the notion of justification is epistemically transparent, and, as I remarked at the end of Sect. 18.1, this cannot be demonstrated, but only assumed. However—as I said – an assumption may be more or less plausible.

There is an aspect of the notion of C-justification that might seem problematic in this respect: C-justifications for universal quantification and implication are functions or general methods, and one may wonder whether such general methods *are* epistemically transparent. This question concerns not only C-justifications, but also Prawitz's grounds and Heyting's proofs; I see two reasons for considering the transparency assumption plausible in this respect. First, it seems to be a matter of fact that humans are in general capable of recognizing laws and rules—entities having a functional nature. Second, it is important to stress that the subject I have made reference to in the preceding pages is an idealized one, essentially in the same sense as a language user is idealized in linguistics, i.e. having no limits of memory, attention, and so on: a subject whose cognitive capacities and performances in any given occasion are taken as representative of the ones of an arbitrary member of the same species. Now, if such an idealized subject is not able, when acquainted with a general method m, to recognize it as a method with such and such properties, the sole conclusion it is natural to draw is that m is *not* such a method. How could it be a method with such and such properties if *nobody* were capable to acknowledge that it is? Of course, it is possible that I am not capable to realize that something is such a method, because of the limits of my IQ, memory, attention, and so on; but these are precisely the factors from which we make abstraction when we make reference to an idealized subject.

The computational nature of justification I have tried to catch with my definition entails a major departure from the neo-verificationist tradition which Prawitz adheres to. The logical background of that tradition is Curry-Howard's Isomorphism, one of whose consequences is that the process of reduction of an intuitionistic derivation in natural deduction to its normal form can be seen as the process of computation of a term in a typed λ-calculus; since grounds are admissible denotations of λ-terms, the possession of a ground for A can be equated with having constructed a term that denotes that ground—and such possession can, according to Prawitz, be assumed to be epistemically transparent: it manifests itself in the naming of that ground.[41] I agree with this idea, except for the crucial case of atomic statements. The structure of a term denoting a ground for an atomic statement $P(n)$ does not reveal anything about the nature of its denotation; as a consequence, merely having a name t that, as a matter of fact, denotes a ground g for $P(n)$ does not enable one to see that one has a ground for $P(n)$, unless one *sees* that t denotes g, and that g is a ground for $P(n)$. This problem is largely disregarded in the neo-verificationist tradition, perhaps because of the assumption that atomic statements are decidable. However, on the one hand, this assumption is not plausible in empirical domains; on the other hand, and more to the point, the decidability of an atomic statement is logically independent of the epistemic transparency of its grounds.

[41] Prawitz (2009, p. 187).

I have therefore chosen a different strategy, consisting in giving an explicit account of the data and algorithms that must be available to a knowing subject for him to be able to see that he has a justification for $P(n)$. Since I have characterized C-justifications for $P(n)$ as stemming from the composition of C-authorizations to use the name n and C-authorizations to use the predicate P, I have been confronted with the following question: what does allow one to see that n denotes the object o? This has led me to characterize objects from an internalist standpoint, i.e. as they are given to the computational apparatus of a subject[42]; as a consequence, C-objects are not individuals, but classes of terms. This choice seems to me necessary in order to flesh out the idea that objects are the inputs of epistemically transparent computational processes. Computations are triggered by symbols, i.e. by terms of some representational system or, more generally, of some language internal to the computing apparatus; an individual of the external world can be seen as the input of a computational process only modulo an interpretation associating a term to that individual. The problem is that such interpretation may, in general, be out of the epistemic control of the subject performing the computation itself, in the sense that the subject may not know which individual the term is a name of.[43] My internalist notion of C-object is intended to offer a solution to this problem, since C-objects are obviously recognizable.

Moreover, this internalist characterization of justifications seems to yield a way out of a difficulty typical of verificationist approaches in general—a difficulty stated by Prawitz in the following terms:

> Unfortunately, the defined concepts [ground, inference, deduction] do not make justice to deductions outside of mathematics. [...] But as Dummett illustrates by the example of Euler's proof concerning the bridges in Königsberg, we may deduce in applied mathematics that a person during a walk has crossed some of the bridges of Königsberg twice, if she has crossed every one of them at least once. We may have a ground for asserting that the person crossed every one of the bridges at least once in the form of a relevant observation, but applying Euler's proof to this empirical ground will not be a deduction in my sense, because obviously it cannot yield an operation that transforms the given observation to an observation that the person has crossed one bridge twice. (Prawitz 2011, p. 24)

If justifications for atomic statements are not defined as observations one can make in certain external conditions, but as mental states individuated by their internal structure (i.e. by available data and algorithms), without reference to external conditions, the difficulty is avoided, since the method provided by Euler's proof does transform every justification for the antecedent into a justification for the consequent.

Finally, let me stress an important advantage of the notion of (empirical) justification I am proposing. A typical criticism to which inferentialist notions of justification are exposed is the skeptical conclusion that no belief is inferentially justified. According to the inferentialist approach, a subject s is inferentially justified in believing that B if and only if s is justified in believing that A, A is true, and B is inferred from

---

[42] "Internalist" is therefore meant here in the purely methodological sense in which it is used, for instance, by Chomsky in (2000).

[43] This difficulty concerns Prawitz's notion of open ground as well, and seems to be another reason against its epistemic transparency.

A. The typical skeptical argument has two main steps: (i) the observation that the inference of *B* from *A* is not necessary, i.e. that the possibility cannot be ruled out that *A* is true and *B* false; and (ii) an argument to the conclusion that, in order to rule out such a possibility, it is necessary to use the very form of inference that is questioned. The notion of empirical justification I propose is *not* inferentialist: a cognitive state is not something to be believed, nor the premise of an inference. It is rather a notion of justification based on explanation: a subject is justified in believing that *A*, in this sense, if and only if *A* is the best explanation of the data available to the subject. In this case the skeptic has no room to make the first step. The only possibility would be to say that something *seems* to explain the data, but does not really explain them; but this is just what cannot be done: in Lipton's words,

> We do not know how to make the contrast between understanding and merely seeming to understand in a way that makes sense of the possibility that most of the things that meet all our standards for explanation might nonetheless not really explain. (Lipton 1991, p. 25)

Given the central role played by explanation in the definition I have proposed of the concept of *C*-justification, the absence of any gap between apparent explanation and real explanation is another reason for the plausibility of the assumption that *C*-justifications are epistemically transparent.

References

Bierwisch, M. (1992). From concepts to lexical items. *Cognition*, *42*(1–3), 23–60.

Casalegno, P. (2002). The problem of non-conclusiveness. *Topoi*, *21*(1–2), 75–86.

Chomsky, N. (2000). *New horizons in the study of language and mind*. Cambridge: Cambridge University Press.

Cozzo, C. (2015). *Necessity of thought*. This volume.

Dummett, M. (1973). *Frege: Philosophy of Language*. London: Duekworth.

Evans, G. (1982). *The varieties of reference*. Oxford: Oxford University Press.

Gentzen, G. (1936). The consistency of elementary number theory. In: M.E. Szabo (Ed.), *The Collected Papers of Gerhard Gentzen* (pp. 132–213). Amsterdam: North-Holland Publishing Company.

Heyting, A. (1974). Intuitionistic views on the nature of mathematics. *Synthese*, *27*(1–2), 79–91.

Lipton, P. (1991). *Inference to the best explanation*. London: Routledge.

Marr, D. (1982). *Vision*. New York: W. H. Freeman and Company.

Nelson, D. (1949). Constructible falsity. *The Journal of Symbolic Logic*, *14*, 16–26.

Nelson, D. (1959). Negation and separation of concepts in constructive systems. In A. Heyting (Ed). *Constructivity in Mathematics* (pp. 208–225). Amsterdam: North-Holland Publishing Company.

Prawitz, D. (1970) Constructive semantics. In *Proceedings of the 1st Scandinavian logic symposium Åbo 1968* (pp. 96–114). Filosofiska Föreningen och Filosofiska Institutionen vid Uppsala Universitet.

Prawitz, D. (1973). Towards a foundation of general proof theory. In P. Suppes (Ed.), *Logic, Methodology and Philosophy of Science IV* (pp. 225–250). Amsterdam: North-Holland Publishing Company.

Prawitz, D. (1977). Meaning and proofs: on the conflict between classical and intuitionistic logic. *Theoria*, *43*(1), 2–40.

Prawitz, D. (1980). Intuitionistic logic: a philosophical challenge. In G.H. von Wright (Ed.), *Logic and Philosophy* (pp. 1–10). Leiden: Nijhoff.
Prawitz, D. (2002). Problems for a generalization of a verificationist theory of meaning. *Topoi, 21*(1–2), 87–92.
Prawitz, D. (2009). Inference and knowledge. In M. Pelis (Ed.), *The Logica Yearbook 2008* (pp. 175–192). London: College Publications.
Prawitz, D. (2010). Validity of inferences. In *Proceedings from the 2nd launer symposium on analytical philosophy on the occasion of the presentation of the launer prize at Bern 2006*.
Prawitz, D. (2011). To explain deduction, unpublished manuscript.
Prawitz, D. (2012). Truth as an epistemic notion. *Topoi, 31*(1), 9–16.
Usberti, G. (1995). *Significato e Conoscenza*. Milano: Guerini e Associati.
Usberti, G. (2004). On the notion of justification. *Croatian Journal of Philosophy, 4*(10), 99–122.
Usberti, G. (2012). Anti-realist truth and truth-recognition. *Topoi, 31*(1), 37–45.
van Fraassen, B. (1980). *The scientific image*. Oxford: Oxford University Press.
Williamson, T. (2000). *Knowledge and its limits*. Oxford: Oxford University Press.

Chapter 19
Dag Prawitz's Published Books and Papers, by Year (Selected)

Translations, reprints and new editions are listed under the original editions.

Papers and doctoral dissertation

1960

1. "A mechanical proof procedure and its realization in an electronic computer" (together with H. Prawitz and N. Voghera), *Journal of the Association for Computing Machinery* 7, pp 102–128. Reprinted (with a new commentary) in: *Automation of Reasoning 1, Classical Papers on Computational Logic*, pp 202–228, J. Siekmann and G. Wrightson (eds), Springer Verlag, 1983.
2. "An improved proof procedure", *Theoria* 26, pp 102–139. Reprinted (with a new commentary) in: *Automation of Reasoning 1, Classical Papers on Computational Logic*, pp 162–201, J. Siekmann and G. Wrightson (eds), Springer Verlag, 1983.

1964

3. "Normal deductions" (Abstract of paper read at a meeting in the Association for Symbolic Logic, New York 1964), *Journal of Symbolic Logic* 29, p 152.

1965

4. *Natural Deduction. A Proof-Theoretical Study*, Almqvist and Wiksell, Stockholm. Russian translation, Moscow, 1997. Reprinted (with a new preface and an errata list) by Dover Publications, New York, 2006.

1967

5. "A note on existential instantiation", *Journal of Symbolic Logic* 32, pp 81–82.
6. "Completeness and Hauptsatz for second order logic", *Theoria* 33, pp 246–58.

H. Wansing (ed.), *Dag Prawitz on Proofs and Meaning*, Outstanding Contributions to Logic 7, DOI 10.1007/978-3-319-11041-7_19

1968

7. "A survey of some connections between classical, intuitionistic and minimal logic" (together with P.-E. Malmnäs), in: *Contributions to Mathematical Logic*, pp 215–29, H.A. Schmidt et al. (eds), North-Holland.
8. "Propositions", *Theoria* 34, pp 134–46.
9. "A discussion note on utilitarianism", *Theoria* 34, pp 76–84.
10. "Utilitarism och alternativen till handlingar" (Eng. "Utilitarianism and the alternatives of actions"), in: *Sanning Dikt och Tro. Till Ingemar Hedenius* (Eng. *Truth, Fiction, Faith. To Ingemar Hedenius*), pp 251–254, A-M Henschen Dahlquist et al. (eds), Bonniers.

1969

11. "Hauptsatz for higher order logic", *Journal of Symbolic Logic* 33, pp 452–57.
12. "Advances and problems in mechanical proof procedures", in: *Machine Intelligence 4*, pp 59–71, B. Meltzer et al. (eds), Edinburgh.

1970

13. "Some results for intuitionistic logic with second order quantifiers", in: *Intuitionism and Proof Theory, Proc. of the Summer Conference at Buffalo*, pp 259–69, J. Myhill et al. (eds), North-Holland.
14. "A proof procedure with matrix reduction", in: *Symposium on Automatic Demonstration*, pp 207–14, M. Landet et al. (eds), Springer Verlag.
15. "Constructive semantics", in: *Proceedings of the 1st Scandinavian Logic Symposium Åbo 1968*, pp 96–114, Filosofiska studier 8, Filosofiska Föreningen och Filosofiska Institutionen vid Uppsala Universitet, Uppsala.
16. "On the proof theory of mathematical analysis", in: *Logic and Value*, pp 169–80, T. Pauli (ed.), Filosofiska studier 9, Filosofiska Föreningen och Filosofiska Institutionen vid Uppsala Universitet Uppsala.
17. "The alternatives to an action", *Theoria* 36, pp 116–26.

1971

18. "Ideas and results in proof theory", in: *Proceedings of the 2nd Scandinavian Logic Symposium*, pp 237–309, J. Fenstad (ed.), North-Holland. Italian translation: "Idee e risultati nella teoria della dimostrazione", in: *Teoria della dimostrazione,* pp.127–204, D. Cagnoni (ed.), Feltrinelli, 1981. Croatian translation: "Ideje i rezultati teorije dokaza", in: *Novija Filozofija Matematike*, Z. Sikc (ed.) pp 193–241, Nolit, Beograd 1987.

1972

19. "The philosophical position of proof theory", in: *Contemporary Philosophy in Scandinavia*, pp 123–34, R.E. Olson et al. (eds), The John Hopkins Press. Japanese translation in: *Readings in the Philosophy of Mathematics: After Gödel*, I. Takashi (ed.), 1995, pp 161–180.

1973

20. "Towards a foundation of general proof theory", in: *Logic, Methodology and Philosophy of Science IV*, pp 225–50, P. Suppes et al. (eds), North Holland.
21. "Logik som filosofisk disciplin" (Eng. "Logic as a philosophical discipline"), in: *Mening og Handling* (Eng. *Meaning and Action*), pp 128–135, Institutt for filosofi, Universitetet i Oslo.

1974

22. "On the idea of a general proof theory", *Synthese* 27, pp 63–77. Also published in: *Bollettino della Unione Matematica Italiana*, pp 108–121, Bologna 1974. Reprinted in: *A Philosophical Companion to First-Order Logic*, pp 212–224, R.I.G. Hughes (ed.), Hackett, 1993. Italian translation: "Sull'idea di una teoria generale della dimostrazione", in: *Teoria della dimostrazione*, pp 205–220, D. Cagnoni (ed.), Feltrinelli, 1981.

1975

23. "Comments on Gentzen-type procedures and the classical notion of truth", in: *Proof Theory Symposium Kiel 1974*, pp 290–319, A. Dold et al. (eds), Springer Verlag.
24. *ABC i symbolisk logik* (Eng. *The ABC of Symbolic Logic*), Filosofiska studier nr 23, Uppsala (mimeographed).
    - New printed edition, Thales, Stockholm, 199.
    - Second revised edition, Thales, Stockholm, 2001.
    - Third revised and extended edition, Thales, Stockholm, 2010.
25. "Causality and action", in: *Kausalitet, Nordisk seminar om kausalitet i Oslo 1975* (Eng. *Causality, Nordic seminar on causality in Oslo*), pp 133–146, Dagfinn Føllesdal et al. (eds), Inst. for filosofi, Univ. i Oslo, Oslo.

1976

26. "Comments [to C. Lejewski, "Ontology and Logic"]", in: *Philosophy of Logic*, pp 43–48, S. Körner (ed.), Blackwell, Oxford.

1977

27. "Meaning and proofs: On the conflict between classical and intuitionistic logic", Theoria 43, pp 2–40. Hungarian translation in: A matematika filozófiája a 21. század küszöbén, pp 123–63, C. Ferenc (ed.), Budapest 2003.
28. "Logisk intuitionism, sanning och mening" (Eng. "Logical intuitionism, truth, and meaning"), *Norsk filosofisk tidsskrift*, pp 139–72.

1978

29. "Om moraliska och logiska satsers sanning" (Eng. "On truth in ethics and logic"), in: *En filosofibok* (Eng. *A Book of Philosophy*), pp 144–55, L. Bergström et al. (eds), Bonniers.
    - Portuguese translation: "Sobre a verdade das proposiçoes morais edas proposiçoes da lógica", *Analytica, Revista de Filosofia* 11, pp 127–141, 2007.

1979

30. "Proofs and the meaning and completeness of the logical constants", in: *Essays on Mathematical and Philosophical Logic*, pp 25–40, J. Hintikka et al. (eds), D. Reidel, Dordrecht. German translation: "Beweise und die Bedeutung und Vollst"andigkeit der logischen Konstanten", *Conceptus*, XVI, pp 3–44, 1982.
31. "Decision theory and ethics in the application of science" in: *Abstracts, 6th International Congress of Logic, Methodology and Philosophy of Science*, pp 244–245, Hannover.

1980

32. "Intuitionistic logic: A philosophical challenge", in: *Logic and Philosophy*, pp 1–10, G. H. von Wright (ed.), Martinus Nijhoff Publishers, The Hague.
34. "Rationalitet och kärnkraft" (Eng. "Rationality and atomic power"), *Filosofisk tidskrift* 1, pp 1–23.
35. "Beweis, Beweisbarkeit" (Eng. "Proof, Provability"), in: *Handbuch wissenschaftstheoretischer Begriffe* (Eng. *Handbook of Concepts from Philosophy of Science*, pp 90–95, J. Speck (ed.), Vandenhoeck and Ruprecht, Göttingen.

1981

36. "Philosophical aspects of proof theory" in: *Contemporary Philosophy. A new survey, vol 1*, pp 235-77, Martinus Nijhoff Publishers, The Haugue.
37. "Validity and normalizability of proofs in 1st and 2nd order classical and intuitionistic logic", in: *Atti del congresso nazionale di logica*, pp 11–36, Bibliopolis.

1985

38. "Remarks on some approaches to the concept of logical consequence", *Synthese* 62, pp 153–71.
39. "I fondamenti della matematica oggi, Intervento di Dag Prawitz", in *Atti degli incontri di logica matematica IV-VII*, C. Bernardi and P. Pagli (eds), pp 297–307, Dipartimento di Matematica, Siena.
40. "Normalization of proofs in set theory", in: *Atti degli incontri di logica matematica*, pp 357–71, Siena.
41. "Värdenihilism—en vanföreställning" (Eng. "Value nihilism—a fallacy)", *Tvärsnitt* 7, pp 11–15.
42. "Några filosofiska synpunkter på rationell argumentation inom juridiken" (Eng. "Remarks on rational argumentation in law from a philosophical point of view"), in: *Rationalitet och empiri i rättsvetenskapen* (Eng. *Rationality and empiricism in jurisprudence*), pp 24–29, Juridiska fakulteten i Stockholm; skriftserien, 6.

1987

43. "Some remarks on verificationistic theories of meaning", *Synthese* 73, pp 471–77.
44. "Dummett on a theory of meaning and its impact on logic", in: *Michael Dummett, Contributions to Philosophy*, pp 117–65, B.M. Taylor (ed.), Martinus Nijhoff Publishers, Dordrecht.

1989

45. "Von Wright on the concept of cause", in: *The Philosophy of Georg Henrik von Wright*, P. Schilp et al. (eds), pp 417–44, Open Court.
46. "Kunskap och bevis: Översikt över en aktuell diskussion" (Eng. "Knowledge and proof: A survey of a contemporary discussion"), in: *Kungl. Vitterhets Historie och Antikvitets Akademiens* ⇒ *Arsbok 1988* (Eng. *The 1988 Yearbook of The Royal Swedish Academy of Letters, History and Antiquities*, pp 102–108.

1990

47. "Att överleva eller att må väl: om olika mål för miljövård" (Eng. "To survive or to live well: about different goals in environmental conservation"), in: *Vad tål naturen?* (Eng. *What Does Nature Bear*), L. Lundgren (ed.), pp 45–52, Naturvårdsverket (Rapport 3738).

1991

48. "Psykoanalytisk sanning - natur eller humanvetenskaplig?" (Eng. "Psychoanalytical truth—a concept of the natural sciences or of the humanities?"), in: *Psykoanalys och kultur* (Eng. *Psychoanalysis and Culture*), H. Reiland and F. Ylander (eds), pp 127–138, Natur och Kultur, Stockholm.

1992

49. "Turing och Witgenstein—två verklighetsuppfattningar" (Eng. "Turing and Wittgenstein—two concepts of reality"), *Dialoger* 22.23, pp 47–51.

1993

50. "Remarks on Hilbert's program for the foundation of mathematics", in: *Bridging the Gap: Philosophy, Mathematics, and Physics*, pp 87–98, G. Corsi et al. (eds), Kluwer Academic Publishers.
51. "Michael Dummetts språkfilosofiska program" (Eng. "Michael Dummett's program for philosophy of language"), in: *Huvudinnehåll*, pp 147–156, Nya Doxa.

1994

52. "Meaning and experience", *Synthese* 98, pp 131–41.
53. "Meaning theory and anti-realism", in: *The Philosophy of Michael Dummett*, B. McGuiness et al. (eds), pp 79–89, Kluwer Academic Publishers.
54. "Medvetandets substans" (Eng. "The substance of the mind"), in: *Om själen* (Eng. *On the soul*), A. Ellegård (ed.), pp 127–132, Natur och Kultur.

55. "Book reviews. The Logical Basis of Metaphysics, by Michael Dummett", *Mind* 103, pp 373–376.

1995

56. "Quine and verificationism", *Inquiry* 37, pp 487–94.
57. "Förord (Eng. "Preface"), in: Michael Dummett, *Metafysik och mening* (Eng. *Metaphysics and meaning*), Thales.

1997

58. "Progress in philosophy", in: *The Idea of Progress*, A. Burgen et al. (eds), pp 139–53, de Gruyter, Berlin.

1998

59. "Truth and objectivity from a verificationist point of view", in: *Truth in Mathematics*, H.G. Dales et al. (eds), pp 41–51, Clarendon Press, Oxford.
60. "Truth from a constructive perspective", in: *Truth in Perspective: Recent Issues in Logic, Representation and Ontology*, C. Martinez et al. (eds), pp 23–35, Ashgate, Aldershot.
61. "The significance of philosophical logic", in: *In Search of a New Humanism*, R. Egidi (ed.), pp 157–161, Kluwer Academic Publishers, Dordrecht.
62. "Comments on the papers", *Theoria* 64 (Special issue on the philosophy of Dag Prawitz), pp 283–337.

2001

63. "A note on Kanger's work on efficient proof procedures", in: *Collected Papers by Stig Kanger with Essays on his Life and Work*, G. Holmström-Hintikka et al. (eds), pp 43–52, Kluwer Academic, Dordrecht.

2002

64. "Meaning and Objectivity", in: *Meaning and Interpretation*, Konferenser 55, Kungl. Vitterhets Historie och Antikvitets Akademien, D. Prawitz (ed.), Almqvist and Wiksell International, pp 101–114.
65. "Problems for a Generalization of a Verificationist Theory of Meaning", *Topoi* 21, pp 87–92.

2003

66. "Sanningens återkomst: Sanningen finns men frågan är hur" (Eng. "The return of truth: There is the truth but the question is how"), *Axess 2003*, pp 230–244.
67. "Sanningen i vitögat" (Eng. "Face the truth", *Forskning and Framsteg*, nr 1 jan-feb 2003, pp 32–37.

2005

68. "Logical Consequence from a Constructivist Point of View", in: *The Oxford Handbook of Philosophy of Mathematics and Logic*, S. Shapiro (ed.), pp 671–695, Oxford University Press.

2006

69. "Meaning Approached via Proofs", *Synthese* 148, pp 507–524.
70. "Skäl och goda grunder" (Eng. "Reasons and good grounds"), *Filosofisk tidskrift* 27, pp 3–14.

2007

71. "Pragmatist and Verificationist Theories of Meaning", in: *The Philosophy of Michael Dummett*, The Library of Living Philosophers, vol XXXI, R.E. Auxier and L.E. Hahn (eds), pp 455–481, Open Court, Chicago.

2008

72. "Proofs Verifying Programs and Programs Producing Proofs: A Conceptual Analysis", in: *Deduction Computation Experiment. Exploring the Effectiveness of Proofs*, R. Lupacchini and G. Corsi (eds), pp 81–94, Springer, Milano.

2009

73. "Logical Determinism and the Principle of Bivalence", in: *Philosophical Probings. Essays on von Wright's Later Work*, F. Stoutland (ed.), pp 111–135, Automatic Press, United Kingdom.
74. "Inference and Knowledge", in: *The Logica Yearbook 2008*, M Pelis (ed.), pp 175–192, College Publications, London.

2010

75. "Assertions in the Context of Inference", in: *Judgements and Propositions. Logic, Linguistic and Cognitive Issues*, S. Bab and K. Robering (eds), pp 89–98, Logos Verlag, Berlin.

2011

76. "Proofs and Perfect Syllogisms", in: *Logic and Knowledge*, C. Cellucci et al. (eds), pp 385–402, Cambridge Scholars Publishing, Newcastle upon Tyne.

2012

77. "Truth and Proof in Intuitionism", in: *Epistemology versus Ontology*, P. Dybjer et al. (eds), Springer, pp 45–67.
78. "The Epistemic Significance of Valid Inferences", *Synthese* 187, pp 887–898.
79. "Truth as an Epistemic Notion", *Topoi* 31, pp 9–16.

2013

80. "Validity of Inferences", in: *Reference, Rationality, and Phenomenology: Themes from Føllesdal*, M. Frauchiger (ed.), Ontos Verlag, pp 179–204.

2014

81. "An Approach to General Proof Theory and a Conjecture of a Kind of Completeness of Intuitionistic Logic Revisited", in: *Advances in Natural Deduction. A Celebration of Dag Prawitz's Work*, Luiz Carlos Pereira, Edward Hermann Haeusler, and Valeria de Paiva (eds.), pp 269–279, Trends in Logc vol. 39, Springer, Dordrecht.
82. "A short scientific autobiography", this volume.
83. "Explaining deductive inference", this volume.

Edited books and special issues of journals

En filosofibok, (Eng. *A book of philosophy*), Bonniers, Stockholm 1978 (together with L. Bergström and H. Ofstad).

Annals of Pure and Applied Logic vol. 63, number 1, 1993. Special issue, A selection of papers presented at the 9th International Congress of Logic, Methodology and Philosophy of Science, August 7–14, 1991, Uppsala, Sweden (together with D. Westerståhl).

Logic, Methodology and Philosophy of Science IX, North Holland, Elsevier 1994 (together with B. Skyrms and D. Westerståhl).

Logic and Philosophy of Science in Uppsala, Papers from the 9th International Congress of Logic, Methodology and Philosophy of Science, Kluwer Academic Publishers, Dordrecht, 1994 (together with D. Westerståhl).

Theoria, vol 60, part 3, 1994. The Rolf Schock Prize in Logic and Philosophy 1993, A Special Issue on the Philosophy of W.V. Quine.

Synthese vol 106, No 1, 1996, Varia, Including a Symposium on Descartes and Contemporary Philosophy of Mind (together with J. Hintikka).

Meaning and Interpretation, Konferenser 55, Kungl. Vitterhets Historie och Antikvitets Akademien, Almqvist and Wiksell International, 2002.

MIX
Papier aus verantwortungsvollen Quellen
Paper from responsible sources
FSC® C105338
www.fsc.org

If you have any concerns about our products,
you can contact us on
ProductSafety@springernature.com

In case Publisher is established outside the EU,
the EU authorized representative is:
Springer Nature Customer Service Center GmbH
Europaplatz 3, 69115 Heidelberg, Germany

Printed by Libri Plureos GmbH
in Hamburg, Germany